RESIDENTIAL ARCHITECTURE:

DESIGN AND DRAFTING

Ernest R. Weidhaas

**Professor Emeritus of Engineering Graphics
The Pennsylvania State University**

Mark D. Weidhaas, P.E.

**Structural Engineer
Morgan & Associates, Inc.
Colorado Springs, Colorado**

Delmar Publishers

an International Thomson Publishing company

Albany • Bonn • Boston • Cincinnati • Detroit • London • Madrid
Melbourne • Mexico City • New York • Pacific Grove • Paris • San Francisco
Singapore • Tokyo • Toronto • Washington

NOTICE TO THE READER

Cover illustration by Frank Mahoney

Delmar Staff

Acquisitions Editor: Sandy Clark
Developmental Editor: Peg Gantz
Senior Project Editor: Christopher Chien

Production Coordinator: Jennifer Gaines
Art and Design Coordinator: Mary Beth Vought
Editorial Assistant: Christopher Leonard

COPYRIGHT © 1999
By Delmar Publishers
a division of International Thomson Publishing Inc.

The ITP logo is a trademark under license.

Printed in the United States of America

For more information, contact:

Delmar Publishers
3 Columbia Circle, Box 15015
Albany, New York 12212-5015

International Thomson
  Publishing Europe
Berkshire House 168-173
High Holborn
London, WC1V 7AA
England

Thomas Nelson Australia
102 Dodds Street
South Melbourne, 3205
Victoria, Australia

Nelson Canada
1120 Birchmont Road
Scarborough, Ontario
Canada, M1K 5G4

International Thomson Editores
Campos Eliseos 385, Piso 7
Col Polanco
11560 Mexico D F Mexico

International Thomson
  Publishing GmbH
Konigswinterer Strasse 418
53227 Bonn
Germany

International Thomson Publishing
  Asia
221 Henderson Road
#05-10 Henderson Building
Singapore 0315

International Thomson Publishing—
  Japan
Hirakawacho Kyowa Building, 3F
2-2-1 Hirakawacho
Chiyoda-ku, Tokyo 102
Japan

2 3 4 5 6 7 8 9 10 XXX 04 03 02 01 00 99

Library of Congress Cataloging-in-Publication Data

Weidhaas, Ernest R.
   Residential architecture : design and drafting / Ernest R. Weidhaas, Mark D. Weidhaas.
      p. cm.
   Includes index.
   ISBN 0-8273-7848-3
   1. Architectural drawing—Technique. 2. Architectural design—Technique.
3. Architecture, Domestic—Designs and plans.
I. Weidhaas. Mark D. II. Title.
NA2708.W45 1997
720'.28'4—dc21
                                          97-31678
                                          CIP

Contents

SECTION IV *Working Drawings*

SECTION V   *Light Construction*

Appendices

Preface

To Instructors

To an instructor, a textbook can be a valuable assistant, relieving much repetitive explanation; to a student, it can be an inspiration, offering encouragement in addition to facts and methods. In *Residential Architecture: Design and Drafting,* these principles are recognized. This text is a comprehensive reference for new teachers and a dependable tool for experienced instructors. It will challenge advanced students as well as guide students who are new to the field of architectural design and drafting.

This is a book on architectural **drafting**. However, when architectural drafters draw a line not previously supplied in the form of a sketch, they are also engaged in **design**. Therefore, this book puts equal emphasis on architectural design and architectural drafting. Much of the material included here is concerned with **residential** drawing and design. Because students are more familiar with houses than with any other structures, their design and construction seem the proper introduction to such a vast field.

This book is arranged in eight sections, each of which addresses specific architectural design and drafting concepts. The chapters within each section provide helpful material for all students, from beginner to advanced. Each chapter is self-sufficient, so the chapters may be arranged or omitted according to the available class time. The introduction to each chapter shows the relative importance of that chapter and how it fits into the overall process of drafting and design.

Within each chapter, separate, self-contained boxed articles provide topic-related information on **applied academics**, **architecture-related careers**, and **computer and internet** skills. These articles can be integrated into the chapter to enhance the information taught there or they can be used as independent activities. They can also serve as a springboard to activities that build upon specific student interests.

The descriptive material has been arranged in an interesting, readable format using step-by-step illustrations whenever applicable. Illustrations have been designed for maximum clarity and accuracy. For greater realism, pictorials have been drawn in perspective. A variety of photographs taken throughout the country are included to illustrate practical applications of building principles. Each chapter provides helpful material for a range of students from beginner to advanced. Beginning students will appreciate having the *reasons* behind the *rules;* advanced students will find challenges to stimulate their creativity.

A contemporary home for Family A is designed from preliminary program to finished rendering, allowing students to follow the design and drafting process in reality as well as in theory. The plans for this house are included in the relevant chapters. The plans for a split-level modular home for Family M (dimensioned in metric) are presented in Chapter 14, and a two-story traditional home for Family Z as well as an A-framed solar home for Family S are given in Appendix A.

Also included are the plans for a five-story commercial building of bolted and welded steel construction. Photographs of this building being erected are in Chapter 46. Chapters on accessibility (including the ADA), acoustics, and fire protection follow. Although these generally are considered commercial construction topics, an awareness of the principles underlying these subjects can be of value in residential construction, too.

Author Ernest R. Weidhaas, professor emeritus of Engineering Graphics at the Pennsylvania State University and a registered professional engineer, prepared this book with the welcome contributions of co-author Mark D. Weidhaas, architectural engineer. Daniel Engstrom and Ronda Engstrom researched and wrote the boxed articles and designed the integrated chapter objectives and end-of-chapter questions.

This edition includes new sections on computer-aided drafting, reproduction processes, drafting media, postmodern architecture, room design (including applicable ADA accessibility guidelines), plumbing, alternate heating and cooling methods, pest protection, earthquake protection, hurricane protection, radon control, and construction administration.

The authors would be most grateful to receive suggestions and correspond with readers interested in this book. You can contact them c/o Drafting Editor, Delmar Publishers, 3 Columbia Circle, P.O. Box 15015, Albany, NY 12212-5015.

To Students

You are about to be introduced to architectural design and drafting, a satisfying and creative profession: a profession of ideas and planning; of intelligent design, detailing, and specification writing; and finally, of actual construction—when all will see the product of your talent and imagination.

Are you wondering if architecture should be your career? The authors sincerely hope that this book will help to answer that question for you. If your answer is "yes," you can look forward to an interesting life of constant and increasing challenges with the enormous satisfaction of a job—a big job—well done.

Although the primary aim of this book is to present information vital to the profession of architectural design and drafting, it may also lead you into one of the allied fields of work, many of which are described in this book. But remember that neither this book nor any other can provide all of the extensive information you will need to become a registered architect or architectural engineer.

If you decide that your interests lie in another direction, your time and effort will not be wasted. What you learn will be useful in many ways. You will have learned the value of creative imagination, the importance of careful investigation before making a final decision, the need for a knowledge of construction materials and methods, and an awareness of safety. All of this will be helpful to you through your life—especially as a homeowner.

The design and drafting of a building are not easy. Even experienced architects must work hard and long and will have moments of doubt, but they have come to realize that the greater the demands upon you, the greater the rewards.

Ernest R. Weidhaas
Mark D. Weidhaas

How to Use This Book

There are a number of features in this text designed to aid learning. **Objectives** and **Key Terms** enable you to recognize what you are expected to learn in each chapter, and **end-of-chapter questions and activities** measure your mastery of the material.

At the beginning of each chapter, learning **Objectives** identify the knowledge or skills you will develop as you study the material. Read the objectives before you begin studying the chapter and keep them in mind as you read through the material. When you have finished the chapter, review them to insure that you have met each objective.

Key Terms, listed at the beginning of each chapter, are important words and phrases that you will encounter as you study the chapter. Each key term is highlighted in color at the first significant use in the text, and is included in the glossary or index at the back of the book.

CHAPTER 2

Architectural Drafting Tools

OBJECTIVES

By completing this chapter you will be able to:

- Identify various drafting tools utilized in architectural drafting.
- Describe different types of drafting media and explain when each is appropriately used.
- Classify types of drafting machines according to their designed functions.
- List various types of document storage and reproduction methods for architectural drafting.

KEY TERMS

T Square	Templates
Triangle	Blueprint
Scale	Whiteprint
Drafting media	

To do any job properly and efficiently, you must have the proper tools. This is true of all trades and professions—whether carpentry or drafting, surgery or architecture. Keep this in mind when you select your basic drawing tools or add more advanced and special equipment. If you are interested in becoming an architectural drafter, purchase tools of good quality, since they represent an investment that you can carry with you into your career.

BASIC EQUIPMENT

You should be thoroughly familiar with the following basic tools of the drafter.

Drafting Table or Drawing Board

Although the majority of architectural drafting work is mounted with tape directly to a drafting table (Figure 2–1), a drawing board does have the advantage of portability. Select a smooth-surfaced board without any warpage. The ends should be true and square. Solid basswood, hollow basswood, and metal-edged basswood drawing boards are available.

Figure 2–1 Drafting table. (Courtesy ALVIN.)

REVIEW QUESTIONS

1. What is architectural styling?
2. Distinguish between the two major architectural styles.
3. List four styles classified as traditional.
4. List four styles classified as contemporary.
5. Differentiate between the various colonial styles of architecture.

ENRICHMENT QUESTIONS

1. Walk around your local neighborhood and describe five different houses according to their style.
2. Explain why different architectural styles are dominant in various parts of the United States.
3. Prepare a report on the life and accomplishments of one or more of the following architects. Include examples of structures built and proposed by each. Show how the work of each was influenced by predecessors.
   - a. Frank Lloyd Wright
   - b. Walter Gropius
   - c. Le Corbusier
   - d. Mies van der Rohe
   - e. Eero Saarinen

PRACTICAL EXERCISES

1. Sketch an example of each type of traditional and contemporary architectural style.
2. Prepare a display that exhibits various examples of one architectural style.
3. Build a model of a typical residence of one of the following styles:
   - a. English
   - b. Colonial
   - c. Ranch
   - d. Split-level
   - e. Modern
   - f. Solar
   - g. Underground
   - h. Postmodern
4. Sketch or computer-produce several alternate exterior styles for your original house design.

Review Questions are written to insure that you have adequately read the chapter and understand the material. Most of these questions come directly from the text itself or are explicitly referred to in the reading. These questions can be completed in little time without the use of a computer or drafting tools.

Enrichment Questions are written so that you have to comprehend and apply the material in the chapter and deduce your own answers. Some of these questions require you to use the chapter information and develop appropriate answers, while others require you to produce a simple drawing and explain the solution in words. Access to a CAD program or drafting tools is necessary to complete most of these problems. Enrichment questions require more time to complete than the review questions.

Practical Exercises enable you to synthesize the chapter material and other information in the text. They require considerable time to complete, although some can be done outside of class. You must have access to drafting tools or a CAD program, and occasionally you will need reference books from your school library or a local library.

Each chapter has separate boxed articles that relate to the information you are studying. There are three types of articles: **Career Profiles**, **Real World Applications**, and **Archinet**.

CAREER **PROFILES**

Name: Howard Abrams   *Age:* 63   *Gender:* Male
Occupational Title: Zoning/Code Enforcement Officer
Employer: Indiana Borough, Indiana, PA 15701
Years of experience: 11½ years
Educational background (degree held, school/university): Bachelor of Arts, Penn State University
Approximate starting salary: $25,000 plus benefits

JOB OVERVIEW:
   I cannot describe a typical day on my job as no day is the same. My responsibilites include: issuing building permits, checking construction to make sure it meets code, inspecting unrelated individuals' dwelling units, checking for garbage and litter and much more. I must also attend borough council, planning commission and zoning hearing board meetings. Being a member of the Indiana Borough Planning Commission for twenty years provided me with experience needed to perform my present job.

Career Profiles describe either a person who is employed in an architecture- or drafting-related profession or the career activities and educational requirements for a profession that is related to architecture or drafting.

Read-World Applications allow you to apply the skills you are learning to a variety of both academic and job-related areas. These applications cover a variety of skills, including math, social studies, science, communications, and computers. Many of them give you an opportunity to practice skills you will need on your job, whether or not you choose to enter an architecture-related field.

REAL WORLD **APPLICATIONS**

Quality Presentations

As a professional, you will be responsible for presenting your ideas and findings to a variety of audiences. They may be corporate clients, prospective homeowners, or your supervisor. To ensure a quality presentation, keep in mind these important skills:

1. Be confident. You are trying to convince your audience that what you have to say is important.

2. Use eye contact. This shows that you are interested in the members of your audience and that you know the material.

3. Use correct language. Slang, jargon, and clichés are not appropriate.

4. Speak clearly. Enunciate each word and take your time.

5. Use your hands to emphasize points. Be aware of what your hands are doing at all times or they may distract your audience.

6. Posture is important. Stand up straight with both feet firmly on the floor.

7. Practice, practice, and practice again.

   Choose a drafting tool from this chapter. Prepare a three- to four-minute persuasive presentation for your class. Your goal is to try to convince them to purchase the tool while informing them how to use it correctly and safely.

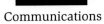

ARCHINET http://www.

Interviewing an Architect

In the same way that you have heard dentists advertise the need for their services, as an architect, you also will be required to sell your services. Knowing why an architect is essential for a building project will enable you to successfully gain clientele. The American Institute of Architects (AIA) has prepared some advice for prospective clients on why and how to choose an architect.
   Use this Internet address, sponsored by the AIA, (http://www.aia.org/begingd.htm) to access information. Now answer these questions.

1. Why should an architect be hired?
2. How can a client save money by hiring an architect?

   Many clients will want to interview several architects before choosing one. It is important that you be prepared for their questions. Review the information on interviewing an architect provided on this Internet page. Which three questions do you feel would be most influential in a client's decision making process? Practice answering these questions by asking a classmate to role play as a prospective client.

Archinet articles encourage you to explore the Internet to discover more about a variety of architecture-related web sites. (A computer with Internet access must be available in order to get full benefit of these articles.)

An icon identifies a specific skill that relates to each article. These icons call out the basic academic skills you will use in each article. These will show you how the academics you're learning in school are used in industry. There are six icons:

Math

Social Studies

Communications

Computers

Job Skills

Science

Acknowledgments

The authors and Delmar Publishers would like to thank the following for their contributions to the text:

Portions of Chapter 6 courtesy of Professor Raymond J. Masters, the Pennsylvania State University, and Professor Richard W. Quadrel, Rensselaer Polytechnic Institute.

"Future Trends" section of Chapter 6 courtesy of Patrick Vernon, State College (Pennsylvania) Area School District.

Portions of Chapters 14 and 50 courtesy of the *Construction Specifications Institute.*

Floor plans for the split-level home in Chapter 14 courtesy of Professor Emeritus M. Isenberg, the Pennsylvania State University.

"Moisture Control" section of Chapter 35 courtesy of B. E. Beneyfield, Moisture Control Consultant.

"Solar Stills" information in Chapter 37 courtesy of Horace McCracken, Solar Equipment Consultant.

Chapter 47 courtesy of Professor John N. Grode, The Behrend College of the Pennsylvania State University.

"Section 08800 Glazing" of Chapter 50 courtesy of Jack Risheberger & Associates, Registered Architects and Engineers.

Floor plans for the single-story contemporary home used throughout the text courtesy of Mr. and Mrs. Donald W. Hamer, State College, Pennsylvania.

Floor plans for the A-framed solar house in Appendix A courtesy of Professors Richard E. and Mary J. Kummer, the Pennsylvania State University.

The authors and Delmar Publishers would like to thank the many professionals who reviewed the manuscript. Their feedback, assistance, and creative suggestions are greatly appreciated.

Gareth Abell
Sandpoint High School Drafting Department
Sandpoint, Idaho

Wes Barker
Sherman High School Drafting Department
Sherman, Texas

Daniel E. Engstrom
Indiana Junior High School Technology
  Department
Indiana, Pennsylvania

John Hammer
Rhinelander High School Technology Department
Rhinelander, Wisconsin

John W. Hughes
Lebanon Union High School Vocational
  Department
Lebanon, Oregon

Lawrence D. Huntzinger
Western Oklahoma State College Technology
  Department
Altus, Oklahoma

Ernest L. Johnson
Savannah Technical Institute Drafting
  Department
Savannah, Georgia

Brian Matthews
Wake Technical Community College
Raleigh, North Carolina

Joe M. Pool
Pace High School Drafting Department
Pace, Florida

Arnold P. Ruskin
Torrey Pines High School Applied Technology
  Department
Encinitas, California

William Scott Thomas
Deer Valley V-Tech Center Drafting Department
Glendale, Arizona

Introduction to Architectural Drafting

SECTION I

CHAPTER 1

Architectural Styling

OBJECTIVES

By completing this chapter you will be able to:

- Differentiate between the principal characteristics of major traditional and contemporary architectural styles.

- List some of the contributions of the world's leading architects.

KEY TERMS

Styling	Victorian
Traditional	Contemporary
English	Modern
Georgian	Ranch
Regency	Split-level
Colonial	Postmodern

One of the most exciting results of the study of architecture is seeing your dreams take shape. However, there's more to drawing a floor plan than sketching some lines on paper or commanding lines to move on a computer monitor to form a layout that seems pleasing to you. You must decide upon the style of your home, consider the site where your home will be located, plan for financing, make a number of basic decisions regarding the size and shape of rooms, number of floors, type of construction and energy system, and, possibly, include special provisions for accessibility and safety. You'll also need to know about the tools and skills needed to transfer your vision to paper or printout. By the time you have finished studying this book, you will have the basic knowledge needed to create a plan of your own dream home, the first step in transforming your dream into reality.

Why is a working knowledge of architectural styling important to you? The first decision you must make when you are planning to build a house is its **styling** (exterior appearance). Do you prefer traditional or modern? Ranch house or split-level? The style you choose influences the interior layout. Some styles require a one-floor house, others one and one-half or two floors. Certain styles call for a symmetrical design with the features evenly balanced on each side of a central hall. Understanding the advantages of a variety of styles helps you decide which style is most appropriate for you. We shall first discuss the styles in more common use, and then show you how to pick the style that will best suit your requirements.

There are two general classifications of popular styles: *traditional* and *contemporary*. Let us take each in turn.

TRADITIONAL

A house built in the **traditional** style today is a copy, with certain modifications, of a kind of house built previously. There are many dwellings built in the traditional style, but the trend has been toward contemporary. Many people feel that some traditional styles overly limit freedom of layout. Also, many traditional styles are of European, not truly North American, origin. However, the traditional house has stood the test of years and will always have a place in the hearts of those who prefer tradition.

English

English houses (Figures 1–1 through 1–5) are fashioned after the type built in England before the eighteenth century. Historical subdivisions of English architecture are Old English (Figure 1–2), Tudor, and Elizabethan. Each has its own particular characteristics, but all have common features as well. For example, the interior layout is informal and unsymmetrical due to the lack of exterior symmetry. Walls are of stone, brick, or stucco, and are sometimes half-timbered. The

Figure 1–1   *Architect's presentation drawing for a Tudor estate residence.* (Courtesy Scholz Design, Toledo, Ohio.)

Figure 1–2   *Old English styling.*

Figure 1–4   *Regency styling.*

Figure 1–3   *Georgian styling.*

Figure 1–5   *Careful exterior planning transformed this simple gabled residence into an attractive English design.*

REAL WORLD ━━━ APPLICATIONS

Becoming an Architect

If you are considering a career in architecture, you have probably wondered what is required to become an architect. First, the term "architect" is legally restricted to usage by registered architects. Becoming registered is a lengthy and challenging endeavor. However, if pursued, the outcome is a rewarding and respected career.

For architectural registration, most states require graduation from a National Architectural Accrediting Board (NAAB) accredited university and three years of practical experience under the supervision of a professional architect, plus the successful completion of a comprehensive written examination. Architectural registration is possible without a degree, but extensive and varied practical experience is required as a substitute for formal schooling.

A Bachelor of Architecture program is a five-year university program. Typical specialized courses for architectural majors include:

🏛 Architectural history

🏛 Architectural graphics

🏛 Computer applications

🏛 Building materials

🏛 Architectural design (five years)

🏛 Structural design

🏛 Electrical design

🏛 Mechanical design

🏛 Landscape planning

🏛 Construction management

🏛 Professional practice

In addition, nearly all baccalaureate programs include common, basic courses such as orientation, English, speech, mathematics, chemistry or physics, engineering mechanics, social sciences, humanities, physical education, and electives.

Students obtaining a bachelors degree in a related field may pursue a Master of Architecture degree from a two-year program. Persons holding degrees in another discipline may enroll in a three- or four-year Master of Architecture program. A Master's degree in architecture usually emphasizes urban planning or regional planning. You can be assured of a high-quality baccalaureate or master's program if you choose a program accredited by the NAAB. A list of universities offering accredited architectural programs (approximately 100 in the United States) can be obtained at no charge from NAAB, 1735 New York Ave. N.W., Washington, DC 20006. Phone: 202-783-2007.

gable ends may be of dark-stained, hand-hewn beams. If the second floor overhangs the first, carved drops may be used at the corners. Fenestration (the arrangement of windows in a wall) is completely random; occasionally a window may appear to be built right through a chimney. The windows are casements (side hinged) and are made of small diamond panes after the prototype, a style that developed because large sheets of glass were not manufactured at the time. Roofs are steeply pitched, the eaves and ridges being at various levels.

The fireplace used to have a more important function in the home than it does today, since it furnished all cooking and heating facilities. Massive chimneys were usually topped off with chimney pots (vertical extensions of tile or brick). Although some features of the English house may be used in constructing small homes, the style is at its best in larger houses.

Although you will find many examples of English homes in the United States today, most of them were built more than fifty years ago. The English style is not popular for new homes (although occasionally individual features are copied) because builders today cannot build in stone, brick, or stucco at a reasonable price. Moreover, the trend in this country is away from continental ideas and toward American Colonial and modern styles.

Georgian

As you will see in the section on architectural history, the Renaissance was characterized by the rebirth of the architecture of the classical civilizations. **Georgian** architecture (1714–1760) was also developed from classical principles of formality and symmetry (Figure 1–3). Many classical details were used, such as pedimented doorways (triangular areas above the doorways), elaborate cornices (parts of a roof that extend beyond the walls), and pilasters (bas-relief columns).

The Georgian house (so called for the kings under whom it flourished) is a large one, two or more floors high, with a gently sloping hip roof. The front of the house must be strictly symmetrical. The front entrance is at dead center, with windows equally spaced on either side, the second-floor windows directly above the first-floor windows. Symmetry is carried through even to the chimneys—one at either end. If only one chimney is required, a false one is included so that symmetry will be preserved. We often find pilasters, sidelights, and columned porches two stories high. The interior plan is, of course, a center-hall one, with bedrooms on the second floor. These restrictions of symmetry do not affect the design of the rear of the house, however.

We shall soon see how the Georgian house served as the inspiration for the various American Colonial styles—becoming adapted to wood instead of stucco and stone.

Regency

The king of England at the time of the American Revolution was George III. His son was appointed regent by the English Parliament to reign in his father's place when George III became too old to rule. This period in English history, known as the *Regency,* gave birth to a particular architectural style.

The **Regency**-style house (Figure 1–4) is similar to the Georgian but has cleaner lines and finer details. Exterior walls are usually brick, often painted white. The Georgian formality and hip roof still prevail. Other typical Regency details are long shutters at the first-floor level, curved copper bay or porch roofs, curved side wall extensions, and fancy iron-work tracery around porches.

Colonial

The term **colonial** is loosely applied to any style developed by colonizers. Colonial architecture, or to be more specific, English Colonial architecture, consists of Early American (before 1720) and American Colonial (after 1720). American Colonial is a modification of Georgian and consists of various regional types, such as New England, Southern, and Dutch.

Even though there were no trained architects in North America before the Revolution, many well-designed houses were built. The reason is that a workable knowledge of architecture was part of every gentleman's training. Each landowner designed his own house (Jefferson designed his own and several others) with the aid of English architectural handbooks which simplified the use of the Renaissance orders. If these handbooks were followed, the landowner could not go wrong. For materials, wood was plentiful in all the colonies, although brick was often used in the South and stone in the middle colonies.

New England Colonial   The New England house (Figure 1–12, page 8) was simple and unpretentious, but of such good design and proportion that it has been copied through the years. Since this style is modeled after the Georgian, the front elevation is symmetrical. Wood was plentiful in the North and was used even for classical columns and pilasters.

Exterior finish was a narrow clapboard with vertical boards covering the corners so that the clapboards did not have to be mitered (beveled at 45 degrees). The roof pitch was steep to shed the heavy snows, and sometimes the eaves at the rear dropped a floor lower than the eaves at the front of the house. In early houses, one chimney was located at the center to serve as many rooms as possible, and in later work there were two chimneys, one at either end. The first windows were glassless and closed by wooden shutters. These were replaced by diamond-paned casements imported from England, and eventually by the double-hung (nicknamed *guillotine*), rectangular-paned window. This house was called a *salt box* (Figure 1–6) because it looked like the salt boxes sold in early general stores.

Garrison   Like all colonists, the new arrivals in North America tried to duplicate the type of house they had known at home. One of the first breaks with this tradition was the garrison house

(Figure 1–7), modeled after the blockhouses the colonists used to fight off Indian raids. These blockhouses had overhanging second stories, which made them difficult to scale and easier to defend. In imitation of these forts, some houses were built with an overhanging second floor, although they were otherwise similar to the New England Colonial. The ends of the heavy timbers projecting down below the overhang were frequently carved in some fashion. An English version (with diamond pane windows) of a garrison house is shown in Figure 1–10.

Cape Cod   The Cape Cod section of Massachusetts has become known for its charming small homes. Although a true Cape Cod house (Figures 1–8 and 1–9) has very definite characteristics, nearly any small house having a steeply pitched gable roof with the eave line at the top of the first story may be called Cape Cod. Actually, a true reproduction will have the following details: double-hung, small-paned windows with shutters, shingle or clapboard walls, a

Figure 1–6   New England "salt box."

Figure 1–8   A "3/4" cape (1 + 2 windows).

Figure 1–7   American garrison.

Figure 1–9   A "full" cape (2 + 2 windows) with multiple dormers.

Figure 1–10   *Front elevation of the English garrison Z residence. See Appendix A for plans.*

Figure 1–11   *Fine contemporary residence with traditional flairs.* (Courtesy Scholz Design, Toledo, Ohio.)

Figure 1–12 Representative New England colonial design. (Courtesy Scholz Design, Toledo, Ohio.)

Figure 1–13 Architect's rendering of a proposed Victorian home. (Courtesy Scholz Design, Toledo, Ohio.)

wood-shingled roof (or an imitation), a main entrance in the center of the front elevation, and a massive center chimney above the multiple fireplaces warming the first floor rooms (or, again, an imitation massive chimney). Small dormers at the front or a shed dormer (jokingly called a *dustpan*) at the rear adds greatly to the usability of the second floor. An American architect associated with the Cape Cod house is Royal Barry Wills.

Southern Colonial   It is true that better examples of architecture can be found in the larger, more expensive houses than in the small ones. This is natural, since only the wealthy could afford the expense of the details of styling. This distinction was particularly true in the South where some very lovely plantation houses were built (Figure 1–14). They differed from the New England Colonial mainly in that a flat porch, two stories high and supported by columns, was used to shade the front windows. Brick was the usual material. Today the Southern Colonial style is not as popular as the colonial styles of the North because the high porch looks ungainly on any but a very large house.

Dutch Colonial   Dutch and German settlers in New Amsterdam (New York), New Jersey, and Pennsylvania built houses with steeply pitched roofs, called *gambrel* roofs (Figure 1–15). This type of roof was invented by a French architect named Mansard, who designed a double-pitched roof, permitting the attic to be used as another floor. The tale goes that this was done originally to evade the heavier tax on two-story houses. When this roof runs on four sides, it is called *mansard* after the originator; when it is on only two sides, *gambrel* or *Dutch Colonial.* The more authentic example of Dutch Colonial has a slightly curved projecting eave with a continuous shed dormer window. Stone construction should be used.

French Colonial   The best examples of authentic French Colonial architecture can be found in the old French Quarter of New Orleans. Here, buildings are crowded together, all with common characteristics, but each different. The hallmark of this style is a flat facade relieved only by wrought-iron balconies. The balconies contain fancy scroll work and may be supported by delicate iron columns or trellises. The plastered fronts are tinted pink, yellow, or green.

French provincial (rustic) styles are identified by their mansard roofs, rough stone finish, and round-headed dormer windows. The larger French chateau (stately) styles are identified by their steep hip roofs, stucco finish, and casement windows (Figure 1–16).

Spanish   A Spanish house (Figure 1–19) should appear to have adobe walls, in keeping with the material originally available. Roofs are tiled and low-pitched. A close reproduction should be built around a patio and have open-timbered ceilings of rough-hewn logs.

Victorian   Queen Victoria reigned in Great Britain for sixty-three years (1837–1901), and ornate houses built during this period were called Victorian. Fanciful wood trimming on Victorian houses was called *gingerbread.* Although difficult to maintain, remaining Victorian houses are now cherished, and some Victorian features are now being revived in contemporary buildings (Figure 1–13).

Figure 1–14   *Southern colonial.*

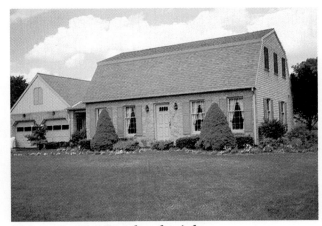

Figure 1–15   *Dutch colonial.*

Figure 1–16 Contemporary residence in the style of a French chateau. (Courtesy Scholz Design, Toledo, Ohio.)

CONTEMPORARY ARCHITECTURE

Many terms such as *modernistic, futuristic,* and *functional* have been used to distinguish nontraditional architecture from traditional. The two terms most widely accepted are **contemporary** and **modern.** To be perfectly correct, *contemporary* indicates any building erected at the present time regardless of style. However, since present-day construction consists mainly of ranch, split-level, and modern, we will restrict this term to these three types. Occasionally *contemporary* is used synonymously with *modern.*

Ranch

The popularity of the **ranch** house (Figure 1–17) today is partly due to its blend of past and present. Most contemporary conveniences can be incorporated into a ranch house design without appearing too extreme. Also, the demand for light construction and low, land-hugging designs has increased, as has the desire for one-floor plans with no stairs to climb.

Nearly any one-story house is called a ranch house today, although it should be rambling,

have an informal plan and a low-pitched roof, preferably hipped. The ranch is well suited to large, flat lots and may be built in the high, medium, and even low price ranges.

To permit better use of the basement level, ranch houses may be built with a basement partially above ground and partially below ground. This style is termed *raised ranch.* When the entrance to a raised ranch is positioned at mid-height between these two floor levels, the style is termed *split entrance.*

Figure 1–17 Ranch-styled residence.

Split-level

A **split-level** house (Figure 1–18) is arranged so that floor climbing is limited to half-flights of stairs. It is a kind of combination of ranch and two-story construction. It might be laid out with bedrooms and bath on the highest level; living, dining and kitchen half a floor lower; and garage and *free room* half a floor lower still (under the bedroom level). A free room is usually a recreation or family room built in "found" or "free" basement areas. The levels may be split between ends or between front and back depending on the terrain. Although the majority of split-levels are built on a more or less stock plan, infinite variations are possible (Figure 1–20). These should offer a real challenge to the designer.

Modern

A truly modern building (Figure 1–21) should be "honest." That is, it should have a fresh ap-

Figure 1–18   Traditional split-level residence.

proach to requirements such as shelter, light, and circulation. The modern architect feels that drawing on the styles of the past to any great extent is "dishonest." He or she designs each house to suit the requirements of those living in it without being fettered by historical

Figure 1–19   Contemporary residence constructed in the Spanish tradition. (Courtesy Scholz Design, Toledo, Ohio.)

Figure 1–20 Contemporary split-level residence.

Figure 1–21 Modern-styled residence.

styling, symmetry, or useless decoration. Some fine examples are homes that apply new technological advances without attempting to hide them behind decorative facades. The requirement for large areas of solar collectors affects the design of solar homes (Figures 1–23 and 1–25), and the requirement for earth covering affects the design of underground homes (Figure 1–24).

A key word used in modern architecture is *functional*. This means that every element of the house should have a function—a reason for existing. The expression "form follows function" is a simplification of the idea that exterior appearance, or form, should be subordinated to the functional and structural aspects.

But the modern house should not be merely a "machine for living," since it also has the function of meeting the family's aesthetic needs. This one factor—a sensitivity to aesthetic need—distinguishes the truly great modern design from the mediocre. Even a mundane structure such as a water tank, when designed by an inspired architect-engineer, can be a community asset rather than a visual pollutant. The 10-million-gallon tank in Figure 1–28 is braced by

Figure 1–22 Contemporary residence with classical detailing. (*Courtesy Scholz Design, Toledo, Ohio.*)

REAL WORLD APPLICATIONS

Biography of an Architect

 Who are the famous architects of the modern era? When were they born? What is their cultural background? How and at what stage in their life did they become interested in architecture? Did they have formal training? What were some of their greatest accomplishments? What events in their life significantly influenced their architectural designs? Answer these questions about one of the architects of the modern era mentioned in this chapter.

Your first step will be to go to the library and use the card catalog or computerized card catalog to find books about the architect you have chosen. You may need to look in general architectural books to find the information. Read the relevant sections of the books you select. As you read, keep in mind the questions you are answering. Also, do not forget to take notes as you read. These notes will be the basis for the biographical essay you will write. In taking these notes, follow these guidelines:

1. Briefly summarize the main points and list only the important details. Paraphrase, but do not copy the written text.

2. When you do quote directly, use quotation marks. Select these quotes carefully; they should include essential information or should be selected for their unique or distinctive language.

3. Write as concisely as you can. Use abbreviations.

4. Write a descriptive title or heading at the top of each note card.

5. Mark your notes to emphasize important information. You may underline, highlight, star, check, and indent.

star-shaped steel members and illuminated indirectly at night. The slender steel tower shown in Figure 1–26 encloses a 320,000-gallon water tank which is set in a reflecting pool. A sixteen-bell carillon is installed in the cage at the top of the spire.

Since sensitivity and originality are objectives of the modern architect, there can be no

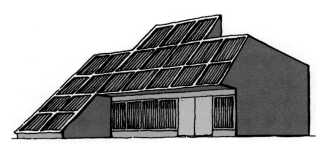

Figure 1–23   Solar house.

Figure 1–24   Underground house.

Figure 1–25 *Passive solar living area.* *(Courtesy Scholz Design, Toledo, Ohio.)*

typical example of modern. However, the style is often marked by a simple (but interesting) design; both native and machine-produced materials; low structures with flat, shed, or clerestory roofs; provision for indoor-outdoor living; and great expanses of glass. Some architects who stand out in modern architecture are Alvar Aalto, Le Corbusier, Walter Gropius, Victor Gruen, Philip Johnson, Louis Kahn, Ludwig Mies van der Rohe (Figure 1–27), Pier Luigi Nervi, Richard Neutra, Oscar Niemeyer, Ioh Ming Pei, Eero Saarinen, Edward Durell Stone (Figure 1–26), Louis Sullivan, Frank Lloyd Wright (Figures 1–29, 1–30, and 1–31), and Minoru Yamasaki (Figure 1–32).

Postmodern

Postmodern architecture applies the ornamentation of past classical designs to the clean lines of contemporary buildings, often in a tongue-in-cheek fashion. For example, classical columns might be used for decoration without supporting any lintels.

A leading postmodern architect is Robert Venturi of Philadelphia. In response to Mies van der Rohe's dictum "Less is more," Venturi proposed "Less is a bore." One of the earliest examples of major postmodern structures is the AT&T Headquarters in New York City, designed by Philip Johnson and John Burgee. Constructed in

Figure 1–26   Water tank and carillon on the SUNY campus in Albany, NY. Edward Durell Stone, architect. (Courtesy SUNY Albany.)

Figure 1–27   Farnsworth House at Plano, Illinois. Mies van der Rohe, architect. (Photographer: Bill Hedrich, Hedrich-Blessing.)

the 1980s, this 645-foot tower is finished with a Chippendale roof (Figure 1–33). A more recent example is the Palmer Museum of Art at The Pennsylvania State University designed by Charles Moore. Built in 1993, this museum combines ancient masonry forms with current technology such as neon lighting (Figure 1–34). Other noted postmodern architects are Peter Eisenman, Michael Graves, Robert Stern, and Stanley Tigerman.

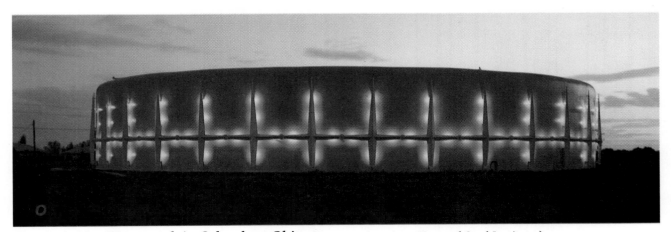

Figure 1–28   Water tank in Columbus, Ohio. (Courtesy American Iron and Steel Institute.)

Figure 1–29 *The Solomon R. Guggenheim Museum, New York. Frank Lloyd Wright, Architect.* (Photograph by David Heald. © The Solomon R. Guggenheim Foundation, New York. Used by permission.)

Figure 1–30 *Fallingwater at Bear Run, Pennsylvania. Frank Lloyd Wright, architect, 1936.* (Courtesy The Western Pennsylvania Conservatory.)

Figure 1–31   Seamless corner window (first used in Wright's Fallingwater). (Courtesy Pella Corporation.)

Figure 1–32   Manufacturers and Traders Trust Building, Buffalo, New York. Minoru Yamasaki and Associates, architects. (Courtesy Balthazar Korab Ltd. Photography, Troy, Michigan.)

Figure 1–33   AT&T Headquarters, New York City. Philip Johnson and John Burgee, architects. (Courtesy Hedrich-Blessing Photographers, Chicago, Illinois.)

Figure 1–34   Interior entrance elevation of Palmer Museum of Art, The Pennsylvania State University. (Photographer: Howard P. Nuernberger.)

REVIEW QUESTIONS

1. What is architectural styling?
2. Distinguish between the two general classifications of architectural styles.
3. List four styles classified as traditional.
4. List four styles classified as contemporary.
5. Differentiate between the various colonial styles of architecture.

ENRICHMENT QUESTIONS

1. Walk around your local neighborhood and describe five different houses according to their style.
2. Explain why different architectural styles are dominant in various parts of the United States.
3. Prepare a report on the life and accomplishments of one or more of the following architects. Include examples of structures built and proposed by each. Show how the work of each was influenced by predecessors.

   a. Frank Lloyd Wright   d. Mies van der Rohe

   b. Walter Gropius   e. Eero Saarinen

   c. Le Corbusier

PRACTICAL EXERCISES

1. Sketch an example of each type of traditional and contemporary architectural style.
2. Prepare a display that exhibits various examples of one architectural style.
3. Build a model of a typical residence of one of the following styles:

   a. English   e. Modern

   b. Colonial   f. Solar

   c. Ranch   g. Underground

   d. Split-level   h. Postmodern

4. Sketch or computer-produce several alternate exterior styles for your original house design.

Architectural Drafting Tools

OBJECTIVES

By completing this chapter you will be able to:

🏛 Identify various tools used in architectural drafting.

🏛 Describe different types of drafting media and explain when each is appropriately used.

🏛 Classify types of drafting machines according to their designed functions.

🏛 List various types of document storage and reproduction methods for architectural drafting.

KEY TERMS

T Square	Templates
Triangle	Blueprint
Architect's scale	Whiteprint
Drafting media	

To do any job properly and efficiently, you must have the proper tools. This is true of all trades and professions—whether carpentry or drafting, surgery or architecture. Keep this in mind when you select your basic drawing tools or add more advanced and special equipment. If you are interested in becoming an architectural drafter, purchase tools of good quality, since they represent an investment that you can carry with you into your career.

BASIC EQUIPMENT

You should be thoroughly familiar with the following basic tools of the drafter.

Drafting Table or Drawing Board

Although the majority of architectural drafting work is mounted with tape directly to a drafting table (Figure 2–1), a drawing board does have the advantage of portability. Select a smooth-surfaced board without any warpage. The ends should be true and square. Solid basswood, hollow basswood, and metal-edged basswood drawing boards are available.

Figure 2–1 *Drafting table. (Courtesy ALVIN.)*

T Square

Make certain that the blade of the **T square** (Figure 2–2) is perfectly straight (except for a desirable bow away from the drawing surface) and free of nicks. For accurate work, the blade must be fastened securely to the head.

Triangles

Clear-plastic **triangles** (Figure 2–3) are used in sets of two: 30°–60° triangle and 45° triangle. As with the T square blade, the outer edges should be flat and nick-free. Some drafters prefer an adjustable triangle (Figure 2–4), which permits lines to be drawn at any angle.

Scales

The architectural drafter normally uses an **architect's scale,** although on occasion an engineer's or mechanical engineer's scale is needed. Either triangular or bevel scales are used according to preference (Figure 2–5). The architectural scales are:

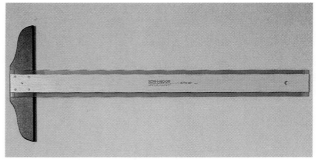

Figure 2–2   T square. (Courtesy Koh-I-Noor Inc.)

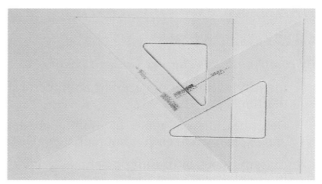

Figure 2–3   Triangles. (Courtesy Koh-I-Noor Inc.)

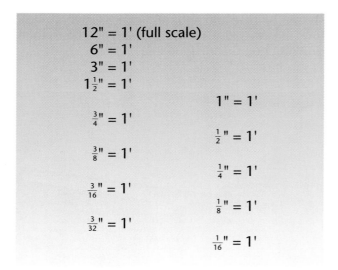

12" = 1' (full scale)
6" = 1'
3" = 1'
$1\frac{1}{2}$" = 1'

1" = 1'

$\frac{3}{4}$" = 1'

$\frac{1}{2}$" = 1'

$\frac{3}{8}$" = 1'

$\frac{1}{4}$" = 1'

$\frac{3}{16}$" = 1'

$\frac{1}{8}$" = 1'

$\frac{3}{32}$" = 1'

$\frac{1}{16}$" = 1'

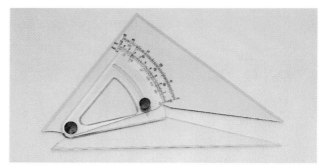

Figure 2–4   Adjustable triangle. (Courtesy Koh-I-Noor Inc.)

Notice that there are really *two* architectural scale systems. One is based on the full scale and proceeds to smaller scales; the other is based on a $\frac{1}{16}$" = 1' scale and proceeds to larger scales. Combined, these two architectural scale systems offer a wide range of scale choice.

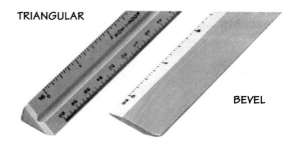

TRIANGULAR

BEVEL

Figure 2–5   Scales. (Courtesy Koh-I-Noor Inc.)

REAL WORLD APPLICATIONS

Scale Conversions

A line is drawn that is $4\frac{3}{8}''$ long. Without using an architect's scale, how long is the line it represents if the scale is $\frac{1}{4}'' = 1'0$? The ability to manipulate fractions is an important skill in architectural drafting because nearly all projects are drawn to a smaller scale, and therefore sizes must be transferred to an appropriate measurement. Follow these steps to solve this problem:

1. Since each $\frac{1}{4}''$ represents one foot, the problem can be solved by determining the number of $\frac{1}{4}''$ intervals in the $4\frac{3}{8}''$ line. That is, divide $\frac{1}{4}''$ into the $4\frac{3}{8}''$ line:

$$\frac{4\frac{3}{8}}{\frac{1}{4}}$$

2. Convert the $4\frac{3}{8}$ mixed numerator to a fraction of $\frac{35}{8}$:

a. $4 \times \frac{8}{8} = \frac{32}{8}$

b. $\frac{32}{8} + \frac{3}{8} = \frac{35}{8}$

$$\frac{\frac{35}{8}}{\frac{1}{4}}$$

3. To divide by a fractional denominator, invert the denominator and multiply:

$$\frac{35}{8} \times \frac{4}{1}$$

4. Simplify:

$$\frac{35}{\overset{}{\underset{2}{8}}} \times \frac{\overset{1}{4}}{1} = \frac{35}{2} = 17\frac{1}{2}$$

$17\frac{1}{2}$ intervals or $17\frac{1}{2}$ feet

5. Check this answer with your architect's scale.

Complete the table by calculating the distance of each line when used with the appropriate architectural scale. Check your answers with your architect's scale.

Line Length	Scale		
	$\frac{1}{4}'' = 1'0$	$\frac{1}{16}'' = 1'0$	$3'' = 1'0$
$5\frac{1}{2}''$			
$2\frac{3}{8}''$			
$7\frac{3}{4}''$			
$10\frac{3}{16}''$			
$3\frac{13}{32}''$			

Drawing Sets

A satisfactory drawing set (Figure 2–6) should contain as a minimum: dividers, pencil compass, pen compass, and drafting pen or technical fountain pen. Many sets also contain mechanical drafting pencils. The instruments in the better sets are constructed of stainless steel or nickel silver; less costly sets are of chrome-plated brass.

Drafting Pencils

One refillable drafting pencil with an assortment of different grade leads is sufficient for the classroom. A professional drafter, however, has a number of refillable pencils—each color-coded to one lead grade so that no time will be lost in switching leads. Drafting pencil leads are graded from 9H (hard) to F (firm) to 6B (black), as shown in Table 2–1. Some drafting pencils are

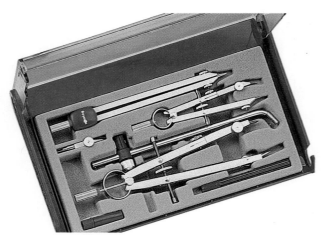

Figure 2–6 *Drawing instrument set.* (*Courtesy Koh-I-Noor Inc.*)

designed to hold leads that require sharpening, but thin drafting pencils hold leads of such small diameter that no sharpening is needed (Figure 2–7). Thin leads are available in diameters of 0.3 mm, 0.5 mm, 0.7 mm (all recommended for architectural drafting), and 0.9 mm.

TABLE 2–1   *Grades of Pencil Leads*
9H
8H
7H
6H
5H
4H
3H
2H ⎫ Used for most
H ⎬ architectural
F ⎭ work
HB
B
2B
3B
4B
5B
6B

Figure 2–7 *Thin-lead drafting pencil.* (*Courtesy Koh-I-Noor Inc.*)

Lead Pointers

A mechanical lead pointer (Figure 2–8) provides an easy method to obtain perfectly formed conical points. The abrasive liners can be replaced quickly and easily.

Erasers

A wide variety of pencil erasers (Figure 2–9) is available. A satisfactory eraser should be capable of completely removing pencil and ink lines without roughing the surface of the paper or leaving colored marks. The artgum eraser is designed to remove smudges rather than lines. Electric erasing machines (Figure 2–10) are helpful when a great amount of erasing is needed.

Lettering Guides

A number of devices to assist in drawing guidelines for lettering are available. The RapiDesign guide illustrated in Figure 2–11 is used for $\frac{1}{16}$", $\frac{3}{32}$", $\frac{1}{8}$", or $\frac{5}{32}$" guidelines. Some drafters prefer the Ames lettering instrument or Braddock-Rowe lettering triangle.

Technical Fountain Pen

Some drafters prefer a technical fountain pen (Figure 2–12) to a standard ruling pen since they can obtain a variation in line weight by

Figure 2–8 *Lead pointer.* (*Courtesy Staedtler, Inc.*)

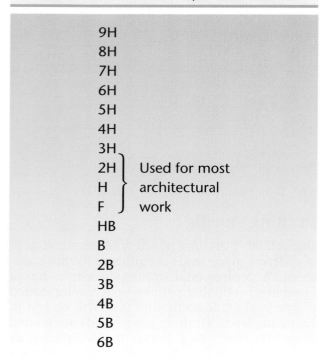

Figure 2–9   *Erasers.* *(Courtesy Staedtler, Inc.)*

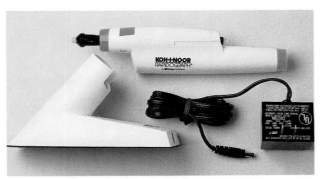

Figure 2–10   *Rechargeable electric eraser.*
(Courtesy Koh-I-Noor Inc.)

changing the speed of the stroke. This is most useful in preparing presentation drawings.

Drafting Media

Different **drafting media** are used for different stages of architectural design. Preliminary architectural designs (called *schemes*) are sketched on inexpensive *tracing paper*, but finished working drawings are drafted on *tracing vellum* or on *drafting film.*

Tracing vellum is a durable medium with good translucency. It is manufactured from 100 percent rag stock treated with high-impact

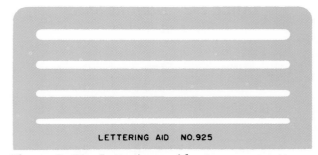

Figure 2–11   *Lettering guide.* *(Courtesy Berol®
Corporation.)*

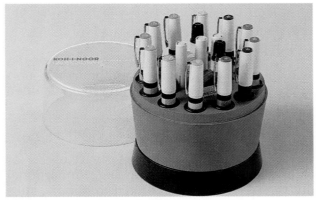

Figure 2–12   *Technical fountain pen set.* *(Courtesy Koh-I-Noor Inc.)*

resins to prevent graphite from penetrating so that erasures are ghost-free. If lead still remains after erasing, it is called "ghosting." This is undesirable because it appears in reproductions.

Drafting film is a polyester film commonly called by the trade name *Mylar.* It may be used with graphite lead, plastic lead, or ink. It has excellent transparency and is dimensionally stable, unaffected by temperature or humidity. It has superior erasability, ideal for continuous revisions. Drafting film is available in single or double matte finish. (Double matte permits drafting on both sides of the film.)

All drafting media are available in cut sheets and rolls. Sheets are sized in multiples of $8\frac{1}{2}$" × 11" and 9" × 12" as shown in Table 2–2. Both sets of sizes are approved by the American National Standards Institute and designated by letters A through E. Rolls are available in widths from 18" to 48" in 6" increments and in lengths of 20 and 50 yards.

TABLE 2–2 Drafting Sheet Sizes in Inches

Designation	Size	Size
A	$8\frac{1}{2} \times 11$	9×12
B	11×17	12×18
C	17×22	18×24
D	22×34	24×36
E	34×44	36×48

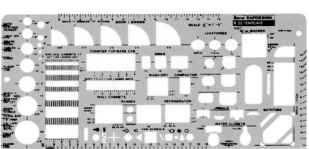

Figure 2–14   Circle Template. (Courtesy Berol® Corporation.)

ADVANCED AND SPECIAL EQUIPMENT

The following equipment, although not normally found in a school drafting room, is often used in a professional drafting office.

Proportional Dividers

The instrument in Figure 2–13 is very useful for enlarging or reducing a drawing. Proportional dividers are usually calibrated to obtain ratios from 1:1 up to 10:1.

Templates

The architectural drafter uses a large variety of **templates** to speed the work. A circle template (Figure 2–14) may be used to supplement the compass. The architectural template (Figure 2–15) simplifies the drawing of kitchen and bathroom fixtures, and the furniture template (Figure 2–16) helps in the preliminary design stage to draw furniture the proper size. Many other specialized templates are available, such as the computer work station template (Figure 2–17), which gives the shape and various sizes of work station furniture and peripherals.

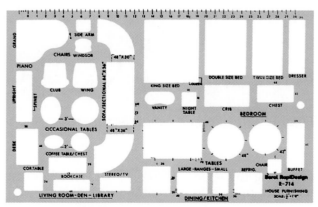

Figure 2–15   Architectural template. (Courtesy Berol® Corporation.)

Figure 2–16   Furniture template. (Courtesy Berol® Corporation.)

Figure 2–13   Proportional dividers. (Courtesy ALVIN.)

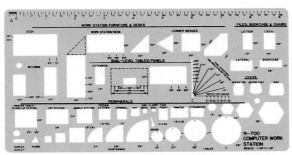

Figure 2–17 *Computer work station template.*
(Courtesy Berol® Corporation.)

Lettering Sets

Although pencil lettering is best drawn freehand, ink lettering may be improved by the use of lettering stencils (Figure 2–18). These stencils can provide a variety of sizes and styles of letters. Figure 12–13 (page 137) shows an architectural alphabet formed with the aid of a lettering stencil.

Parallel-rule Drafting Table

A parallel-rule drafting table (Figure 2–19) or drawing board may be used in place of the standard drawing board and T square. The straightedge can be moved up and down, and it is designed to remain perfectly horizontal, thus allowing accurate work.

Drafting Machines

Drafting machines may be used in place of the standard drawing board, T square, triangles, scales, and protractor. As with the parallel-rule drawing board, a straightedge remains hor-

Figure 2–18   Lettering set. *(Courtesy Koh-I-Noor Inc.)*

Figure 2–19   Parallel-rule drawing board.
(Courtesy ALVIN.)

izontal. In addition, though, there is a vertical edge. Both edges are graduated to act as scales and may be rotated to any position by pressing a release button. The X-Y plotter (Figure 2–20) and arm type (Figure 2–21) perform similar functions.

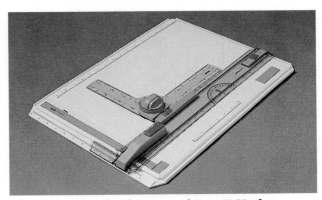

Figure 2–20   Drafting machine, X-Y plotter type. *(Courtesy ALVIN.)*

Figure 2–21   Drafting machine, elbow type.
(Courtesy VEMCO.)

REAL WORLD APPLICATIONS

Quality Presentations

 As a professional, you will be responsible for presenting your ideas and findings to a variety of audiences. They may be corporate clients, prospective homeowners, or your supervisor. To ensure a quality presentation, keep in mind these important skills:

1. Be confident. You are trying to convince your audience that what you have to say is important.

2. Use eye contact. This shows that you are interested in the members of your audience and that you know the material.

3. Use correct language. Slang, jargon, and clichés are not appropriate.

4. Speak clearly. Enunciate each word and take your time.

5. Use your hands to emphasize points. Be aware of what your hands are doing at all times or they may distract your audience.

6. Posture is important. Stand up straight with both feet firmly on the floor.

7. Practice, practice, and practice again.

Choose a drafting tool from this chapter. Prepare a three- to four-minute persuasive presentation for your class. Your goal is to try to convince them to purchase the tool while informing them how to use it correctly and safely.

Drafting Tables

Automatic drafting tables (Figure 2–22) are used for large-scale commercial drafting. Drafting table surfaces can be adjusted for height and tilted up to 90 degrees by means of electrical, mechanical, hydraulic, or pneumatic actuators. A drafting table with an endless belt drafting surface is useful for very large drawings. A drawing taped to the belt can be moved horizontally or vertically to a convenient position for the drafter with the unused portion of the drawing rolled to the back side of the drafting surface.

Reproduction Processes

Working drawings are drawn on translucent tracing vellum or film so that prints can be produced easily. An early reproduction process produced prints having white lines on a dark blue background. They were called **blueprints**. Although this process has been replaced by the

Figure 2–22 *Adjustable drafting table. (Courtesy Mayline Company.)*

whiteprint process, the term *blueprint* is still used to refer to any type of print.

Whiteprints **Whiteprints** are reproductions having dark lines on a white background. Most whiteprints are reproduced using the *diazo* process, often called *Ozalid,* a trade name formed from *diazo* (a chemical compound) spelled backwards. There are two basic steps in the diazo process (Figure 2–23).

1. *Exposing.* The translucent tracing is placed over a sheet of diazo-coated paper, and both are fed into a machine that exposes them to ultraviolet light. Where light passes through the tracing, the coating on the print paper is deactivated and will be white; where light is prevented by a pencil line from reaching the print paper, the coating will remain as an exact image of the line.

2. *Developing.* The print paper, separated from the tracing, is exposed to ammonia vapor to develop the images into dark lines. The dark-

ness of any printed line depends upon the darkness of the tracing line and the length of exposure time. The ammonia developer requires venting to the outside for safety, but several patented developers now available require no venting.

Microforms Plans or their prints may be stored flat in drawers or rolled and placed in tubes. For a large architectural office, this requires considerable storage space, so the plans may be reproduced to a size smaller than a postage stamp, called a *microcopy* or *microfilm.* For viewing, the microcopy is used to project a magnified image onto a screen, or a full-size reproduction can be made. A microcopy can be mounted in a hole in a keypunch card called an *aperture card.* This permits information to be located using a sorting machine. A number of microcopies can also be reproduced on one sheet called a *microfiche.*

Computer-aided Drafting Stations

Modular drafting and computer stations (Figure 2–24) are designed to be easily expanded as new technical systems become available. Components of the station are interchangeable so that drafting machines, computer terminals, keyboards, digitizers, printers, and plotters can be added, removed, or repositioned.

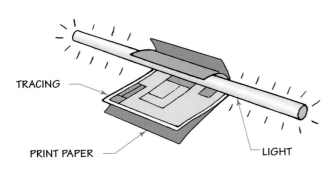
TRACING
PRINT PAPER
LIGHT
1) EXPOSING

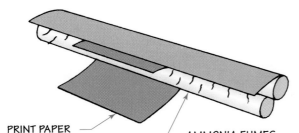
PRINT PAPER
AMMONIA FUMES
2) DEVELOPING

Figure 2–23 *Whiteprinting process. (Courtesy Bruning Division of Addressograph-Multigraph.)*

Figure 2–24 *Computer-aided drafting station. (Courtesy Mayline Company.)*

REAL WORLD  APPLICATIONS

Tools Over Time

 The evolution of drafting tools has changed the field of architectural drafting. Research the historical changes of one of the drafting tools in this chapter and prepare an executive report about it. An executive report is a one-page summary of a document. Corporate executives' time is too valuable to be spent reading lengthy reports. The summary presents the most important information and significant findings in a concise manner.

Report on one of the following tools: scales, compass, pencil, erasers, technical fountain pens, drafting media, drafting machines, templates, reproduction processes, or another tool of your choice.

REVIEW QUESTIONS

1. Draw a T square and label the blade and the head.
2. Draw a 30°–60° triangle and label the three angles.
3. List the minimum drawing equipment needed for a satisfactory set.
4. Which pencil leads are most commonly used for architectural drafting?
5. List three different types of drafting media.
6. What is the difference between a blueprint and a whiteprint?
7. Distinguish between a lettering guide and a lettering stencil.
8. What is the difference between microfiche and microfilm?

ENRICHMENT QUESTIONS

1. Interpret what 3" = 1' means in words.
2. What is one advantage and one disadvantage of using templates in architectural drafting?
3. If you had a choice between using a drafting machine or a parallel rule, which would you use and why?

PRACTICAL EXERCISES

1. Document what you feel a "typical" architectural drafting table may look like in the next twenty-five years. Indicate *why* you feel these changes will take place.

2. Measure a room in your house or school. Using $\frac{1}{4}$" = 1'0" scale, redraw it onto drafting paper.

3. Do you feel that drafting tools have become unnecessary because of the improvements made in computer-aided drafting and design (CADD)? Why or why not?

CHAPTER 3

Sketching

OBJECTIVES

By completing this chapter you will be able to:

- 🏛 Demonstrate the proper technique used to sketch various lines and shapes.
- 🏛 Explain the importance of proportion when sketching.
- 🏛 Produce a sketch of a room in a house.
- 🏛 Compare and contrast basic elements of pencil sketching and computer-assisted sketching.

KEY TERMS

Sketching	Construction lines
Graph paper	Proportion

Each profession has a symbol that has been generally accepted as the hallmark of the field. Often the symbol is an instrument or device commonly used by the members of that profession: a palette for art, a baton for music, a transit for surveying, a stethoscope for medicine. For architects, the pencil has been the symbol, and for drafters, the T square. A pencil represents freehand sketching at the initial design stage, and the T square represents the accurate drafting of the final plans. A successful architectural drafter certainly must be proficient in using the tools of the trade: the pencil and the T square. Chapters 3 and 4, then, share with you the sketching and drafting methods that other architectural drafters are now using to do their jobs quickly and accurately.

SKETCHING

Freehand **sketching** is *not* difficult! An extremely steady hand or other special skill is not necessary. All that is needed is an understanding of what *is* required and what is *not* required to produce an excellent sketch. Just remember that you control your hand; your hand does not control you.

Lines

A freehand sketch should not look like an instrumental drawing with perfectly straight and accurate lines. In fact, the unevenness of a properly sketched line is more attractive and interesting than a mechanically perfect line. The weight, direction, and proportions of sketched lines *are* important, however, and the following rules should help. But please remember to concentrate on the desired *results* rather than on the rules.

1. A soft pencil (such as an F grade) is best for sketching. The point should be long and tapered as shown in Figure 3–1. The point is slightly rounded rather than needle-sharp. After sketching each line, twist your pencil slightly to avoid developing a flat portion. This will reduce the number of necessary sharpenings.

2. As you might expect, a right-handed person sketches short horizontal lines from left to

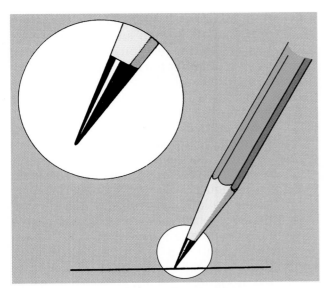

Figure 3–1   Lead point for sketching.

right and a left-handed person sketches from right to left. Short vertical and inclined lines are sketched from top to bottom as shown in Figure 3–2. If the angle of a line seems awkward for you, turn your paper to a more comfortable position. This is the reason that drafters don't tape sketches to a drawing board as they do when drafting with instruments.

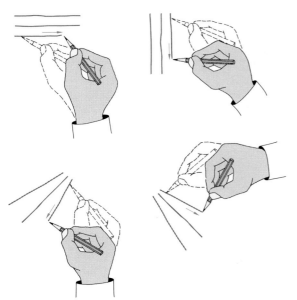

Figure 3–2   Sketching short lines.

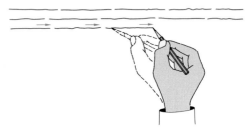

Figure 3–3   Sketching long lines.

3. To sketch a line, rest your hand on the drawing surface and pivot only your fingers. Most persons cannot sketch a line longer than 1" without sliding their hand. Therefore, sketch lines longer than 1" in short intervals, with a small gap left between each interval, as shown in Figure 3–3. Do not omit these gaps, for they add a professional touch to a sketch.

4. The correct *direction* of a sketched line is most important. Horizontal lines should be horizontal and not inclined. Vertical lines should be vertical and not leaning. The easiest way to accomplish this is to sketch on **graph paper** or on tracing paper placed over graph paper. Graph paper with $\frac{1}{4}$" grids is commonly used (Figure 3–4, top). Special tracing vellum with grid lines that disappear when reproduced is also available. Isometric paper may be used for pictorial sketching (Figure 3–4, center).

   Another advantage of tracing paper is that alternate schemes can be sketched and studied. Details that are to remain unchanged can be traced easily from the initial scheme. Often you will find it easier to correct mistakes by retracing than by erasing.

5. When graph paper is not used, the sketch is blocked in using very light **construction lines** (Figure 3–4, bottom). These construction lines give you a chance to study your sketch for the proper line direction and proportion before you darken it with the final outlines.

6. Good **proportion** is important to all design and immediately distinguishes an excellent sketch from a mediocre one. Proportion is simply a matter of relating one length or area to another—for example, the relation of the

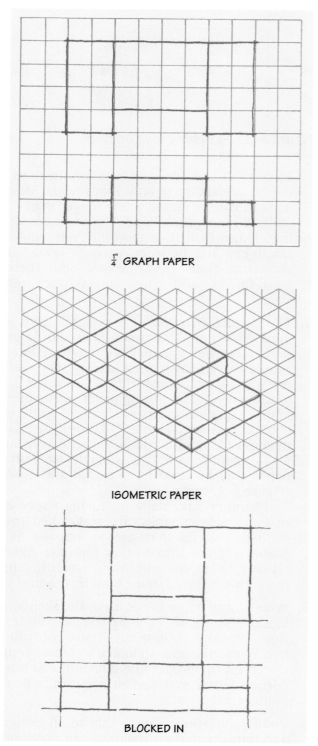

Figure 3–4   *Top: Sketching on $\frac{1}{4}$" graph paper. Center: Sketching on isometric graph paper. Bottom: Blocking lines to aid direction and proportion.*

width of an object to its height (is the width twice the height or two and one-half times the height?), or the relation of one portion of an object to another portion. The overall proportions of an object are especially important, for they will determine the proportions of all smaller elements.

Often areas can be compared with a simple geometric form such as a square, triangle, or circle. Rectangles can be divided easily by eye into several squares. Examples are shown in Figure 3–5.

Many building components occur so often that it is worthwhile to study and sketch them to become familiar with their proportions: 2" lumber, 8" × 16" concrete block modules, $2\frac{2}{3}$" × 4" × 8" brick modules as shown in Figures 3–6 and 3–7 (page 34). In Figure 3–8, notice how only a slight change in the proportion of an object—a door—will change the type of door. With such basic materials properly proportioned, other elements will fall into place more easily.

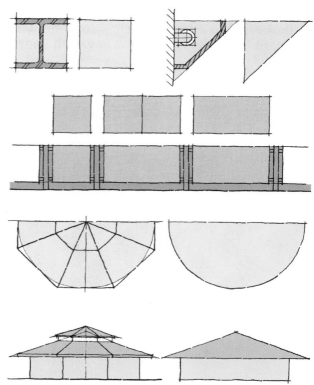

Figure 3–5   *Proportioning by comparison with geometric forms.*

CAREER PROFILES

Architecture as a Career

Interested in a career as an architect? If so, you may wonder who employs architects? What kind of respect do architects enjoy? What general skills are needed? In what settings do architects work? For some answers, continue reading.

In 1997 there were an estimated 100,000 persons employed as architects. The majority held jobs in architectural firms having fewer than five employees. Others were self-employed, with a few working for contractors, real estate developers, and government agencies.

Architecture and engineering are professions on a par with medicine, dentistry, law, theology, science, and education. The architect occupies a respected position in the spectrum of occupations ranging from design-drafter to architect.

Architects design and supervise construction of residential and commercial buildings, schools, churches, theaters, hospitals, stadiums, and monuments. Occasionally, architects will design entire communities. In addition to designing buildings, an architect may be responsible for the selection of the site, preparing the cost analysis, and aiding in long-term planning for land development. Performing these tasks requires numerous skills including design, engineering, managerial, communication, and supervisory. Successful architects must be both artistic and practical. They must have the ability to work smoothly with a variety of clients and contractors.

Working conditions of architects are comfortable, with most of their time being spent in the office. However, this does not mean they sit at their desks all day. They may meet with clients, develop reports and drawings, and consult with other architects and engineers. In addition, they often go to construction sites to check on the progress of their projects. While architects normally work a forty-hour week, this may be extended and include evenings and weekends when meeting a deadline.

Interns at architectural firms earn approximately $25,000 annually. Principals or partners of firms earn significantly more on average—$50,000 annually. Architects' fees range from 3 percent of the construction budget for a large building to 10 percent for a modest building.

For more information about the architect's work, working conditions, training, job outlook, and salary, access the Occupational Outlook Handbook Home Page at:
http://www.bls.gov/ocohome.htm

From this page, go to "Outlook for Specific Occupations" and then to "Professional Specialty Occupations." Architect is one of the occupations listed on this page.

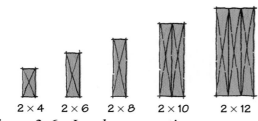

Figure 3–6   Lumber proportions.

2 × 4   2 × 6   2 × 8   2 × 10   2 × 12

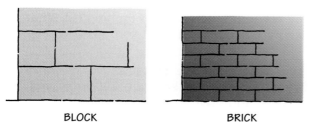

BLOCK   BRICK

Figure 3–7   Masonry proportions.

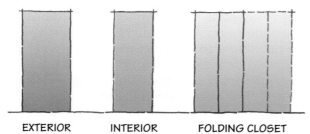

EXTERIOR   INTERIOR   FOLDING CLOSET

Figure 3–8   Door proportions.

Circles and Arcs

Large circles are best sketched by inscribing them inside a square of light construction lines as shown in Figure 3–9. The center lines deter-mine the tangent points. Small circles can be sketched using only the center lines.

Ellipses

When a circle is inclined to the picture plane, it will appear as an ellipse. Consequently, ellipses often must be drawn in pictorial sketches. As with circles, the best method is to sketch first the center lines with light lines and then the enclosing rectangle with very light construction lines. Then draw tangent arcs at the ends of the major and minor axes, and complete the ellipse by connecting these tangent arcs. See Figure 3–10.

Figure 3–11 shows these techniques in several sketches.

COMPUTER-AIDED SKETCHING

A computer-aided sketch substitutes a stylus for a sketching pencil and a digitizer tablet for graph paper on a drawing board. (A puck, which requires centering crosshairs rather than pointing, may seem awkward at first, but results are identical.) But the different techniques needed for freehand sketching and instrumental drafting tend to blend together in computer-aided sketching and drafting. For a computer-aided sketch is merely your first digitized layout, which will be revised as often as needed to form your final design. It is the built-in accuracy of computer-produced lines combined with availability of grids that quickly transform your initial sketch into a finished drawing. Straight lines, circles, arcs, and free-form sketched lines can all be produced with computer-aided drafting programs.

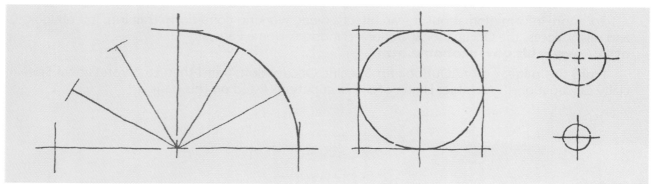

Figure 3–9   Sketching circles.

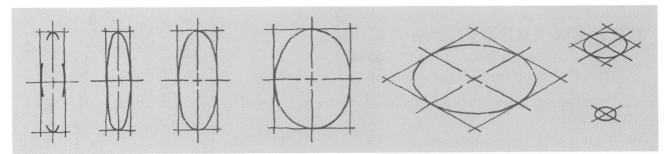

Figure 3–10   *Sketching ellipses.*

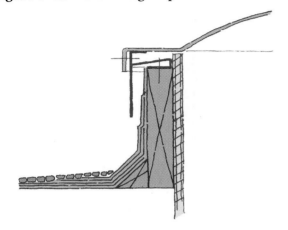

INSTALLATION DETAIL

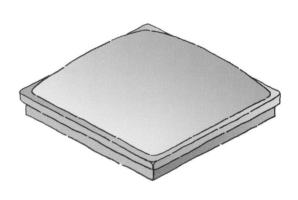

PLASTIC DOME ISOMETRIC

Figure 3–11   *Pencil sketching technique.*

REAL WORLD APPLICATIONS

Sketching Handbook

Inevitably, whatever your chosen career, teaching others will be an important element of your job. For example, a junior high school art teacher might approach you as an architect to develop a brief instructional handbook for teaching sketching to seventh-grade students. You agree on a voluntary basis and begin the task.

As you prepare this booklet, keep your instructions clear and basic. Remember that seventh-graders have a much narrower vocabulary than the professionals with whom you are accustomed to working. It should be written in a step-by-step format. Include examples of sketched lines, circles, shapes, and items that will be of interest to this particular audience. Include hands-on exercises for students to complete so they can apply the information covered. Also, include a brief introduction on the relevance of sketching to drafting, architecture, and other related careers. Finally, provide the students with a brief overview of what an architect does in a typical day or on a typical project.

REVIEW QUESTIONS

1. What is the purpose of leaving line gaps while sketching?

2. Describe the technique that should be used while sketching to ensure the pencil remains pointed.

3. What is meant by proportion?

4. Give the pencil stroke direction for a right-handed person for the following:

   a. Horizontal lines

   b. Vertical lines

   c. Lines inclined to the lower right

   d. Lines inclined to the lower left

5. What are two advantages of sketching with graph paper?

6. What is one advantage and one disadvantage of computer-aided sketching versus pencil sketching?

ENRICHMENT QUESTIONS

1. What are three situations in which sketching would be better than board drafting or CADD?

2. On an "A" size sheet of paper, lay out twelve $2\frac{1}{4}$" square areas and sketch by hand or with the computer the patterns shown in Figure 3–12.

3. Why is sketching an important part of architectural drawing?

Figure 3–12   Patterns for Enrichment Question 2.

PRACTICAL EXERCISES

Figure 3–13 *Pattern for Practical Exercise 1.*

1. Draw or computer-produce a low wall 2' high × 6' long) constructed of 12"-square concrete screen blocks patterned as shown in Figure 3–13. Use $\frac{1}{4}$" mortar joints and 1" = 1'0 for the drawing scale.

   a. Sketch by hand or with the computer on $\frac{1}{4}$" graph paper.

   b. Sketch by hand or with the computer on unruled paper.

   c. Use your computer's sketch command.

2. Imagine yourself glued to the ceiling of a room in your house and prepare a sketch by hand or with the computer of what you would see.

3. Refer to a standard structural steel reference and sketch by hand or with the computer cross sections of the following structural steel members:

   a. W 12 × 40

   b. S 8 × 18.4

   c. C 8 × 11.5

   d. L 4 × 4 × $\frac{1}{4}$

   e. HP 8 × 36

4. Use an architectural reference guide to sketch by hand or with the computer four different styles of bow and bay windows.

CHAPTER 4

Drafting

OBJECTIVES

By completing this chapter you will be able to:

🏛 Communicate design intent by means of instrumental architectural drafting.

🏛 Use the appropriate drafting tools to:

◼ Draw parallel, perpendicular, and inclined lines

◼ Construct tangential lines and arcs to each other

◼ Draw irregularly shaped arcs

◼ Divide a line into equal and proportional parts

◼ Construct an equilateral triangle, square, pentagon, hexagon, and octagon

◼ Draw conical sections: ellipses, parabolas, and hyperbolas

◼ Construct a cylindrical helix

KEY TERMS

Ellipse   Catenary
Parabola   Helix
Hyperbola

DRAFTING BASICS

Drafting usually refers to drawing with the aid of drafting instruments such as the T square and triangles, discussed in Chapter 2. With the help of these drafting instruments, drawings can be produced quickly and accurately. In some instances, they are even used as patterns for finished products.

Straight lines are drawn with the aid of a straightedge, and circular lines with the aid of a compass. Here are some techniques used by experienced drafters.

Drafting Pencil

The point of your drafting pencil should be long and tapered, as shown in Figure 4–1, and slightly sharper than the pencil point used for sketching. There is another difference between sketching and drafting. In drafting, the pencil is

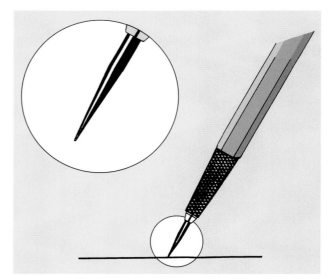

Figure 4–1 Lead point for drafting.

rotated slightly *during* the drawing of each line (rather than *after* the line is drawn). This keeps the point uniform so that you can produce lines of constant weight.

Horizontal Lines

A right-handed person draws horizontal lines from left to right using the top edge of the T square, parallel rule, or drafting machine as a guide. When using a T square, first press its head firmly against the working edge of the drawing board or table with your left hand (Figure 4–2). Then you move your left hand to the middle of the blade of the T square to hold it down (Figure 4–3). Your left hand is now in a position to also hold a triangle pressed against the top of the blade, if you wish to draw vertical lines. Left-handers, in general, reverse this rule. Draw horizontal lines from right to left holding your T square with your right hand with the head of the T square at the right.

Vertical Lines

A right-handed person draws vertical lines upward using the left edge of a triangle (Figure 4–3). Left-handers draw vertical lines upward by using the right edge of the triangle.

You should realize that these recommendations for the direction of drawing lines are not arbitrary rules. Rather, they are determined by the direction of light that illuminates the drawing surface. Right-handed drafters prefer light from their left rather than their right because this reduces objectionable shadows from their right hands. Consequently, they draw horizontal lines from left to right, and vertical lines using the left edge of a triangle. Also, all drafters prefer light from the front rather than the rear since this reduces objectionable shadows from the body. Consequently, draw horizontal lines using the top edge of the T square. You must, however, draw vertical lines upward to prevent your hand from blocking your vision.

Inclined Lines

Right-handers draw most inclined lines from left to right using the top edge of a triangle (Figure 4–4).

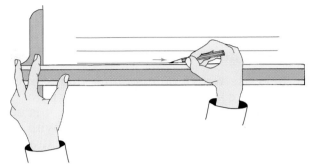

Figure 4–2   *Drafting horizontal lines (right handed).*

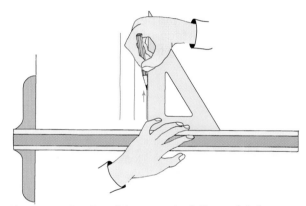

Figure 4–3   *Drafting vertical lines (right handed).*

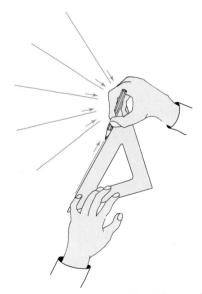

Figure 4–4   *Drafting inclined lines (right handed).*

Left-handers draw most inclined lines from right to left.

Circles and Arcs

Drafters use wedge points rather than conical points for compass leads. A wedge point is needed to keep the compass sharp, for it is not possible to rotate the compass lead (with respect to the direction of the line being drawn) as with drafting pencils. A wedge point is formed by sanding just one flat surface. That flat surface may face inward or outward depending on the design of the compass and the radius to be drawn. As shown in Figure 4–5, the lead should be perpendicular rather than inclined to the surface of the paper. Also, the needle point should extend slightly farther than the lead point.

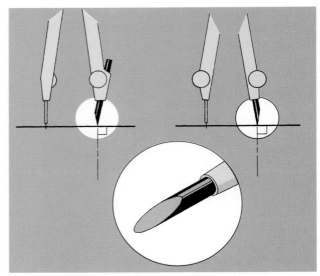

Figure 4–5   Lead point for compasses.

CAREER PROFILES

Architectural Design-Drafter

Plans for virtually every structure ever built were first prepared by an architectural design-drafter. That includes the building you are in right now, and all the buildings you can see from your present location. Large and small buildings—office buildings, industrial structures, public buildings, and private residences—all were completely described by design-drafters before their foundations were laid. A simple residence may be described by a comparatively small set of plans, but a large commercial building may require dozens of plans and specifications requiring hundreds of hours of drafting on a board or CAD.

The level of skill required to produce acceptable architectural plans can vary widely. As a newly-employed apprentice drafter, you will probably be given the simpler tasks by the chief drafter—perhaps a sketch of a building detail. Using a drafting machine or a CAD system, you will be expected to convert the sketch into an appropriately scaled, finished plan, complete with dimensions, notes, materials, and symbols. In all likelihood, you'll discover that even your "simple" introductory task will require decisions on your part to complete missing information. Other design-drafters will probably be eager to help you and answer questions, but you must be careful not to divert them from their assigned work too often.

Take every opportunity to visit buildings under construction—those you have planned and others. Even better, part-time employment as a builder will improve the design decisions you make as a design-drafter.

GEOMETRIC CONSTRUCTIONS

Inclined Lines

Lines inclined 30, 45, and 60 degrees to the horizontal can be drawn using the 30–60 and 45-degree triangles. Lines inclined 15 and 75 degrees can be drawn using both triangles in combination (Figure 4–6). Any angle can be measured or drawn using the adjustable triangle or protractor.

Applications: Pictorial drawings, auxiliary views, section symbols.

Parallel Lines

A line parallel to a given line can be drawn by aligning any one edge of a triangle with the given line, and then sliding another edge of the triangle along a second triangle (or along the T square) to the desired position (Figure 4–7). Or an adjustable triangle can be used.

Application: A basic construction used in all forms of projection.

Perpendicular Lines

A line perpendicular to a given line can be drawn by aligning one leg of a triangle with the given line, and then sliding the hypotenuse of the triangle along a second triangle (or along the T square) to the desired position (Figure 4–8). Or an adjustable triangle can be used.

Application: A basic construction used in all forms of projection.

Line Tangent to Arc

A line can be drawn through a point tangent to an arc by aligning one leg of a triangle to the arc and through the point. Slide the hypotenuse of the triangle along a second triangle (or along the T square) so that the second leg is aligned with the center of the arc. The intersection of

Become very familiar with the building codes governing your design area. *Sweet's Architectural Catalog* is a set of product descriptions by manufacturers of architectural products. Indexed by firms, products, and trade names, *Sweet's* is very helpful in selecting materials and equipment. Also located in every architectural office are *Architectural Graphic Standards* and *Time Saver Standards*. These thick volumes will answer many of your design questions. Use the Internet to search for specific information provided by manufacturers, suppliers, or public institutions. A good place to begin is the Architecture on the Web homepage, which has extensive links to other Internet sites that deal with architecture. The address is: **http://www.arch.su.edu.au/~white_k/arch/title.html**

In all probability, you will develop a specialty depending upon your skill and interest. This may be as an illustrator, checker of shop drawings, electrical drafter, HVAC specialist, structural drafter, or architectural designer. Consider writing specifications, an important component of the legal contract between the owner and contractor. But always remember you are working on a team.

The academic preparation of an architectural design-drafter can be quite varied. The drafter may have been introduced to the field through one or several architectural graphics or engineering graphics courses offered at a high school, trade school, community college, technical institute, or university. The drafter may have an associate or baccalaureate degree in a technical field. Occasionally, the degree may even be in a non-related field.

The Occupational Outlook Handbook on the Internet will give you more information on the drafter's work, working conditions, job outlook, salary, and training. You can access it at: **http://www.bls.gov/ocohome.htm**

Click on "Outlook for Specific Occupations" and then on "Technicians and Related Support Occupations" and you will see a listing for drafters.

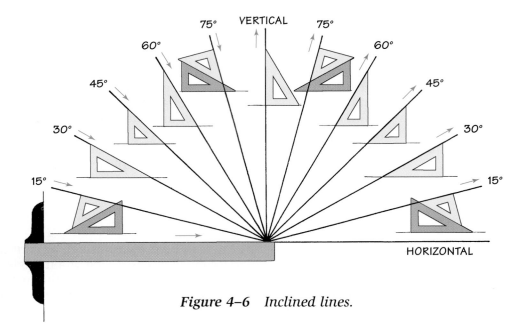

Figure 4–6   Inclined lines.

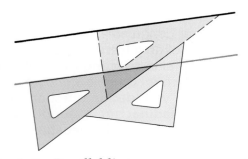

Figure 4–7   Parallel lines.

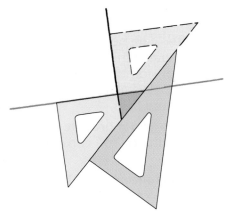

Figure 4–8   Perpendicular lines.

the second leg and the arc locates the point of tangency (Figure 4–9). Slide the triangle back to its original position, and draw the required tangent line. Or use an adjustable triangle.

Application: Accurate location of tangent points.

Dividing a Line

A line can be divided into any number of equal or proportional parts by the principle of proportional triangles. Construct a triangle with the given line as one leg and an easily divisible construction line as a second leg. Project the divisions, parallel to the third leg, back to the given line (Figure 4–10).

Application: Layout of repetitive elements such as stairs.

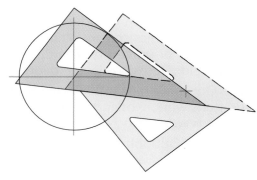

Figure 4–9   Line tangent to arc.

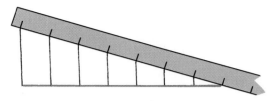

Figure 4–10   Dividing a line.

Arc Tangent to Perpendicular Lines

An arc of desired radius can be drawn tangent to two perpendicular lines by striking an arc of the desired radius from the intersection of the perpendicular lines to locate the points of tangency. Strike arcs from the points of tangency to locate the center of the desired arc (Figure 4–11).
Application: Rounded corners.

Arc Tangent to Two Lines

An arc of desired radius can be drawn tangent to any two lines by drawing construction lines parallel to the given lines, and at a distance equal to the radius, to locate the center of the arc. Drop perpendiculars from the center to the given lines to locate the points of tangency (Figure 4–12).
Application: Straight forms having fillets or rounds.

Arc Tangent to Arcs

An arc of desired radius can be drawn tangent to any two arcs by striking construction arcs from the centers of the given arcs using a radius equal to the given arc radii plus (or minus, if desired) the desired radius. This locates the center of the desired arc. Draw construction lines from the center of the desired arc to the center of each given arc to locate the points of tangency (Figure 4–13).
Application: Curved forms having fillets or rounds.

Arc Tangent to Line and Arc

An arc of desired radius can be drawn tangent to any line and any arc by combining the principles of the two previous geometric constructions (Figure 4–14).
Application: Combined straight and curved forms.

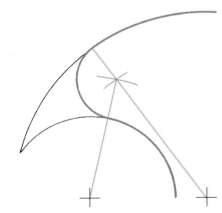

Figure 4–13   Arc tangent to arcs.

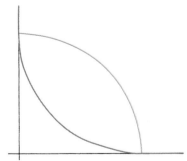

Figure 4–11   Arc tangent to perpendicular lines.

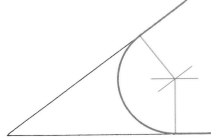

Figure 4–12   Arc tangent to two lines.

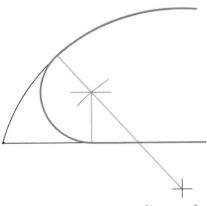

Figure 4–14   Arc tangent to line and arc.

Equilateral Triangle

Given a side, an equilateral triangle can be constructed by drawing 60-degree lines inward through the ends of the side (Figure 4–15).
Application: A basic proportional element.

Square

Given a side, a square can be constructed by drawing 45-degree construction lines inward through the ends of the side. Draw adjacent sides at 90 degrees to the given side (Figure 4–16).
Application: A basic proportional element.

Pentagon

Given a side, a pentagon can be constructed by drawing 54-degree construction lines inward to the ends of the side to locate the center of the circumscribing circle. Draw this circle and step off the remaining sides (Figure 4–17). This con-

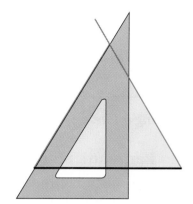

Figure 4–15   Equilateral triangle.

struction can be used for *any* regular polygon by using an angle of

$$90° - \frac{180°}{N}$$

when *N* equals the required number of sides.
Applications: Plan shapes, polyhedra.

REAL WORLD   APPLICATIONS

Arc Lengths

¼"=1'-0   Your clients would like to outline their driveway in lights. Since their driveway is curved, as shown in the figure, they must know the length of the outside circular arc to determine the number of lights they will need. Since the circumference of a circle is π times the diameter, the formula for calculating the length of a circular arc is:

$$ARC = \frac{\alpha \pi D}{360}$$

where: ARC = circular arc length
α = central angle
D = diameter

Insert the appropriate numbers from the figure and calculate the outside length of the curved driveway. In this case the central an-

gle is 180 degrees. If the lights are to be spaced 2 feet apart and are 6 inches in diameter, how many lights will the client need?

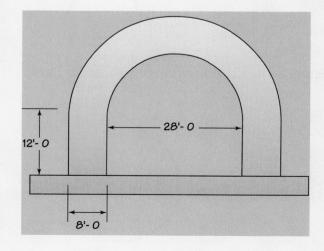

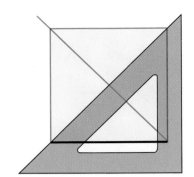

Figure 4–16   Square.

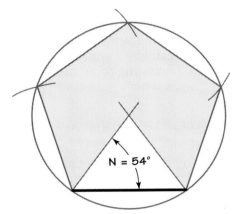

Figure 4–17   Pentagon.

Hexagon

Given a side, a hexagon can be constructed by drawing 60-degree construction lines in both directions through the ends of the side to locate the center of the hexagon. Draw a construction line through the center of the hexagon and parallel to the given side to locate adjacent corners. Draw 60-degree lines inward through these corners to locate the remaining corners (Figure 4–18).
Applications: Plan shapes, polyhedra.

Octagon

Given a side, an octagon can be constructed by drawing 45-degree sides outwardly. Construct a square through the given side and the far ends of the 45-degree sides. Swing arcs with centers at the corners of the square, and radii equal to half the length of the diagonal of the

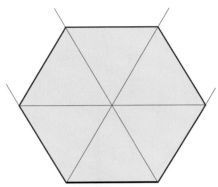

Figure 4–18   Hexagon.

square. Add the remaining sides to complete the octagon (Figure 4–19).
Applications: Plan shapes, polyhedra.

Regular Polygon

Given a side, a regular polygon of any number of sides (N) can be constructed by drawing a semicircle on the given side as illustrated. Divide the semicircle into N equal parts, and draw an adjacent side through the second outermost division. Draw perpendicular bisectors of these sides to locate the circumscribing circle. Draw radial construction lines through the divisions of the semicircle to locate the remaining corners of the polygon (Figure 4–20).
Applications: Plan shapes, polyhedra.

Conic Sections

When a right circular cone is cut by a plane surface, the intersection is called an **ellipse** when the plane is inclined at an angle smaller than the base angle of the cone; a **parabola** when

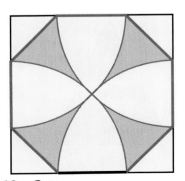

Figure 4–19   Octagon.

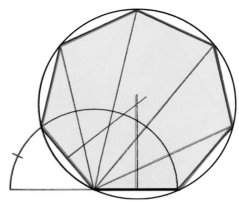

Figure 4–20   Regular polygon.

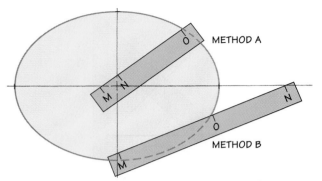

Figure 4–22   Ellipse (trammel method).

inclined at the same angle; and a **hyperbola** when inclined at a greater angle (Figure 4–21).

Ellipse (Trammel Method)

Given the major and minor axes, an ellipse can be constructed by using a trammel (a paper straightedge). On the trammel, mark off half of the major axis (M) and half of the minor axis (N) from a common origin (O). Then place the trammel in any convenient direction so that the major axis mark falls on the minor axis and the minor axis mark falls on the major axis. The origin mark locates one point on the ellipse. Rotate the trammel to obtain as many points as desired (Figure 4–22). Using an irreg-

ular curve, complete the ellipse by drawing a smooth curve through the origin marks. Two methods are illustrated.

Often only one quadrant of the ellipse need be plotted, for you can mark your irregular curve and use it for the remaining three quadrants.
Application: Circles which project as ellipses.

Ellipse (Concentric Circle Method)

Given the major and minor axes, an ellipse can be constructed by drawing two concentric circles having diameters equal to the major and minor axes. Draw any convenient number of radial construction lines. Draw construction lines parallel to the major axis through the intersection of the radial lines and the inner cir-

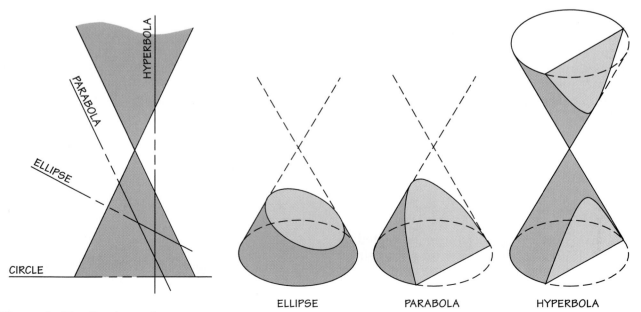

Figure 4–21   Conic sections.

cle. Draw construction lines parallel to the minor axis through the intersection of the radial lines and the outer circle. Using an irregular curve, complete the ellipse by drawing a smooth curve through the intersections of the construction lines (Figure 4–23).

Often only one quadrant of the ellipse need be plotted, for you can mark your irregular curve and use it for the remaining three quadrants.

Application: Circles which project as ellipses.

Parabola of Given Rise and Span

Given the rise and span, a parabola can be constructed by drawing a rectangle having the span as a base and an altitude equal to the rise. Divide the altitude and half the base into the same number of equal parts. The intersections of the construction lines locate points on the parabola as illustrated (Figure 4–24).

Applications: Approximate catenary curve, sound- and light-reflecting surfaces, bending moment at any point of a uniformly loaded beam.

Parabola Tangent to Two Lines

Given two lines, a tangent parabola can be constructed by extending the lines to their intersection. Divide the line extensions into the same number of equal parts. Draw construction lines as illustrated (Figure 4–25). The parabola will be tangent to these construction lines.

Applications: Arch forms, warped roof surfaces.

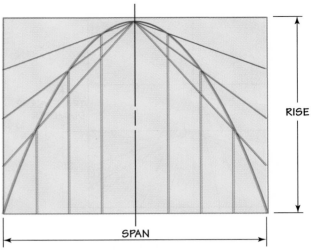

Figure 4–24   Parabola of given rise and span.

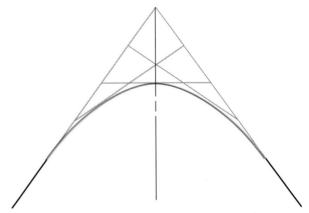

Figure 4–25   Parabola tangent to two lines.

Hyperbola of Given Diameter, Rise, and Span

Given the transverse diameter, rise, and span, the hyperbola can be constructed by drawing a rectangle having the span as a base and an altitude equal to the rise. Divide the altitude and half the base into the same number of equal parts. Intersections of the construction lines locate points on the hyperbola as illustrated (Figure 4–26).

Application: Warped roof surfaces.

Catenary

Given any three points, a catenary curve can be drawn by hanging a fine chain through the three points marked on a vertical drawing board. Prick the desired number of guide points

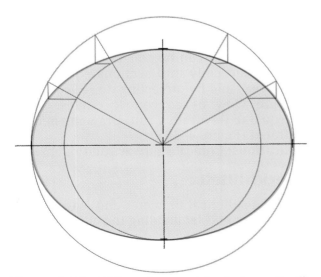

Figure 4–23   Ellipse (concentric circle method).

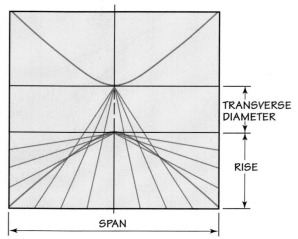

Figure 4–26 *Hyperbola of given diameter, rise, and span.*

through the links of the chain. A **catenary** is not a conic section but can be approximated by a parabola as illustrated (Figure 4–27). For accurate results, use formulas or tables found in standard references such as *Marks' Handbook*.
Application: Cable-supported roofs, uniform cross-sectional arches.

Cylindrical Helix

Given the cylindrical diameter and lead, a **helix** can be generated by dividing the circumference of the cylinder and the lead into the same number of equal parts. Project the circumference marks to the adjacent view until they intersect the lead lines. Draw the helix through these intersections as illustrated (Figure 4–28).
Applications: Spiral stairways and ramps.

Conical Helix

Given the conic diameter, conic altitude, and lead, a helix can be generated by dividing the

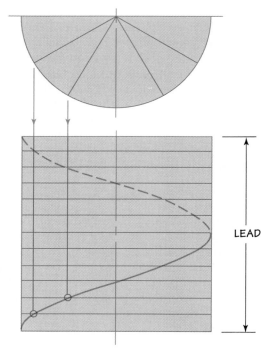

Figure 4–28 *Cylindrical helix.*

base of the cone and the lead into the same number of equal parts. Project the base marks to the edge view of the base in the adjacent view, and then to the vertex of the cone until they intersect the lead lines. Draw the helix through these intersections as illustrated (Figure 4–29). The plan view of the conical helix is called the *spiral of Archimedes*.

Conical Helix of Uniform Slope

Given the conic diameter, conic altitude, rise, and run, a helix can be generated by dividing the cone vertically into sections equal to the desired rise, and by projecting these sections to the plan view as concentric circles. Strike arcs equal to the desired run. Draw the helix as illustrated (Figure 4–30).
Applications: Conical stairways and ramps.

Spherical Helix

Given the spherical diameter and lead, a helix can be generated by dividing the circumference of the sphere and the lead into the same number of equal parts. Project the circumference marks to the adjacent view until they intersect the lead lines. Draw the helix through these

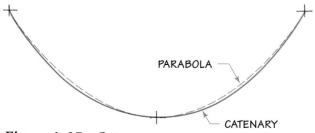

Figure 4–27   *Catenary.*

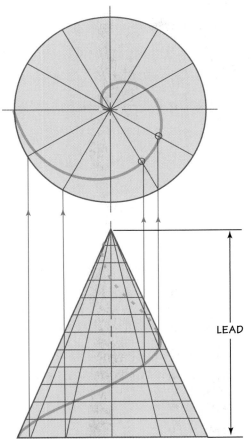

Figure 4–29   *Conical helix.*

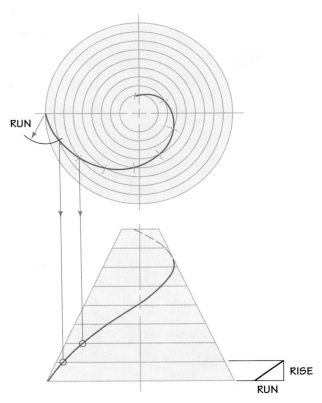

Figure 4–30   *Conical helix of uniform slope.*

intersections as illustrated (Figure 4–31).
Application: Ornamental space curve.

Spherical Helix of Uniform Slope

Given the spherical diameter, rise, and run, a helix can be generated by dividing the sphere vertically into sections equal to the desired rise, and by projecting these sections to the plan view as concentric circles. Strike arcs equal to the desired run and draw the helix as illustrated (Figure 4–32). Notice that the desired slope cannot be maintained beyond the tangent point of the slope line on the spherical surface.
Applications: Spherical stairways and ramps.

COMPUTER-AIDED DRAFTING

A computer-generated and plotter-produced structural plan is shown in Figure 4–33. Other commercially produced computer-generated

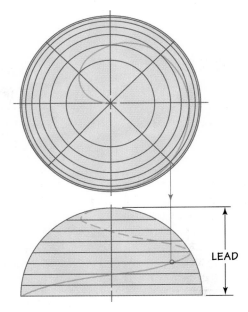

Figure 4–31   *Spherical helix.*

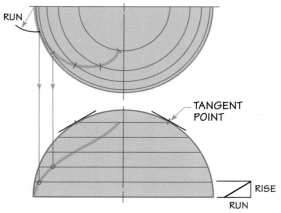

Figure 4–32   Spherical helix of uniform slope.

drawings appear throughout this book. As you study these plans, notice that they are virtually indistinguishable from hand-drawn and stencil-lettered drawings, but the required equipment, production methods, and necessary skills are quite different. Whichever method you use to generate your drawings, a working knowledge of the principles and techniques of drafting is extremely important.

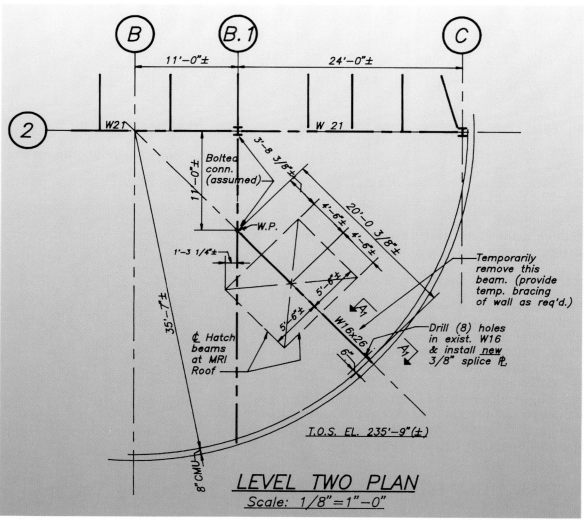

Figure 4–33   A computer-generated structural plan.

REVIEW QUESTIONS

1. Give the proper stroking, depending on whether you are right-handed or left-handed, for the following lines drawn with instruments.

   a. Horizontal lines

   b. Vertical lines

   c. Lines inclined to the lower right

   d. Lines inclined to the lower left

2. Why is a compass lead point sharpened in wedge shape?

3. What is the process used to draw parallel lines using two triangles?

4. What is the process used to draw perpendicular lines using two triangles?

5. When is a line tangent to an arc?

6. Ellipses, parabolas, and hyperbolas are all based on what solid geometric figure?

7. How does a circumscribed circle differ from an inscribed circle?

ENRICHMENT QUESTIONS

1. Write a set of instructions that would guide a drafting tool novice through the tasks shown in Figures 4–7, 4–8, and 4–9.

2. Why must a pencil be rotated as a line is being drawn?

3. Describe the procedure used to accurately divide a line into equal segments. Given an instance when this would be necessary.

4. Draw an ellipse that has a major axis of $4\frac{3}{16}$" and a minor axis of $2\frac{3}{4}$".

PRACTICAL EXERCISES

1. Locate the instruction sheet from a compass and describe the proper procedure used to sharpen a compass lead.

2. On an "A" size sheet of paper, lay out twelve $2\frac{1}{4}$" blocks and use drafting instruments to draw the patterns shown in Figure 4–34. Estimate all dimensions to obtain similar proportions.

3. Draw or computer-produce a random, broken-course-and-range stone wall as illustrated in Figure 4–35. Include an area 18" high × 3'0 wide to a scale of $1\frac{1}{2}$" = 1'0.

4. Using the table placement and dimensions in Figure 4–36, lay out a 21' × 23' dining room to contain the maximum number of tables. Scale of $1\frac{1}{2}$" = 1'0.

   a. Sketch on graph paper.

   b. Draw with drafting instruments.

   c. Computer-produce.

5. Draw or computer-produce a low wall (2' high × 6' long) constructed of 12"-square concrete screen blocks patterned as shown in Figure 4–37. Use $\frac{1}{2}$" mortar joints. Scale: 1" = 1'0.

6. Draw or computer-produce the plan of a 33'-long serpentine wall using the dimensions given in Figure 4–38. Scale $\frac{1}{4}$" = 1'0.

Figure 4–34 *Patterns for Practical Exercise 2.*

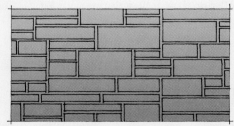

Figure 4–35 *Random broken-course-and-range stone wall for Practical Exercise 3.*

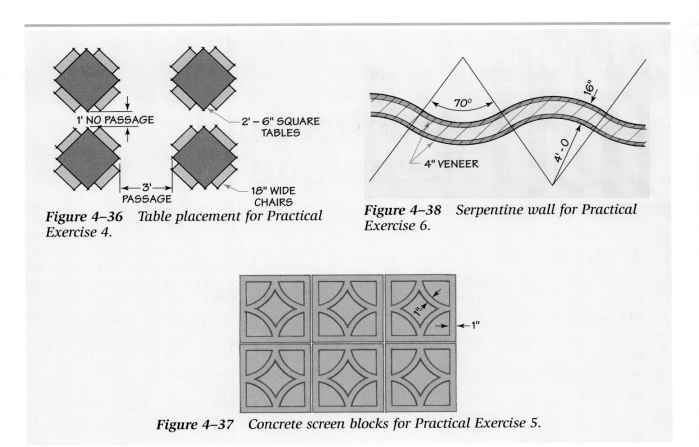

Figure 4–36   *Table placement for Practical Exercise 4.*

Figure 4–38   *Serpentine wall for Practical Exercise 6.*

Figure 4–37   *Concrete screen blocks for Practical Exercise 5.*

CHAPTER 5

Architectural Technique

KEY TERMS

Linework
Line technique
Architectural symbols

The technique of linework on architectural drawings is completely different from that used on other forms of mechanical drawing. In fact, the difference is so great that it is difficult for a drafter trained in another field—engineering drafting, for example—to make the switch to architectural drafting, since doing so means "unlearning" many accepted rules and practices. The reason for this dissimilarity is the difference in the goals of the two fields. Whereas engineers must turn out a perfectly exact indication of their needs to be used by persons trained in their own fields, architects want to produce an artistically complete picture to be read by trained workers and laypersons alike. Basically, architects must sell their product. And to do this, their plans must be appealing and warm as well as accurate.

LINE CHARACTER

Remember that you are not only a drafter, but an artist as well. Don't be afraid to let your lines run past each other at their intersections (Figure 5–1). A better end result will be obtained if the mind and hand are not cramped by trying to stop at a given point.

Use a soft pencil. Many architects never touch a pencil harder than 2H grade. It is impossible to obtain line quality in a drawing done with a hard pencil. The two skills to be mastered with a soft pencil are the slight twisting to keep the point sharp and the extra care required to keep the paper clean. Recommended pencil grades are shown in Figure 5–2.

POOR   GOOD   BEST

Figure 5–1   Line technique at corners.

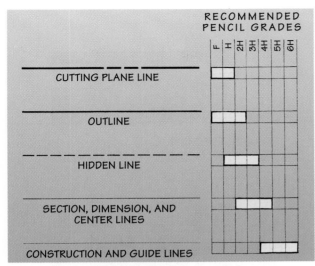

Figure 5–2 Architectural line weights.

Lines (other than dimension lines and a few others) often are *not* drawn uniform in weight. Indeed, they may nearly fade out in the middle. The ends of the lines are accentuated and should come to a distinct stop rather than just fade away.

All instruments and materials should be kept in perfect condition. A violinist might as well try to play with boxing gloves on as a drafter try to draw with a blunt pencil point. A good drafter who is forced to use poor materials and still produces an acceptable drawing does so in spite of them, not because of them. No good artistic work is sloppy.

Keep your drawings clean and smudge-free. These suggestions will help:

1. All your drafting equipment should be cleaned frequently. Soap and water, if thoroughly dried off, will not injure triangles and T squares. Use a clean cloth or paper towel to clean your desktop before beginning work.

2. Wash your hands. Try to keep your hands off the face of your drawing. Never allow anyone else to touch your drawing. When lettering, place a clean paper shield under your hand.

3. Do not sharpen your pencil over the desktop. Be certain that all graphite particles fall only in a waste receptacle. Blow or wipe off your pencil point immediately after sharpening. Wipe off each drafting instrument regularly.

4. Cover the unworked-on portion of your drawing with a clean paper shield so that it is not smudged and dirty before you start to draw. Cover any completed portion of your drawing with a clean paper shield also.

5. Do not slide your triangles and T square across your drawing; lift them slightly.

6. Keep all lines (especially construction lines) sharp and accurate to reduce the amount of erasing to a minimum.

7. Either blow or brush off any excess graphite from your drawing after drawing each line.

8. Some drafters find that a special cleaning powder (known under various trade names, such as Dry-Clean Pad, Draft-Clean Powder, Dust-it, Scumex) will aid in keeping their drawings smudge-free.

9. As a last resort, change pencil grades. A harder pencil grade will not smudge as readily as a soft pencil grade.

LINE TECHNIQUE

Architectural drafters develop their own styles of linework just as they develop their own styles of lettering. **Linework** consists of a combination of light and dark lines. As shown in Figure 5–3, avoid lines of all one, monotonous weight. Some of the most successfully used styles of **line techniques** are illustrated in Figure 5–4.

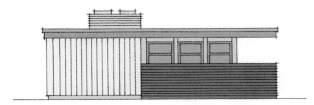

LINES TOO LIGHT FOR PROPER REPRODUCTION

Figure 5–3 Poor line techniques.

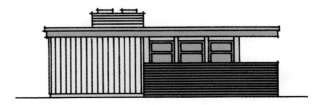

LINES HEAVY, BUT NO CONTRAST

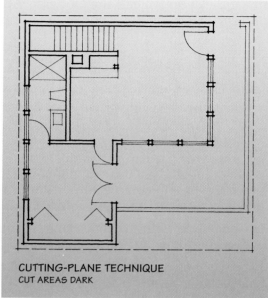

CUTTING-PLANE TECHNIQUE
CUT AREAS DARK

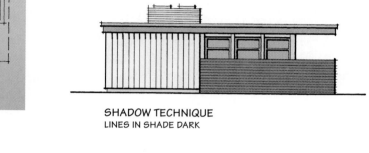

SILHOUETTE TECHNIQUE
OUTLINES DARK

SHADOW TECHNIQUE
LINES IN SHADE DARK

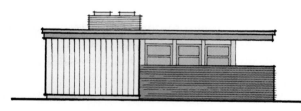

DISTANCE TECHNIQUE
NEAR LINES DARK; FAR LINES LIGHT

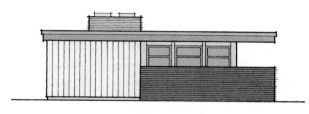

MAJOR–FEATURE TECHNIQUE
MAJOR ELEMENTS DARK

Figure 5–4   Good line techniques.

Cutting-plane Technique

This technique is used for section views. The lines formed by the cutting plane are darkened.

Distance Technique

It is possible to show depth in an architectural drawing by emphasizing the lines closest to the observer. Even if the plan in Figure 5–4 had been omitted, you would be able to visualize the shape of the building by this technique.

Silhouette Technique

The silhouette is emphasized by darkening the outline. One of the oldest techniques, it is still used today.

Shadow Technique

Recessions and extensions can be shown by darkening the edges away from the light source. The light is usually assumed to be coming from the upper left.

Major-feature Technique

This is a commonly used technique. The major elements are outlined, and the elements of lesser importance are drawn in with finer lines. The diagrams in this text are drawn using this technique.

Obviously, all of the above techniques cannot be used simultaneously. You will develop your own favored style, but remember to remain flexible enough to be able to adapt your personal

REAL WORLD   APPLICATIONS

Interviewing Techniques

It's your big day! Your hard work in school has paid off. After sending out resumés to various architectural firms, you have been called for an interview as one of the five finalists at one firm. This firm is one of your top choices for employment. You are excited, yet understandably nervous. Before going for the interview, review the interviewing skills you have learned.

Some of these skills include:

1. Greet the interviewers with a firm handshake and a pleasant smile.

2. Show confidence in your abilities.

3. Speak clearly.

4. Provide direct answers to questions.

5. Demonstrate your ability to cooperate and work on a team.

6. Present yourself as knowledgeable about architecture, yet recognizing your limitations.

7. Listen attentively. Never respond to a question you did not hear clearly by saying, "Huh?"; rather, respond by asking, "Could you please repeat that?"

Finally, remember to ask the interviewers questions about the firm itself. Not only are they interviewing you, but you are interviewing them to see if this is a company you wish to join.

Find a partner and practice interviewing each other. Record the positive and negative aspects of your interview so you can be ready when the big day really comes.

---

technique to the standards set up by a particular office.

ARCHITECTURAL SYMBOLS

A system of **architectural symbols** to indicate certain materials and features has developed through the years. Properly used, these symbols complement the architectural linework and form an attractive and useful language.

Figure 5–5 shows the symbols most often used on architectural sections, and Figure 5–6 shows those used on architectural elevations. Notice that most materials have different symbols for section and elevation views. Also remember that all section and elevation symbols should be drawn lighter than the outlines.

Figure 5–7 shows some common structural shapes used in architectural design. Either the W-shape or S-shape beam may be specified for house girders. The W shape is often used as

a column in industrial buildings, but pipe columns are used in residences. The angle sections may be used to support masonry over wall openings; they may be obtained with equal or unequal legs. Both angle and channel sections are generally used as elements of built-up sections for commercial structures. Steel is also obtainable in round bar shapes, square bar shapes, and rectangular plates.

Wall symbols are shown in Figure 5–8, together with the accepted dimensioning practices: to the outside face of the studs for frame walls, and to the outside of the masonry for masonry walls. Notice that no section symbol is specified for the frame wall. A wood symbol or poché (darkening of wall by shading or light lines) may be used.

All fixed equipment supplied by the builder should be included in the plans, whereas equipment furnished by the owner is omitted. Figure 5–9 shows an assortment of symbols that may be used. An invisible line may represent an

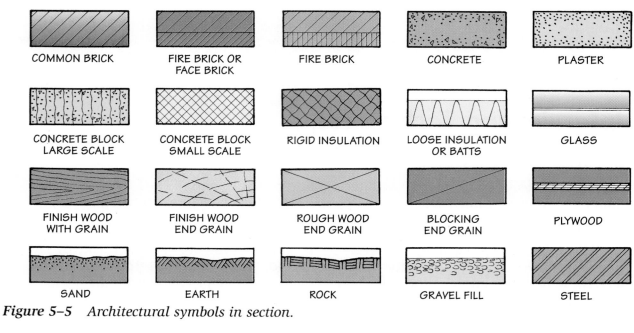

Figure 5–5   *Architectural symbols in section.*

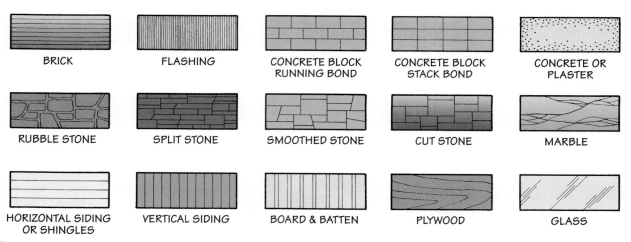

Figure 5–6   *Architectural symbols in elevation.*

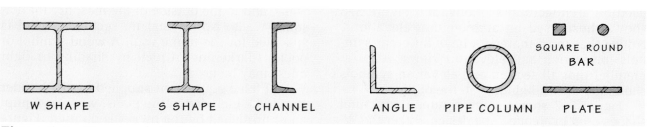

Figure 5–7   *Structural steel shapes.*

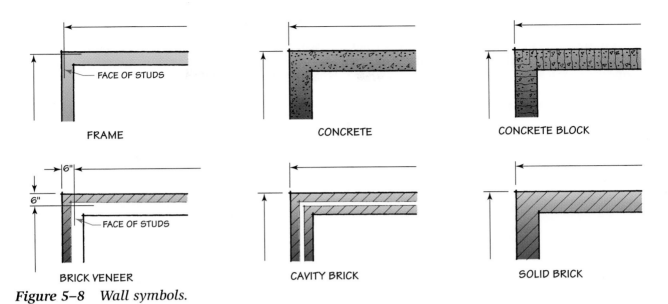

FACE OF STUDS

FRAME

CONCRETE

CONCRETE BLOCK

6"

6"

FACE OF STUDS

BRICK VENEER

CAVITY BRICK

SOLID BRICK

Figure 5–8 *Wall symbols.*

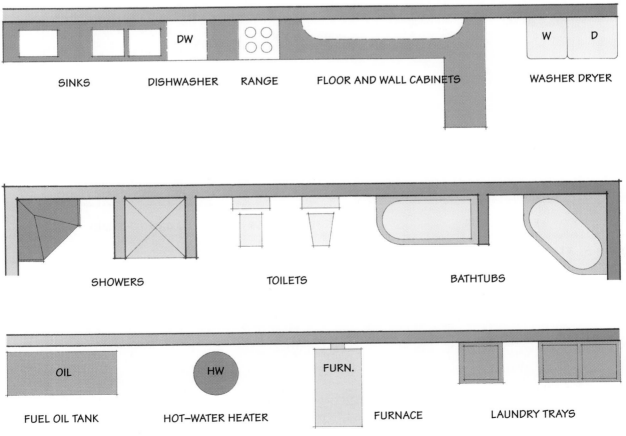

DW

W   D

SINKS   DISHWASHER   RANGE   FLOOR AND WALL CABINETS   WASHER DRYER

SHOWERS   TOILETS   BATHTUBS

OIL   HW   FURN.

FUEL OIL TANK   HOT–WATER HEATER   FURNACE   LAUNDRY TRAYS

Figure 5–9 *Fixed equipment symbols.*

invisible object (like the dishwasher shown built in under the counter) or a high object (like the wall cabinets that are *above* the plane of the section).

The conventions used to indicate windows and doors in a frame wall are shown in Figure 5–10. Although the doors are shown opened a full 90 degrees, an angle of 30 degrees may also be used. These same window and door conventions may be adapted to other kinds of walls, as shown in Figure 5–11. Notice the definite contrast in line weight between the walls and conventions.

COMPUTER-GENERATED SYMBOLS

CAD software suppliers offer menus of numerous architectural, structural, mechanical, and

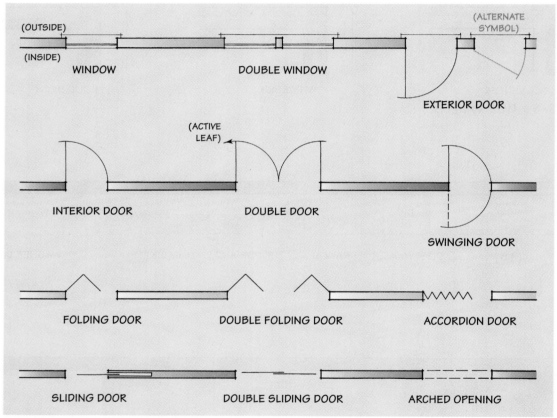

Figure 5–10   Frame wall openings.

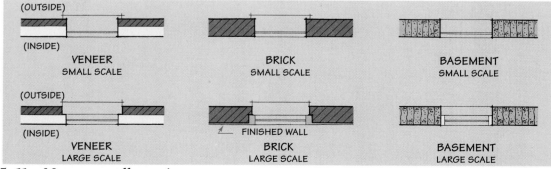

Figure 5–11   Masonry wall openings.

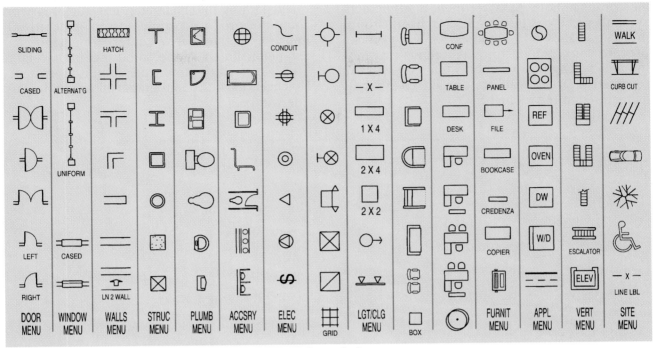

Figure 5–12   *Computer-generated architectural symbols.* *(Courtesy ASG.)*

REAL WORLD ───── APPLICATIONS

Architectural Symbols Brochure

After two years on the job, you discover that many of your clients are confused by architectural symbols. Based on this observation, your firm decides to produce a pamphlet that describes twenty of the most commonly used architectural symbols. It is your job to oversee this project. It is to be a professionally-designed brochure that clearly explains the symbols. It will be word-processed.

While completing this assignment, adhere to the following page layout guidelines.

1. First decide on a general format. How will the leaflet be folded? Where on each page will the symbols be located?

2. Review a number of fonts (letter styles) and choose one that is readable.

3. Pay attention to the headlines you select. They should be in larger size font and in bold type.

4. Lay out the page so that a reader will naturally follow the main elements in the path of a backwards N (if more than one column is used on a page) or a Z.

5. Keep it simple. Cluttered brochures make the information confusing.

electrical symbols. An assortment of equipment symbols produced by an ASG (Archsoft Group) program is shown in Figure 5–12. You also can prepare your own customized menus by drafting and storing those symbols that you will use repeatedly. Or you can draft or sketch symbols as required at the moment.

Regarding line character, it is customary on computer-produced drawings to generate all lines (visible, invisible, center, dimension, and extension lines) at the same medium weight. This saves both time and computer storage. However, most CAD programs allow you to vary line widths.

REVIEW QUESTIONS

1. What is the difference between line work used on engineering drawings and architectural drawings?

2. What are the advantages of using:
   a. Hard pencil grades   b. Soft pencil grades

3. List five different types of line techniques used by architectural drafters.

4. What is the purpose of using architectural symbols instead of labeling each feature?

ENRICHMENT QUESTIONS

1. Draw or computer-produce a legend that lists and illustrates:
   a. Section symbols   d. Wall symbols
   b. Elevation symbols   e. Fixed symbols
   c. Structural steel shapes f. Frame wall openings

2. Why are different line techniques used in elevation views?

3. Lay out the 10'-6" × 8'-0 awning window wall shown in Figure 5–13. Use appropriate pencil or computer technique. Scale $\frac{1}{4}$" = 1'-0.

4. Today, many architectural symbols are not drawn using instruments; rather, they are drawn using templates. Locate an architectural template and draw the same items listed in question #1.

5. What are some advantages and disadvantages of using templates versus computer-produced symbols?

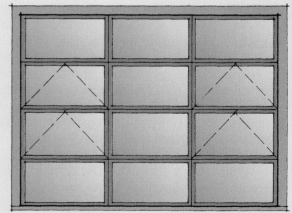

Figure 5–13   Awning window wall for Enrichment Question 3.

PRACTICAL EXERCISES

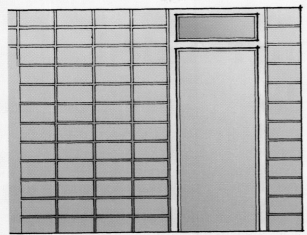

Figure 5–14 Stacked bond wall with door and transom for Practical Exercise 1.

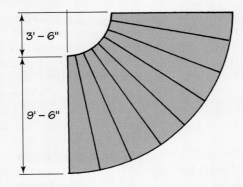

Figure 5–15 Spiral concrete stairway for Practical Exercise 2.

1. Draw or computer-produce the elevation of a stacked bond wall containing a 2'-8" × 8'-0 door and transom as indicated in Figure 5–14. Masonry units are 8" × 8" × 16". Use appropriate pencil or computer technique. Scale $\frac{1}{4}$" = 1'-0.

2. Draw or computer-produce an elevation of the helicoidal concrete stairway shown in Figure 5–15. Use appropriate pencil or computer technique. Scale $\frac{1}{4}$" = 1'-0.

3. Draw or computer-produce the cross section of a small theater using the dimensions indicated in Figure 5–16. Use appropriate pencil or computer technique. Scale $\frac{1}{8}$" = 1'-0.

4. Locate a Standard and Poor's set and find an entrance doorway with side windows. Use the indicated dimensions and draw or computer-produce it. Scale $\frac{1}{2}$" = 1'-0.

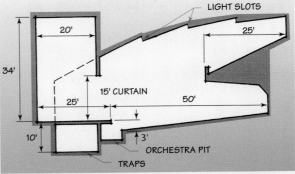

Figure 5–16 Sketch of theater for Practical Exercise 3.

CHAPTER 6

Computer-Aided Drafting

OBJECTIVES

By completing this chapter you will be able to:

- Define terminology related to computers and computer-aided drawing (CAD).

- Describe the general function of the components of a computer and how they work together.

- Describe the benefits and shortcomings of solid modeling.

- Identify various types of architectural databases.

- Predict possible future advances in computer technology.

KEY TERMS

Central Processing
Unit (CPU)
Random Access
Memory (RAM)
Peripherals
Monitor
Keyboard
Digitizer
Printer

Plotter
Wireframe
Primitives
Lightpen
Joystick
Mouse
Trackball

BACKGROUND

Graphic communication (drawing and reading pictures) is one of the earliest forms of human communication, and today's drafting tools have evolved through a continuously changing process that constantly seeks to improve graphic communication.

While drafting tools have a long history of use, some of the most important innovations in design-drafting have been made in just the past few decades. American Herman Hollerith is credited with the first successful computer design. His machines used electrically charged wires that could pass through holes punched in a card to complete a circuit, and was used to tabulate the results of the 1890 census. The company he founded was renamed International Business Machines (IBM) in 1924. Newly invented transistors replaced vacuum tubes, and the first fully transistorized computers were manufactured in 1958 by Control Data Corporation. American students Steven Jobs and Stephen Wozniak introduced the Apple II personal computer in 1977. IBM manufactured its PC in 1981, and Apple its Macintosh desktop computer in 1984. Today, courses in computer-aided design and drafting (CADD) are included in the curriculum for those studying design and drafting.

THE COMPUTER WORKSTATION

Almost without exception, the principal component of today's drafting room is the computer workstation, which usually includes the following components:

- **Central processing unit (CPU).** This is the computer hardware that processes information (the computer's "brain").

- **Random access memory (RAM).** This is the computer's main memory for storage and retrieval of data.

🏛 *Storage devices* such as *floppy disks* or *hard disks.*

🏛 **Peripheral** *equipment* such as the **monitor,** also called cathode ray tube (CRT), **keyboard, digitizer** (a tablet that senses the location of a puck), and **printer** or **plotter.**

🏛 *Computer programs* that instruct the workstation to perform drafting functions.

In practice, the monitor produces plans, elevations, sections, and pictorials on its display (Figure 6–1). Because the monitor does not produce a permanent record, it is called a *soft-copy* device. Printers and plotters produce permanent records and are called *hard-copy* devices. (A printer draws on paper by means of ink jet or laser beams, while a plotter draws by moving pens.) These devices, when coupled with the computational and storage facilities of the computer workstation, present a wide range of possibilities for architectural design and drafting. And if the workstation has sufficient processing capabilities, even dynamic animations such as *walk-throughs* and *fly-*

Figure 6–1   *A workstation monitor displaying an exterior perspective. (Courtesy Tri-Star Computer.)*

throughs of the design are possible. When an office has several workstations, they are often *networked* together, allowing all workstations to communicate with one another and share computing, storage, printers, and plotters (Figure 6–2).

The architectural drafter typically issues commands to a computer by using an alphanumeric

Figure 6–2   *Networked multiple workstations. (Courtesy Hewlett Packard Company.)*

keyboard and either a digitizer or a mouse. An alphanumeric keyboard is similar to a typewriter and is used to transmit alphabetical and numerical commands. A digitizer is a device for producing graphics information in the form of precise coordinates. A mouse is also capable of providing coordinate information, although it is less precise than a digitizer. For this reason, a mouse is often considered to be a pointing device for selecting options from menus, each of which indicates a different function to the computer. However, when reinforced by software functions such as *snap-to-grid* or *join-to-line*, the coordinate information provided by a mouse can become as accurate as that from a digitizer. Other menu options such as *draw, arc, erase,* and *zoom-in* are quite common.

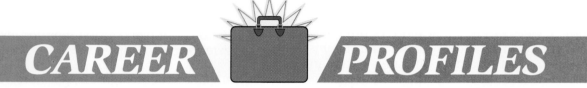

CAREER PROFILES

Name: John Heffley   *Age:* 29   *Gender:* Male

Occupational Title: Mechanical drafter & project manager/Mechanical contractor

Employer: Dutchland Incorporated, Gap, PA 17527

Years of Experience: 5 years as an architectural draftsman. 1 year in mechanical drafting. 2 years as a project manager.

Educational Background: Studied construction technology, carpentry, and architectural drafting in high school and then went to technical school for architectural and mechanical drafting.

JOB OVERVIEW:

Drafters generally work in air conditioned offices, working with computers, reference books, catalogs, and design sketches from designers. Occasionally site visits are required to collect information on conditions at the job site.

A majority of my day is spent in the office working with computers and making phone calls. Drafting and design consists of compiling plans and details of the components needed to construct the project. A drafter primarily works with architects, engineers and designers. In the project management capacity, the job includes working with customers, engineers, job supervisors, subcontractors, material suppliers, and state, county, and township representatives.

The variety of responsibilities keeps my work very interesting. There are always many tasks bidding for my time, so I need to be good at prioritizing my day. This is probably one of the greatest challenges of my job.

Various work experiences have helped me develop my drafting skills and knowledge of the construction industry. I previously worked for a lumber company, serving customers in the construction industry. My second job was working for an architectural design firm drafting plans for commercial buildings. The contractor I currently work for designs and builds water and wastewater facilities such as water storage tanks, pump stations, and wastewater treatment plants.

I continue to enjoy learning about all aspects of construction. Many avenues can be taken in the construction industry, including designing, estimating, sales, purchasing, and marketing.

POPULAR SYSTEMS

A complete computer graphics system used by an architectural office would include a workstation with a high-resolution graphics monitor, a line printer or plotter, and facilities for storing and retrieving graphics databases on either magnetic or optical storage devices. The workstation would contain a selection of programs (software) providing numerous computer-aided drafting functions. The following description of a typical application will illustrate how such a system is used to produce a plan.

Computer-assisted Plans

You sit at a design console consisting of a monitor, an alphanumeric keyboard, and a digitizer (Figure 6–3). While there are many types of digitizers, each using different technologies for generating coordinate information, most digitizers resemble ordinary drafting boards. Under the surface of this drawing tablet is a fine mesh of horizontal and vertical wires. When a special stylus, resembling an ordinary pen (Figure 6–4), or a small pad with crosshairs etched on a transparent plastic window, comes in close proximity to the tablet, the computer electrically senses its exact X and Y location. Small switches or buttons on the stylus allow the designer to select and store that location. Lines can be described to the computer simply by registering the coordinates of their endpoints. These lines can also be commanded, through software, to be pre-

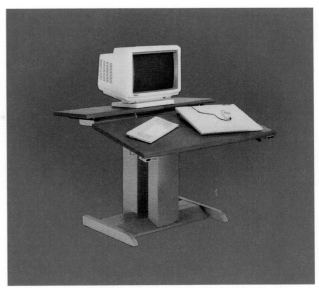

Figure 6–3   A digitizer, puck, monitor, and keyboard. (Courtesy Mayline Company.)

cisely horizontal, vertical, or at an angle, and of a specified length. Other geometric shapes, such as circles and arcs, can be described by locating centers and extreme points. In addition to this method, most systems allow you to bypass the digitizer and enter exact dimensions using the alphanumeric keyboard.

For example, to design a plan composed of repeated elements such as motel rooms, you might first sketch a typical room, then place the sketch on a digitizing tablet and digitize each corner of the room. The computer, as it receives

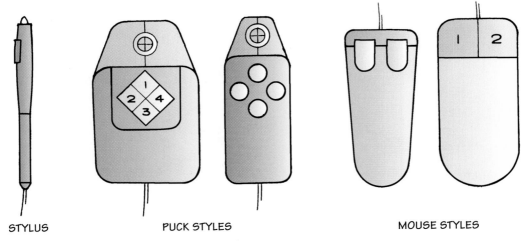

STYLUS   PUCK STYLES   MOUSE STYLES

Figure 6–4   Typical stylus, puck, and mouse styles.

this information, can be commanded to correct the sketch, making all lines straight and true and all arcs circular. The work is continually displayed and updated on the monitor. If a mistake occurs, you can select an editing function from a menu on the monitor and indicate if lines need to be added, deleted, or revised.

After the room has been described to the computer, you can use a combination of software commands or menu items to store and retrieve the drawing, to scale it up or down in size, to move it to another location, to have it repeated at various locations and orientations, or have it combined with previously described elements to create a complete plan. This plan can then be stored for future revisions, and even transmitted via a network to other workstations where allied professionals might perform additional design such as structural, electrical, plumbing, or heating, ventilating, and air conditioning (HVAC). Hard copies can be obtained

from a line printer (Figure 6–5). An example of such a drawing is shown in Figure 6–6.

When a project progresses into the working drawing phase, you can call up libraries of stan-

Figure 6–5   *A laserjet printer.* (Courtesy Hewlett Packard Company.)

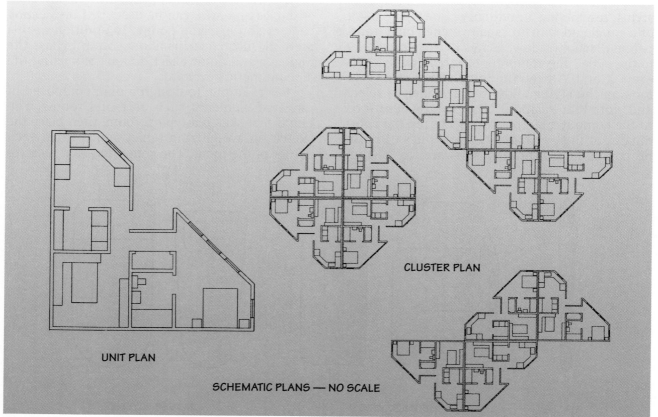

CLUSTER PLAN

UNIT PLAN

SCHEMATIC PLANS — NO SCALE

Figure 6–6   *A computer-produced study plan.* (Courtesy Professor Raymond J. Masters, The Pennsylvania State University.)

dard building details stored on magnetic or optical media. Commercial databases of such items are commonly available and contain thousands of standard building details in electronic form. These details can be displayed on a monitor so that the designer can edit portions and change dimensions to suit the specific needs of a project. The final detail can be added to the plan and appears on the plotted version of the working drawings. A typical computer-produced detail is shown in Figure 6–7.

You can also enter drawings and lettering into a computer directly (without digitizing) by *optical scanning* of previously drawn plans. An optical scanner uses a movable head having a light beam and a sensor of reflected light that senses dark lines and thus reads, records, and displays drawings and lettering. The heads can be either handheld or automatic (called desktop).

Computer-assisted Pictorials

Computer-assisted drafting is not limited to plans, elevations, and details. A computer can also generate perspectives of a building (Figure 6–8) as viewed from any vantage point, outside or inside the building. To calculate and plot a perspective image, a computer first requires information on the three-dimensional nature of the building. Normally, this information is in the form of X, Y, and Z coordinates that describe the corners, edges, and other elements of the building. Many manufacturers of building products such as windows, doors, and curtain walls provide numerical descriptions of their products to simplify this task. Some manufacturers also have software available that allows you to import the pre-drawn product into your drawings.

Building coordinates can be described in several ways. One of the most commonly used techniques incorporates a system of geometric descriptions that defines a building as a combination of planes, rectangular prisms, cylinders, and so on. Such systems save considerable time by eliminating the tedious task of describing every coordinate point since a computer can automatically generate coordinates from a minimum of information about an object. For

REAL WORLD APPLICATIONS

Converting to CAD

Your first job! You were hired three months ago at a small architectural firm as a drafting technician. All architectural plans are drawn by hand. Although most of your training has involved using CAD, you have the skills necessary to perform this job successfully. In your opinion, however, this process is not cost effective and is an inefficient use of time.

One day during lunch your boss asks what you think of the job. You respond favorably, but mention the advantages of using CAD instead of drawing plans by hand.

Intrigued, your boss asks you to develop a proposal for a complete computer workstation. It should include all necessary equipment ranging from the power surge protector to the computer desk.

Prepare a maximum two-page proposal that delineates all costs and includes a rationale for each major piece of equipment. Additionally, the proposal should specify at least three benefits of CAD as justification for shifting from hand-drawn plans to CAD. This proposal *must* be word-processed and professionally prepared.

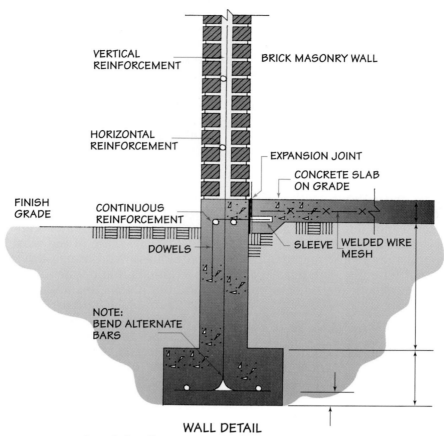

VERTICAL
REINFORCEMENT

BRICK MASONRY WALL

HORIZONTAL
REINFORCEMENT

EXPANSION JOINT

CONCRETE SLAB
ON GRADE

FINISH
GRADE

CONTINUOUS
REINFORCEMENT

DOWELS

SLEEVE

WELDED WIRE
MESH

NOTE:
BEND ALTERNATE
BARS

WALL DETAIL

Figure 6–7 *A computer-produced detail. (Courtesy Professor Raymond J. Masters, The Pennsylvania State University.)*

Figure 6–8 *A monitor displaying an interior perspective. (Courtesy Tri-Star Computer.)*

example, the six faces, twelve lines, and twenty-four endpoints of a rectangular prism can be calculated knowing only the length, width, height, and the location of one corner. Another technique allows the designer to determine X, Y, and Z coordinates by cross-referencing digitized plans and elevations. There are also three-dimensional digitizers that can be used when a scale model is available.

These X, Y, and Z coordinates, their connectivity, and even the material of the building element they are defining can be stored in a permanent file and then retrieved by programs that can calculate and plot the appropriate perspective view. Figure 6–9 is an example of a perspective plot from such a program. Because every line in the building is visible, this type of drawing is called **wireframe.** More advanced programs can calculate and draw only those portions of the building that are visible to the

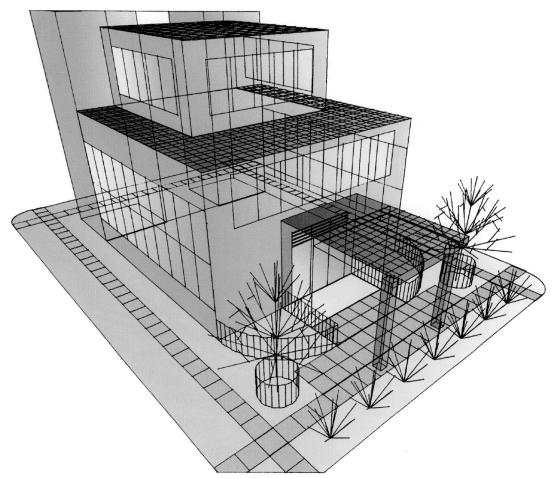

Figure 6–9 *A computer-produced wire-frame perspective.* (Courtesy Professor Raymond J. Masters, The Pennsylvania State University.)

observer. Such programs are described as having *hidden-line removal* capabilities and produce *solid models.* An example of a solid-model perspective plotted with the hidden lines removed is shown in Figure 6–10. Many programs with solid modeling capability also contain functions that light models to produce shadows and permit surfaces to have both texture and color. The images produced by the best of such programs are often nearly indistinguishable from photographs of actual buildings.

Solid Modeling

Computers produce solid-model perspectives in a variety of ways. Complex images may be built from a number of **primitives,** much like a child's castle constructed from building blocks.

Primitives consist of basic geometric shapes such as rectangular prisms, cylinders, and spheres. The primitives may undergo a number of logical operations: the joining of two shapes is called a *union;* the volume that is shared by two overlapping shapes is called the *intersection.* Primitives can also be *subtracted* from each other to produce a resultant object (Figure 6–11). For example, a rectangular block with a hole through it is modeled as a cylinder subtracted from a rectangular prism. This process of creating complex forms by performing logical operations on primitives is called *constructive solid geometry.*

Other solid modeling methods do not deal with three-dimensional primitives but synthesize the third dimension through some operation on two-dimensional shapes. *Extrusion* is a

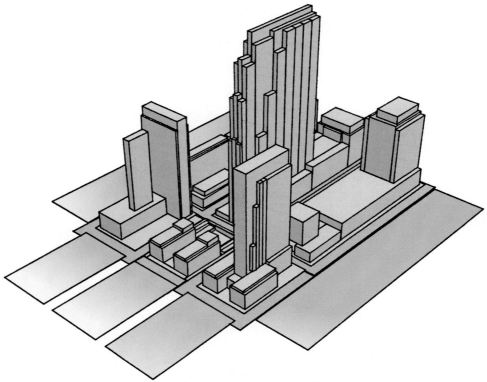

Figure 6–10 *A computer-produced solid-model perspective of Rockefeller Center, New York City.*
(Courtesy Professor Richard W. Quadrel, Rensselaer Polytechnic Institute.)

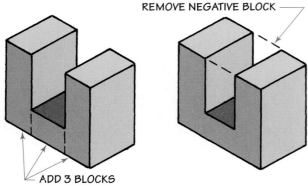

Figure 6–11 *Twin towers formed by adding or subtracting primitives.*

technique that allows a designer to "sweep" a two-dimensional shape (such as a floor plan) into the third dimension (the building's height) to produce a three-dimensional solid. This process is called *sweep geometry.*

Boundary geometry produces a solid image by describing the locations of two-dimensional shapes (planes such as walls, floors, and roofs) that make up its surface. Ideally, a designer would have a variety of techniques available through the workstation's software and be able to select the most appropriate method.

Interactive Systems

Today's computer graphics systems are often called *interactive systems,* allowing the user to communicate instantaneously with the software through the use of dynamic input devices such as a lightpen, joystick, mouse, or trackball.

A **lightpen** is a penlike device containing a photoelectric sensor that senses positions on a monitor without the need for a digitizer.

A **joystick** is a computer input device that a user can move forward and back and left to right, providing two directional parameters. Rotating the stick provides a third parameter. These three parameters can be sensed and interpreted by a program as X, Y, and Z coordinates, and used, for example, to change viewing locations. Thus, the movement of a joystick could change the observer's position for a perspective drawing. As the user moves the joystick, the computer-drawn perspective on the monitor changes to simulate an observer approaching or circling the building. If the displayed perspective changes fast enough to simulate true animation, the display is called *real time*.

A computer **mouse** is a small, hand-size, electromechanical device that can be moved around a work surface, often on a special rectangle called a *mouse pad.* Two common technologies using a computer mouse are rollerball and optical, with rollerball being more common. As the mouse is moved around the work surface, differential rotations of the rollerball are translated into X and Y coordinates, appearing on the monitor and displayed as a cursor or arrow. A button on the mouse can be used to indicate to the software when a particular location is meaningful. Although mouse information can be processed into exact coordinate values by the software, it is not as accurate as a digitizer. Therefore it is categorized as a *pointing device* rather than a *positioning device.*

Trackballs employ technologies similar to the mouse. Once again, the differential rotations of a trackball are translated into X and Y locations. However, in use, the trackball is stationary and the rollerball is manually moved. Trackball accuracy is the same as a mouse.

Although the older, traditional drafter's tools will continue to be around for a while, computer technology (Figure 6–12) has become more affordable and is commonplace in an architectural office. The computational power and large storage capacities of these systems will continue to influence not only the production of architectural drawings but also the entire architectural design process itself.

ARCHITECTURAL DATABASES

Some computer systems are customized to produce multiview and pictorial drawings of buildings. Other systems are often tailored to employ analytical programs to solve structural, mechanical, and electrical problems in buildings. Both types of systems require geometric information about the building to perform their tasks, although in many cases the graphic functions and analytical (numerical) functions are kept entirely separate.

By creating a database containing the building representation, both graphic and analytical data can be stored in a common location. The

Figure 6–12   *A workstation monitor displaying an elevation design.* (Courtesy Tri-Star Computer.)

building representation contains geometric information (such as the location of walls and roofs) and additional information (such as insulation, fire-hour ratings, and sound absorption). The efficiency of a computer system is increased when both graphic programs and analytical programs can draw their information from a common database. Consequently, changes to this database from the graphic side (such as moving a wall) automatically changes the data for a subsequent engineering analysis. As updates are made to the building design, the database also changes and, in turn, produces different analytical results as well as new drawings reflecting the designer's changes.

All of the building elements must be linked geometrically for the computer to understand how the building's pieces fit together. Once the building geometry has been described, the computer can integrate various programs for solving some of the nongraphical problems that might include some of the following topics.

Architectural Engineering

A common area of computer application is the engineering of structural, mechanical, and electrical systems. It is possible for an engineer to describe, graphically or numerically, a design concept for computer analysis. After the appropriate loading conditions have been keyed into the computer, it can analyze the structure and produce stress diagrams. A structural engineer can then determine the most efficient solution for the building framing to minimize the weight, difficulty of construction, and cost of the project. After the final design changes have been completed, the computer will produce the shop fabrication drawings. Heating, ventilating, and air conditioning systems can be analyzed in a similar fashion. Plumbing and electrical systems can be sized and optimized and pictorials produced.

Code Checking

The process of checking a building design against state building codes is one of the most time-consuming, yet most important, tasks in the design process. A computer can be used to determine if a given building design meets the code requirements for accessibility and fire safety. For example, the computer can check the distance between each room and the nearest exit for safe evacuation. It can also check widths and clearances for maneuvering wheelchairs. Any violations of the codes can be brought to the attention of the architect for redesign.

Site Analysis

Site features, prevailing winds, and solar orientation can be studied by a computer to determine the placement of the building to minimize energy losses and conflicts with natural systems. Cut-and-fill studies can be performed to reduce the cost of site development. Roadway alignment, parking, and storm runoff can be checked to ensure feasible, economical, and safe solutions. Figure 6–13 shows a computer-generated foundation plan.

Estimating

Architectural drafters are often required to prepare quantity takeoffs, schedules for items such as doors and finishes, cost estimates, and specifications for the building construction contracts. Computers can simplify those processes that involve careful measuring of material quantities as shown on architectural drawings. If the drawings have been developed through a computer program and the information stored in a database, then it is possible for quantities to be automatically calculated, schedules produced, and current cost indexes added to produce a cost estimate.

Specifications

Computerized specification writing was one of the first applications of computers to the building design field. A computer can maintain a master file of current product specifications and produce basic specifications by request from the architect. The documents are then assembled and printed within minutes. Word processing allows fast additions to, and deletions from, the text and automatically formats the specifications to produce a final copy.

Architectural Management

Problems associated with the management of multiple design projects may be solved through

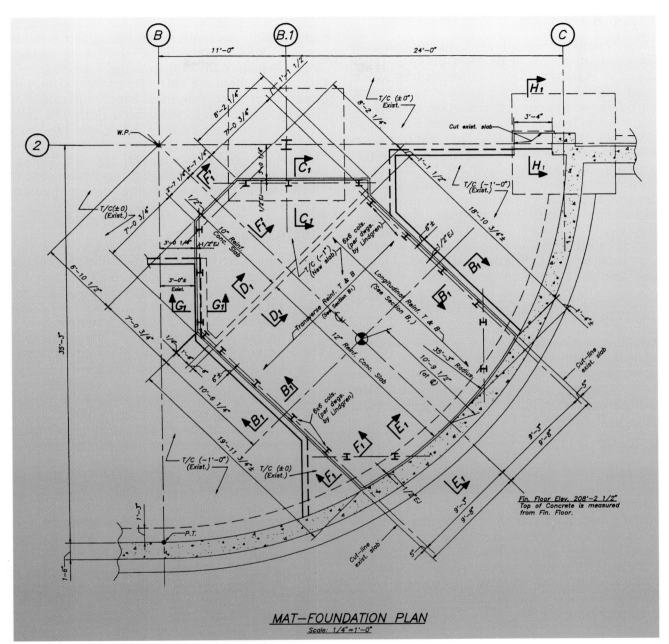

MAT—FOUNDATION PLAN
Scale: 1/4"=1'-0"

Figure 6–13 *A computer-generated foundation plan.*

computer techniques. To increase the efficiency of an office, a computer can store information on these projects, and, by analyzing their requirements, help allocate the necessary labor and resources to each project at the most appropriate time. Other computational techniques such as the *critical path method* are useful for scheduling and controlling the progress of a building project from its inception through construction. A particularly useful feature of this type of management is the ability to control the cash flow of a project in order to reduce cost.

Many contractors use laptop computers at the job site or portable computers connected with the home office's central computer by modem.

Video Animation

Most computer graphics displays produce images that have four or more times as much detail as standard television screens. However, with additional display hardware called *video encoders,* which convert computer displays to standard video, most workstations can record directly to a standard VCR. This feature allows a designer to create an animated presentation of a design on standard videotape for review by their clients. These animations are also valuable for clients to present to financial institutions or to potential tenants. They offer a unique opportunity to experience a design visually before any actual construction occurs, and improve the design process by eliminating misunderstandings and ambiguity in the design team.

FUTURE TRENDS

Looking to the future, it is evident that advances in computer graphics will occur at an increasing rate. Many believe that speech processing, creative design, and virtual reality offer the greatest potential for further development and applications.

Speech Processing

We can expect to see more user-friendly computer systems that will recognize human voice and respond by synthetic voice. Rather than requiring manual operation of keyboards, the user "dictates" to the computer vocally, and the computer "learns" to recognize the user's speech patterns. Some systems now available commercially permit voice-actuated telephone dialing, preparation of architectural specifications, and manufacturing control.

Creative Design

One of the most controversial areas of architectural research is the use of computers in creative design. Despite considerable investigation in this area, researchers have yet to produce a comprehensive computer program to model the creative design process. There are probably two major reasons for this. First, the human design process is not clearly understood. Much of the process depends upon intuition, pattern recognition, and cognitive synthesis, traits that are not readily programmable. Secondly, current computer systems process information serially, which does not necessarily pattern the human thought process.

Researchers are currently trying to develop special types of computers capable of *artificial intelligence* (also called *cybernetics*). The construction of this computer will allow it to learn through experience, recognize patterns, draw inferences, and perform other "creative" tasks. Computer systems displaying this "intelligence" will have a major impact upon the creative design process.

Virtual Reality

A virtual reality (VR) system is one that places the user in a computer-generated environment that appears to be quite realistic. The person experiencing the simulation may actually believe that he or she is within that environment.

The most important use of VR systems will be in scientific and design fields. Virtual environments can be used for purposes such as commercial training (to guide assemblers by overlaying instructions upon real work areas), military training (such as military pilot training using flight simulators), and education (virtual classrooms where each student can be assigned different lessons).

In the area of design and modeling, the designer will be able to show a client how a concept actually looks. An architect designing a building to meet specific codes (such as accessibility codes), will be able to set up parameters in the system that satisfy the code requirements. The architect will be able to "walk" through a simulated virtual model of the proposed design to check the building in its proposed environment. Any environment can be simulated, and the spatial relationships of various design elements can be experienced firsthand. A novice designer can safely and easily test out a new design. This technology will allow more "what if" thinking.

Computing Power

Now available are low-cost, high-performance computer systems capable of generating all the geometric elements that go into making up the fine details of the virtual world. Gone are the "steppy"-looking graphics of past years. Now, curves are smooth, straight lines are really straight, and simulated objects can be shown with natural-looking textures. This can all be accomplished in real time, or quickly enough that the user will not notice a delay in updating the image or any lag in interaction.

Computer Storage Capacity

The storage capabilities of a modern computer far exceed those of just a few years ago. Tremendous increases in storage capabilities, through such developments as optical discs and magneto-optical drives, have made complex virtual environments possible. This lowers the load on computing power, since objects need not be recalculated as frequently as before, and also gives a better virtual environment with more varied objects, as the data for each object can now be readily stored. Nanotechnology research, currently under way, will provide an even greater level of storage.

Software Advances

Also readily available are imaging, animation, and rendering software packages that make possible the affordable creation of virtual worlds by most computer programmers. These software packages now produce results that previously required the attention of senior programmers at advanced research centers.

Lightweight Head-mounted Displays

Head-mounted displays are beginning to take the form of eyeglasses. The 1" high resolution screens invented by Tectronic have made possible the eye phones used by NASA and others. These offer a sensible, hygienic approach to solving the problems that were associated with earlier versions of the head-mounted displays. We can now obtain lightweight, truly stereoscopic head-mounted displays with sufficient image performance to allow the user to ignore the technical aspects of the image generation process and concentrate on what is occurring on the screen. The best of these displays offer resolution approaching the limits of the human eye, have a wide field of view, and are sufficiently lightweight so as not to cause eye fatigue when used for extended periods of time (Figure 6–14).

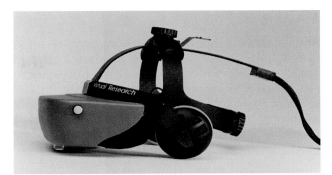

Figure 6–14 *A lightweight virtual reality head mount.* (Courtesy Virtual Research.)

Head Trackers

Head and body movement trackers allow the computer to adjust what is being viewed to the direction in which the user's head is pointing, thus allowing the user to "walk" through the virtual world. This creates a more natural viewing environment than that available from television, large-screen displays, or laser-projected imaging.

Hand Trackers

There are new and innovative interactive devices such as a data glove that sends the positions of your hand and all your fingers to the computer. Also, a three-dimensional mouse provides hand location while one or more buttons can be used to interact with the VR environment. Hand trackers and devices to sense the position and motion of the hand and fingers allow the user to place a "virtual hand" within the simulated environment and manipulate objects with ease. Work is in progress using a number of techniques to provide feedback on what is being felt by your hand when interacting. One technique uses air pressure to inflate microbladders in the glove proportional to the pressure being applied by the glove's user. This technology, called *haptic interface,* permits a user to feel and distinguish between wood, masonry, and plastic materials.

Body Trackers

Whole-body sensor suits can be used to control remote robots. These devices sense the position and movement of your body and all appendages to a high degree of accuracy.

Far Future

Sometimes science fiction stories feature a cyborg—half human and half machine. Such a creation seems to lessen humanity, but as more and more is learned about the human brain and ways of interfacing with it, ultimately a melding of the brain and a computer will occur. Such a hybrid will have the best of both worlds, for he or she will have human emotions, human intuition, a computer's analytical skills and extensive memory, and the ability to communicate directly with other humans without having to go through such limiting factors such as telephones or even speech.

VR technology excites the imagination and opens some doors once closed to all but the most esoteric of researchers. As new opportunities arise, VR usage will certainly expand.

REVIEW QUESTIONS

1. Define the following computer-related terminology by identifying its purpose as it relates to CAD.

    a. CPU

    b. Monitor

    c. Keyboard

    d. Digitizer

    e. Hard Drive

    f. Plotter

2. Differentiate between hard-copy and soft-copy.

3. Give an example of an "X,Y, Z" position and explain what it means.

4. Differentiate between a wire-frame model and a solid model.

5. For what is the constructive solid geometry (CSG) used?

6. When is an extrusion used as it relates to CAD?

7. List two benefits of interactive systems.

8. What is an architectural database?

9. When is code checking done? Why?

10. List five practical applications for virtual reality.

ENRICHMENT QUESTIONS

1. Provide specific examples when a wire-frame model would be more beneficial than a solid model and vice versa.

2. Explain how a digitizer gathers and sends information to the CPU.

3. Beside the items mentioned in the chapter, what other type of information would be useful in an architectural database?

4. What advantages would estimating the cost of a project have if it is done using CAD?

5. List some ways that the computer can act as a consultant.

6. Explain what would happen if you "toured a virtual kitchen."

7. Which of the futuristic equipment listed in this chapter will have the most significant impact on CAD?

PRACTICAL EXERCISE

1. Design a CAD workstation that a person could use at home. Be sure to specify make, model, and specifications for each piece of equipment and software.

7

Architectural Design With CAD

OBJECTIVES

By completing this chapter you will be able to:

🏛 Utilize a CAD program to draw:

 ■ Lines, arcs, and circles

 ■ Sections and section symbols

 ■ Dimensions and lettering

🏛 Differentiate between computer input devices and explain the advantages and disadvantages of each.

🏛 Demonstrate how to produce three-dimensional drawings using a CAD program.

🏛 Produce drawings on CAD, output them to paper, and save the files.

GETTING STARTED

Introduce yourself to a new computer or a new CAD (computer-aided drafting) program by first reading the introductory pages of the manufacturer's hardware and software manuals. In addition, refer carefully to training material that has been prepared to teach you the unique features of the computer system at your training location. Do not hesitate to question your instructor or other experienced users.

A tutorial diskette or CD-ROM will most likely also be available to help you learn to perform basic CAD operations. Some software is accompanied by comprehensive multimedia tutorials complete with graphics and sound. Re-

sults of your progress are recorded each session as you work.

Portions of this chapter are based upon AutoCAD, a dominant CAD software in North America. Most other systems are quite similar.

Setting Up

Turn on your computer workstation (Figure 7–1) and follow any additional instructions that apply at your facility. (The international symbol for "on" is a line; "off" is a circle.) CAD laboratories are organized differently, both in connected hardware and installed software, but typically an introductory display appears on your monitor. This shows the software logo, trade

Figure 7–1   A CAD station with a digitizing tablet. (Courtesy Mayline.)

name, some legal information, and possibly a telephone number to call your regional dealer for assistance.

Your monitor is now energized and ready to accept the commands of your keyboard and your pointing device. Your pointing device might be a *puck* or *stylus* when used with a digitizer tablet, or it might be a *mouse*, *trackball*, or *jcystick* when used without a tablet.

Following the display of loading information, the monitor may show:

> Enter your login name:

Using your keyboard, type your name and press the "Enter" (or "Return") key. This is usually a large key in a reversed L shape and contains a bent arrow symbol (↵). Remember to press the "Enter" key following *any* response or command from your keyboard. You may also be asked:

> Enter your password:

This information is required at some facilities for invoice accounting and to prevent unauthorized use.

Before continuing, the following terms should be familiar to you:

Digitizer and Puck

A *digitizer* tablet is an electronically sensitized drawing surface that senses the location of a *puck* (or stylus) on its surface. The puck looks like a mouse but has a clear plastic crosshair mounted at the upper end. The computer registers the location by recording its coordinates and displaying a cursor on your monitor.

The puck contains four or more buttons. The "pick" or "select" button is commonly used to pick a point on your digitizer as well as to pick a menu option or dialog box selection. For some operations, the pick button is held down as you slide or "drag" the puck along the surface of the digitizer. The pick button is the top (and lowest numbered) button on a puck. If your digitizer uses a *stylus* (looks like a pen), the pick button is built into the point of the stylus and is operated by merely pressing the stylus on the surface of the digitizer.

Mouse

If your computer has a mouse instead of a digitizing tablet, it will have two or three buttons. The first button (leftmost) is the pick button.

Cursor

The *cursor* is a horizontal and vertical set of crosshairs on your monitor that moves about the drawing area reflecting the movement of the mouse, the arrow keys on your keyboard, or the puck on the digitizer tablet. The cursor changes to an arrow to highlight and select commands in the menu areas and dialog boxes. Also, the cursor can change into a "pick box" (small square) for selecting lines and other objects of your drawing.

Drawings formed on your monitor are called "soft copy"; when reproduced by your printer or plotter, they are called "hard copy." The cursor itself, however, never appears on hard copy.

Using Your Keyboard

A computer *keyboard* is similar to a typewriter keyboard but has a number of additional keys, such as function keys. Commands are typed by typing their full command name or an abbreviated shortcut.

Start AutoCAD using the method described by your instructor.

THE MONITOR

Returning to your monitor (computer screen), AutoCAD displays information along three side areas of your monitor: (1) a *command prompt* area and *status line* at the bottom of your monitor, (2) one or more *toolbars* at the left, and (3) a *menu bar* at the top (Figure 7–2). The remaining area of the monitor can be used for drawing.

Command Prompt

The lower *command prompt* area provides three lines of space in which your commands are displayed. Here AutoCAD prompts you to provide additional information regarding your commands. You will soon learn to pay close attention to the command prompt area, for this is

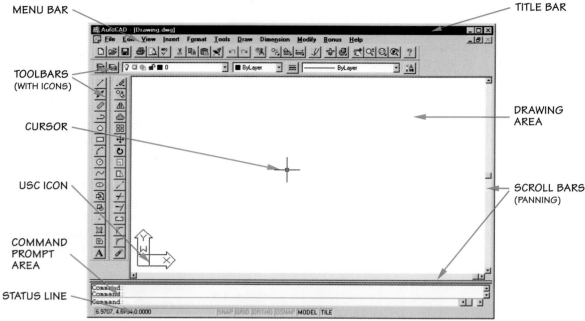

Figure 7–2   The components of the AutoCAD® screen.

where the AutoCAD program most often communicates with you.

Status Line

The *status line* provides a single line of space which shows the status of your drawing, such as the coordinates of the cursor.

For practice, type **Line** at the keyboard. The CAD program is prompting you to draw a line as shown in the command prompt area. In this instance, the first prompts to your command will be:

        From point:

You can reply by picking the starting point with your mouse or entering the starting point's coordinates using your keyboard. This prompt is followed by:

        To point:

Then pick the line's ending point. Press the Enter key to end the **Line** command.

Toolbars

The left-hand *toolbars* show "pictures" (called icons) of commands, in logical order. Some

icons show a tiny triangle. The triangle indicates that there are additional icons hidden underneath (called a "flyout"). Each icon is a CAD command. To find out the meaning of an icon, move the cursor over the icon and wait. A second or two later, the meaning of the icon is displayed in a small yellow box (called a "tool tip") and a sentence-long description is displayed on the status line.

Menu Bar

Move the cursor to the menu bar. Now select any heading by clicking it with your cursor. Notice that when you click with your mouse, the heading becomes highlighted. Click your mouse and the menu selection drops (pulls down) into the main drawing area. Figure 7–3 shows the **Draw** pull-down menu.

The small triangle (▶) after some selections indicates that additional sub-menu selections are available. In Figure 7–4, **Circle** ▶ has been selected and additional sub-menus have appeared.

To remove these menus, click your cursor on the title bar area (topmost) or press the keyboard's "Escape" (ESC) key.

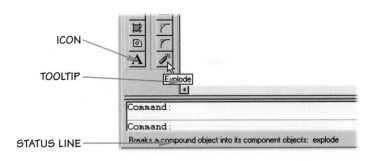

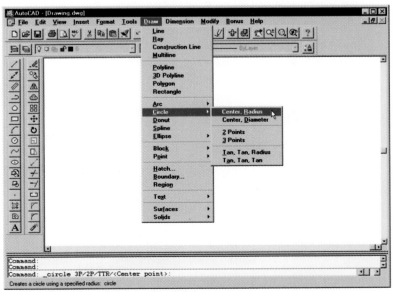

Figure 7–3   AutoCAD® **Draw** menu commands.

Figure 7–4   AutoCAD® **Circle** submenu commands.

Many CAD programs and suppliers offer hundreds of menus and dialog boxes arranged in a branching order from general headings of menus to sub-menus and to specialized dialog boxes. Dialog boxes replace subcommand menus whenever a large number of prompts and commands are offered. They give you the opportunity to interact with your computer program, which explains why they are called "dialog." Prompts on dialog boxes can be answered in any sequence you desire. All are constantly being updated and expanded, so you must often refer to your facility's training bulletins to determine what is available at your workstation.

DIGITIZER TABLET MENUS

In addition to the keyboard, side menu, and pull-down menus, a *digitizer tablet menu*, also called a *template*, offers yet another method of entering commands. The digitizer tablet is covered by a menu card which identifies commands by name and icon (symbol). A standard menu is shown in Figure 7–5. The tablet menu consists of basic command areas to the left and

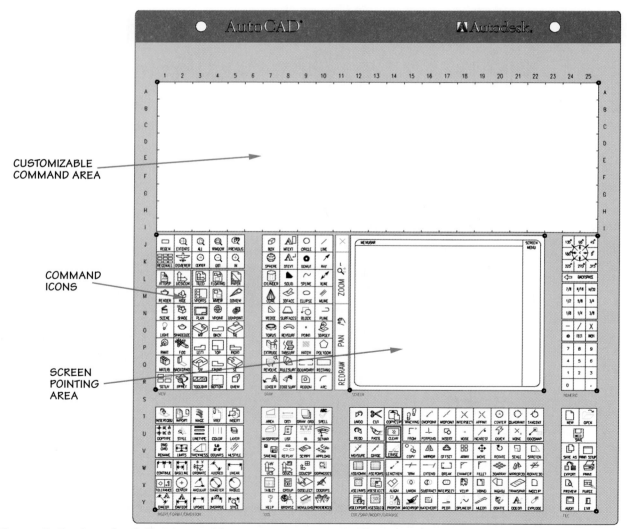

Figure 7–5 *Standard digitizer tablet menu.* (Courtesy Autodesk Inc.)

REAL WORLD APPLICATIONS

Software Updating

As a drafting technician in a small architectural firm, you are aware that the CAD software is at least two years out of date. A newer version, or possibly a different program altogether, would make your work much easier and would provide many new drawing features. For instance, your current software does not include any detail templates, nor does it support more than three layers. These are essential features for the new software. Therefore, in your opinion, purchasing such software would prove beneficial although it may be more expensive.

During a meeting with your supervisor, you express these thoughts. No one is denying that the software could be better, but no one wishes to spend the time researching different software programs. As a result, your supervisor agrees to allow you one hour per day for the next ten days to research software. After the ten-day period, you will have an additional five days to prepare a detailed report. The only requirement that your supervisor stipulated was to include at least five different programs in your research.

under the screen pointing area. The small area to the right provides English numerals, but can be configured for metric units if you desire. The upper area is usually used for custom-designed commands.

By moving your puck over the screen pointing area of a digitizer tablet, you cause the screen's cursor (Figure 7–6) to also move; by clicking your puck over the desired command areas of a digitizer tablet, you issue commands which duplicate the menu bar commands and keyboard commands. Whichever method or combination of methods you use is purely a matter of personal preference.

To activate the standard tablet menu, you must configure the digitizer tablet to define the menu area and the screen pointing area. This is accomplished by clicking **Tools** on the menu bar, then the **Tablet: Config** subcommand, and completing additional prompts. If you later want to use the entire tablet area for digitizing (for example, to trace and digitize an existing drawing), you must press function key F4. This increases your pointing area tenfold. Press F4 a second time to return to using the command overlay.

DRAWING WITH CAD

To begin drawing, you create lines to make up your drawing. Later, you may wish to edit these lines, section, or "poché" portions of your drawing, and finally dimension and add notes to complete the drawing.

GRIDS

CURSOR

Figure 7–6   Screen cursor and grids.

Drawing Lines

CAD gives you the opportunity to draw lines in several different ways. You can choose to draw a single line, or a series of separate lines, or a continuous connected line (called a "polyline"). Curved lines, arcs, and circles can be drawn as well as triangles, squares, or other polygons. You can even use your mouse to create a freehand sketch.

To draw a single line, type **Line** at the command prompt. Now look at the command prompt area. It is asking you to select a starting point by prompting:

        From point:

Now move the cursor (Figure 7-6) and click your mouse to the desired starting point within the drawing area (point 1 in Figure 7-7). The command prompt area now asks you to select another point by prompting:

        To point:

As you move your cursor to the next desired point (point 2 in Figure 7-7), you'll notice a line trailing along. This trailing line helps you to visualize your drawing intention, and this is called "dragging" a line. Press the Enter key to end the **Line** command. Type **U** (for Undo) on your keyboard to remove the lines.

Zooming and Panning

If you discover that your starting point or scale selection causes your drawing to fall off the screen, use the **Zoom** or **Pan** commands to change the view.

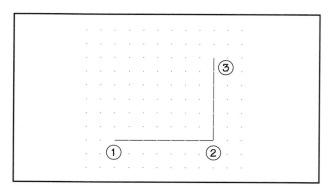

Figure 7-7   Drawing lines.

The **Zoom** command is similar to the zoom lens of a camera. It lets you enlarge your view (zoom in) to focus upon a specific detail, or expand your field of vision (zoom out) to see details that previously had been outside the boundaries of the viewing area. For example, you might wish to zoom in to draw a customized symbol accurately to a large scale and then return to the default size to place the symbol.

To zoom in to your drawing, select the **Zoom** command and provide the diagonally opposite corners of a rectangle surrounding the area to be enlarged (called a "window"). These corners can be chosen by moving the cursor and clicking your mouse at the corners or by entering the coordinates of the corners using your keyboard. You can also zoom by entering a scale of the amount of magnification. Any value greater than 1 will produce a size increase; any value less than 1 will produce a size reduction. For example, "2" doubles the size of the drawing, and ".5" reduces the drawing to half size.

The **Pan** command lets you slide your drawing from side to side or up and down, but *without* changing the magnification, as you would pan with a camera. To pan a drawing, select the **Pan** command and specify the amount of panning by choosing any point and commanding its displacement. As an alternative to the **Pan** command, you may use the scroll bars.

Drawing Aids

As a CAD drafting project for this chapter, you draw a 10" × 20" footing section. To help you draw precisely, all CAD programs include several drawing aids often known as **Ortho**, **Grid**, and **Snap**.

To make lines precisely horizontal and vertical, turn on the **Ortho** mode. This is accomplished by pressing the appropriate function key (the F8 key in AutoCAD) or by double-clicking the word **ORTHO** on the status line. With Ortho turned on, lines that you place are precisely horizontal (0 or 180 degrees) and vertical (90 or 270 degrees).

Another valuable drawing aid is **Grid**, which is a pattern of tiny dots (Figure 7-6) placed in the drawing area of your screen. In AutoCAD, press F7. These dots only appear on the screen; they are not reproduced by your printer or plotter. The dots are spaced a consistent number of units apart starting at the origin, which is near the lower left corner of the drawing area. The default spacing is 0.5" but you may change the spacing with the **Grid** command. The "default" is the option that is used when you don't chose another option. The default is always shown between two angle brackets, such as < 0.5000 >.

Another drawing aid is called **Snap**. You can think of Snap as resolution. For example, setting the snap spacing to 1" forces you to draw in one-inch increments. In AutoCAD, press function key F9 to turn on Snap, which is 0.5" by default. You may change the snap spacing with the **Snap** command.

Drafting Example

Continuing with our example drafting project, assume that our 20" wide footing requires a space around it twice as large (2 × 20") to ac-commodate dimensioning and notes. To set the area of the grid, use the **Limits** command, setting the limits to 0,0 (lower left corner) and 80,80 (upper right corner). To see the all of the grid, use the **Zoom All** command (Figure 7–8).

Before starting to draw, ensure Snap (0.5"), Grid (1"), and Ortho modes are turned on by looking at the status line. Now draw the footing with the **Line** command. When the **Line** command asks you for points, respond with the following x,y coordinates:

> From point: **20,10**
> To point: **40,10**
> To point: **40,20**
> To point: **20,20**
> To point: **c**

By typing **C**, you complete the line. In other words, AutoCAD draws a line to the point where you began the **Line** command, then automatically exits the **Line** command.

Notice that this drawing is being made in *units* rather than *inches*. Since each unit can

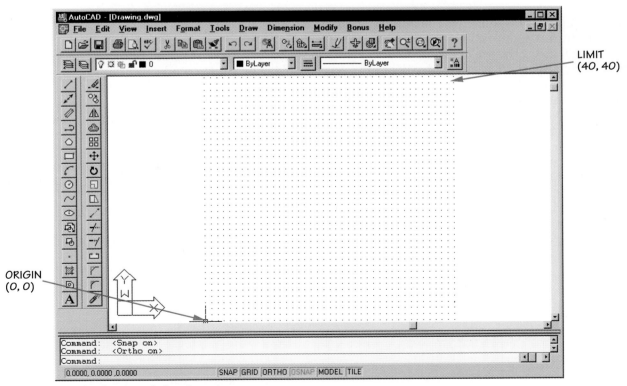

Figure 7–8   The limits of the drawing.

represent any desired size (inch, foot, millimeter, meter, and so on), the appropriate scale is not specified until your drawing is sent to your printer or plotter.

Continue developing the outline of your footing section by adding two vertical upward lines spaced 10 units apart to represent a 10" wide concrete foundation wall resting upon its footing. Also add three horizontal lines spaced 4 units apart and to the left from the foundation wall to represent a 4" thick concrete floor slab and 4" gravel fill. A vertical line $\frac{1}{2}$ unit to the left of the foundation wall represents the left edge of a $\frac{1}{2}$" expansion joint (see Figure 7–9).

The **Line** command can be used for all these lines, or after drawing on one side of the 10" wall, you can use the **Copy** command to duplicate it or an **Offset** command to designate the 10" distance for the duplicated line location.

There are many different options the drafter can use to do the same thing. Experiment with various options to find the preference that will be most efficient for you.

Linetypes

At this stage, all of your lines have been drawn as minimum weight outlines (called "Continuous" lines in CAD). In addition to these outlines, invisible lines (called "Hidden" or "Dashed" in CAD), center lines, property lines, cutting-plane lines, and phantom lines can be specified. All may be varied in linetype and color. All types of lines are usually drawn the same width on your monitor (although polylines can have a designated width that will be visible on the screen), but different weight pens

can be specified later for plotting different types of lines. Plotters accomplish this by associating a different pen thickness with each color of line chosen on the monitor.

Lines such as the outlines you have already drawn can be changed or modified into different linetypes, colors, or layers.

Layers

Layers (overlays) are useful in organizing your drawing and is a tool that allows you to turn off or freeze designated layers while you modify objects on other layers. Layers of a drawing are used to produce electrical, plumbing, air conditioning, and structural plans that coordinate with architectural plans. Computer-produced architectural renderings in color are produced using a different layer for each color. Although there is almost no limit to the number of layers you can display, you can draw on only one layer at a time. Layers are accessed through a dialog box and toolbars that prompt you to select the type and color of lines and lettering (Figure 7–10).

Drawing Circles and Arcs

To draw a circle, select the **Circle** command. You will have several options to define your circle. The most common option is to specify the circle's center and any one point on the circumference.

As with the **Circle** command, the **Arc** command has many options to define an arc. The most useful option permits you to specify the arc's center, then its starting point, and counterclockwise to its ending point. Arcs are usually

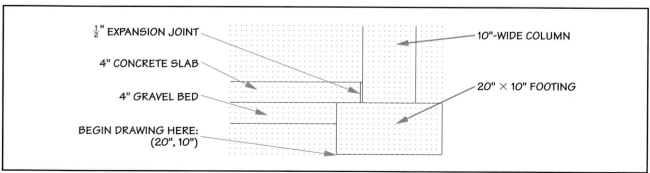

Figure 7–9 Adding additional lines.

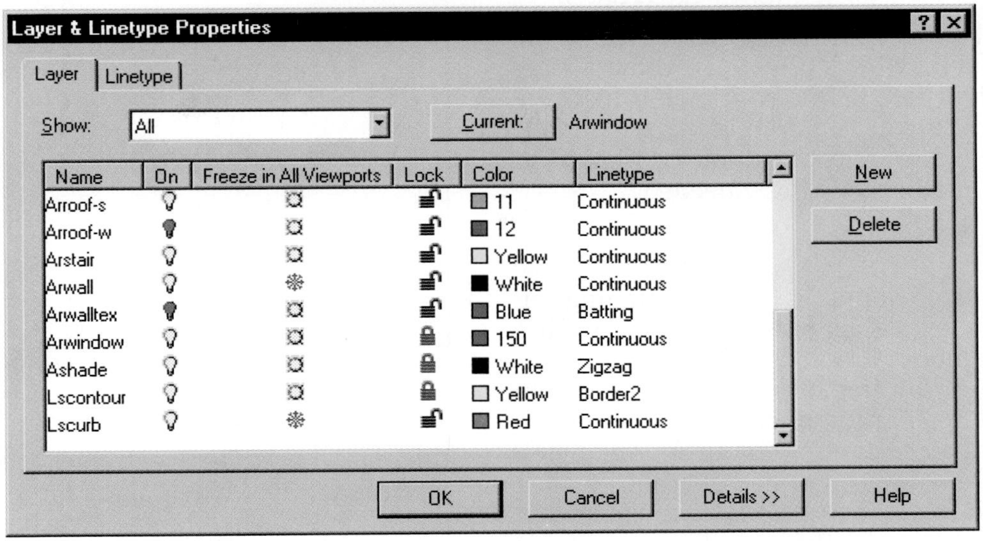

Figure 7–10   *AutoCAD® Layer and Linetype Properties dialog box.*

drawn counterclockwise by CAD unless specified as a negative angle.

Geometric Shapes

Geometric shapes such as regular polygons and ellipses are obtained by selecting the **Polygon** or **Ellipse** commands.

Polygons with 3 to 1024 sides can be produced either by specifying the length of the polygon's side, or by specifying a polygon circumscribed about an imaginary circle of a given radius, or specifying a polygon inscribed within an imaginary circle of a given radius.

Ellipses can be produced by several methods, but the **Center**, **Axis**, **Axis** subcommand is commonly used to specify the location of the ellipse's center, location of either end of its major axis, and location of either end of its minor axis.

Symbols

Symbols for architectural features can be added in a variety of ways. A number of suppliers offer menus of architectural, mechanical, electrical, and structural symbols. Architectural menus include standard symbols of walls, windows, doors, stairs, appliances, and fixtures. Specialized menus, such as piping menus, are also available.

Sketching

Free-form sketched lines can be produced with the **Sketch** command, which lets you draw freehand with your puck or mouse just as you would sketch with a pencil. Actually, however, CAD is recording your sketched line as a series of short, connected straight lines called a "polyline." Before sketching, you will be prompted to:

```
Record increment:
```

This asks you to specify the length of each of these short increments. If you specify too large an increment, your sketched line will appear jagged rather than smooth, but if you specify too small an increment, you will be using excessive memory storage as well as slowing the speed of "regenerating" or refreshing your drawing. See the plot plan of the A residence (Figure 25–2, page 292) for an example of computer-generated sketched lines.

In addition to the **Record increment** option, you may be prompted to **Erase**, **Quit**, or **Exit**. **Erase** lets you correct or erase your polyline, which at this point is only temporarily stored. **Quit** terminates the **Sketch** command *without* storing your polyline, while **Exit** terminates your command and stores your polyline.

For example, break lines at the top of the foundation wall and to the left of the floor slab can be added by sketching. The break lines shown in Figure 7–11 were drawn using the Sketch command, which lets you draw a free-form sketched line. After you have drawn the symbol once, it can be saved (called a "Block" in CAD) so it can be recalled later within your drawing or another drawing if needed.

Sectioning

Sectioning or pochéing is called "hatching" in CAD. Some available sectional symbols are shown in Figure 7–12.

To section an area, it must be completely surrounded by border lines. For example, to place the gravel sectioning in our footing detail (Figure 7–13), you need to first add lines (called "Construction lines" in CAD) to form connected lines which surround the gravel area. (This is

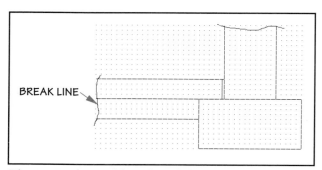

Figure 7–11   Adding break lines.

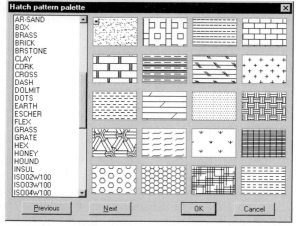

Figure 7–12   AutoCAD® Hatch Pattern Palette dialog box of sectional symbols.

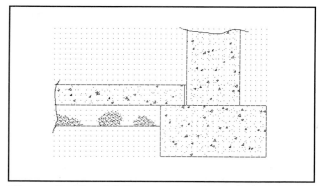

Figure 7–13   Adding sectional symbols.

the reason that the break lines were added before sectioning.) Then place the sectioning and erase those surrounding construction lines, which need not remain on your drawing.

You select a sectioning symbol from a dialog box by clicking the symbol with your cursor. Notice that each symbol on a dialog box has a different code name or number. Advanced users are able to custom-design their own symbol menus.

Corrections

Errors made in giving commands as well as errors made on your drawing can be corrected or erased.

If you type an incorrect command name, your CAD program will prompt you to try again. For example, if you wish to enter the **Line** command by typing on your keyboard (rather than using your mouse) and you mistyped and spelled "Linne" and then depressed the Enter key before noticing your mistake, the command and prompt area would show:

```
Command: Linne

Unknown command "LINNE".
      Press F1 for help
```

This prompts you to either correctly retype the command, or to press the F1 function key to access **Help** and a complete list of approved commands.

To correct a *typing* error before you have depressed the Enter key, press the keyboard's Backspace key to erase your typed letters. In the previous example, backspace twice to remove the last two letters and type an "e" again to obtain the correct command (**Line**).

To cancel a command you have entered in error, press the keyboard's ESC key (located along the top row). The current command will be canceled and you will be back at the **Command**: prompt.

You can erase a *drawing* error by selecting the **Erase** command. Your cursor (crosshairs) will change into a small square, called a "pick box." Using your mouse, move the pick box until some part of it touches the item to be erased and pick the item by pressing the left mouse button. The item becomes highlighted, which lets you review your selection before clicking your mouse's right button, causing that item to disappear. If you find that you inadvertently erased the wrong item, you can restore the item with the **Undo** command.

If you need to erase a large quantity of lines, they can be surrounded by a rectangle called a "window box." The outline of the window box is selected by clicking your cursor to two opposite corners of the desired rectangle. If the window box is drawn beginning from the left and going to the right, all lines that are *completely* within the window box are highlighted for your inspection before erasure. However, if the window box is drawn from right to left (called a "crossing window"), all lines that are *completely or partially* within the window box will be selected.

Dimensioning

The steps needed to dimension most CAD drawings are:

1. Select the object to be dimensioned by picking it or its end points.

2. Move the dimension line away from the object.

3. Add the dimension itself on or over its dimension line.

For instance, to dimension the 20" wide footing width (Figure 7–14), you must first access the Dimension menu and then specify the type of dimension by selecting the Linear subcommand. You can now dimension the horizontal line by moving the cursor and clicking your

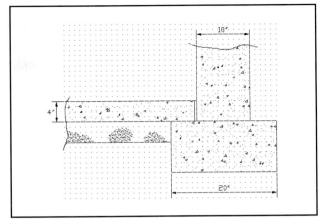

Figure 7–14   Adding dimension lines and dimensions.

mouse at the line's beginning point (coordinates X = 20, Y = 10) and then its ending point (coordinates X = 40, Y = 10).

Additional prompts ask you to specify the location of the dimension line. In this example, you could specify 5 units downward from the object line, or you could pick that location with your cursor.

The CAD program now automatically calculates the length of the dimension and prompts in the command and prompt area:

```
Dimension text <20>:
```

You accept the calculated dimension by pressing the keyboard's Enter key, or if you want to enter a different response, type the desired length using your keyboard and then press **Enter**.

Extension lines are added, and the dimension itself can be located *above* its dimension line or *in* a space provided in a break at the middle of the dimension line. If necessary, the dimension can be offset rather than centered (for example, to avoid a conflict with a center line). The dimension routine also provides dimension line arrowheads or "tic" marks and extension line offsets as accepted in full-scale drafting practice. ("Offset" means the space between extension line and object as well as the distance the extension line continues beyond the dimension line.)

Note that when your drawing is produced at another scale, the size of dimensions, arrowheads, and extension line offsets must be scaled

REAL WORLD ⬛ APPLICATIONS

File Organization

One of the advantages to using CAD for preparing architectural plans is that it is easier to make revisions. The files can be recalled, modified, and saved under a new file name. Drawing with paper and pencil means erasing parts of the drawings or starting over again. However, many CAD users also spend much time looking through old computer files in search of a particular drawing. If care is given to naming the files in the beginning, this process becomes much easier.

When naming files, computers running DOS and Windows versions 3.X and earlier allow you to enter names not exceeding eight characters followed by a three-character extension. Computers running Windows 95/NT, UNIX, or Macintosh allow more descriptive file names, sometimes up to 255 characters long. The file name and extension are separated by a period (.). The extension identifies the application that created that file. Some common CAD extensions include the following. What are some extensions of other programs that you are using—CAD or otherwise?

🏛 AutoCAD and AutoCAD LT	.DWG
🏛 AutoSketch	.SKD
🏛 Drawing Interchange Format	.DXF
🏛 Encapsulated PostScript	.EPS
🏛 CADKey	.PRT
🏛 MicroStation	.DGN
🏛 TurboCAD	.TCW

File names should identify the type of drawing and sometimes the version of the drawing. Many times these need to be abbreviated to fit within the eight-character limit. Be careful not to name files with generic names such as "DRAWING.DWG." A better name would be "FELEV-1.DWG." The reference to "FELEV" in the file name means that this is a *front elevation* view. The "1" indicates that this is the first version of this project. Notice that this file name does not contain any information about who the client is. This is because the directory (folder) that contains the files for this project is "MILLE598." The "MILLE" stands for the client's name, in this case Miller. If there is more than one client with the same last name, use their first initial after the last name. The "598" refers to the month and year in which the project was completed. In this case it is May 1998.

The key thing to remember when naming and organizing files is that they should be consistent. Develop a set of "rules" that you will use when saving files. This standard then dictates how to save files in the future.

up or down to the desired size. For example, $\frac{1}{8}$" high full-scale dimensions appear as only $\frac{1}{64}$" high when added to a drawing being produced at $\frac{1}{8}$ scale. You can use the **Dimscale** command to adjust the size of dimensions, arrowheads, and extension line offsets to appropriate size.

Choose **Architectural** in the Units section of your Units Control dialog box (Figure 7–15) to show linear dimensions in feet and inches, and choose **Deg/Min/Sec** in the Angles section to show angular dimensions in degrees and minutes.

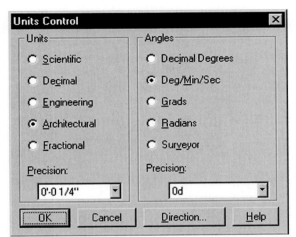

Figure 7–15   AutoCAD® Units Control dialog box.

Lettering

To add notes and titles to your drawing (Figure 7–16), select the **Text** command (lettering is called "Text" in CAD). The CAD software prompts you to justify, to select lettering style or font, to select a starting point for your lettering, to select lettering direction (called "Rotation angle"), and choose lettering height, width, and slant (called "Obliquing angle"). Some of the prompts and their defaults follow.

The **Justify** prompt asks you to indicate the position of your lettering in relation to the start-ing point that you specify. For example, the lettering may be located to the right, left, or middle of the starting point. Also, the lettering can be produced over, centered, or under the starting point.

The **Style** default is the lettering style previously selected and entered. Numerous styles of Gothic and Roman lettering can be selected from storage. Commercial Gothic is a popular selection, but some architectural offices have designed their own architectural alphabet. Figure 12–12 (page 136) shows examples of each.

The **Start Point** default is the lower left corner of your line of lettering. This can be selected with your mouse or by typing coordinates on your keyboard. Replace this default by the **Right** option to select the lower right corner, which is useful for a leader such as shown in Figure 7–16.

The **Rotation angle** default is a horizontal line of lettering, shown as < 0 > degrees. A line of lettering inclined to the horizontal (say vertical lettering reading from the right) can be obtained by typing in the angle of rotation (90).

The **Height** option prompts you to specify a lettering height suitable for your drawing. The **Width** default provides standard lettering proportions, but you may condense or extend these proportions if you desire. The default value is < 1.00 > .

The **Obliquing angle de**fault provides upright letters for all fonts except italic. The default value is shown as < 0.0 > . To command lettering that slants 10 degrees backward, you would type "–10."

Finally, type your text (letters, numerals, and punctuation symbols) using your keyboard. Errors are corrected by backspacing.

THREE-DIMENSIONAL DRAWING

For many years, two-dimensional drawings and three-dimensional architectural models have been the principal methods to design and record the exterior and interior of buildings. Three-dimensional objects, such as buildings, can be shown in great detail on two-dimensional paper

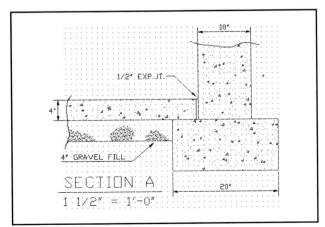

Figure 7–16   Adding notes and titles to complete your drawing.

either by multiview projection or pictorial drawing. Full-size or scaled models offer the opportunity to study a three-dimensional object from different angles. Computer-aided design and drafting programs offer the advantages of both drawings and models, for they can be used to obtain a detailed two-dimensional representation of an object and also to view the object from different three-dimensional directions. As discussed in Chapter 6, computer-generated pictorials may show hidden as well as unhidden lines (called "wire frame") or omit hidden lines.

Traditional Isometric Drawing

Conventional isometric, oblique, and perspective drawings are two-dimensional representations of three-dimensional objects, and they can be drawn using an AutoCAD program. For example, to produce a traditional isometric drawing, move the cursor and click your mouse on the **Drawing Aids** command under the **Tools** heading on the menu bar. The dialog box offers an isometric grid and snap. You have the choice of three isometric grids: the **Left** default grid lines are vertical and 30 degrees to the left, the **Right** grid lines are vertical and 30 degrees to the right, and the **Top** grid lines are 30 degrees to the right. Select one of these commands and draw in isometric using your isometric grids to click the coordinates of the corners of your object. Notice that you need to switch to a different set of isometric coordinates (press F5 in AutoCAD) whenever you want to draw in a different isometric plane.

Although the preceding instructions create a traditional isometric drawing, it is seldom used because you have not utilized the full capabilities of CAD to create a three-dimensional pictorial which can be viewed from many directions.

Wire-frame Pictorial

Up to now, you have been working in only the X and Y coordinate plane of the World Coordinate System (WCS). For three-dimensional work, however, you must also include the Z axis. In WCS the Z axis is mutually perpendicular to the X and Y axes on the drawing

surface and extends from that surface up toward you (Figure 7–17). Objects may be constructed by specifying the X, Y, and Z coordinates of the first corner, then drawing a line to the X, Y, and Z coordinates of the second corner, and so on.

One method of introducing yourself to three-dimensional commands is to draw a simple block 4" wide by 2" deep by 1" high. Enter the coordinates of all eight corners and connect them with lines to produce all twelve edges. Although you have entered all corners and edges, your screen shows only a 2" by 4" rectangle! The 1" height information has been entered into the drawing, but you don't see it because the 1" vertical lines appear as dots at the corners of the 2" by 4" rectangle. They are perpendicular to the screen coming right at you.

Your location in relation to the screen is called your *viewpoint*, and in WCS the default value of your viewpoint is X = 0, Y = 0, and Z = 1, which means you are located directly over the origin. To see all three dimensions of the block, move your viewpoint away from the origin in either the X or the Y direction. That is, your viewpoint needs any X or Y coordinate other than zero. This is accomplished by changing the viewpoint, which can be moved

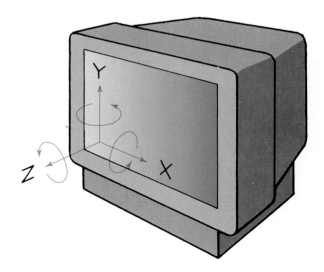

Figure 7–17   Positive directions of X, Y, and Z coordinates and positive rotations in WCS.

about the object (from 0 to 360 degrees) or moved vertically up and down (from 0 to +90 or –90 degrees) resulting in a variety of axonometric projections.

Although the viewpoint command presents a large number of viewing directions, it does not control the *distance* between you and the object. Your axonometric projection can be transformed into a perspective projection through the **Dview** command, which lets you specify this distance. You will be prompted to provide the X, Y, and Z coordinates of both the station point (called "camera point") and the object (called "target point"), resulting in the opportunity to specify an infinite variety of wire-frame perspectives.

Hidden-line Removal

You can transform a wire-frame pictorial (*all* lines showing) into a solid-looking model (only visible lines showing) by "hiding" the invisible lines. The CAD program automatically erases all lines which fall behind surfaces in front (Figure 7–18).

Solids Modeling

Another method of obtaining building pictorials is through the use of *primitive* solids such as cubes, rectangular blocks (called "boxes"), prisms (called "wedges"), cylinders, tori, and cones. These primitive forms can be joined together (called a "union") or subtracted from one another (by specifying a negative primitive),

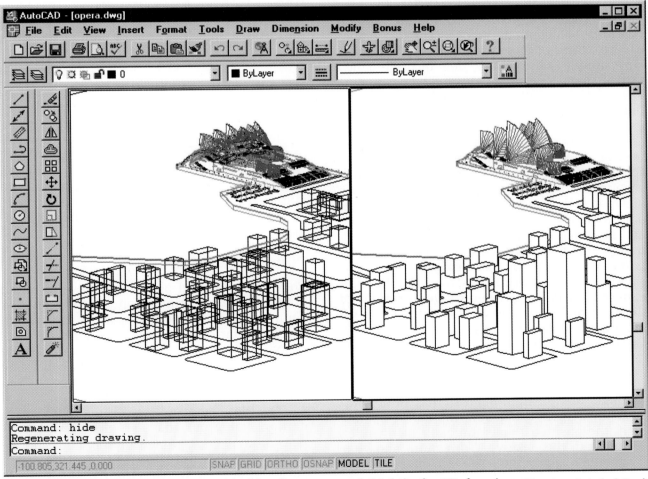

Figure 7–18   *Wire-frame (left) and hidden lines removed (right) of a 3D drawing. (Courtesy Autodesk Inc.)*

resulting in the desired shape. Intersecting portions of two primitives also can be used to define a solid.

Additional methods of obtaining building pictorials are through the use of *extrusions* and *revolutions*. Using extrusion commands, a two-dimensional shape such as a floor plan can be extruded or thickened into a three-dimensional solid by sweeping or moving it up to the building's height. Using revolution commands, round objects can be constructed by specifying the two-dimensional shapes and the angle of rotation.

User Coordinate System

The User Coordinate System (UCS) is a coordinate system that you define at any angle or location within the three-dimensional space of the WCS. The World Coordinate System (WCS) is the default of the UCS. The UCS planes can be rotated or moved to any desired position. For example, an auxiliary view of the inclined face of an object can be produced by positioning a UCS plane parallel to the inclined face. Or a building can be produced and viewed from several positions simultaneously using UCS planes and the **Viewports** command.

The **Viewports** command can divide the screen into two, three, or four sections (each section is a viewport). Figure 7–18 shows two viewports. Using four UCS planes, you can produce a plan view, front elevation, end elevation, and pictorial drawing. Although you can work in only one viewport at a time, the other viewports instantly display any changes.

When moving or rotating the UCS planes, it is necessary to know the positive directions of the X, Y, and Z axes as well as the positive direction of rotation about each axis. Two *right-hand rules* are used:

1. To find the positive direction of each axis, point your thumb, index finger, and middle finger of your right hand so that all are mutually perpendicular. Your thumb then points in the positive X axis direction, your index finger in the positive Y direction, and your middle finger in the positive Z direction.

2. To find the positive direction about an axis, curl the fingers of your right hand and extend your thumb. Point your thumb in the positive direction of any axis, and your fingers will indicate a positive direction of rotation.

Rendering

To render a pictorial, you can remove hidden lines using the **Hide** command (Figure 7–18). Another method is to add lighting materials and shading to produce a "photo-realistic" rendering (Figure 7–19).

A pictorial produced by the **Render** command can be viewed on your screen and printed by a color printer.

SHUTTING DOWN

When you have finished working on a drawing, you may wish to save it for future revision or addition, produce a hard copy, or erase it completely.

Saving Your Work

The **Save** command (**QSave** in AutoCAD) stores your drawing under its file name. Experienced drafters save their drawings at quarter-hour intervals so that no more than 15 minutes of work is lost in the event of a power interruption or program crash. Many CAD programs have an automatic save option that allows you to specify the interval of elapsed time between automatic saves.

Plotting or Printing

The **Plot** command plots your drawing on an ink-jet plotter or prints it on a laser printer. A pen plotter creates a drawing using pens which produce your drawing lines. An ink-jet plotter sprays colored ink in fine jets from replaceable cartridges. Printers create drawings consisting of a series of tiny dots which form lines at 75 to 1200 dots per inch (DPI) resolution.

The **Plot** command results in a Plot Configuration dialog box that lists the default (current) plotting or printing device and possible alternate connected devices. You will be asked questions such as which portion of the drawing to reproduce, the scale desired, line weight, and line color.

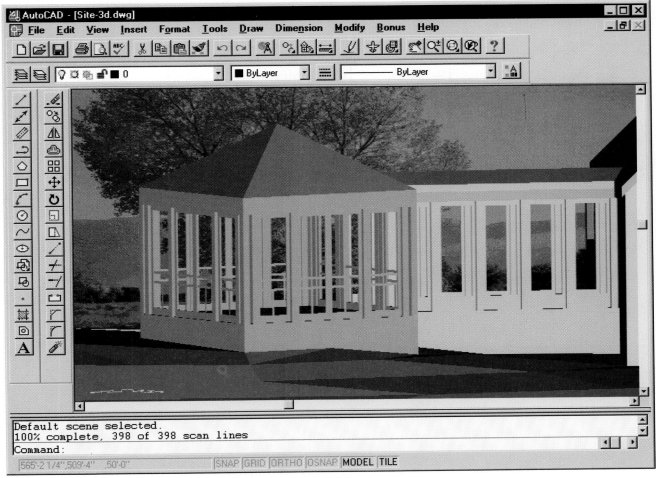

Figure 7–19   *3D drawing rendered.* (*Courtesy Autodesk Inc.*)

Quitting

The **Quit** command allows you to completely disregard all work not previously saved.

To avoid destroying files or damaging equipment, do not turn off your computer workstation without the approval of the CAD laboratory's supervisor.

REVIEW QUESTIONS

1. List two places to receive information about a new software program.

2. Differentiate between the following input devices:

   a. Puck
   b. Mouse
   c. Digitizer tablet
   d. Keyboard

3. What is the purpose of the command prompt area?

4. What does it mean that an object is "highlighted"?

5. How does the "status line" assist the CAD operator?

6. Differentiate between the purposes of a dialog box and a pull-down menu.

7. Define digitizing.

8. List and describe three drawing aids used by CAD.

9. What must be done to an area to section before hatch lines can be added?

10. Describe two methods that can be used to input a dimension line.

11. Differentiate between zoom and pan.

12. Differentiate between the WCS and the UCS. When is each used?

13. List the proper steps to be used when shutting down your computer and quitting the CAD program.

ENRICHMENT QUESTIONS

1. Delineate the steps needed for a novice computer user who wants to use CAD to draw an arc, circle, box, or freehand sketch.

2. Explain why computers display their drawings in units rather than inches.

3. Produce a chart that shows various linetypes drawn by CAD compared to those done by hand.

4. How would you explain the purpose of CAD layers?

5. Identify advantages and disadvantages of a wire-frame versus a solid model.

PRACTICAL EXERCISES

1. Produce specialized symbols you will need for your original house design.

2. Computer-produce lettering that would exemplify each of these font styles:
   a. Rotation angle = 75°
   b. Angle of rotation = 65°
   c. Height = 1.5"
   d. Width = 1.25"
   e. Oblique angle = −15°

3. Computer-produce a floor plan for a bedroom. Include all dimensions, furniture, windows, and doors.

4. Computer-produce Figures 4–36, 4–37, and 4–38 (page 53).

5. Use Figure 9–15 (page 116) and computer-produce each of them. The scale is one block = 1'-0.

6. Computer-produce the kitchen for the Z residence that is shown in Appendix A.

7. Choose three windows and three doors from Appendixes B and C respectively and computer-produce each. Save them as a symbol.

Drawing Conventions

SECTION II

CHAPTER 8

Projections

OBJECTIVES

By completing this chapter you will be able to:

- Distinguish between a perspective and a parallel projection.

- Distinguish between an orthographic and an oblique projection.

- Distinguish between a multiview and an axonometric projection.

KEY TERMS

Perspective projection
Parallel projection
Oblique projection
Orthographic projection
Multiview projection
Axonometric projection

The ability to communicate complex ideas is one of the talents that distinguishes humans from other forms of life. The first methods of communication were spoken languages and picture languages. The picture languages have developed through the years into a great number of written languages and one universally accepted graphic language. This universal graphic language is based upon a theory of *projections*. That is, it is assumed that imaginary sight lines, called *projectors,* extend from the eye of the observer to the object being described. The projectors transmit an image of the object onto an intervening transparent surface called the *picture plane*. This image is called a *projection* of the object.

TYPES OF PROJECTIONS

Perspective Projection

When the projectors all converge at a point (the observer's eye) as shown in Figure 8–1, the re-

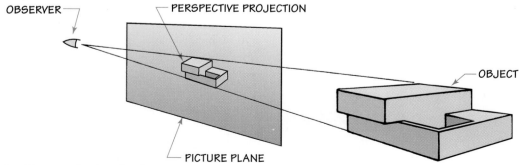

Figure 8–1   Perspective projection.

sulting projection of the object on the picture plane is called a **perspective projection.** Perspective projections are often used by architects to present a realistic picture of a proposed building. As shown in Figure 8–6 and Table 8–1, there are three types of perspective projection: *one-point* (Figure 8–2), *two-point* (Figure 8–3), and *three-point* perspective. Detailed information can be found in Chapter 38.

Parallel Projection

When the projectors are all parallel to each other (as if the observer had moved to infinity), the resulting projection is called a **parallel projection.** Parallel projectors angled (oblique) to the picture plane result in an **oblique projection** as detailed in Chapter 11, but for most architectural drafting, the projectors are assumed to be perpendicular to the picture plane. Since *ortho* is a Greek prefix meaning "at a right angle," this is called an **orthographic projection.** There are two kinds of orthographic projection

depending upon the relation of the object to the picture plane. These are called **multiview projection** (Figure 8–4) and **axonometric projection** (Figure 8–5).

Multiview Projection

In multiview projection, the object is positioned so that its principal faces are parallel to the picture planes. This is the type of projection most useful to architects because principal lines and faces appear true size and shape on the picture plane. The term *multiview* is used because more than one view is required to show all three principal faces. *Multi* is a Latin prefix meaning "many." See Chapter 9 on multiview projections for more detail.

Axonometric Projection

In axonometric projection, the object is tilted with respect to the picture plane so that all faces and axes are visible but not in true shape. *Axono*

Figure 8–2 *A one-point perspective view. (Courtesy California Redwood Association.)*

Figure 8–3 *A two-point perspective view. (Courtesy California Redwood Association.)*

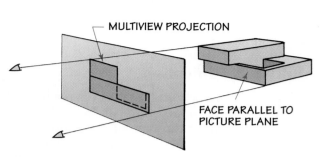

Figure 8–4 *Multiview projection.*

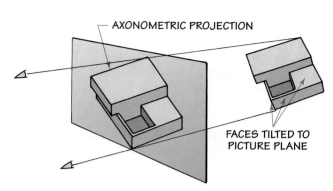

Figure 8–5 *Axonometric projection.*

is a Greek prefix meaning "axis." Axonometric projections are easier to draw than perspectives and, consequently, are used often. See Chapter 11 on pictorial projections for more detail.

The different features of each type of projec-tion are illustrated in Figure 8–6 and Table 8–1. Study them carefully and refer back to them of-ten as you read related Chapter 9 on multiview projections, Chapter 11 on pictorial projections, and Chapter 38 on perspective projections.

TABLE 8–1 *Types of Projections*

			Type	Relation of Projectors to:		Relation of Object Faces to Picture Plane
				Each Other	*Picture Plane*	
Perspective			One-point	Converging	Many angles	One face parallel
			Two-point	Converging	Many angles	Vertical faces oblique
			Three-point	Converging	Many angles	All faces oblique
Parallel	Orthographic	Multiview	First-angle	Parallel	Perpendicular	Parallel
			Third-angle	Parallel	Perpendicular	Parallel
		Axonometric	Isometric	Parallel	Perpendicular	Three equally oblique
			Dimetric	Parallel	Perpendicular	Two equally oblique
			Trimetric	Parallel	Perpendicular	Three unequally oblique
	Oblique		Cavalier	Parallel	Oblique	45°
			Cabinet	Parallel	Oblique	Arc tan 2

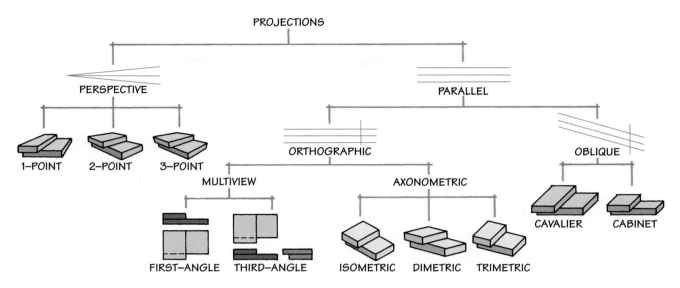

Figure 8–6   *Types of projections.*

Interviewing an Architect

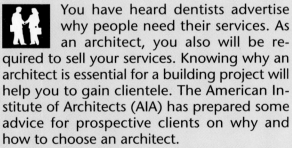
You have heard dentists advertise why people need their services. As an architect, you also will be required to sell your services. Knowing why an architect is essential for a building project will help you to gain clientele. The American Institute of Architects (AIA) has prepared some advice for prospective clients on why and how to choose an architect.

Use this Internet address, sponsored by the AIA, to access this information: **http://www.aia.org/begingd.htm** Now answer these questions.

1. Why should an architect be hired?

2. How can a client save money by hiring an architect?

Many clients interview several architects before choosing one. It is important that you be prepared for their questions. Review the information on interviewing an architect provided on this Internet page. Which three questions would be most influential in a client's decision-making process? Practice answering these questions by asking a classmate to role-play as a prospective client.

REVIEW QUESTIONS

1. What is the theory of projections?

2. What is a picture plane?

3. Give the relationship of the projectors to each other and to the picture plane in the following types of projection:

    a. Multiview

    b. Axonometric

    c. Oblique

    d. Perspective

PRACTICAL EXERCISES

1. According to Figure 8–6, what would be one advantage and one disadvantage of a perspective projection as opposed to a multiview projection?

2. What is the relationship between the observer and the picture plane in an orthographic projection? An oblique projection?

CHAPTER 9

Multiview Projections

OBJECTIVES

By completing this chapter you will be able to:

- 🏛 Differentiate between third- and first-angle projections.

- 🏛 Draw and describe the use for various architectural lines.

- 🏛 Explain types of sections used in architectural drafting.

KEY TERMS

Multiview projection	Dimension line
Third-angle projection	Extension line
First-angle projection	Construction line
Line weight	Guideline
Outline	Rule of configuration
Hidden line	
Cutting-plane line	Section
Section line	
Center line	

All buildings and nearly all elements of a building are three-dimensional, but they are designed and specified by means of two-dimensional plans. A three-dimensional object can be described on a flat, two-dimensional plan by any of the types of projection discussed in Chapter 8, but the type of projection most useful, and therefore most commonly used, is **multiview projection.** Multiview projection is so popular because the exact shape of each face of a building and the elements of a building can be shown without distortion. Also, the length of every line can be shown true size or to a convenient scale.

The "Glass Box"

The easiest way to understand multiview projection is to imagine an object placed inside a "glass box" so that all six faces (front, rear, plan, bottom, and both ends) are parallel to the faces of the glass box. This is illustrated in Figure 9–1, where an object (a clay model of a building) has been surrounded by imaginary, transparent planes. Now, if projectors were dropped perpendicularly from the object to each face of the glass box (Figure 9–2), a number of projection points would be obtained which could then be connected to give a true-size and -shape projection of the six principal faces of the object. If the

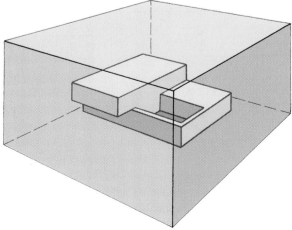

Figure 9–1   The "glass box."

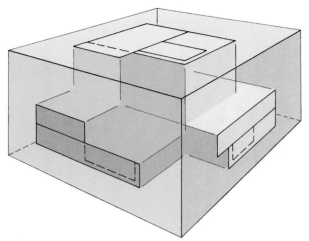

Figure 9–2 *Projecting to the "glass box."*

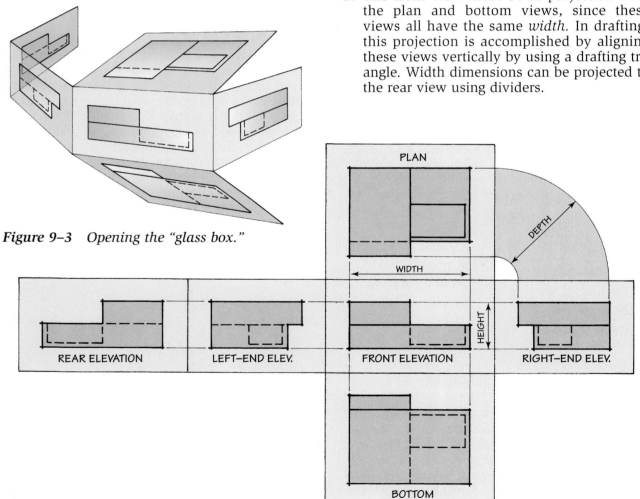

glass box is then unfolded, as shown in Figure 9–3, all six faces can be illustrated upon a single sheet of paper as shown in Figure 9–4. Note the terms *height, width,* and *depth. Height* is a vertical distance, *width* is an end-to-end distance, and *depth* is a front-to-rear distance.

Study Figure 9–4 and notice that all adjacent views must be in projection. For example:

1. The front view must be in projection with the rear and end views. These four views all have the same *height* and are often called *elevations* (such as front elevation, rear elevation, right-end elevation, and left-end elevation). In drafting, the projection between elevations is accomplished by aligning these views horizontally using a T square.

2. The front view must be in projection with the plan and bottom views, since these views all have the same *width.* In drafting, this projection is accomplished by aligning these views vertically by using a drafting triangle. Width dimensions can be projected to the rear view using dividers.

Figure 9–3 *Opening the "glass box."*

PLAN

DEPTH

WIDTH

HEIGHT

REAR ELEVATION   LEFT–END ELEV.   FRONT ELEVATION   RIGHT–END ELEV.

BOTTOM

Figure 9–4 *Standard arrangement of the principal views.*

REAL WORLD APPLICATIONS

Preparing a Memo

 During the final review of a house you have designed, several changes were requested by your client. These changes will require that your drafter modify the plans, elevations, and sections of the house. Send an in-house memo to your drafter stating the required changes and the requested completion date. Write this memo with a word processor using these basic guidelines:

1. Include the date.
2. Clearly identify the recipient and sender.
3. State the subject of the memo.
4. Briefly state the purpose of the memo in the first sentence.
5. Write concisely, but be sure to include *all* necessary information.

When memos are written well, they make a good impression on the receiver. The figure shows an example of a typical memo format.

Name of Organization
DATE: Memo Date
TO: Name, Title
FROM: Name, Title, Signature
SUBJECT: STATED CLEARLY IN FULL CAPS
Introductory Paragraph (Single Spaced)
Body Paragraph (Single Spaced)
Concluding Paragraph (Single Spaced)

3. Also notice that four views (plan, bottom, and both ends) have a common element of *depth.* In drafting, depth measurements may be transferred by dividers, by scale, by using a 45 degree miter line, or by drawing 90 degree circular arcs (Figure 9–5).

In architectural drafting, views of an entire building are often so large that each view requires an entire sheet of paper. In such cases, titles or other identifications are used to clarify the relationships between views.

Number of Views

Six views are seldom drawn. A simple architectural detail usually requires only two or three views, but a complex building might require a great number of views in addition to sections and details. Partial views may also be drawn. The governing rule is to draw as many views as necessary to describe the object clearly and accurately—no more, and certainly no fewer.

Third- and First-angle Projections

The frontal and horizontal planes of the glass box in Figure 9–2 can be extended to divide space into four sectors, known in geometry as *quadrants* (Figure 9–6). The object can be placed in any quadrant and projected to the projection planes. The horizontal plane is folded clockwise into the frontal plane as shown by the arrows in Figure 9–6. This results in four alternate arrangements of views, called first-, second-, third-, and fourth-angle projections.

Third-angle projection produces the relationship between views described previously in which the plan view is *above* the front elevation. **First-angle projection** produces a slightly different relationship between views, in that the plan is *below* the front elevation. In first-angle projection, the picture plane is *beyond* the object rather than *between* the object and observer. Second- and fourth-angle projections produce overlapping views and are not used.

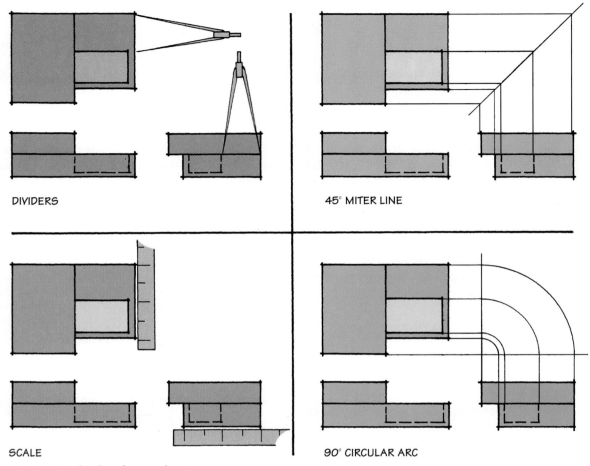

DIVIDERS

45° MITER LINE

SCALE

90° CIRCULAR ARC

Figure 9–5 *Methods of transferring measurements.*

Third-angle projection is used for most architectural and technical drafting in this country. Occasionally, however, first-angle projection is used in architectural drafting when it is more convenient to place a plan below an elevation. In Figure 29–10 (page 352), notice that the plan and front elevation of each fireplace is in first-angle projection, whereas the front and end elevations are in third-angle projection.

Language of Lines

Nine types of lines constitute the basic "alphabet" of drafting. They are all illustrated in Figure 9–7. Notice that these lines are drawn using five different line weights. **Line weight** refers to the blackness and thickness of a line and ranges from an extremely heavy cutting-plane line to barely visible construction lines and guidelines. A heavy line is obtained by using a soft pencil (such as an F grade), a slightly rounded point, and some hand pressure. A light line is obtained by using a hard pencil (such as a 6H grade), a sharp point, and less pressure.

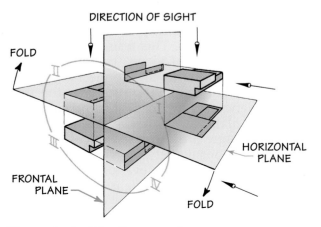

DIRECTION OF SIGHT

FOLD

FRONTAL PLANE

HORIZONTAL PLANE

FOLD

Figure 9–6 *The four quadrants.*

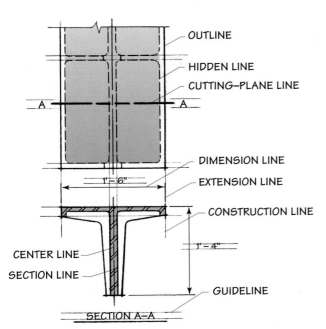

Figure 9–7   *The language of lines.*

Outline   **Outlines** (also called *visible* lines) are heavy lines used to describe the visible shape of an object including edges, edge views of planes, and contours of curved surfaces.

Hidden Line   **Hidden lines** (also called *invisible* lines) are outlines that cannot be seen by the observer because they are covered by portions of the object closer to the observer. The locations of such hidden edges are indicated when necessary to describe the object accurately. The dashes of hidden lines are about $\frac{1}{8}$" in length and $\frac{1}{32}$" apart. Hidden lines are medium-weight lines.

Cutting-plane Line   The **cutting-plane line** represents the edge view of a cutting plane sliced through the object to reveal inner features. It is drawn as the heaviest-weight line so that the location of a section can be identified easily.

Section Line   **Section lines** are used to cross-hatch any cut portion of an object. A number of sectioning symbols are shown in Figure 5–5. Section, center, dimension, and extension lines are all drawn the same light weight.

Center Line   **Center lines** indicate axes of symmetry. Most center lines consist of alternating $\frac{1}{8}$" short dashes and 1" long dashes spaced about $\frac{1}{32}$" apart. In small-scale drawings, spaces may be omitted.

Dimension Line   **Dimension lines** are used to indicate the direction and limits of a linear dimension. Some types of arrowheads used with dimension lines are illustrated in Figure 13–1 (page 141).

Extension Line   **Extension lines** serve as an extension of a feature on the object so that dimensions can be placed *next* to a projection rather than crowded *on* the projection.

Construction Line   **Construction lines** are extremely light lines barely visible to the eye. They are used to lay out a view or to project between views.

Guideline   Horizontal and vertical **guidelines** are construction lines used to guide hand lettering. See Figure 12–5 (page 133).

"Reading" a Drawing

In addition to drawing the projections of an object from a mental picture, an architectural drafter must be able to "read" drawings. That is, given a projection of an object, the drafter must be able to visualize its shape and features. The following rules may help:

1. The same features must always be in projection in adjacent views. Consequently, a point or line in one view may be projected to and read in an adjacent view to help understand what it represents.

2. Read views simultaneously rather than one at a time. Staring at a single view usually will not be particularly helpful. Your eyes should project a feature back and forth between views until you are able to visualize the feature and eventually the entire object.

3. There is a rule called the **rule of configuration** which states that the configuration (shape) of a plane remains about the same in all views, unless the plane appears in its edge view. For example, a five-sided surface will always have five sides—not four or six—unless it appears on edge.

Sectioning

Architectural components are seldom solid objects as discussed in the preceding sections. Rather, they consist of complex assemblies that require sectional views to describe them adequately. A **section** is an imaginary cut through a component (part) or an assembly of components. All the material on one side of the cut is removed so that the interior can be studied. Often, sections are drawn through entire structures, walls, floors, roofs, foundations, structural assemblies, stairs, and fireplaces. The scale of sectional views is often increased to further clarify the details. Cutting-plane lines are used only when needed to show where the cut was taken. Sight direction arrows are added to the ends of cutting-plane lines only when needed to show the direction of sight.

Full Section   A full section is a cut through the entire building or component. As shown in Figure 9–8, when a horizontal cutting plane is passed through an entire building, a floor plan results. The horizontal cutting plane is assumed to be located 4' above the floor (midway between floor and ceiling).

Cutting planes can be vertical as well as horizontal. A vertical cut through the long dimension of a building is called a *longitudinal section,* and through the short dimension of a building, a *transverse section.* Both cuts are helpful in analyzing the building's structure and detailing.

Offset Section   The cutting plane can be *offset* (bent) to permit it to cut through all necessary features. For example, although a horizontal cutting plane through an entire building is usually assumed to be about 4' above floor level, it would be offset *upward* to cut through a high strip window as shown in Figure 9–9, and offset *downward* to cut through a lower level of a split-level house as shown in Figure 9–10. Usually such offsets need not be indicated by a cutting-plane line.

Half Section   A half section is a cut to remove only *one-quarter* of a symmetrical component. Thus both the exterior and interior can be shown in one view as indicated in Figure 9–11.

Broken-out Section   A broken-out section has the advantage of permitting the drafter to select the most critical area for sectioning and still present the exterior appearance of the component—all in one view. See Figure 9–12.

Revolved Section   A revolved section is a section that has been revolved 90 degrees and drawn on the exterior view of a component. Like a broken-out section, this permits the showing of a greater amount of information in a small space. See Figure 9–13.

Revolved partial sections are used to indicate the sectional profile of a special column, jamb, or molding. See Figure 9–14.

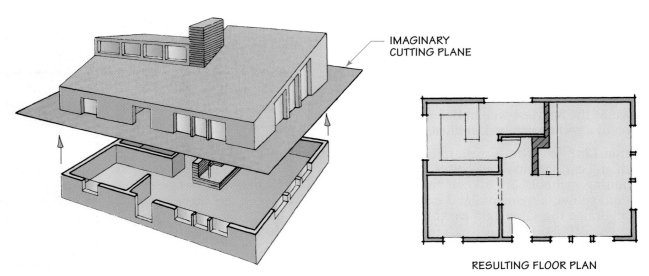

IMAGINARY CUTTING PLANE

RESULTING FLOOR PLAN

Figure 9–8   *Imaginary cutting plane used to obtain a full section.*

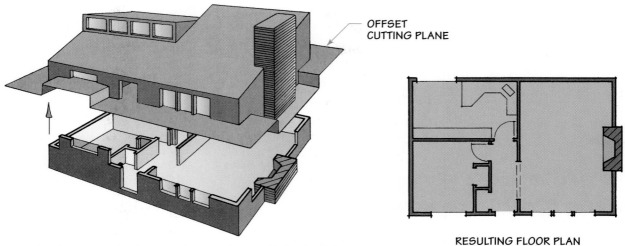

OFFSET
CUTTING PLANE

RESULTING FLOOR PLAN

Figure 9–9   *Cutting plane offset to show all desired features.*

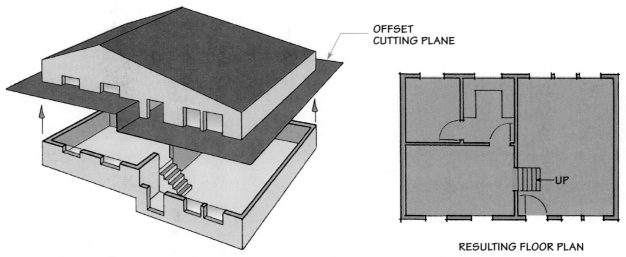

OFFSET
CUTTING PLANE

RESULTING FLOOR PLAN

Figure 9–10   *Cutting plane offset to obtain a floor plan of a split-level house.*

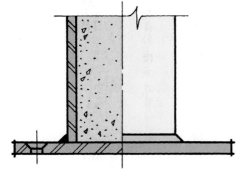

Figure 9–11   *A half-section of a welded steel column.*

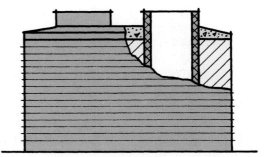

Figure 9–12   *A broken-out section of a chimney cap.*

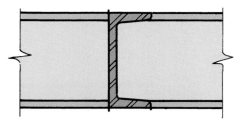

Figure 9–13   *A revolved section of a steel channel.*

Removed Section   A removed section is a revolved section that has been removed to another location and often drawn to a larger scale. A cutting plane should be used to indicate where this sectional cut was taken.

Computer-aided Projections   Computer graphics systems can be used to produce the several types of lines and sections required for multiview projection. Proper line weights are ob-

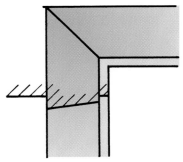

Figure 9–14   *A revolved partial section of door trim.*

tained by using a "Polyline" command or by assigning each type of line to a different "layer." Chapter 7 contains operating details on producing computer-aided multiview projections. However, all rules of reading a drawing apply equally to hand-drawn and computer-drawn projections.

American Institute of Architects

Architects, along with those in most related fields, are considered professionals. As a professional it is important to join professional organizations. Most of these organizations are national or international with state affiliations and local chapters. They provide benefits such as resources, professional development through conferences, licensing boards, and an opportunity to meet other professionals, often referred to as networking. One of the most important advantages of membership is the status that comes from the professional affiliation.

Most registered architects are members of The American Institute of Architects (AIA), and subscribe to AIA's *Code of Ethics and Professional Conduct*. More information about this organization can be gained by accessing the AIA homepage on the Internet. The address is:

http://www.aia.org

Answer these questions about the AIA by using the information on this homepage.

1. What is the mission of the AIA?
2. What resources are available through AIA?
3. What is the membership fee? How often do you need to renew your membership?
4. What is required to become an architect?
5. Describe the AIA's membership structure?
6. Where is your local chapter's headquarters?
7. What is included in AIA's *Code of Ethics and Professional Conduct*?

REVIEW QUESTIONS

1. What is a multiview projection?

2. Distinguish between first-angle and third-angle projections.

3. Describe four methods that can be used to transfer distances between the plan view and the elevations.

4. What type of line weight is used for each of these types of lines? Label the heaviest line as a five and the lightest line as a one.

   a. Hidden
   b. Visible
   c. Cutting-plane
   d. Section
   e. Center
   f. Dimension
   g. Extension
   h. Construction
   i. Guideline

5. Identify one advantage of:

   a. Half-sections
   b. Offset sections
   c. Broken-out sections
   d. Revolved sections
   e. Partial sections
   f. Removed sections

ENRICHMENT QUESTIONS

1. Describe when it would be appropriate to use only two views in a multiview projection. Sketch an example.

2. Explain the rule that indicates how many views in a multiview projection to draw.

3. Explain the rule of configuration and sketch an example to demonstrate it.

4. According to Figure 9–9, the cutting plane does not pass through the object on a level plane. Explain why it is necessary to adjust the height of the cutting plane for certain objects.

PRACTICAL EXERCISES

1. Draw or computer-produce multiview projections of the models shown in Figure 9–15 as assigned:

   a. Sufficient views to describe the model

  b. Front elevation, end elevation, and plan

  c. Plan and all elevations

2. Draw or computer-produce half-sections of a $3\frac{1}{2}$" concrete-filled steel pipe column at its base and cap.

3. Draw or computer-produce an elevation and a revolved section of the following:

  a. W8 × 31 steel beam

  c. L6 × 4 × $\frac{1}{2}$ steel angle

  b. C8 × 13.75 steel channel

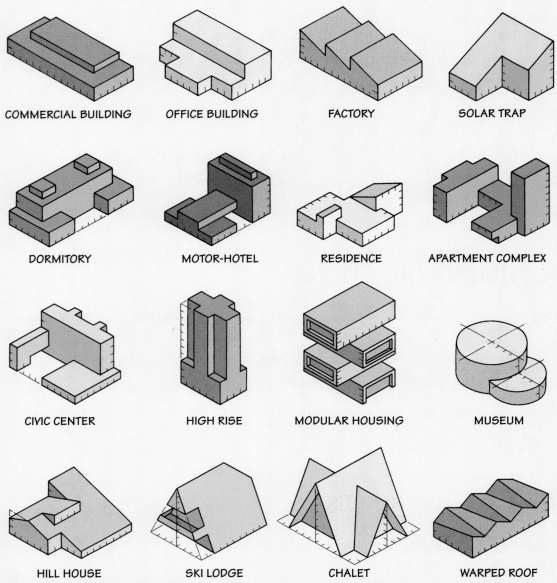

COMMERCIAL BUILDING　　OFFICE BUILDING　　FACTORY　　SOLAR TRAP

DORMITORY　　MOTOR-HOTEL　　RESIDENCE　　APARTMENT COMPLEX

CIVIC CENTER　　HIGH RISE　　MODULAR HOUSING　　MUSEUM

HILL HOUSE　　SKI LODGE　　CHALET　　WARPED ROOF

Figure 9–15   Mass models for Practical Exercise 1.

CHAPTER 10

Auxiliary Views

OBJECTIVES

By completing this chapter you will be able to:

- 🏛 Explain the need for primary auxiliary and secondary auxiliary views.
- 🏛 Create and place a primary auxiliary view by following the given steps.
- 🏛 Create and place a secondary auxiliary view by following the given steps.

KEY TERMS

Primary auxiliary view
Secondary auxiliary view
Related views

Occasionally in architectural drafting, a view which is not a principal view is required. Such views are called *auxiliary views* and may show the true size and shape of an inclined or oblique surface or the true length of an inclined or oblique edge. Auxiliary views are classified as primary auxiliary views and secondary auxiliary views. A **primary auxiliary view** is a view that is perpendicular to only one of the three principal planes of projection and is inclined to the other two. A **secondary auxiliary view** is an auxiliary view that is obtained by projection from a primary auxiliary view.

Primary Auxiliary View

A primary auxiliary view is obtained by projection from a principal view. Common examples in architectural drafting are the auxiliary views needed to show the true size and shape of each face of a building that has walls or wings that are not at a 90 degree angle to each other. An example is shown in Figure 10–1. The procedure used to draw a true-size and -shape elevation of the inclined wall 1-2-3-4 is as follows:

Step 1. Draw the edge view of a projection plane parallel to the edge view of plane 1-2-3-4. This is usually called a *reference line.* Since the auxiliary view is to be projected from the plan, label the reference line *P/A* (*P* for *plan* and *A* for *auxiliary elevation*).

Step 2. All points on the inclined face are projected from the adjacent view (the plan view in this example) to the auxiliary elevation view. Projection lines are *always* drawn perpendicular to their reference line. In this example, points 1 and 4 project along the same projector. Also, points 2 and 3 project along the same projector.

Step 3. Locate points 1, 2, 3, and 4 in the auxiliary elevation view by transferring distances from a related view (the front elevation in this example). **Related views** are two views that are adjacent to the same view. Dividers are often helpful in transferring these measurements. Connect the points in the auxiliary view in the proper order.

Secondary Auxiliary View

Secondary auxiliary views are auxiliary views projected from a primary auxiliary view. Although not commonly used in architectural drafting, they are occasionally required for accurate shape description or to solve structural problems. An example is shown in Figure 10–2,

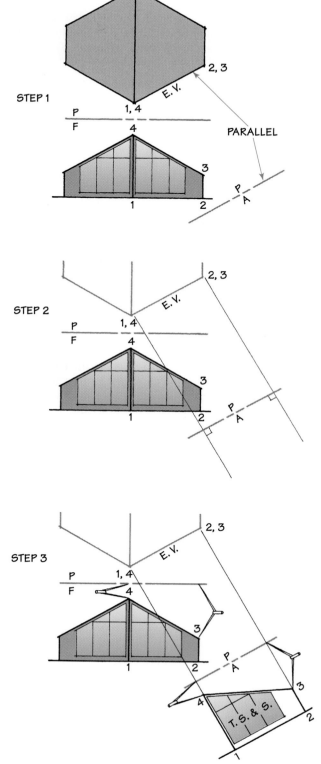

Figure 10–1   *Drawing a primary auxiliary view.*

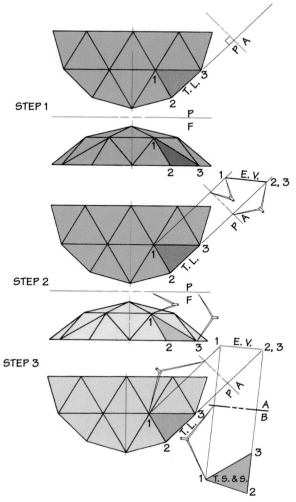

Figure 10–2   *Drawing a secondary auxiliary view.*

which illustrates the procedure to find the true size and shape of face 1-2-3 of the geodesic dome so that a pattern can be made. Two auxiliary views are required, because an edge view of face 1-2-3 must be drawn before the true-size and -shape view can be found.

Step 1. Analyze the problem. Before the true size and shape of face 1-2-3 can be found, the edge view must be drawn. To find the edge view of any plane, the point view of a line in that plane is found. The point view of any line can be found by projecting parallel to the true-length view of that line. Since line 2-3 is true length in the plan view, project parallel to it by drawing a perpendicular *P/A* reference line.

REAL WORLD APPLICATIONS

Calculating Areas

Not only is it important to know square footage for roof calculations, but also when ordering materials such as carpet, linoleum, cabinets, brick, drywall, siding, and paint. In order to calculate the square footage of a home or roof, it is important to be familiar with the formulas for computing rectangular, triangular, and circular areas. The formulas to calculate each of these areas are shown in the figures.

When determining the dimensions of a roof, measurements must be taken from a surface that is a true shape. The dimensions provided on a floor plan and elevation are typically not sufficient to ascertain the true shape of a pitched roof. This is especially true if the roof is pitched on a plane that is neither parallel to the elevation view nor the floor plan. As discussed in this chapter, if all of the lines in a surface are true length, then together the shape they form is true shape. Once an auxiliary view, and thereby the true shape, is determined, the square footage can be calculated. For a roof, the square footage is necessary when purchasing shingles, roofing felt, plywood, or other roofing material.

What is the area of the following items in the two-story residence shown in Appendix A?

a. Dining room

b. Paul's bedroom

c. Study

d. Total closet space

e. Above-ground surface area of right elevation

f. Garage roof

h. House roof

Finally, what is the area of a circular entrance way measuring 6'-8" in diameter?

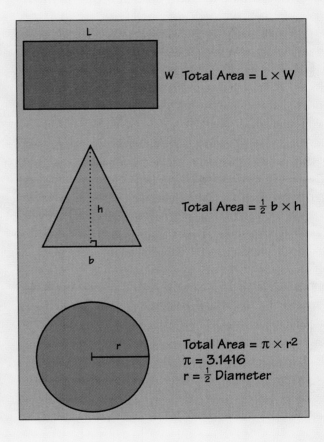

Total Area = L × W

Total Area = ½ b × h

Total Area = π × r²
π = 3.1416
r = ½ Diameter

Step 2. Project face 1-2-3 to the primary auxiliary view so that it appears as an edge view.

Step 3. Draw an *A/B* reference line parallel to the edge view of face 1-2-3. Project face 1-2-3 to the secondary auxiliary view so that it appears true size and shape.

Computer-produced Auxiliary Views   Although a number of special programs are available for computer-producing auxiliary views, they can also be produced with standard commands. For example, an auxiliary view showing the true size and shape of the window wall in Figure 10–1 is produced by rotating the screen's UCS (User Coordinate System) so that a UCS plane is parallel to the window wall plane rather than in the default WCS (World Coordinate System). In addition to the "Grid" command, use the "Snap" command to position points accurately on your auxiliary view and obtain useful coordinate readouts.

Refer to Chapter 7 for operating details on computer-producing auxiliary views.

REVIEW QUESTIONS

1. When would it be necessary to utilize a primary auxiliary view in architectural drafting?

2. What makes primary auxiliary and secondary auxiliary views different from each other?

3. When drawing projection lines, they must be _____ to each other and _____ to reference lines.

4. What do the abbreviations WCS and UCS represent?

ENRICHMENT QUESTIONS

1. How would producing an auxiliary view enhance a set of working drawings for the house shown in Figure 10–3?

2. Describe how CAD software could enhance the production of auxiliary views.

Figure 10–3 *A primary auxiliary view is required to show the shape of this wall.*
(Courtesy Pella Corporation.)

PRACTICAL EXERCISES

1. Using primary auxiliary views, find the true size and shape of each roof section of the residence shown in Figure 10–4. Calculate the number of squares of roofing required (1 square = 100 sq. ft.).

2. Draw or computer-produce a secondary auxiliary view of the roof section of the pavilion shown in Figure 10–5. Calculate the number of squares of roofing required for the entire roof.

3. Draw or computer-produce a secondary auxiliary view of a typical wall of the pavilion shown in Figure 10–5. Calculate the square feet of insulation required for one wall section.

4. Using hand-drawn or computer-produced auxiliary views, find the following data for the cable-supported roof shown in Figure 10–6 so that proper angle brackets can be specified.

   a. The angle between each cable and the mast

   b. The angle between each cable and the roof

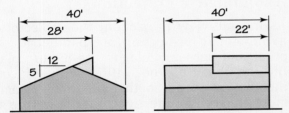

Figure 10–4   Drawing of residence for Practical Exercise 1.

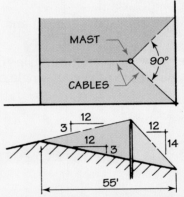

Figure 10–6   Drawing of cable-supported roof for Practical Exercise 4.

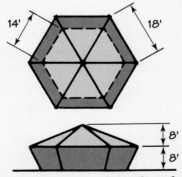

Figure 10–5   Drawing of pavilion for Practical Exercises 2 and 3.

CHAPTER 11

Pictorial Projections

OBJECTIVES

By completing this chapter you will be able to:

- 🏛 Identify and create various types of axonometric projections.

- 🏛 Differentiate between a cabinet and cavalier drawing.

- 🏛 Indicate how an axonometric drawing changes when the projection angle changes.

KEY TERMS

Axonometric projection
Isometric projection
Isometric drawing
Oblique projection

In addition to multiview projections, pictorials are often used by architects because they better describe the actual appearance of an object. In pictorials, all three principal faces of an object can be shown in one view, and such pictures are easily understood by persons not trained in reading multiview projections.

Pictorial projections can be classified as *perspective* and *parallel.* Perspective projections are more realistic, but parallel projections are easier to draw. Perspective drawings are de-scribed in Chapter 38. Parallel pictorial projections are classified as *axonometric* and *oblique* as follows.

Axonometric Projection

As in multiview projection, the projectors in **axonometric projection** are parallel to each other and perpendicular to the picture plane. But the object has been tilted with respect to the picture plane so that all three principal faces are seen in one view, but not in true size or true shape. When the object is tilted so that all three principal faces are equally inclined to the picture plane, the axonometric projection is called an *isometric projection.* When only two faces are equally inclined to the picture plane, a *dimetric projection* results. When *no* two faces are equally inclined, a *trimetric projection* results. A form of isometric projection, called *isometric drawing,* is by far the most popular pictorial method.

Isometric Projection   An **isometric projection** can be obtained by revolving an object 45 degrees about a vertical axis as shown in Figure 11–1. Then the object is tilted forward so that all principal edges form equal angles with the picture plane. This angle is approximately 35°-16′. Dimetric and trimetric projections can be obtained by similar methods, but this is a cumbersome procedure and, consequently, is seldom used.

Isometric Drawing   **Isometric drawing** differs from isometric projection in that the principal edges are drawn true length rather than

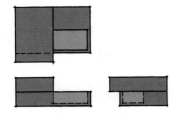

STEP 1   A MULTIVIEW PROJECTION OF AN OBJECT

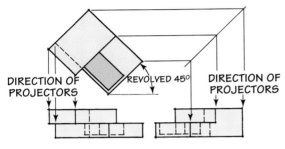

DIRECTION OF PROJECTORS   REVOLVED 45°   DIRECTION OF PROJECTORS

STEP 2   THE OBJECT REVOLVED

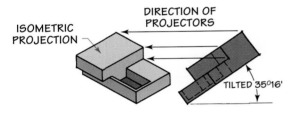

ISOMETRIC PROJECTION   DIRECTION OF PROJECTORS

TILTED 35°16'

STEP 3   THE OBJECT REVOLVED AND TILTED

Figure 11–1   *The theory of isometric projection.*

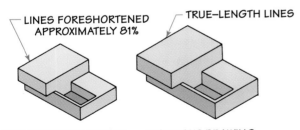

LINES FORESHORTENED APPROXIMATELY 81%   TRUE–LENGTH LINES

ISOMETRIC PROJECTION   ISOMETRIC DRAWING

Figure 11–2   *Comparison of isometric projection and isometric drawing.*

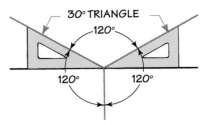

30° TRIANGLE   120°   120°   120°

Figure 11–3   *The isometric axes.*

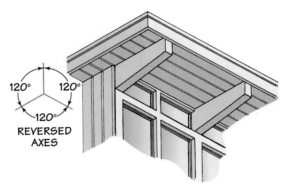

120°   120°   120°   REVERSED AXES

Figure 11–4   *A reversed isometric drawing.*

foreshortened (Figure 11–2). Therefore, isometric drawings can be drawn directly and quite quickly. The principal edges appear as vertical lines or as lines making an angle of 30 degrees to the horizontal (Figures 11–3 and 11–4).

It is important to realize that in isometric drawing, only isometric lines (principal edges) are drawn true length. Consequently, nonisometric lines cannot be obtained by direct measurement but by offset measurement. *Offset measurement* is simply the procedure of boxing in a shape so that the position of any point can be measured along isometric lines as shown in Figures 11–5 and 11–6.

Circles in isometric drawing appear as ellipses. But rather than plotting these ellipses by offset measurement, an elliptical template can be used. For large ellipses, a four-center approximation can also be used (Figure 11–7). An example of an approximation for a semicircle is shown in Figure 11–8. The procedure is as follows:

Step 1. A square is assumed to be placed tangent to the semicircle at the three intersections with its center lines (points 1, 2, 3).

Step 2. The tangent square and center lines of the circle are drawn in the isometric drawing.

Step 3. Perpendiculars are erected to the sides of the isometric square at points 1, 2, and

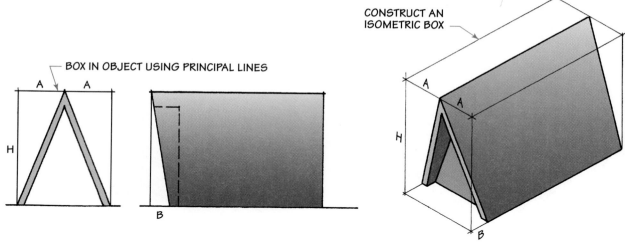

Figure 11–5   *Drawing nonisometric lines by offset measurement.*

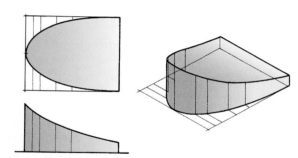

Figure 11–6   *Plotting irregular curves using offset measurements.*

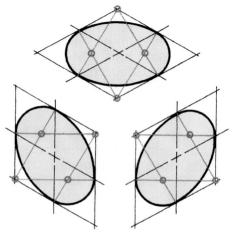

Figure 11–7   *A four-center ellipse can be drawn in all principal planes.*

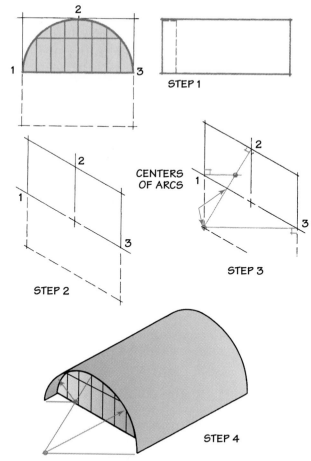

Figure 11–8   *Plotting a semicircle in isometric drawing by the approximation method.*

3. The intersections of these perpendiculars are the centers of two arcs tangent to the isometric square at 1, 2, and 3.

Step 4. Using a compass, draw two arcs tangent to the isometric square.

Oblique Projection

An **oblique projection** (Figure 11–9) is obtained by parallel projectors that are oblique rather than perpendicular to the picture plane. The projectors can be assumed to be at any angle to the picture plane, but it is most common to project them to produce receding lines that will appear at an angle of 45 degrees to the horizontal. Usually these 45 degree lines are drawn to the right.

The angle of the projectors also determines the amount of foreshortening of the receding lines. For example, the receding lines can be reduced to half size (a *cabinet* drawing) or drawn to full scale (a *cavalier* drawing, Figure 11–10). The proportions $\frac{2}{3}$ and $\frac{3}{4}$ are also used.

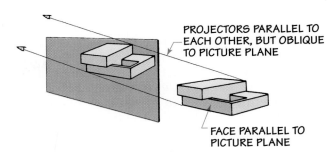

Figure 11–9 *Oblique projection.*

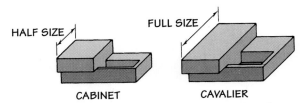

Figure 11–10 *Comparison of cabinet and cavalier drawings.*

REAL WORLD   APPLICATIONS

Adding Angles

¼"=1'-0 In junior high mathematics, you probably learned how to add and subtract angles. It seems simple, but a review is always helpful. Angles are important in architecture and the ability to add them quickly is beneficial.

Angles are measured either using degrees (°), minutes (') and seconds ("), or as a decimal. The first method is commonly used in architectural drafting so it will be the one described here. First, remember that there are 60 seconds in a minute, and 60 minutes in a degree. Let's add two angles that use this system. Begin by adding the numbers as you normally would.

$$14° \ 24' \ 16''$$
$$+13° \ 40' \ 50''$$
$$27° \ 64' \ 66''$$

Notice that 66" = 1' 6". Now add 1' to 64'. The result is 27° 65' 6". Notice that 65' = 1° 5'. Now add 1° to the 27°. The final total is 28° 5' 6".

Always remember to not leave any number greater than 59' or 59". Here are some practice problems.

33° 45' 33"	0° 12' 49"	29° 30' 15"
+18° 33' 12"	+10° 14' 14"	+ 50° 29' 45"

A major advantage of oblique drawing is that one face of the object is parallel to the picture plane and therefore remains in its true size and shape. Consequently, it is common sense to position the object so that the most irregular outline is the face parallel to the picture plane. The faces not parallel to the picture plane are distorted and must be constructed using offset measurements. For example, circles in the parallel face can be drawn as circles (Figure 11–11), but circles in the receding faces must be plotted by offsets or by the four-center ellipse method.

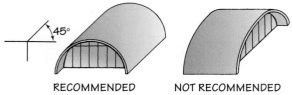

RECOMMENDED   NOT RECOMMENDED

Figure 11–11 Positioning curved faces parallel to the picture plane.

A disadvantage of oblique drawing is that receding lines do not converge and, consequently, appear to the eye to be distorted. The distortion can be minimized by positioning the object, when possible, so that the largest dimensions are parallel to the picture plane rather than receding.

Computer-produced Pictorials

Isometric and oblique pictorials as described in this chapter can be CAD-produced by replacing the default standard horizontal and vertical drawing grid by isometric or oblique drawing grids. In either case, only *one* of the three pictorial grids can be displayed and utilized at a time, and only *two* coordinates of any point on an object can be entered on each grid.

An improved method of isometric drawing permits the drafter to enter all *three* coordinates of each point on an object and then to move the viewing location, resulting in a variety of axonometric drawings. Also, by reducing the distance

REAL WORLD APPLICATIONS

Word Structure and Vocabulary

What do the words "axonometric," "isometric," "dimetric," and "trimetric" have in common? Yes, they have the same word root—"metric." One way to improve your vocabulary is to learn about root words, prefixes, and suffixes. The words listed above have the same root but different prefixes. Use a dictionary to define "metric" and then each of the prefixes. Put these together to write a complete definition for each word. Do the same for the following architectural and drafting terms. Some of these have suffixes so you will need

to look up the suffix separately from the root. Sometimes the definition will indicate where the parts of the word originated. Get in the habit of memorizing prefixes, roots, and suffixes and your vocabulary will quickly expand.

1. Anthropometrics
2. Microcopy
3. Microfilm
4. Subfloor
5. Construction
6. Gigabyte
7. Multitasking
8. Convection
9. Conduction
10. Perspective

between the viewer and object, an infinite variety of wire-frame and solid-model perspectives can be produced. A computer-generated, three-point architectural perspective is shown in Figure 11–12.

Refer to Chapter 7 for operating details on providing various computer-generated pictorials.

Figure 11–12   A computer-produced three-point architectural perspective. (Courtesy Tri-Star Computer.)

REVIEW QUESTIONS

1. What are the two classifications of pictorial projections and the advantages of each?

2. What are the three types of axonometric projections?

3. List the differences between an isometric projection and an isometric drawing.

4. Draw and describe what is meant by "offset measurement."

5. What is meant by an oblique projection?

6. What is the difference between a cabinet and a cavalier drawing?

ENRICHMENT QUESTIONS

1. The models in Figure 9–15 (page 116) are drawn isometrically. If more of the front of the chalet or ski lodge were to be seen, how will the isometric axes change?

2. Redraw the chalet and the ski lodge to show more of the front of these buildings.

3. How will a circle appear on an oblique drawing if it is located in the front of the object? Why?

4. Using CAD to produce isometric or oblique drawing is advantageous because it can produce wire-frame or solid models. What is the difference between a wire-frame and a solid model produced using CAD?

5. Why is the most irregular face of an object positioned parallel to the picture plane in oblique drawings?

6. Why are hidden lines usually omitted from pictorial drawings?

PRACTICAL EXERCISES

1. Draw or computer-produce pictorials of the 2" × 4" wood joints shown in Figure 11-13:

   a. Isometric drawing

   b. Oblique drawing (choose appropriate projection angle)

   c. Cavalier drawing

   d. Cabinet drawing

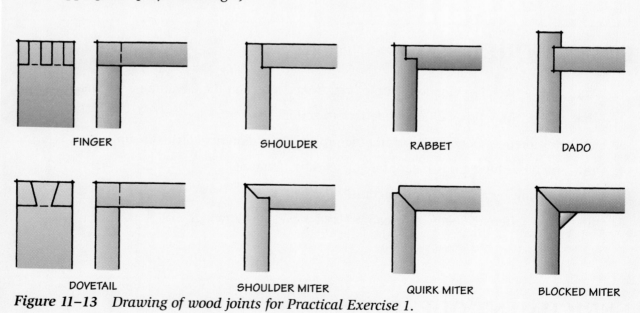

Figure 11–13   Drawing of wood joints for Practical Exercise 1.

2. Draw or computer-produce the civic center shown in Figure 9–15 according to the specifications given below. Each mark will represent $\frac{1}{2}$". State what happened to each drawing when the angles changed.

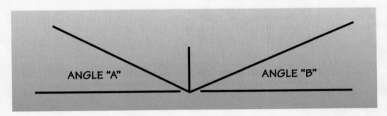

Drawing #1:   Angle "A" = 30°   Drawing #4:   Angle "A" = 45°
              Angle "B" = 30°   Angle "B" = 15°

Drawing #2:   Angle "A" = 15°   Drawing #5:   Angle "A" = 15°
              Angle "B" = 30°   Angle "B" = 45°

Drawing #3:   Angle "A" = 30°
              Angle "B" = 15°

CHAPTER 12

Architectural Lettering

OBJECTIVES

By completing this chapter you will be able to:

- 🏛 Letter an architectural drawing in a clear, professional architectural style.
- 🏛 Communicate design intent through a uniform and concise system of lettered drawings.
- 🏛 Discuss the advantages of ink lettering stencils and appliqué lettering.
- 🏛 Construct a title block.

KEY TERMS

Serifs	Proportion
Guidelines	Spacing
Form	Presentation drawing
Stability	

The ability to letter is as important to a drafter as the ability to paint is to an artist. An artist who has an idea for a painting must be able to put that idea on canvas in a neat, orderly fashion using a definite technique. Similarly, an architectural drafter who has ideas for a building must be able to put those ideas on vellum in a clear, orderly manner using a definite system of notation. The ability to do a professional job of lettering is considered so important by most employers that they require a lettering sample to be submitted at the time of application for employment.

LETTERING STYLES

Of the many styles of lettering, *Old English, Roman,* and *Gothic* are best known.

Old English lettering (or **Text**) is shown in Figure 12–1. Although it is attractive, it is not widely used today because it is difficult to letter and read. Most high school and university diplomas use Old English lettering because of its elegant appearance.

Roman letters have strokes of different widths. Notice in Figure 12–2 that the horizontal lines are thin and the verticals thick. This is because the early pen points (quills) were flat and made lines that varied in width depending upon the direction of stroke. Also notice the small lines at the ends of every stroke. These are called **serifs**. Roman lettering is used extensively by book and magazine publishers, but not by engineers or architects. You may find Roman lettering on monumental structures cut into stone or metal plaques.

ABCabcdef 123

Figure 12–1 Old English lettering. (Courtesy Paratone, Inc.)

ABC alpha 123

Figure 12–2 Roman lettering. (Courtesy Paratone, Inc.)

ABCdefg 123

Figure 12–3 Gothic lettering. (Courtesy Paratone, Inc.)

Gothic lettering (Figure 12–3) differs from Roman lettering in two important aspects: (1) All strokes are exactly the same width, and (2) no serifs are used. Of the many types of Gothic lettering, only *one* type has been approved by ANSI (American National Standards Institute) for use on engineering and architectural drawings. See ANSI (formerly American Standards Association), American National Standard Drafting Manual: Y14.2M—1992. This one approved type of Gothic lettering, called *Commercial Gothic,* is used on all engineering drawings. On architectural drawings, it is often altered slightly to suit the taste of the draftsman or the style of a particular office. Let us first study the standard Commercial Gothic lettering and then look at the usual methods by which it can be altered to better suit architectural needs.

Commercial Gothic   Figure 12–4 shows the American Standard Commercial Gothic lettering in the three sizes used on architectural drawings: $\frac{1}{4}$" for important titles and drawing numbers, $\frac{1}{8}$" for lesser headings, and $\frac{3}{32}$" for dimensioning and notes. The form and proportion of each letter should be studied carefully since this alphabet is used universally.

LETTERING SECRETS

There are six lettering "secrets" which have been collected by professional drafters who use Commercial Gothic lettering. Practice your lettering with these secrets in mind.

Guidelines

A professional drafter always uses guidelines when lettering. **Guidelines** are very light lines (usually drawn with a 4H pencil) that aid in forming uniformly sized letters. Guidelines are *not* erased, since they are drawn so lightly that they are not objectionable. They should be visible to you when lettering, but invisible when you hold the drawing at arm's length. Some drafters prefer to use non-photo reproducible blue lead.

The horizontal guidelines used for capital letters (like *ABC*) are a base line and a cap line.

Figure 12–4 USASI Commercial Gothic lettering.

REAL WORLD  APPLICATIONS

From Cave Drawings to the ABCs

 Imagine humans had never progressed past cave drawings or pictograms. You would not be practicing lettering but rather "picturing." Fortunately, we now have a sophisticated alphabet with which we communicate. The development from cave drawings to what we know as the ABCs of today has been a slow but fascinating progression. The Phoenicians are credited with creating the first true alphabet. Describe this first alphabet and how it was developed. What roles did other cultures play in the alphabet's history? Write a brief report that traces the history of the alphabet. Use these questions for guidance.

Often, the first step in writing a research report is choosing a topic of interest to you. In this case, though, the topic is assigned. Go to the library and select a number of books and articles to read. You may want to begin reading with a general reference book such as an encyclopedia. After some initial reading, limit your subject by developing a thesis statement. A thesis statement is a sentence that defines the focus of your report. An important step in writing a research paper, often skipped by students, is preparing a preliminary bibliography. This is a list of books and articles on your topic which you plan to utilize. These are found by using the card catalog, computer databases, *Reader's Guide to Periodical Literature,* and other reference publications. The bibliography should include the author, year of publication, title, and publisher for books. For articles, the bibliography should include the author, year of publication, title of the article, name of journal, volume number, issue number, and page numbers. Sample bibliographic entries for a book and article are shown in the figure. Take special note of the punctuation and format in these examples.

Book

Hutchinson, John & Karsnitz, John R. (1994). *Design and Problem Solving in Technology.* Albany, NY: Delmar.

Journal Article

Staff. (1990). Underground cities: Japan's answers to overcrowding. *The Futurist,* 24(4), 29.

Now begin taking notes, which will lead to writing the report. As you read the selected sources take relevant notes. There are three ways to take notes: summarize, paraphrase, and quote directly. Use all three. Summarizing is reducing the information to a few important points. Paraphrasing is rewriting the information in your own words. When you quote directly from a source use quotation marks and include the page number from which the quote was taken as it will need to be footnoted. Once you have gathered information in an organized manner, write a working outline. As you discover missing points in this outline, do further research to complete it. Rewrite your outline once research is complete and begin writing your first draft. Revise this draft at least two times. It may be beneficial to have a friend help with the second revision. Now compose your final report, but remember to proofread before submitting it.

Figure 12–5 Guidelines.

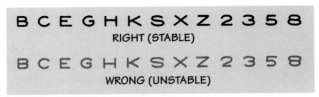

Figure 12–7 Stability of letters.

The horizontal guidelines used for lowercase letters (like *abc*) include also a waist line and a drop line (Figure 12–5). Lowercase lettering, however, is seldom used in architectural work. If you wish to simplify the task of measuring and drawing guidelines, the Rapidesign guide, Ames lettering device, or Braddock-Rowe lettering triangle may be used.

Either vertical or inclined guidelines are also used. These are spaced at random and are used to keep all letters vertical, or slanting at a uniform slope of $67\frac{1}{2}$ degrees.

Form

The exact **form** of every Commercial Gothic letter should be memorized and used. This task is much simplified if you notice that all capital letters (except S) are based upon *straight* and *circular* lines. The numerals (and the letter S) are based upon *straight* and *elliptical* lines. Some of the common mistakes made in forming letters are shown in Figure 12–6.

Stability

You must also remember that *letters and numerals should appear stable* whenever possible. **Stability** means that the letters and numerals should be able to "stand on their own two feet." To prevent any possibility of appearing *unstable* or top-heavy, the letters B, C, E, G, H, K, S, X, and Z and the numerals 2, 3, 5, and 8 are drawn with their lower portions slightly larger in area

than their upper portions. Examples of stability in lettering are shown in Figure 12–7.

Proportion

Of all the lettering secrets, this is the most important one to the beginner: *Make your letters much wider than you think they should be.* Notice in Figure 12–4 that nearly all letters are as wide as they are high. Thus the O and Q are perfect circles. The M and W are even wider than they are high. Also notice that the letters are somewhat wider than the numerals.

Letters that are narrower than standard are called *condensed,* and letters wider than standard are called *extended.* Condensed lettering is used only when it is absolutely necessary to fit many letters into a small space. Slightly extended lettering, on the other hand, is often used, since it is more readable and better looking than standard lettering. All of the notes on the illustrations in this book are lettered in extended lettering.

Density

Black lines should be used for lettering. This is necessary for two reasons: to improve the appearance of the lettering and to improve its readability so that it will show up well when reproduced. If your lettering is not black enough, simply use a softer pencil (such as H or F) and *bear down* harder on the paper. Of course you must still remember to keep a sharp point. Professional drafters sharpen their pencils after every two or three words or use fineline drafting pencils that need no sharpening.

Spacing

Proper **spacing** of letters to form words, and words to form sentences, is a must. The best lettering has the *letters close together* to form words but the *words far apart* to form sentences. The spacing of letters is not measured

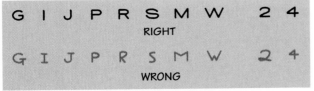

Figure 12–6 Common lettering errors.

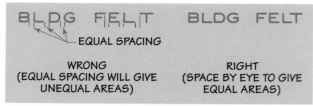

Figure 12–8 Spacing of letters.

Figure 12–9 Spacing of words.

directly, but is done by the eye so that the areas between letters are visually equal. Notice at the left of Figure 12–8 that the spaces between the letters in the words BLDG FELT were carefully measured, with the result that the space between the L and T appears too large. The second part of the figure shows proper optical spacing, all letters having the same *area* between them.

Figure 12–9 shows a simple method of spacing words: Imagine the letter *O* (a circle) between each word.

Remember to use the six lettering secrets when you practice your lettering:

1. Guidelines
2. Straight lines and circular lines
3. Stable lettering
4. Fat lettering
5. Black lettering
6. Close spacing

ARCHITECTURAL LETTERING

Unless beginning architectural drafters have training in letter design, they should use the Commercial Gothic letter forms without change. Indeed, even some drafters with many years' experience feel that these forms cannot be improved. However, most architectural drafters are not content to use Commercial Gothic lettering because it is, after all, the standard for *engineering* rather than for *architectural* drawing. Since there exists no standard architectural alphabet, most architects take great pride in developing their own style (see Figure 12–10). It has been said that there are as many styles of lettering as there are architects! Architects will, of course, stick to their own style, and that style will appear uniformly on all of their drawings. Architects want to *sell* their work, so the lettering used should be as attractive as possible. However, no matter how fanciful their style in lettering titles and headings, they always use straightforward Commercial Gothic numerals for dimensioning.

Figure 12–11 shows a legible and attractive alphabet. You may wish to practice these letters first and then revise some letters until you find a type of lettering that feels right to you. During this test period, keep these rules in mind:

1. Rapidity of execution is an important factor. Time is money; do not get in the habit of drawing excessively time-consuming letters. This means that the letters should be single-stroke and that stylized portions of individual letters should be drawn in a free and natural fashion. (For example, you should not use the elliptical form of C and D in Figure 12–11 if it does not seem natural to you.)

2. Accentuate the ends of the strokes. Though this detail comes naturally to some, if you find that you cannot easily produce attractive results with these accents after a fair trial, then do not attempt them further. Also, you may want to try chisel-pointed lead to accentuate either horizontal or vertical strokes.

3. In most cases, only vertical capitals are used. Lower and upper drop lines are useful in uniformly ending those lines that drop below or above the normal guides. Each drop line is one-third the capital height. The capital letters G, Q, R, T, and Y may drop down, and the capital letter L may extend upward.

4. Lettering should be legible. Architectural lettering can be as fanciful as you please—as long as it is easily read.

ABCDEFGHIJKLMNOPQRSTUVWXYZ
1234567890

ABCDEFGHIJKLMNOPQRSTUVWXYZ
1234567890

ABCDEFGHIJKLMNOPQRSTUVWXYZ
1234567890

ABCDEFGHIJKLMNOPQRSTUVWXYZ
1234567890

ABCDEFGHIJKLMNOPQRSTUVWXYZ
1234567890

Figure 12–10   Examples of architectural lettering by practicing architects. (Courtesy William B. Meister, Jack Risheberger, William L. Cunningham, James E. Black, and Richard I. Whidet.)

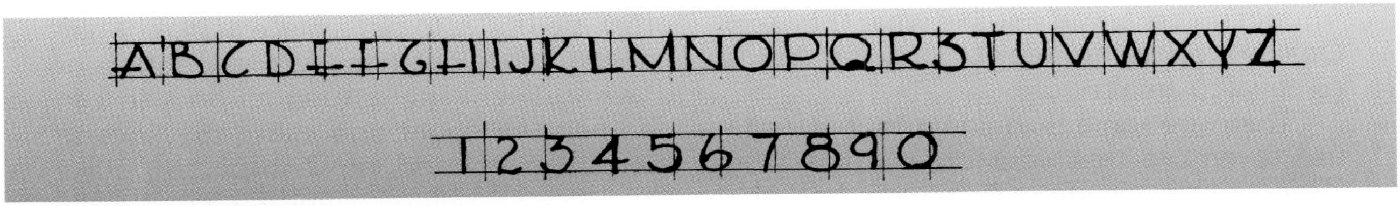

ABCDEFGHIJKLMNOPQRSTUVWXYZ
1234567890

Figure 12–11   Architectural lettering.

COMPUTER-PRODUCED LETTERING

Examples of computer-produced Commercial Gothic and architectural lettering are shown in Figure 12–12. Standard menus of alphabets and numerals are usually used, but some commercial offices prefer to custom-design their own unique style. Lettering, dimensions, notes, and titles are added to a computer drawing by

ABCDEFGHIJKLMNOPQRSTUVWXYZ
1 2 3 4 5 6 7 8 9 0

ABCDEFGHIJKLMNOPQRSTUVWXYZ
1234567890

Figure 12–12   Examples of computer-produced lettering.

typing on a keyboard. The drafter also selects the lettering style (font), starting point, lettering direction, and lettering height, width, and slant. Computer-produced lettering was used for many illustrations in this book. Some typical computer-produced notes are shown in Figure 12–19.

Refer to Chapter 7 for operating details on providing computer-generated lettering.

PRESENTATION DRAWING

The large majority of architectural working drawings are drawn in pencil or computer produced. Occasionally, though, ink drawings are used for special requirements. A **presentation drawing** (a display drawing to show the prospective client) may be done in ink or

REAL WORLD APPLICATIONS

Producing Clean Drawings

When clients look at your work for the first time, they quickly judge it. Like it or not, if drawings have smudges, clients will not like what they see. Quality work is important in any job, but is essential to architecture.

There are some techniques that you can use to ensure that your drawings will be clean and smudge-free. First, work from the top of the paper to the bottom or from left to right if you are right-handed, or right to left if you are left-handed. This way your hand does not slide on recently drawn lines, causing smudges. Second, add all lettering last. This ensures that all lines have been drawn and no dimensions or notes need to be erased because of miscalculations. Another method is to put a piece of plain white paper under your hand while lettering or drawing lines. The natural oil on skin can dampen the paper and cause the lines to bleed. Finally, and most important, take your time. Rushing to finish a job usually ends up taking more time in the end. In many cases, mistakes are made because care is not taken as the drawing is being produced and additional time is required to correct the problems.

ABCDEFGHIJKLMNOPQRSTUVWXYZ

DINING ROOM

Figure 12–13   *The architectural alphabet, obtained from a standard lettering stencil.*

even a combination of pencil and ink. Also, drawings to be printed in newspapers, magazines, or books are best reproduced when drawn in ink.

Lettering Stencils

Although the ink lettering on presentation drawings may be drawn freehand, a lettering stencil is usually used. Some of the popular trade names are LeRoy, Wrico, and Varigraph. These stencils may be obtained in a variety of stock sizes and styles, or they may be ordered in custom-made styles. Often the stock Commercial Gothic stencil is adjusted slightly to obtain lettering with an architectural flavor. Figure 12–13 shows an architectural alphabet obtained by using a standard LeRoy extended lettering stencil.

Pressure-sensitive Transfer (Appliqué)

Prepared lettering sheets may be obtained from which drafters can transfer individual letters to their drawings merely by rubbing the letter as shown in Figure 12–14. Some popular trade names are ACS Instant Lettering, Paratone Alphabets, and Mico/Type.

Instrumental Lettering

Drafting instruments are seldom used for lettering because they are too slow. Occasionally, however, a large presentation drawing may require a special title. Figure 12–15 shows one of the many possible styles of single-line lettering, and Figure 12–16 shows one style of double-line (boxed) lettering. The boxed letters may be filled in if desired.

1. Transfer of letters is effected by registering guide lines on type sheet with guide on artwork. With a very soft pencil or ballpoint pen rub down letter with light pressure.

2. Carefully lift away type sheet — letter is now transferred — repeat procedure until setting is complete. Finally place backing sheet over setting and burnish for maximum adhesion.

Figure 12–14   *How to use pressure-sensitive transfers.*

ABCDEFGHIJKLMNOPQRSTUVWXYZ

Figure 12–15   *Single-line instrumental lettering.*

Figure 12–16   *Double-line instrumental lettering.*

Title Block

Professional offices use vellum or drafting film with printed border lines and title blocks to save

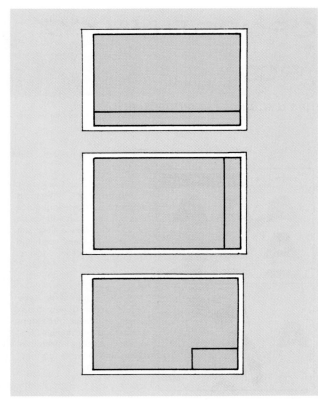

Figure 12–17 Common forms of title blocks.

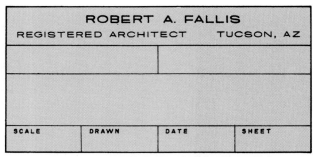

Figure 12–18 Appliqué title block used by professional architects.

is shown in Figure 12–18. There are many other sizes and types of title blocks, but all should contain the following information as a minimum:

1. Name and location of structure

2. Name and address of owner

3. Name and address of architect

4. Name of sheet (such as "First-Floor Plan")

5. Number of sheet

6. Date

7. Scale

8. Draftsman's initials

drafting time. As shown in Figure 12–17, title blocks are usually of a horizontal, vertical, or corner format. A sample 2" × 4" title block

GENERAL NOTES

1. _Design Live Loads_
   Roof (Snow) ——— 40 psf.
   Floors ——— 60 psf.
   Wind —— 18 psf.

2. Assumed maximum safe soil bearing pressure ————————————————— 3000 psf.

3. All footings shall bear on natural undisturbed virgin soil free of organic matter or deleterious fill. Soft or deleterious material below bottom of footing shall be excavated down to clean virgin soil and filled with clean granular material, compacted in place, before pouring footings.

4. Slabs—on—grade shall bear on natural undisturbed soil or on controlled compacted fill. Remove existing fill material and replace with clean granular fill compacted in 12" layers to obtain 95% maximum density at optimum moisture content.

5. Do not raise or lower footings from depths shown on plans without permission of the Structural Engineer.

6. Slabs—on—grade shall be 5" thick(minimum) and reinforced with 6x6-6/6 welded wire mesh unless otherwise noted.

Figure 12–19 Typical computer-produced structural notes.

REVIEW QUESTIONS

1. What sizes of Commercial Gothic lettering are used on architectural drawings?
2. List the six "secrets" of good lettering.
3. Are guidelines erased? Why or why not?
4. Show the difference between a drop line, base line, waist line, and cap line.
5. What is meant by *stability* in lettering? Give some examples.
6. What is the proper method for spacing letters and words?
7. When is ink lettering used in architectural work?
8. List the minimum information contained in every text block.

ENRICHMENT QUESTIONS

1. Why is it important for architectural lettering to be dense black?
2. List some reasons that a specialized font is used for architectural drawings.
3. Design a 2" × 4" title block like the one shown in Figure 12–18. Prepare the information as if you were the architect and the plans were for a fictitious client.
4. Practice writing general notes like those found in Figure 12–19 using your own style of architectural font. Remember the key is legibility, not ornamentation.

PRACTICAL EXERCISES

1. Repeat the five sample alphabets of common architectural lettering practices shown in Figure 12–10. This lettering should be $\frac{1}{4}$" high, a common title height. Remember to use guidelines and take your time. If you are not satisfied with your results, try again on a second line.
2. Repeat the architectural alphabet shown in Figure 12–11 using $\frac{1}{8}$"–high letters.
3. On drawing paper or the digitizer tablet and screen, design a style of lettering to be used on all future plans. Your instructor will indicate his/her approval of each letter type. Use the sizes assigned:

   a. $\frac{1}{4}$"   c. $\frac{3}{32}$"   e. 3 mm

   b. $\frac{1}{8}$"   d. 5 mm   f. 2 mm

4. Use a common lettering guide (Ames) to draw guidelines for an Architectural style text. The guidelines should be for $\frac{1}{8}$", $\frac{3}{32}$", $\frac{1}{4}$", and $\frac{1}{2}$" letters.

CHAPTER 13

Architectural Dimensioning in English Units

OBJECTIVES

By completing this chapter you will be able to:

- 🏛 Dimension an architectural drawing in English units using a clear, professional architectural style.

- 🏛 Dimension a topographical drawing in English units using a clear, professional engineering style.

- 🏛 Demonstrate the principles and advantages of modular coordination.

- 🏛 Dimension a modular drawing in a clear, professional style.

KEY TERMS

Dimensioning
Elevation dimensions
Modular coordination
Unicom

To read architectural plans, you must become familiar with the graphic language used in two different professions: architecture and surveying. The architect uses architectural drawing to provide the instructions for constructing a building, and the surveyor uses topographical drawing to describe the plot of land occupied by the building. The dimensioning practice in each of these fields differs slightly. Let's look at both.

ARCHITECTURAL PLAN DIMENSIONS

Architectural **dimensioning** practices depend entirely upon the method used to construct the building. The masonry portion of a building, therefore, is dimensioned quite differently from the frame or veneered portion. For example, in masonry construction, the widths of window and door openings are shown since these dimensions are needed to lay up the wall. Openings in a frame wall, however, are often dimensioned to their center lines to simplify locating the window and door frames. In masonry construction, dimensions are given to the faces of the walls. In frame construction, overall dimensions are given to the outside faces of studs because these dimensions are needed first. Masonry partitions are dimensioned to their faces, whereas frame partitions are usually dimensioned to their center lines. The thicknesses of masonry walls and partitions are indicated on the plan, but frame wall thicknesses are indicated on the detail drawings where construction details may be shown to a larger scale.

Masonry veneer on a wood-frame wall is dimensioned as a frame wall would be (to the outside faces of the studs) since the wood frame is constructed before the veneer is laid up. Figures 13–2 through 13–4 illustrate these differences.

Some additional rules of architectural dimensioning follow.

1. Dots, small circles, triangles, perpendicular lines, or diagonal lines (as shown in

Figure 13–1) may be used in place of arrowheads. Dots should always be used when dimensioning small distances in tight spaces.

2. Dimension lines are spaced about $\frac{3}{8}$" apart. Often three lines of dimensions are needed on each wall: a line of dimensions close to the wall locating windows and doors, a second line locating wall offsets, and finally an overall dimension.

3. Extension lines may or may not touch the plan, but be consistent. To avoid crossing extension and dimension lines, place the longer dimensions farther away from the plan, as in Figures 13–2 through 13–4.

4. Dimension lines are continuous with the numerals lettered above them. Numerals are placed to read from the bottom and right-hand side of the drawing.

5. Dimensions may be placed on the views, but avoid dimensioning over other features. Several complete lines of dimensions in both directions are ordinarily needed to locate interior features.

6. Dimension numerals and notes are lettered $\frac{3}{32}$" high.

7. Give all dimensions over 12" in feet and inches (to the nearest $\frac{1}{16}$"). The symbols for

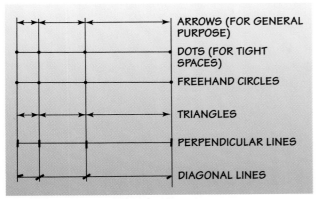

Figure 13–1 *Arrowhead types.*

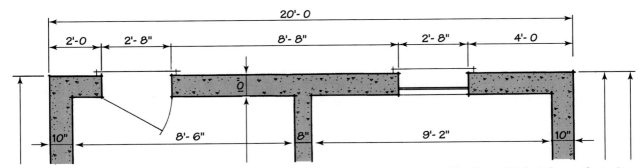

Figure 13–2 *Dimensioning masonry construction (concrete, concrete block, solid brick, and cavity brick).*

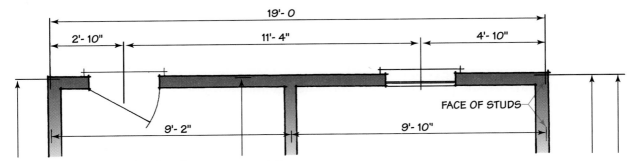

Figure 13–3 *Dimensioning frame construction.*

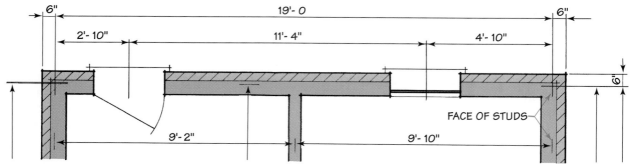

Figure 13–4   Dimensioning veneer construction.

REAL WORLD   APPLICATIONS

Quality Standards

As an architect, maintaining a level of excellence is of utmost importance. Many times clients choose an architect based on a reference from a former client. When your clients are dissatisfied with your work, they may tell more people of their dissatisfaction than when they are pleased. Therefore, when producing drawings it is essential to follow accepted standards. Not only will this enable the contractor to follow your plans easily, but it will allow clients examining your work to see your abilities firsthand. For example, when dimensioning a drawing in English units, certain rules must be followed. Some of these have been identified and illustrated in this chapter.

The idea of standardization began in 1798 with Eli Whitney and his mass production of 10,000 muskets for the United States government. Since that time one of the most influential leaders in quality management and quality control was Dr. W. Edwards Deming. In the book *The Deming Management Method,* Mary Walton outlines and explains Dr. Deming's fourteen points of quality control. The points especially relevant to an architect are summarized below.

1. Create constancy of purpose for improvement of product and service. Dr. Deming's thought was that a company needs to focus on producing quality products, providing jobs, and constantly improving rather than just making a profit.

2. Adopt the new philosophy. As an architect, you should never accept poor workmanship from yourself or any of your fellow workers.

3. Improve training and retraining. Although the field of architecture has been around for thousands of years, standards and expectations, especially in relation to CAD, are constantly changing. Firms must ensure that their workers stay on the cutting edge of technological developments.

4. Demonstrate leadership. Leadership is not condemning workers for their faults but rather assisting them to become more productive.

5. Remove barriers to pride of workmanship. Architects and those working with them desire to do good work. Barriers such as faulty or outdated equipment and software must be eliminated.

feet (') and inches (") are used except for zero inches. For example:

6"	not	0'-6"
1'-0	not	1'-0"
1'-6"	not	1'-6

8. Do not try to "fancy up" dimensions with artistic numerals. *Legibility* is the only concern.

9. Never crowd dimensions.

10. No *usable* dimension is omitted even though the dimension could be obtained by addition or subtraction of other dimensions. Be sure to include overall dimensions, any change in shape of outside walls, all rooms, halls, window locations, and exterior door locations. A common mistake is failing to check cumulative dimensions with overall dimensions. Incorrect dimensions can cause the builder much delay and added expense.

11. All obvious dimensions *are* omitted. For example:

    a. Interior doors at the corner of a room need not be located.

    b. Interior doors centered at the end of a hall need not be located.

    c. The widths of identical side-by-side closets need not be dimensioned.

12. Columns and beams are located by dimensions to their center lines.

13. To free the plan from excessive dimensions, the sizes of windows and doors are given in window and door schedules.

14. House drawings are usually made to a scale of $\frac{1}{4}$" = 1'-0. Larger buildings are usually drawn to a scale of $\frac{1}{8}$" = 1'-0. Details are drawn to larger architectural scales. *Always* indicate the scale used near the drawing or in the title block.

EXAMPLE

Show the dimensions needed on the plan of the simple frame cottage shown in Figure 13–5.

Step 1: Window and door locations. A line of dimensions is placed on every outside wall containing a window or door to locate them for the builder. This dimension line is positioned about $\frac{3}{8}$" beyond the farthest projection (such as a chimney, window, or doorsill). Dimension lines are spaced by eye rather than actually measured.

Step 2: Wall locations. A second line of dimensions is placed on every wall containing offsets to provide the builder with the subtotal of dimensions for each wall.

Step 3: Overall dimensions. Both overall dimensions are placed. Check that all cumulative dimensions equal the subtotal dimensions and that the subtotal dimensions equal the overall dimensions.

Step 4: Partition locations. Lines of interior dimensions are placed to locate partitions and interior features. In this example, only one line of dimensions is required since partition X is located by alignment with the exterior wall. Note that location dimensions for interior doors are not necessary since their positions are obvious. Additional required notes and schedules will complete the dimensioning. See the drawings of the A residence in Chapter 24 for the dimensioning required on a larger plan.

ELEVATION DIMENSIONS

Of the many different methods of indicating **elevation dimensions,** the two following are most often used.

Finish Dimensions

This method indicates the actual dimension of the inside of the room when completely finished. Thus the distance between the *finished floor* and the *finished ceiling* is specified. This method is often used by the designer to quickly specify a desired room height. The height 8'-0 is often used for the first floor, and 7'-6" for the second floor.

Construction Dimensions

This method indicates the dimensions actually needed by the contractor when framing a building. Thus, in platform framing, the distance between the *top of subflooring* and the *top of plate* indicates the exact height to construct the sections of walls and partitions. This method is preferred by builders. The National Lumber Manufacturers Association recommends a first-floor height of $8'\text{-}1\frac{1}{2}$" and a second-floor height of $7'\text{-}7\frac{1}{2}$". These dimensions will result in 8'-0 and 7'-6" room heights after the finished floor and ceiling have been added.

When in doubt as to the correct method of dimensioning an architectural drawing, there is one simple rule to follow: put yourself in the place of the builder and give the dimensions that

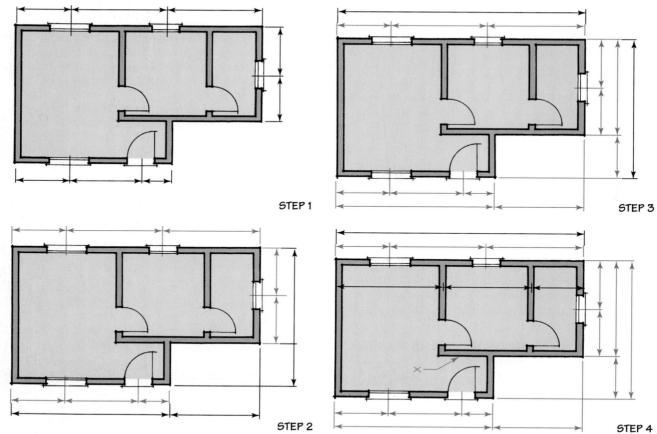

STEP 1

STEP 3

STEP 2

STEP 4

Figure 13–5   Frame cottage plan.

will help build with a minimum amount of calculation. Refer to the elevation dimensions of the A residence in Chapter 27 for a complete example.

SECTIONAL DIMENSIONS

The sectional view provides an opportunity to specify the materials and sizes not shown on the plans and elevations. Since sectional views are drawn to a larger scale, more detailed dimensions may be shown. The dimensions, material, and location of all members are specified, leaving nothing to the imagination of the builder. Nominal ("name") sizes are used for rough material, but actual sizes are used for finish material, as in:

1" × 8" subflooring (Use nominal dimensions)

$\frac{3}{4}$" × 7" fascia   (Use actual dimensions)

Some offices attempt to show rough material by omitting the inch marks from nominal di-

mensions. Thus the dimension 2" × 4" would indicate finished lumber measuring 2" × 4", whereas the dimension 2 × 4 would indicate rough lumber measuring $1\frac{1}{2}$" × $3\frac{1}{2}$". See the A residence sections in Chapter 27 for examples of the dimensions required on sectional views.

TOPOGRAPHICAL DIMENSIONS

A complete study of topographical drawing would be quite lengthy. Fortunately, the architectural drafter is usually interested only in the areas of topography related to the plot plan.

The boundaries of a plot are described by dimensions given in hundredths of a foot (two places beyond the decimal point), such as 151.67'. However, the surveyor will, whenever possible, lay out plots using even lengths (such as 100'). When this is done, the dimension

REAL WORLD APPLICATIONS

Fraction and Decimal Conversion

¼"=1'-0 Many times an architect is expected to use math that was learned in junior high school. One example is the conversion of fractions into decimals and vice versa. When you design a home, it will typically be dimensioned with fractional inches, but some specialty products are specified in decimal inches.

To convert a fraction to a decimal, divide the denominator into the numerator. For instance, to find the decimal equivalent of $\frac{3}{16}$", divide 16 into 3. The result, when rounded to three decimal places is .188". When rounding decimals, remember that if the *fourth digit is five or greater the third digit is increased by one*, but if the fourth digit is less than five the third digit remains the same.

On the other hand, what would you do if you were given .188" and wanted to find the fractional equivalent? Shown in the figure is a decimal inches to fractional inches conversion chart. Find the closest decimal amount and then move across to locate the equivalent fraction. Many of these conversion charts are made to fit into a pocket or wallet. They also make excellent bookmarks in reference material that commonly requires such conversions.

Now try these problems.

a. $\frac{5}{16}$" =   d. .754" =

b. $\frac{5}{8}$" =   e. .212" =

c. $\frac{11}{16}$" =   f. 1.358" =

Decimal Inches to Fractional Inches

Fractional Unit	Decimal Unit	Fractional Unit	Decimal Unit
1/32	0.031	17/32	0.531
1/16	0.063	9/16	0.562
3/32	0.094	19/32	0.594
1/8	0.125	5/8	0.625
5/32	0.156	21/32	0.656
3/16	0.188	11/16	0.688
7/32	0.219	23/32	0.719
1/4	0.250	3/4	0.750
9/32	0.281	25/32	0.781
5/16	0.313	13/16	0.813
11/32	0.344	27/32	0.844
3/8	0.375	7/8	0.875
13/32	0.406	29/32	0.906
7/16	0.438	15/16	0.938
15/32	0.469	31/32	0.969
1/2	0.500	1	1.000

is given simply as 100' rather than 100.00'. Bearings (such as N 5° 10′ 15″ E) are also given to show the compass direction of the boundaries. The bearings are given starting at one corner of the plot and proceeding around the perimeter until the starting point is again reached. Thus two opposite and parallel sides of a plot have opposite bearings (such as N 30° E and S 30° W).

Contour lines are dimensioned by indicating their elevation above sea level or some other datum plane like a street or the floor of a nearby house. The elevations of the land at the corners of the plot and the house are also shown. An engineer's scale is used rather than an architect's scale, 1" = 20' being quite common. The plot plan of the A residence in Chapter 25 shows these required dimensions.

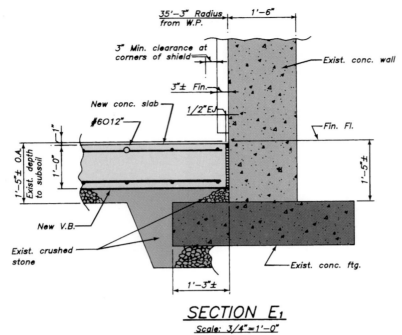

Figure 13-6 *A computer-dimensioned concrete slab detail.*

COMPUTER-AIDED DIMENSIONING

Dimension lines are added to computer-generated plans by selecting the end points of each feature to be dimensioned. The dimension is then automatically calculated. Initially, the dimension line is located directly over the dimensioned feature, but it can be moved to a more appropriate location. Extension lines and arrowheads are then added. The dimension itself can be located either centered or offset above its dimension line, or it can be centered or offset within a break in the dimension line. A computer-dimensioned concrete slab detail is illustrated in Figure 13-6. Refer to Chapter 7 for operating details on providing computer-generated dimensioning.

MODULAR COORDINATION

Module (from the Greek "measure") means a standard unit of measurement. A modular system, then, is a system of design in which most materials are equal in size to an established module or a multiple of that module. Such a system is called **modular coordination** because the materials will fit together—or *coordinate*—with-out cutting. For example, let's plan an open barbecue pit to be built of 8" × 8" × 16" concrete blocks with inside dimensions approximately 3' square. (The actual size of an 8" × 8" × 16" block is $7\frac{5}{8}$" × $7\frac{5}{8}$" × $15\frac{5}{8}$". When laid up with a $\frac{3}{8}$" mortar joint, however, the blocks fit in an 8" module.) If we designed this pit without giving thought to the size of the blocks, we would find that the four corner blocks in each course must be cut from 16" down to 12" (Figure 13-7).

If, however, we planned on an 8" module as shown in Figure 13-8, there would be no cutting required, and we would obtain a larger barbecue pit without using additional blocks.

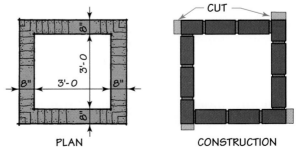

Figure 13-7 *Nonmodular barbeque pit design.*

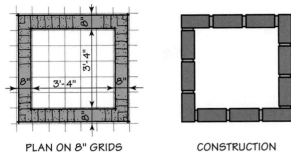

PLAN ON 8" GRIDS   CONSTRUCTION

Figure 13–8   Modular barbecue pit design.

Advantages

From this simple example, it is evident that modular coordination has some definite advantages:

1. It reduces cutting and fitting.

2. It reduces building costs.

3. It standardizes sizes of building materials.

4. It reduces drafting errors by reducing fractional dimensions.

At present, about 20 percent of U.S. architectural firms use modular dimensioning. These are the firms that specialize in masonry or precut lumber buildings.

Size of Module

Any convenient size module may be used. An 8" module was used in the preceding example. A 4' planning module is useful in layout work.

A 20' structural module is often used in steel factory construction. The most useful module, however, is 4", since brick, block, structural tile, window frames, and door bucks are all available in multiples of 4". Countries with metric rather than English measurements use a 100-mm module (4" = 101.6 mm).

Rules of Modular Dimensioning

1. Show light 4" grids on all plans, elevations, and sections drawn to a scale of $\frac{3}{4}$" = 1' or larger. For smaller scales, only a 4' planning grid is shown since it is impractical to show 4" grids.

2. Whenever possible, fit the building parts *between* grid lines or *centered on* grid lines. Occasionally an edge arrangement is necessary (Figure 13–9). Typical walls in plan are shown in Figure 13–10.

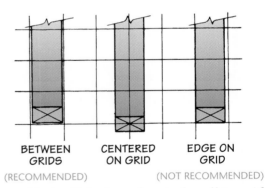

BETWEEN   CENTERED   EDGE ON
GRIDS   ON GRID   GRID
(RECOMMENDED)   (NOT RECOMMENDED)

Figure 13–9   Relation of a stud wall to grid line.

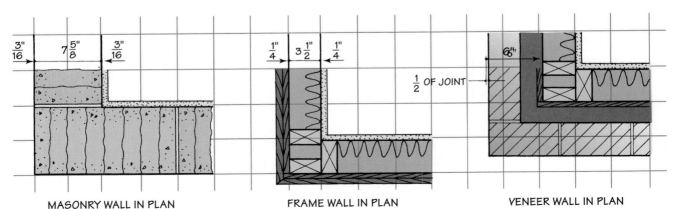

MASONRY WALL IN PLAN   FRAME WALL IN PLAN   VENEER WALL IN PLAN

Figure 13–10   Dimensioning typical walls using modular coordination.

3. To indicate location of building parts, use grid dimensions (dimensions from grid line to grid line). Use *arrows* for grid dimensions as shown in Figure 13–11. These arrows are used to indicate grid dimensions even when the grid lines do not appear on the drawing (as on the 4' planning module).

4. To indicate any position *not* on a grid line, use a dot.

5. The plans and elevations will contain mostly grid dimensions. But since many materials are not sized in 4" modules ($3\frac{1}{2}$" studs, for example), these materials are related to the nearest grid line by means of location dimensions in the section views. Location dimensions will have an arrow on the grid end and a dot on the off-grid end.

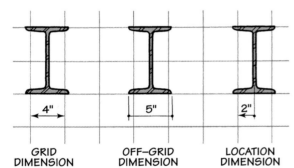

GRID DIMENSION   **OFF–GRID DIMENSION**   **LOCATION DIMENSION**

Figure 13–11   Modular grid dimensions, off-grid dimensions, and location dimensions.

6. The 4" module is three-dimensional, applying to both horizontal and vertical dimensions. Elevation grids are established as follows (Figure 13–12):
   a. The top of the subfloor in wood-frame construction coincides with a grid line.
   b. The top of the slab in slab-on-ground construction coincides with a grid line.
   c. The actual finished floor in all other types of construction is located $\frac{1}{8}$" below a grid line.

For plans containing some elements of modular dimensioning, see the split-level house in Chapter 14 and the South Hills Office Building in Chapter 43.

Computer-aided Modular Drafting

Computer-produced modular dimensioning is similar to non-modular dimensioning, but provision must be made for modular grid lines. This is usually accomplished by commanding a separate layer of 4" grid lines to be pen-plotted with a thin pen width. Some offices prefer to use a "capture" program to plot 4" modular grid coordinates (points) rather than grid lines.

Unicom System

The National Lumber Manufacturers Association recommends a modular system of house construction called **unicom** (for "uniform components"). In the unicom system,

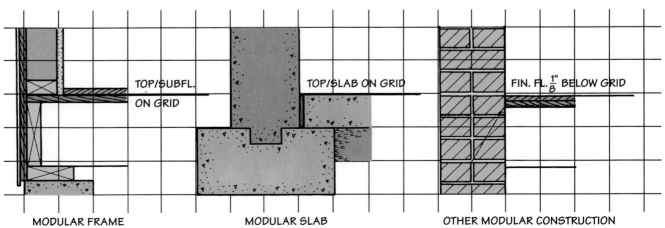

MODULAR FRAME   **MODULAR SLAB**   **OTHER MODULAR CONSTRUCTION**

Figure 13–12   Vertical positioning in modular coordination.

components such as wall, window, and door sections are all based upon a 16" or 24" module, thus requiring suppliers to stock a smaller number of different sized components. Houses may be erected using only these prefabricated components, or framed in the conventional manner.

Unicom Grid The unicom grid is based upon the 4" modular grid but with several variations. Figure 13–13 shows a unicom grid containing three weights of grid lines and a hidden line. The lightweight grids indicate 4" modules, the medium-weight 16" modules, and the heavy-weight 48" modules. The hidden line indicates 24" modules.

Unicom Panels Some typical wall, window, and door panels are shown in Figure 13–14. These panels are multiples of the 16" module and can be combined without cutting into the desired modular design. Floor panels, roof panels, partitions, roof truss components, and stairs are also available to unicom specifications.

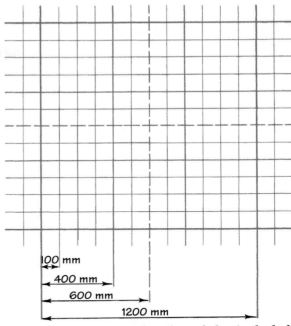

Figure 13–13 *Four related modules included in the unicom grid system.*

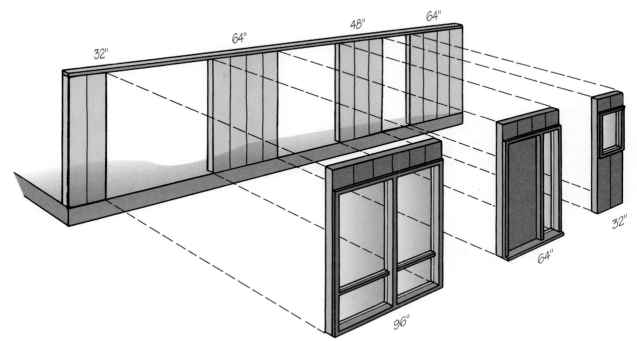

Figure 13–14 *Unicom wall, window, and door components.*

REVIEW QUESTIONS

1. Compare the methods of dimensioning a window in a masonry wall and a window in a wood-frame wall.

2. Give the proper method of indicating the following dimensions on an architectural drawing.
   a. Four inches
   b. Fourteen inches
   c. Four feet
   d. Four feet, four inches
   e. Forty-eight inches

3. Give the recommended:
   a. Distance between dimension lines.
   b. Height of dimension numerals.
   c. Height of architectural notes.

4. What is the purpose of window and door schedules?

5. Give two methods of indicating elevation dimensions. When is each method used?

6. When are nominal sizes and actual sizes used to dimension materials?

7. On a plot plan of a rectangular lot, why do parallel lines have different bearings?

8. What is a modular system?

9. What is a unicom system?

10. What module size is most commonly used?

11. What is meant by bearing?

ENRICHMENT QUESTIONS

1. Overall dimensions are always positioned as the outermost dimension. How does this influence the readability of the drawing?

2. In modular dimensioning:
   a. When are 4" and 4' grids shown?
   b. When are arrows used? When are dots used?

3. In modular elevation dimensioning, what is the position of the floor with respect to a grid line in:

   a. Wood-frame construction

   b. Slab-on-ground construction

   c. Other types of construction

4. Why are interior partitions dimensioned to the center of the wall?

5. In Figure 13–3, the window and door sizes are not dimensioned. How will the builder know how large to make them?

PRACTICAL EXERCISES

1. Using $\frac{1}{4}$" cross-section paper or $\frac{1}{4}$" computer grids, sketch by hand or with a computer a typical exterior corner of a residence constructed of the following materials. Indicate the method of dimensioning and completely note all materials.

   a. Frame

   b. Brick veneer

   c. 10" cavity brick

   d. 8" concrete block

2. Draw or computer-produce the dimensions on your plans of the A residence in Chapter 27.

3. Draw or computer-produce the dimensions on your original house design.

4. Redraw Figure 13–5, but add appropriate dimensions according to your choice.

CHAPTER 14

Architectural Dimensioning in Metric Units

OBJECTIVES

By completing this chapter you will be able to:

- Dimension an architectural drawing or plot of land using SI metric dimensioning practices.

- Dimension an architectural drawing or plot of land using dual (English/metric) dimensioning practices.

- Recognize and specify commonly used metric sizes of materials.

- Convert English units of dimensioning to SI units, and convert SI units to English units.

KEY TERMS

Soft conversion	SI system
Hard conversion	

The metric system was conceived over three hundred years ago by Gabriel Mouton, a Frenchman, who designed a decimal system based upon the circumference of the earth. The unit of length was called the meter, from the Greek "metron" (measure). In 1960 the meter was redefined internationally in terms of the wavelength of a specific color of light. Most industrialized countries have adopted this system, known as the SI (Système International) metric system. France officially adopted metric units in 1793. The United Kingdom started its conversion in 1975, and Canada in 1978. The United States is following on a voluntary—but steady—basis. Due to this voluntary approach, *hard* conversion of building products from English to metric will vary considerably depending upon the publication of technical standards and the speed of manufacturing conversion.

Soft and Hard Conversion

The terms *soft conversion* to metric and *hard conversion* to metric are often used. **Soft conversion** refers to changing the English dimensions of a product to metric dimensions without changing the size of the product. **Hard conversion** refers to changing both the dimensions and the size of a product to a rational metric size. For example, in soft conversion terms, the size of a 4' × 8' plywood sheet would be unchanged but called a 1220 × 2440 mm sheet. In hard conversion terms, it would be changed to a modular size and manufactured as a 1200 × 2400 mm sheet.

THE SI METRIC SYSTEM

Use of SI Units

The **SI system** is based upon the following seven units which are of interest to the architect and builder. Multiples and submultiples are expressed as decimals.

1. Length: meter (m)

2. Time: second (s)

3. Mass: kilogram (kg)

4. Temperature: kelvin (K)

5. Electric current: ampere (A)

6. Luminous intensity: candela (cd)

7. Amount of substance: mole (mol)

Prefixes are used to eliminate insignificant digits and decimals. For example, 3 mm (3 millimeters) is preferred to 0.003 m (0.003 meter). Metric prefixes are shown in Table 14–1.

TABLE 14–1 *Metric Prefixes*

Prefix	SI Symbol	Multiplication Factor
tera	T	10^{12} (1 000 000 000 000)
giga	G	10^{9} (1 000 000 000)
mega	M	10^{6} (1 000 000)
kilo	k	10^{3} (1 000)
hecto	h	10^{2} (100)
deka	da	10^{1} (10)
deci	d	10^{-1} (0.1)
centi	c	10^{-2} (0.01)
milli	m	10^{-3} (0.001)
micro	μ	10^{-6} (0.000 001)
nano	n	10^{-9} (0.000 000 001)
pico	p	10^{-12} (0.000 000 000 001)
femto	f	10^{-15} (0.000 000 000 000 001)
atto	a	10^{-18} (0.000 000 000 000 000 001)

Wherever possible, use multiple and sub-multiple prefixes representing steps of 1000. For example, express length in millimeters, meters, and kilometers. Avoid using the centimeter and the decimeter. Do not use a period after an SI symbol except when it occurs at the end of a sentence (e.g., 2 mm, not 2 mm.). To assist in reading numbers with four or more digits, and to eliminate confusion by the European use of commas to express decimal points, place digits in groups of three separated by a space, without commas, starting both to the left and right of the decimal point (e.g., 12 625 not 12,625). The space is optional with a four-digit number, however (e.g., either 1500 or 1 500).

Computer-aided SI Dimensioning

Figure 14–1 is a computer-produced plan dimensioned in metric units. To produce metric designs on your computer, locate the "NUMERIC" command area on your digitizer tablet (the small area to the right of the screen pointing area), and override the English unit default by configuring to metric units. The standard CAD digitizer tablet menu offers mm, cm, and m units.

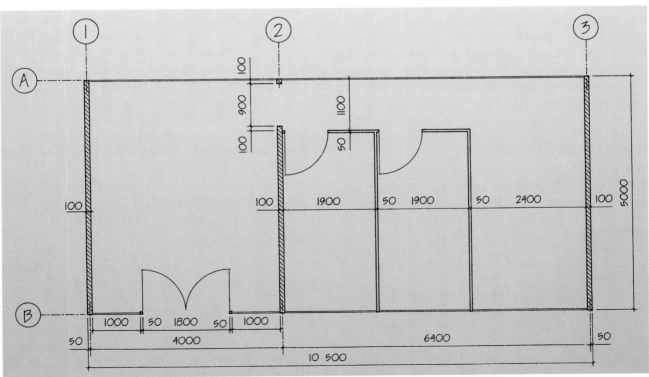

Figure 14–1 *A computer-produced plan dimensioned in metric units (millimeters).*

REAL WORLD

APPLICATIONS

English-Metric Conversions

¼"=1'-0 Conversions from English to metric (SI) units can be made with the help of Table 14–14 (page 161). Retain in all conversions the same number of significant digits so that accuracy is neither sacrificed or exaggerated. With the help of the conversion table, converting $1\frac{1}{16}$" to mm is quite easy. First, convert $\frac{1}{16}$ to a decimal by dividing 1 by 16 (1 ÷ 16 = .06250). As the table shows, 1" = 25.4 mm. Then multiply 1.0625 by 25.4 to obtain the result: 1.0625" × 25.4 mm = 27 mm. Building site or plot plans are generally in decimals, such as 101.24'. Conversion of these dimensions is easy. In this case, multiply the footage by the number of meters in a foot. As Table 14–14 shows, there are .3048 meters in a foot. Therefore, multiplying 101.24 by .3048 will convert this dimension to metric: 101.24' × .3048 m = 30.86 m.

To convert from metric to English, the same calculations are required, but the lower half of Table 14–14 showing metric to English units is used. For instance, to convert 4.7 meters to yards notice that 1 meter equals 1.09361 yards. Multiply 4.7 by 1.09361 to convert this measurement to yards: 4.7 m × 1.09361 yds = 5.1 yds. Be aware that metric units to be converted into feet will provide a decimal number that must be converted to inches. This can be done by multiplying the decimal by 12 since there are 12 inches in 1 foot. For instance, how many feet are in 6.25 meters? Multiply 6.25 m by 3.28084'. The result is 20.51'. Now multiply .51 by 12 to arrive at the number of inches: .51' × 12" = 6.12". Therefore, the final answer is approximately 20'-6". Now try the following conversions.

1. 122.68' = _____ m

2. 32.5 mm = _____"

3. 16" = _____ mm

4. 1'-7" = _____ m

5. 3.11 m = _____'-_____"

Use of Non-SI Units

Some non-SI units are so commonly used that they will continue to be accepted. For example, we will continue to indicate time in the English units of minutes, hours, and days in addition to the SI unit of seconds. Angles will continue to be measured in the English units of degrees, minutes, and seconds in addition to the SI unit of radians.

Also some new non-SI terms will be introduced. For example, the term *metric ton* is not an approved SI unit, but it is now being used to indicate 1000 kg.

Some non-SI symbols will also be used. The SI symbol for liter, for example, is the lowercase *l*. Because this symbol is so easily confused with the numeral 1, the capital *L* is commonly used.

USE OF SI UNITS ON ARCHITECTURAL DRAWINGS

Working Drawings

The preferred SI measurement unit on working drawings scaled between 1:1 and 1:100 is the millimeter. The symbol (mm) is deleted, but the note "All dimensions in millimeters except as noted" should be added. On drawings of large structures scaled between 1:200 and 1:2000, the preferred measurement unit is the meter, taken to three decimal places (e.g., 8.500). Again, the symbol (m) is deleted, but a note, "All dimensions in meters except as noted," is added.

Plot Plans

Surveyors indicate land distances on plot plans in meters (and on maps in kilometers). Surveyors normally measure to an accuracy of about 1 cm, and therefore such distances on plot plans are shown as meters taken to two decimal places (e.g., 8.50). Contour lines are usually established at 0.5-m intervals, 1-m intervals, 2-m intervals, and 5-m intervals. The 0.5-m interval is shown in Figure 14–6.

Angles

According to the National Bureau of Standards, plane angles will be specified in three ways: (1) The SI unit (the radian) will be used in calculations. (2) Engineers will specify angles in degrees with decimal submultiples (e.g., 11.25°) on engineering and construction drawings. (3) Surveyors will continue to specify angles in degrees, minutes, and seconds (e.g., N 10° 12' 30" E) on plot plans. (*Recommended Practice for the Use of Metric (SI) Units in Building Design and Construction,* U.S. Department of Commerce/National Bureau of Standards, reprinted with corrections June 1977.)

Metric Modules

It is anticipated that 100 mm (about $\frac{1}{16}$" less than 4") will be accepted as a basic metric module. Figure 14–2 shows a unicom grid based upon the 100-mm module with multiples of 400 mm (approximately 16"), 600 mm (approximately 24"), and 1200 mm (approximately 48"). Compare this grid with the related grid in the previous chapter. It is expected, then, that hard conversion of building materials will result in metric sizes as discussed in the following sections. (These examples of the modular sizes of building materials should be considered tentative until standards are established and manufacturing changeovers are announced.)

LUMBER

Rough Lumber

U.S. standards for softwood lumber have not yet been established, but the American National Metric Council (ANMC) has recommended that

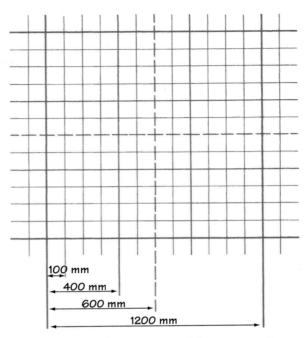

Figure 14–2   The unicom grid system using metric modules.

English lumber sizes be soft-converted to the nearest mm and specified in actual rather than nominal sizes. Thus a 2 × 4 having actual dimensions of $1\frac{1}{2}$" (38.1 mm) × $3\frac{1}{2}$" (88.9 mm) would be specified as 38 × 89. Other similar metric conversions are shown in Table 14–2.

TABLE 14–2 Some Metric Sizes Recommended by ANMC for Rough Softwood		
Nominal English Size	**Actual English Size**	**Actual Metric Size**
1"	$\frac{3}{4}$"	19 mm
2"	$1\frac{1}{2}$"	38 mm
3"	$2\frac{1}{2}$"	64 mm
4"	$3\frac{1}{2}$"	89 mm
6"	$5\frac{1}{2}$"	140 mm
8"	$7\frac{1}{4}$"	184 mm
10"	$9\frac{1}{4}$"	235 mm
12"	$11\frac{1}{4}$"	286 mm

The International Organization for Standardization (ISO), however, has established nominal size standards as shown in Table 14–3. The ISO consists of standards organizations in each country such as the American National Standards Institute (ANSI), the British Standards Institution, and the Canadian Standards Association. The lumber sizes commonly available in European countries are shown in Table 14–4. Notice that lumber in rough widths of 100 mm (4"), 150 mm (6"), 200 mm (8"), 250 mm (10"), and 300 mm (12") are all available in a rough thickness of 50 mm (2"). Only the smaller sizes of 100 mm and 150 mm are also available in a thickness of 38 mm ($1\frac{1}{2}$"). Notice that strapping is commonly available in 19 mm ($\frac{3}{4}$") by 75 mm (3") and 100 mm (4") as well as in 25 mm (1") by 100 mm (4") and 150 mm (6").

Regarding nomenclature, the familiar 2 × 4 may be called a 50 × 100. However, you should realize that the nominal size is a value assigned for the purpose of convenient designation and may exist in name only. Consequently, terms such as *2 × 4* and *2 × 6* may continue to be used for some time. For example, the term *8-penny nail* is still used even though such nails can no longer be purchased for 8 cents per hundred.

Table 14–5 shows the ISO standard lengths of joists and rafters. These lengths are commonly available in metric countries. A 2.4-m length is approximately $1\frac{1}{2}$" ($1\frac{1}{2}$ percent) shorter than an 8' length.

Current joist and rafter spacings of 16" oc (on center) and 24" oc convert to spacings of 400 mm and 600 mm.

Finish Lumber

Finish (dressed) lumber is commonly available in metric countries in 12-mm (approximately $\frac{1}{2}$") and 19-mm (approximately $\frac{3}{4}$") thicknesses. As shown in Table 14–6, the dressed widths are 5 mm smaller than the rough widths shown in Table 14–4.

Panels

Sheets of plywood, fiberboard, and hardboard will be available in a 1200 × 2400-mm size. The 4' × 8' size will also be stocked for some time since the $1\frac{1}{2}$" difference between English and metric sizes can cause serious problems. For

---

TABLE 14–3 *Some Nominal Metric Sizes Recommended by ISO for Rough Softwood*

Thickness		Width	
Nominal Metric Size	Approx. Equivalent English Size	Nominal Metric Size	Approx. Equivalent English Size
19 mm	$\frac{3}{4}$"	75 mm	3"
25 mm	1"	100 mm	4"
38 mm	$1\frac{1}{2}$"	150 mm	6"
50 mm	2"	200 mm	8"
		250 mm	10"
		300 mm	12"

TABLE 14–4 *Some Nominal ISO Sizes Available in Metric Countries for Rough Softwood (with approximate equivalent English sizes)*

	75 mm (3")	100 mm (4")	150 mm (6")	200 mm (8")	250 mm (10")	300 mm (12")
19 mm ($\frac{3}{4}$")	⊟	⊠				
25 mm (1")		⊠	⊠			
38 mm ($1\frac{1}{2}$")		⊠	⊠			
50 mm (2")		⊠	⊠	⊠	⊠	⊠

TABLE 14–5   Lumber Lengths Recommended by ISO

Metric Size	Approx. Equivalent English Size
2.4 m	8'
3.0 m	10'
3.6 m	12'
4.2 m	14'
4.8 m	16'
5.4 m	18'
6.0 m	20'

TABLE 14–7   Some Thicknesses of Plywood Available in Metric Countries

Metric Thickness	Approx. Equivalent English Thickness
12 mm	$\frac{1}{2}$"
16 mm	$\frac{5}{8}$"
19 mm	$\frac{3}{4}$"
25 mm	1"

TABLE 14–8   Concrete Block Modular Widths

Metric Size	Approx. Equivalent English Size
50 mm	2"
75 mm	3"
100 mm	4"
150 mm	6"
200 mm	8"
250 mm	10"
300 mm	12"

example, the smaller metric panel would not fit in a modular building designed to English dimensions, nor would it serve as a suitable replacement panel in the renovation of an old building. Similarly, each of the larger English panels would have to be trimmed to fit in a metric modular building. Some thicknesses of plywood available in metric countries are listed in Table 14–7.

MASONRY

Concrete Masonry

Concrete masonry units, usually called *concrete blocks,* are manufactured in modular sizes based upon the 4" module. The 8" × 8" × 16" stretch unit is most often used. The actual size of such a block is $7\frac{5}{8}$" × $7\frac{5}{8}$" × $15\frac{5}{8}$" to allow for $\frac{3}{8}$" mortar joints. Converted to metric, such units are 200 × 200 × 400 mm nominally. Actual size is 190 × 190 × 390 mm to allow for 10-mm mortar joints. Larger blocks and smaller partition blocks convert to metric sizes as shown in Table 14–8.

Brick Masonry

Four common types of modular brick are shown in Figure 14–3 and Table 14–9. Actual sizes are 10 mm smaller than nominal sizes to allow for mortar joints.

TABLE 14–6   Some Sizes Available in Metric Countries for Finished Softwood (with approximate equivalent English sizes)

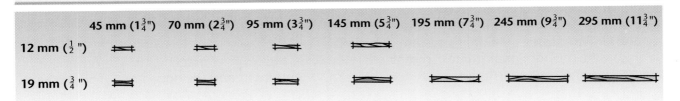

	45 mm ($1\frac{3}{4}$")	70 mm ($2\frac{3}{4}$")	95 mm ($3\frac{3}{4}$")	145 mm ($5\frac{3}{4}$")	195 mm ($7\frac{3}{4}$")	245 mm ($9\frac{3}{4}$")	295 mm ($11\frac{3}{4}$")
12 mm ($\frac{1}{2}$")	▱	▱	▱	▱			
19 mm ($\frac{3}{4}$")	▱	▱	▱	▱	▱	▱	▱

TABLE 14–9 *Brick Sizes*

| | Nominal Metric Dimensions | | | Nominal English Dimensions | | |
| | *(3-course)* | | | *(3-course)* | | |
	Width	*Height*	*Length*	*Width*	*Height*	*Length*
Modular brick	100 mm	200 mm	200 mm	4"	8"	8"
Roman brick	100 mm	150 mm	300 mm	4"	6"	12"
Norman brick	100 mm	200 mm	300 mm	4"	8"	12"
SCR brick	150 mm	200 mm	300 mm	6"	8"	12"

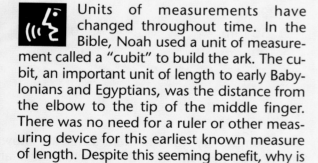

MODULAR   ROMAN   NORMAN   SCR

REAL WORLD   APPLICATIONS

Debating Measurement Systems

Units of measurements have changed throughout time. In the Bible, Noah used a unit of measurement called a "cubit" to build the ark. The cubit, an important unit of length to early Babylonians and Egyptians, was the distance from the elbow to the tip of the middle finger. There was no need for a ruler or other measuring device for this earliest known measure of length. Despite this seeming benefit, why is such a standard of measurement not suitable?

The metric system of measurement greatly improved measurement standards when it was developed in France in 1793. Congress made this new set of standards legal in the United States in 1866. This system, which is based on decimals, has spread throughout the world and is called the International System of Units (SI). However, in the United States the English system, which is based on fractional inches, remains the measurement standard.

Should the metric system be mandated in the United States as it was in Australia? Should companies and industries make changes only if they wish? Should the metric system be the only measurement system taught in schools with the hope that eventually it will infiltrate into all sectors of society?

Choose a position on the place the metric system should have in the United States. Research the topic and prepare to debate your position on these questions with someone in your class. Some key debate points to remember are listed below.

1. Support your position well with data and facts.

2. Speak clearly, passionately, and enunciate each word.

3. Make eye contact with your debate opponent and the audience.

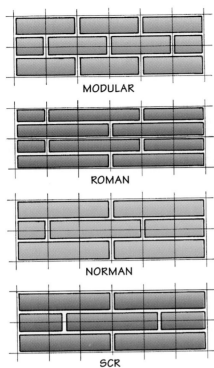

Figure 14–3 Modular coordination of brick types.

Reinforced Concrete

Rebars (reinforcing bars) are available in English sizes from #2 ($\frac{1}{4}$" diameter) to #18 ($2\frac{1}{4}$" diameter) in $\frac{1}{8}$" increments. Hard conversion in most metric countries resulted in rebars sized as shown in Table 14–10.

STEEL

Steel Plate

Availability of steel plate varies depending upon the plate mill, but most mills supply plate in the following English thicknesses:

$\frac{1}{32}$" increments up to $\frac{1}{2}$"

$\frac{1}{16}$" increments over $\frac{1}{2}$" to 2"

$\frac{1}{8}$" increments over 2" to 6"

Although the availability of steel plate in metric sizes also varies depending upon the mill, the following thicknesses are supplied as a minimum:

1-mm increments up to 12 mm

2-mm increments over 12 mm to 20 mm

TABLE 14–10 *Rebar Sizes (diameter)*	
Metric Size	**Approx. Equivalent English Size**
6 mm	$\frac{1}{4}$"
8 mm	$\frac{5}{16}$"
10 mm	$\frac{3}{8}$"
12 mm	$\frac{1}{2}$"
16 mm	$\frac{5}{8}$"
20 mm	$\frac{3}{4}$"
25 mm	1"
32 mm	$1\frac{1}{4}$"
40 mm	$1\frac{1}{2}$"
50 mm	2"

5-mm increments over 20 mm to 60 mm

10-mm increments over 60 mm to 160 mm

Steel Beams

Hard conversion of steel beams to metric sizes occurs only during the last stages of conversion. Soft conversion tables, however, are now available. These tables give the properties of customary beams in metric terms. For example, a W 6 × 15.7 (a W-shape beam sized 6" × 6" weighing 15.7 lb./ft.) is converted to 152 mm × 152 mm having a mass (weight) of 23 kg/m. (*Weight* is a commonly used term, but in technical work it is correct to use the term *mass* to indicate quantity of matter.)

METRIC DESIGN

Ceiling, Head, and Sill Heights

The minimum comfortable ceiling height for a habitable room is 8' (in English units) and 2400 mm (in metric units), as shown in Figure 14–4. Also shown are commonly specified head heights of doors and windows (measured from finished floor to the bottom of rough opening at the head) and windowsill heights (measured from finished floor to the top of rough opening at the sill).

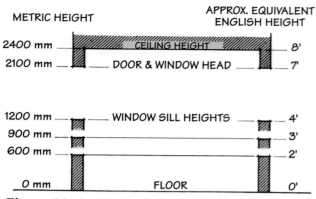

Figure 14–4 Metric ceiling, head, and sill heights.

Width of Door Openings

In most metric countries, the rough opening widths are 600, 700, 800, and 900 mm for interior doors and 900, 1000, 1500, and 1800 mm for exterior doors. These widths are shown in Table 14–11.

Width of Window Openings

In metric countries, the rough opening width for most windows varies from 600 mm to 1800 mm as shown in Table 14–12.

Metric Scales

Common metric scales used in architectural drawing are listed in Table 14–13 together with their equivalent English scales.

TABLE 14–11 *Door Openings*

Metric Width	Approx. Equivalent English Width
600 mm	2'-0
700 mm	2'-4"
800 mm	2'-8"
900 mm	3'-0
1000 mm	3'-4"
1500 mm	5'-0
1800 mm	6'-0

TABLE 14–12 *Window Openings*

Metric Width	Approx. Equivalent English Width
600 mm	2'-0
700 mm	2'-4"
800 mm	2'-8"
900 mm	3'-0
1000 mm	3'-4"
1100 mm	3'-8"
1200 mm	4'-0
1500 mm	5'-0
1800 mm	6'-0

For an example of plans dimensioned in metric units, see the M residence, Figures 14–6 through 14–11.

DUAL DIMENSIONING

Occasionally architectural drawings must be dimensioned in dual units. For example, dual

TABLE 14–13 *Common Architectural Scales*

Metric Scale	Approx. Equivalent English Scale	Used for:
1:1	12" = 1'-0 (1:1)	Full-scale patterns
1:5	3" = 1'-0 (1:4)	Detail sections
1:10	$1\frac{1}{2}$" = 1'-0 (1:8)	Wall sections
1:20	$\frac{1}{2}$" = 1'-0 (1:24)	Structural sections
1:50	$\frac{1}{4}$" = 1'-0 (1:48)	Large-scale plans and elevations
1:100	$\frac{1}{8}$" = 1'-0 (1:96)	Small-scale plans and elevations
1:200	1" = 20' (1:240)	Large-scale plot plans
1:500	1" = 50' (1:600)	Small-scale site plans

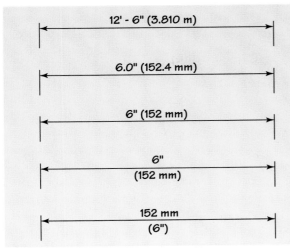

Figure 14-5   Dual dimensioning systems.

TABLE 14-14   *English-Metric Conversions (accurate to parts per million)*

1 inch = 25.4 millimeters
1 foot = 0.304 8 meter
1 yard = 0.914 4 meter
1 mile = 1.609 34 kilometers
1 quart (liquid) = 0.946 353 liter
1 gallon = 0.003 785 41 cubic meter
1 ounce (avdp) = 28.349 5 grams
1 pound (avdp) = 0.453 592 kilogram
1 horsepower = 0.745 700 kilowatt
1 millimeter = 0.039 370 1 inch
1 meter = 3.280 84 feet
1 meter = 1.093 61 yards
1 kilometer = 0.621 371 mile
1 liter = 1.056 69 quarts (liquid)
1 cubic meter = 264.172 gallons
1 gram = 0.035 274 0 ounce (avdp)
1 kilogram = 2.204 62 pounds (avdp)
1 kilowatt = 1.341 02 horsepower

dimensioning is used when a building component has been manufactured to English sizes but must be mated with a component manufactured to metric sizes. Dual dimensioning is simply the placing of the metric counterpart after the English dimension, e.g., 1" (25.4 mm). Dual dimensioning may be shown by any of the methods illustrated in Figure 14-5.

A number of countries are now using a metric system that varies slightly from the SI metric system. For projects in such countries, determine what metric system is used and give dual dimensioning using that country's metric units followed by SI units.

English-metric Conversions

Conversions from English to SI units can be made with the help of Table 14-14. Retain in all conversions the number of significant digits so that accuracy is neither sacrificed nor exaggerated. Conversion is quite easy using a pocket calculator: $1\frac{1}{16}" = 1.0625 \times 25.4 = 27$ mm. Building site or plot plans are generally dimensioned in decimals, such as 101.24'. Conversion of these dimensions is simple: e.g., 101.24' = $101.24 \times 0.3048 = 30.86$ m.

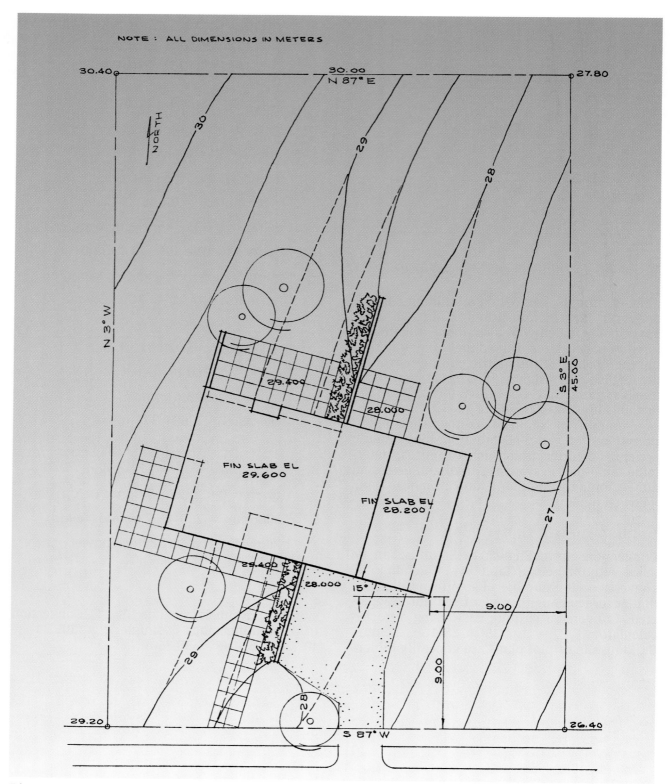

NOTE : ALL DIMENSIONS IN METERS

Figure 14–6 *Plot plan of the M residence.* (*Courtesy Professor M. Isenberg, the Pennsylvania State University.*)

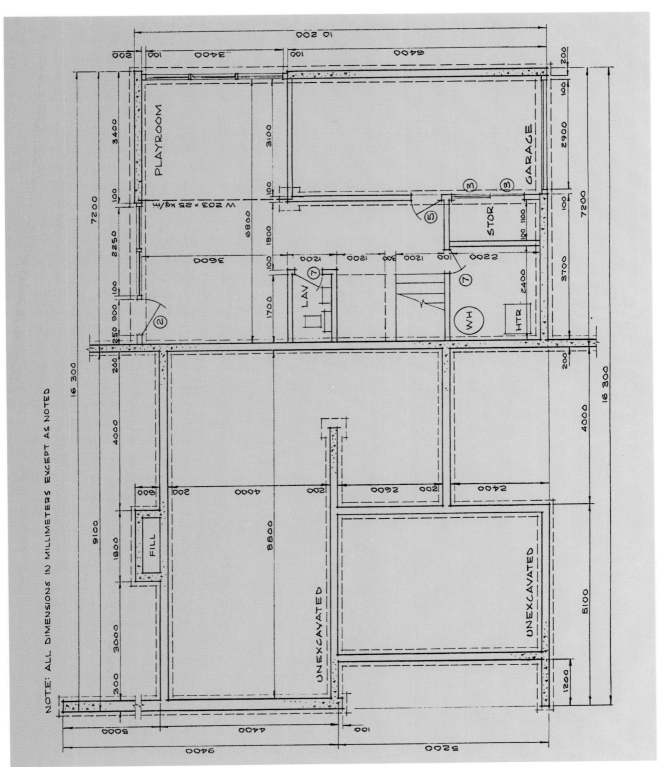

Figure 14–7   *Basement plan of the M residence. (Courtesy Professor M. Isenberg, the Pennsylvania State University.)*

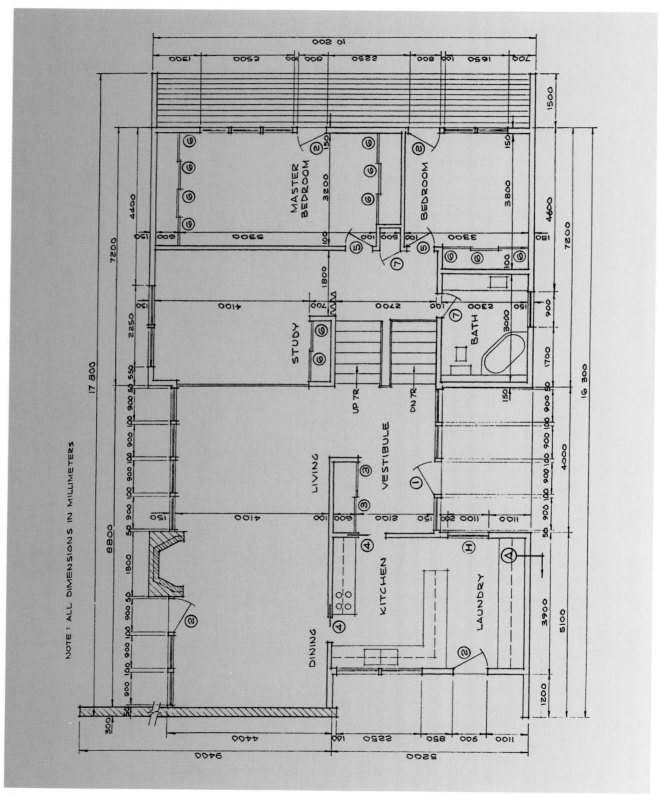

Figure 14–8 *Floor plan of the M residence. (Courtesy Professor M. Isenberg, the Pennsylvania State University.)*

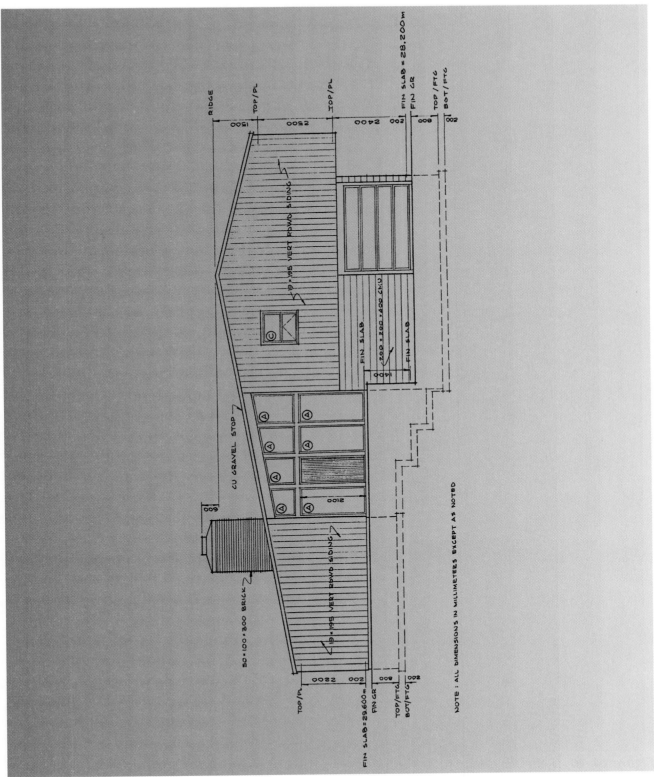

Figure 14–9   *Front elevation of the M residence.* (*Courtesy Professor M. Isenberg, the Pennsylvania State University.*)

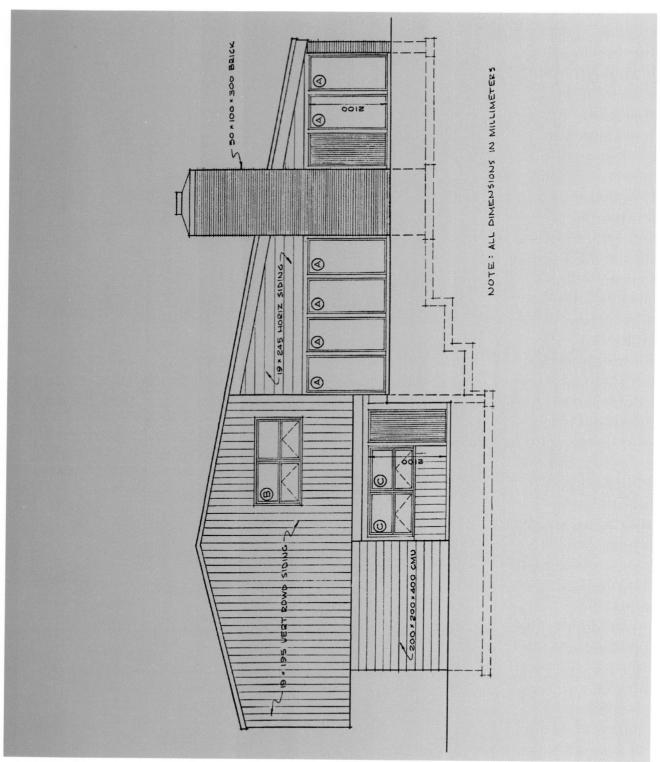

Figure 14–10 *Rear elevation of the M residence. (Courtesy Professor M. Isenberg, the Pennsylvania State University.)*

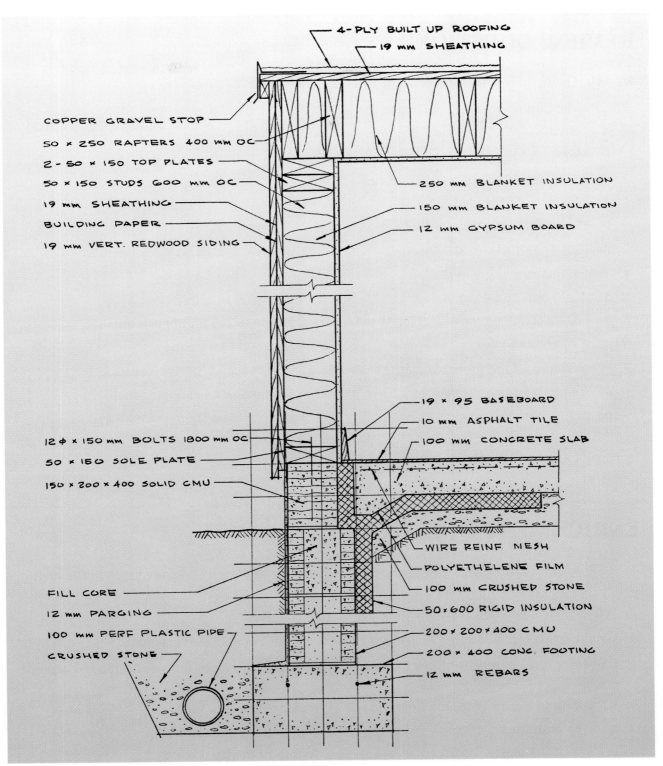

Figure 14–11   *Section A through laundry of the M residence.* (Courtesy Professor M. Isenberg, the Pennsylvania State University.)

REVIEW QUESTIONS

1. What is the difference between soft and hard conversion?

2. Give the SI units for:
   a. Length
   b. Mass
   c. Temperature
   d. Electric current

3. Give the probable hard conversion size of:
   a. 2" × 4" wood studs
   b. 1" × 10" finished wood shelving
   c. $\frac{1}{2}$" × 4' × 8' plywood
   d. 4" × 8" × 16" concrete block
   e. 1" rebar
   f. 3'-wide × 4'-high window opening

4. Which quantity is greater?
   a. A 100-mm or a 4" module
   b. 1200 × 2400-mm or 4' × 8' plywood
   c. A meter or a yard
   d. A kilometer or a mile
   e. A metric ton or an English ton
   f. A kilogram or a pound

5. What would be the probable metric scale of:
   a. A floor plan of a residence
   b. A plot plan of a residence
   c. A floor plan of a shopping mall
   d. A site plan of a shopping mall and its parking area

ENRICHMENT QUESTIONS

1. Identify and explain what you feel are two hindrances to a complete conversion to the metric system for the United States.

2. Why were millimeters used in Figure 14–1?

3. Would you recommend that metric conversions be hard or soft conversions?

4. Call some local hardware stores or lumber yards and see if any of them carry lumber according to metric specifications.

5. According to Figure 14–8, the floor plan of M residence, what is the:
   a. Area of the study
   b. Size of the master bedroom
   c. Width of the stairway
   d. Total area of the floor plan
   e. Width of the fireplace

6. According to Figure 14–11, a sectional view of the laundry of the M residence, what is the:

   a. Thickness of the concrete slab

   b. Thickness of the insulation in the walls

   c. Cross-sectional area of the concrete footing

PRACTICAL EXERCISES

1. Draw or computer-produce the SI dimensioning on your plans of the A residence in Chapter 27.

2. Draw or computer-produce the SI dimensioning of your original house design.

3. Draw or computer-produce the dimensioning of your building using:

   a. English system

   b. SI metric system

   c. Dual dimensioning

4. Draw or computer-produce the following M residence wall sections:

   a. Kitchen

   b. Dining room

Introduction to Architectural Design

SECTION III

CHAPTER 15

Architectural History

OBJECTIVES

By completing this chapter you will be able to:

🏛 Identify the five fundamental structural systems and recognize their place in history and their use in contemporary construction.

🏛 Identify the principal characteristics of the major historical styles of architecture.

🏛 Relate the contributions of past civilizations to the current state of architectural development.

🏛 Discuss architectural trends that will influence future developments in architecture.

KEY TERMS

Lintel	Corinthian
Corbel	Gothic
Arch	Renaissance
Cohesive construction	Classic revival
Truss	Eclecticism
Doric	International style
Ionic	

In this age of new styles, new materials, and new building methods, we are apt to think of the contemporary house as having no ancestry—no history. Actually, the houses we live in today are not completely original designs, but the outgrowth of many centuries of architectural development. No significant architectural style stands alone; rather it is the result of evolution. Every building that has ever been built is a composite idea of all the centuries of ideas that have preceded. It is thus with our present "modern" house, and the trends today will dictate the architecture of the future. We study the history of architecture, therefore, not only to appreciate the many public buildings previously built in classic styles but also to aid in the design of present-day buildings.

In the following discussion, architectural history has been divided into separate chronological groups. But it should be remembered that architectural development is a steady and continuous process. Each age has made its contribution. Then we will see this century in true perspective and realize that no age is an ultimate—but merely one step in this continuing process.

The principal historical styles of architecture are as follows. Each is an evolution of, or has been influenced by, earlier styles.

Egyptian

Greek

Roman

Gothic

Renaissance

Classic Revival, Eclecticism

International, Contemporary

PREVIEW

If a beginning must be found, the architecture of the Egyptians will serve well. The Greeks drew on Egyptian, Assyrian, Persian, and Phoenician ideas. The Romans borrowed the ideas of the Greeks and added to them. History has approved the architecture of the Greeks and Romans to such an extent that many modern public buildings have been built in imitation of it. Later, the Gothic style perfected vaulting, while keeping many of the forms of its predecessors. During the Renaissance, the classic civilizations were studied and revived. But this return to ancient styles led eventually into a period called Eclecticism, which virtually stopped any real architectural development. In reaction to Eclecticism, architects searched for a new and true form of shelter. This was found in the International and the Contemporary.

STRUCTURAL SYSTEMS

The history of architecture is closely tied to the development of methods of spanning the open space between two columns. To accomplish this spanning, five fundamental structural systems have been used (Figure 15–1):

1. Lintel
2. Corbel
3. Arch
4. Cohesive construction
5. Truss

The **lintel** is a horizontal member spanning an opening to carry the weight above. The weight of the lintel and any weight above it causes a vertical downward thrust to the columns. The Egyptians and Assyrians used the lintel almost exclusively. The Greeks developed the column and lintel, in an aesthetic sense, to perfection.

A **corbel** is a block projecting from a wall and supporting a weight. The cantilever is based upon the same principle. The Persians used corbeling extensively.

The **arch** is composed of wedge-shaped blocks, each supporting a share of the load by wedging the adjoining blocks. The weight of the blocks and any weight above causes not only a downward thrust but also outward thrusts. The Assyrians built a few arched structures, but the Romans are credited with much further development. The application of the principle of the arch to the dome is credited to the Roman and Byzantine periods, and the vault reached its height in the Gothic period.

Cohesive construction employs materials that are shaped while plastic and allowed to harden into a homogeneous structure. The Romans used a kind of cohesive construction in their domes. Reinforced concrete is a modern application of this system.

A **truss** is a rigid arrangement of comparatively short members spanning a wide space.

EGYPTIAN ARCHITECTURE

It is a tribute to Egypt that her works dating back to 3500 B.C. are still standing. The surviving architecture, however, does not represent

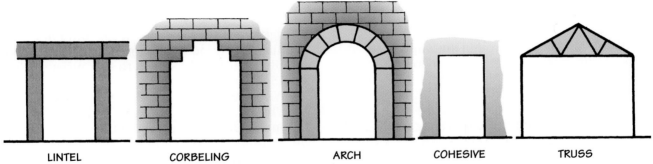

LINTEL   CORBELING   ARCH   COHESIVE   TRUSS

Figure 15–1 Fundamental structural systems.

homes or even palaces, but rather tombs and monuments to the dead. In earliest history, graves were marked by piles of stones called *cairns.* The most noted outgrowth of these cairns are the pyramids built of stone or brick. There are over a hundred pyramids, varying in size and design, each containing intricate passageways designed to confuse anyone entering. Actually, no entrance was found easily, since the tunnels were blocked once the mummy was in place.

The actual construction method of the Egyptians is one of the mysteries of the past. Hundreds of thousands of workers were employed in these huge monuments, but even so it is not clear how stones weighing up to thirty tons were raised. It has been suggested that a temporary sand ramp was erected, and the stone blocks were rolled up this ramp into position. The Egyptian temples of later periods could have been constructed in the same manner, the interior of the temple being filled with sand as the walls and columns were erected, and then dug out at completion. Another theory is that the stones were rocked in place using wooden rockers, which have been unearthed. It has been suggested that the stone was rocked enough for a shim to be inserted under one side, and then rocked again to insert a shim on the other side. This process could continue until the stone reached any required height.

The three largest pyramids are at Giza (Figure 15–2), the most famous being the Great Pyramid of Cheops. This is the only remaining one of the original seven Wonders of the World. The dimensions of the Great Pyramid were originally 764' square × 482' high. It was built of limestone blocks upon a leveled rock plateau and was coated with smooth bands of colored granites.

While only kings built pyramids, lesser rulers built *mastabas.* These were long, rectangular tombs of stone or brick with only a slightly sloping exterior wall and a flat roof of stone slabs. As time passed, these mastabas evolved into tremendous temples (ca. 1500 B.C.) for the deification of the kings. These temples were built on a diminishing plan (see Figure 15–3); that is, the chambers were smaller, darker, and more mysterious the farther they

Figure 15–2   The Sphinx and the Great Pyramid at Giza. (Courtesy G.E. Kidder Smith.)

were penetrated. Also, the floors became higher and the roofs lower as the chambers receded. The exteriors of the temples were windowless oblong boxes fronted by two pylons, somewhat reminiscent of the pyramids, flanking the entrance. Two rows of facing sphinxes might border the approach.

The most noteworthy of these structures is the Karnak Temple, which was built and added to for seven hundred years, finally reaching dimensions of 376' wide × 1,215' long. The first chambers entered, called *hypostyle halls,* were entered at times by the Egyptian upper class for solemn worship. A clerestory allowed light to enter this chamber. In later periods a screen wall (Figure 15–4) was built between the front columns. The wall extended only half the column height to allow light to penetrate to the hieroglyphics on the inner walls and columns. Behind the hypostyle were darkened inner chambers accessible only to the priests (Figure 15–5).

With few exceptions, Egyptian construction was based on the column and lintel. The columns were richly carved and painted; lintels were square, plain, and massive. The entire structure was heavy—much heavier than was actually required for strength. The earliest example of an Egyptian arch dates back to 500

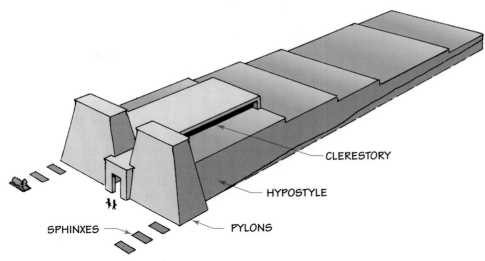

Figure 15–3 Typical Egyptian temple.

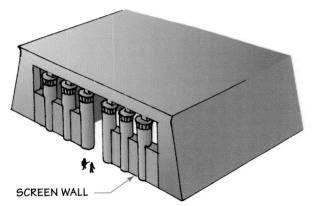

Figure 15–4 A facade of a later period.

B.C. The principle of the 3-4-5 triangle was also known to the Egyptians and used in laying out floor plans. A cord tied in a circle with twelve equally spaced knots was pegged in the earth to produce walls perpendicular to each other. This was a priestly ritual called the *cording of the temple.*

GREEK ARCHITECTURE

The Grecian period (500–100 B.C.) produced some of the most remarkable works of art and architecture known. The Greeks were highly civilized and intelligent, and they built temples of the finest proportion and detail. Their work was influenced by Egyptian, Assyrian, Persian, Phoenician, and Lycian architecture. The

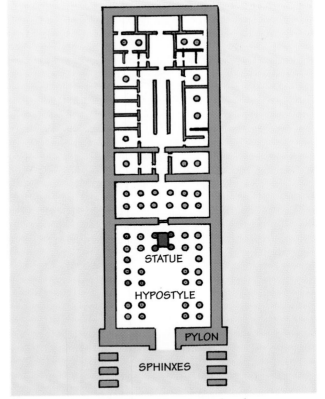

Figure 15–5 An Egyptian temple plan.

Greeks developed an architecture of columned monuments and temples, but the column was used in a manner different from that used in the

Egyptian temples. The Egyptians used columns for internal support, the exteriors being for the most part completely blank walls, whereas the Greeks surrounded their temples with columns, making them the primary elements of exterior design. The number and arrangement of these columns varied greatly, but the plan was the same. The enclosed building itself, called the *cella,* was long, narrow, and windowless. The cella housed the statue of a deity and was open to the public. A small temple might have a single chamber in the cella which would be entered from one end. Columns might be only at that end, or on both ends. A larger temple, such as the Parthenon, might be divided into two chambers and completely surrounded by one or more rows of columns. Interior columns might also be used. Walls and columns were of marble, and the low-pitched roof was wood, covered with marble

or terra-cotta tiles. Color was used on the interior and exterior reliefs, blue and gold being most popular (Figure 15–10). Moldings like the *egg-and-dart* carvings were simple but refined.

There are three styles of Greek columns, each with its own base, shaft, capital, and entablature constituting what is called an *order* (Figure 15–6). These three orders have been called the **Doric**, **Ionic**, and **Corinthian**. The tribes of Dorians and Ionians have given the name to the two styles traditionally Greek, whereas the Romans fully developed the Corinthian order.

The Doric column was four and one-third to seven diameters high with sixteen to twenty elliptical flutes. It had a simple capital, but no base, resting directly on a stepped platform. The finest example of Doric architecture is the Parthenon (Figures 15–7 and 15–8) on the

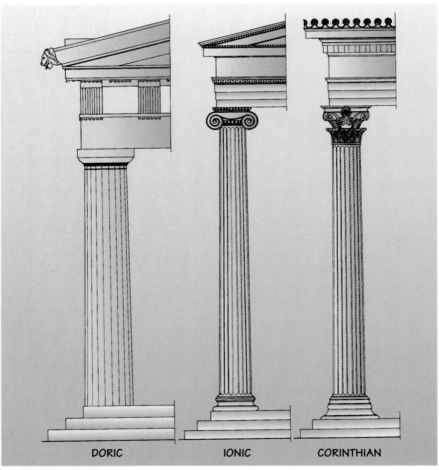

DORIC   IONIC   CORINTHIAN

Figure 15–6 *The Greek orders.*

Figure 15–7   *The Parthenon as it appears today.* (Photographer: Leslie P. Greenhill.)

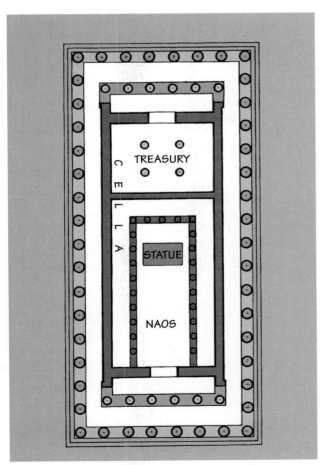

Figure 15–8   *A plan view of the Parthenon at Athens.*

Acropolis of Athens. It was built in 440 B.C. to enshrine the goddess Athena Parthenos and to commemorate the victory of the Athenians over the Persians. The marble blocks were cut to fit together with amazing accuracy and laid without mortar. The details were executed with exquisite refinement, giving rise to the comment that the Greeks "built like Titans and finished like jewelers." Many lines on the Parthenon that look perfectly straight were actually slightly curved to counteract the optical illusion of curvature. Columns have a slight convex curve (only 1" in their 32' height) so that they do not appear concave. Also, the axes of the columns lean a bit inward to prevent a top-heavy look. The steps and eaves curve gently down at the ends—also for visual effect. This can be explained by referring to Figure 15–9. Notice that the horizontal line, which is drawn straight, appears to sag because of the effect of the inclined lines above it. The Greeks took even such details into consideration.

The Ionic column was slender in comparison, being eight to ten diameters in height with

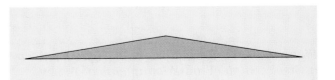

Figure 15–9   *An optical illusion.*

twenty-four flutes cut deeper than in the Doric. It stood on a molded base and was finished by a capital with volutes. The only defect with this capital was that it did not readily lend itself to use as a corner column. To correct this problem, the corner volute was set at 45 degrees so that it could be faced on both sides. The gables on Ionic buildings were frequently sculptured in relief, as were the Doric. Two examples of Ionic buildings are outstanding. One is the Erechtheion, also on the Acropolis, which is noted for its unsymmetrical plan. It enshrined many deities and heroes. Perhaps the best known feature of this temple is the Caryatid Porch, which has robed female figures (caryatids) for columns. An Ionic tomb, the Mausoleum at Halicarnassus, was built by the widow of King Mausolus in 354 B.C. It had a pyramidal roof reminiscent of Egyptian architecture.

The Corinthian order was similar to the Ionic except that the capitals had smaller volutes entwined with rows of acanthus leaves.

The most celebrated example of Greek Corinthian is the circular Monument of Lysicrates in Athens.

ROMAN ARCHITECTURE

Roman architecture was influenced by the Dorians, Phoenicians, Etruscans, and, most of all, the Greeks. The Romans never reached the Greek perfection of design, but they extended the services of architecture to theaters, baths (Figure 15–11), basilicas, bridges, aqueducts, and monuments (Figure 15–12)—all well-engineered structures.

The three Greek orders were adopted as a direct consequence of the Roman conquest of Greece. Not only did the Romans bring home artistic wealth as part of the spoils of war, but they also brought home the desire to make Rome as magnificent as the Greek cities they had destroyed. The Greek orders were modified into what are called the five Roman orders: Tuscan, Doric, Ionic, Corinthian, and Composite. The Tuscan was a simplified Etruscan Doric, with a column seven diameters high. The Doric was similar to the Greek Doric, with a column eight diameters high and often devoid of any fluting. The Ionic was almost identical to the Greek Ionic, with a column nine diameters high. The Corinthian had a column ten diameters high. As previously stated, this

Figure 15–10 *Typical original bright colors on a Greek temple.* From *Monuments of Art* by Lübke.

Figure 15–11 *Roman baths of Dioclesian.* From *Remains of Ancient Buildings in Rome* by DuBourg.

Figure 15–12   Arch of Constantine. *From Remains of Ancient Buildings in Rome* by DuBourg.

order was perfected by the Romans and became the most characteristic order, probably because its ornateness appealed to the Romans. The Composite can be described as a Corinthian column with the volutes enlarged, giving it an Ionic look.

The Romans established sets of rules to simplify the orders which, many feel, were to their detriment. For example, fine elliptical moldings and flutes were cut circular to simplify the construction. However, their tremendous output of architecture probably could not have been accomplished if this kind of standardization had not been introduced. The columns, incidentally, were often cut in one piece (called *monolith*), instead of being built up from smaller sections, as were the Greek columns.

The Romans are credited with further development of the vault and arch. The Etruscans first used vaults; the vaulted great sewer of Rome (500 B.C.) still remains. Roman vaults are basically of three types: barrel, groined, and dome. The barrel vault is a semicylinder, the groined vault is formed from two intersecting barrel vaults, and the dome is a hemisphere supported on a wall of circular plan. The hemisphere supported on a wall of square plan was later developed as the typical Byzantine dome. The vaults were cast over a wooden form, unlike the inclined vault, which required no forms. When hardened, the vault formed a cohesive

curved lintel which did not extend the outward thrust of the stone block construction.

The Romans outdid themselves in the use of the arch, combining it with the column and lintel. This combination has been called the *arch order* (Figure 15–13) and is a characteristic feature of Roman architecture. It is also the poorest feature, because of its inherent inconsistency; either the arch or the lintel has to be redundant. The Romans considered ornateness above function, however, and this sham did not bother them. Lintels were even scored with false joints to make them look like smaller blocks. Much Roman work, especially theaters, used the arch order. False columns were applied to the exterior, a different order for each level.

Many Roman public buildings were built around the basilica located in the Forum. The Forum was to Rome what the Acropolis was to Athens, the basilica being a meeting place and courthouse. The buildings were large and partially or completely roofed, the higher center portion affording a clerestory. Constantine built the first vaulted basilica; previous ones had wooden roofs.

Among the most remarkable Roman structures is the Colosseum in Rome (Figure 15–14), which seated eighty thousand. Erected to house the bloody battles between men and beasts, it could even be flooded to create sea fights. Titus finished the Colosseum in A.D. 80. The plan was a 600' × 500' ellipse; the walls were 153' high arranged in four levels. The three lower levels

Figure 15–13   The arch order.

Figure 15–14 The Colosseum. From *Remains of Ancient Buildings in Rome* by DuBourg.

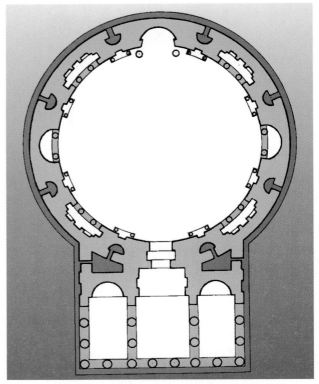

Figure 15–15 A plan view of the Pantheon at Rome.

Figure 15–16 The interior of the Pantheon.
(Photo: Alinari, Florence, Italy.)

were based on the arch orders, using Doric half-columns at the lower level, then Ionic, and then Corinthian. The upper level was composed of Corinthian columns with windows between them. A huge silk awning was stretched from this height over the galleries. The arena itself was un-covered. Ramps led to subterranean vaults which were used to house animals and machinery.

The Circus Maximus outdid the Colosseum in that it measured 400' × 2,100' and seated 260,000. It was used for chariot races.

The Pantheon of Rome (Figures 15–15, 15–16, and 15–17), built in A.D. 138, has been described as the noblest of all circular temples. It is roofed with a dome that admits light through a round opening at the top called the "eye." Rich paneling and statues cover the interior walls.

Less ornamentation was lavished on the Ro-man aqueducts and bridges; instead, the arch was used in its pure form. As a consequence, these are perhaps the Romans' best works. They were built in layers of arches, the greater arches at the bottom. The smaller, topmost arches were adjusted to the exact height required.

Due to Vesuvius's eruption in A.D. 79, we have a good picture of the Roman house—at least the house in the provinces. A small vestibule led to the atrium, open to the sky and containing a pool to catch rainwater. A second vestibule led from the atrium to the inner court, the peristyle. The peristyle (Figure 15–18) was also open, but was a bit larger, with a garden and statues. Both courts, the front living room and the rear family room, were surrounded by smaller chambers, in a manner similar to the Assyrian plan.

Figure 15–17   Exterior of the Pantheon. From *Remains of Ancient Buildings in Rome* by DuBourg.

Figure 15–18   The peristyle of a house at Pompeii. (Photographer: Leslie P. Greenhill.)

GOTHIC ARCHITECTURE

The **Gothic** style developed in Western Europe between 1150 and 1450. The Gothic was an architecture of cathedrals, attacking some of the structural problems of vaulting in new ways. First, the groined vaults were supported entirely by comparatively few piers. The wall between the piers was filled in with marvelous pointed arch windows of stained glass and tracery. Second, the outward thrust of the vaults was countered by an exterior combination of arches and buttresses called *flying buttresses.* These flying buttresses were capped with gables or small steeplelike pinnacles. The third distinguishing Gothic feature is the pointed arch, which was first developed to facilitate the construction of groined vaults on a rectangular plan. The stone vaulted roofs of the cathedrals were covered with a wooden gable roof to protect them from the elements. There are innumerable fine examples of Gothic architecture in use. Outstanding are the Cathedral of Notre Dame at Paris (Figure 15–19), Cologne Cathedral, the Cathedral of Seville, and Westminster Abbey (Figure 15–20).

RENAISSANCE ARCHITECTURE

The decline of feudalism and a new freedom of expression in art, literature, and architecture characterized the period known as the **Renaissance** (1420–1700). The classic civilizations were rediscovered, studied, and imitated. The objection

Figure 15–19   Notre Dame Cathedral, Paris. (Photo: Alinari, Florence, Italy.)

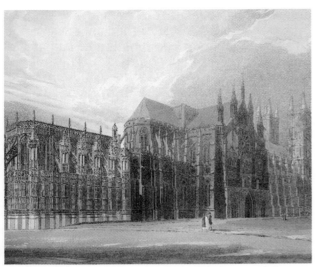

Figure 15–20 Westminster Abbey, London. From *Remains of Ancient Buildings in Rome* by DuBourg.

to the architecture of the Renaissance is that the structural framework was hidden beneath a false classical facade. Also, the classical orders were applied in many instances where they did not fit. Knowledge of the Roman orders was the chief stock-in-trade of the architect of this period.

The first example of this new style was the Cathedral of Florence (1420). The Pantheon in Paris is noted for its strict adherence to classical detail. The largest church in existence today is a Renaissance church: St. Peter's in Rome is 600' long and 405' high. Other notable examples are the Louvre in Paris (Figure 15–21), St. Paul's Cathedral in London, and the Doge's Palace in Venice.

MECHANICAL SYSTEMS

Although the history of architecture is closely allied with that of structural systems, the "me-

REAL WORLD  APPLICATIONS

Historic Landmarks

 Lyndhurst at Tarrytown, N.Y., financier Jay Gould's former home, is not only one of the nation's premier examples of domestic Gothic Revival architecture, but also a house listed in the National Register of Historic Places. The National Register, established by the National Historic Preservation Act of 1966, is a list of buildings, structures, sites, districts, and objects that are considered prominent in American history, architecture, archaeology, and culture. It also includes all national historic landmarks designated under the Historic Sites Act of 1935. The number of preserved buildings and structures has multiplied. Less than two thousand entries were listed in the first issue of the National Register, published in 1969, but by 1989 this number had risen to fifty thousand.

A relatively new strategy that has become increasingly important to historic preservation due to its economic implications is adaptive reuse. Through adaptive reuse strategies, the physical appearance of historic buildings and structures is maintained or restored but they are adapted to fulfill modern needs. Did you ever dine in an old schoolhouse, shop in an old railway station, or visit an office in a former shopping district? If so, you have experienced adaptive reuse.

Locate three buildings in your community that are listed in the National Register. How old are they? What was their original purpose? How has it changed? Are they examples of adaptive reuse strategies? Could you envision ways to adapt them for a different purpose without modifying their physical appearance? Structurally, do you feel they can survive another hundred years?

Figure 15–21   The Louvre, Paris. *(Photo: Alinari, Florence, Italy.)*

chanical" systems (heating, lighting, and sanitation) are interesting studies each in their own right.

Heating

The earliest heating systems were, of course, open wood fires. Holes in the roofs of caves and tents, used to remove smoke, were later replaced by chimneys in buildings. Early buildings in cold climates had a single fireplace or stove for heating and cooking; fireplaces were built of stone or brick with wood chimneys. The first stoves were of brick or tile and later cast iron, and fuel was wood or dried peat. Houses of several rooms had a fireplace or stove located in the kitchen, and hot coals in warming pans were used for sleeping on cold nights. The first built-in house stoves had a stove door on the outside of the building with the remaining three sides within the house, sometimes heating two stories. In 1744 Benjamin Franklin invented an improved iron stove (the Franklin stove) that wasted less heat because the hearth was extended into the room and the stove was connected to the chimney by a funnel-shaped transitional duct.

Central heating systems became common during the nineteenth century, but local heating was still used in many locations. Fuel was coal, oil, or gas with heat distributed by warm air, hot water, or steam. Electric heating and active solar heating developed during the twentieth century.

Lighting

Ancient Egyptians, Greeks, and Romans obtained light by using pottery or bronze dish lamps: wicks were set inside dishes filled with grease or olive oil. Candles made from fats or cooking greases were used for light for many centuries, but in the eighteenth century it was discovered that a flame burns more brightly and with less smoke when it is in a glass chimney. This discovery, combined with increased activity in whaling and the availability of whale oil, created the popular oil lamp. A distilling process to refine oil from coal was discovered by a Canadian geologist, Abraham Gesner, in 1854. This new substance was called coal oil or kerosene and was used both as a lamp fuel and a heating fuel. At about the same time, both natural gas and manufactured gas started to be used for lighting, heating, and cooking.

Natural gas was discovered several thousand years ago and used in temples in China, Japan, India, Russia, and Greece. Ancient Greeks built the temple of Apollo at Delphi over a natural gas crevice. An oracle breathed the fumes (thought to be the breath of Apollo) and predicted future happenings. "Eternal" flames in temples were often fueled by natural gas. In the early seventeenth century it was discovered that gas could be manufactured from coal, but manufactured gas was not used for lighting until the early nineteenth century. The first gas lamp posts were installed along the Pall Mall in London in 1807. The first U.S. city to use manufactured gas street lamps was Baltimore in 1816. The first city to use natural gas street lamps was Fredonia, NY, eight years later. Gas ranges for cooking began being manufactured in the late 1800s. At the same time electricity began to replace gas for lighting, and the electrical convenience outlet made available a wide range of appliances for work and leisure.

Sanitation

Early methods of sanitation usually consisted of the disposal of waste in fields and street gutters, but improved sanitary systems were also designed and built. The Romans built aqueducts of stone, brick, tile, or lead to carry fresh water to buildings and to flush the underground sewers that ran from the buildings to the Tiber River. Ancient cities in India contained paved bathrooms and drains of terra cotta pipes, and other waste was discarded through wall chutes to waste bins in the street, but there was little progress in sanitation during the Dark Ages. The first modern valve toilet (still called a "water closet") was invented in 1596 by the Englishman John Harington, but indoor plumbing was only slowly accepted. Piping began to be installed inside walls only in the early 1900s. Hand-cranked wells and separate outhouses are still in use today.

DOMESTIC ARCHITECTURE

Throughout the previous discussion, private dwellings have not often been used as examples. The reason is that only monumental buildings were of sufficiently durable construction to survive time and the elements. However, we do have examples of comparatively recent private dwellings in all countries. For simplicity, let's look only at North American architecture.

The first shelters of the English colonists in the New World (Plymouth and Jamestown) were lean-tos called *English wigwams.* They were not like Indian tents, however, but were framed with poles and covered with woven twigs called *wattle,* brush, and mud or clay. Sod huts partially buried in the ground with staked walls were also built.

The log cabin was introduced by the Swedes in Delaware. This was the type of dwelling the poorer classes had used in the mother country and was unknown to the Native Americans. Log construction was adopted by other colonists for use in stockades and prisons. In fact, *log house* meant, at first, a jail. Later, the log house and overhanging blockhouse moved west with the frontier.

It should be noted here that traditional North American architecture has its roots in the architecture of the mother countries, particularly England. There was no Native American influence, nor was it the intention of the colonists to found a new architectural style. All they desired was to build a civilization similar to the one they had left. However, available materials and the North American climate soon

dictated certain modifications of the European methods.

Around 1600, the typical English middle-class house was of half-timbered construction. That is, a heavy wooden frame was pegged together and filled in with clay-covered wattle or rolls of clay and straw called *cat and clay.* In the severe New England climate, however, this construction had to be covered with a horizontal sheathing called *weatherboards* (also known in England). Due to the abundance of timber, the clay or brick filling was gradually omitted, being replaced by the hollow frame construction known today. Of course in areas where there was a supply of good building stone (Pennsylvania), or clay for brick making (Maryland), these materials were used.

For the roof covering, thatch was replaced by shingles split from the forests of cedar. In fact, the North Americans abandoned thatched roofs almost two hundred years before the English.

Classic Revival

During the first half of the nineteenth century, the **Classic Revival** held sway. Even the United States turned from its wooden version of the English Georgian (called *Colonial*) to Classic Revival in stone and concrete. Examples of Classic Revival in the United States are the White House (Figure 15–22), the national

Figure 15–22 A view of the south side of the White House (executive mansion), Washington, D.C. (Courtesy Ewing Galloway.)

Figure 15–23 The Senate wing of the Capitol Building, Washington, D.C. (Courtesy Ewing Galloway.)

Capitol (Figure 15–23), several state capitols, and many churches. This artificial return to the past slowed down any real architectural progress.

Eclecticism

The Classic Revival was shortly replaced by **Eclecticism**, a conglomeration of various historical styles, any number of which might be used in a single structure. This searching for ancient methods of enclosing modern structures brought architectural progress to a halt. At first, even the advent of structural steel framing did not change the Egyptian, Classical, and Renaissance coverings. Only when the simple structural lines of the American skyscraper were left undisguised did architecture move on. Today, Eclecticism is synonymous with all that is false in architecture.

International Style

Although the United States led the way in nontraditional design for commercial building, Europe quickly applied this new *functionalism* (the doctrine that a feature should have a *function* and should not be applied for decoration only) to private dwellings. Led by Gropius of Germany

ARCHINET http://www.

Society of Architectural Historians

The study of architecture has been around for thousands of years. The Society of Architectural Historians (SAH) was established to encourage preservation of buildings and the historical accuracy of architecture. The SAH has developed an Internet home page to promote their ideals and organization. The address is: **http://www.upenn.edu/sah**

After accessing this home page, click on Buildings of the United States, which is listed under Home Page Projects. Four volumes have been issued in this series. Are any of these housed in your school or local public library? If they are, write a one-page description about one building of interest to you.

In addition to the home page, the SAH issues the *Journal of the Society of Architectural Historians (JSAH)*. From the home page click on the *JSAH*. There you will find a listing of past journal articles. This shows how the Internet can be used as a resource for bibliographical searches. Choose one interesting article and read it in your school or public library. Make a photocopy of it to share with your class.

and Le Corbusier of France, a type of architecture called the **International Style** developed which was characterized by simple, blocklike exteriors, concrete walls with no roof overhang, and windows located at the corners rather than the sides of walls. Unfortunately, the skillful block proportions used to advantage on North American commercial structures, culminating in buildings like the Empire State Building (1931) and Rockefeller Center (1930–1950), gave these smaller buildings a boxy, factorylike appearance. When the International Style of domestic architecture spread to England and then to the United States, it was not readily accepted. The lack of eaves, the corner windows, and other features of the style were rejected as not being truly functional, but rather false, applied styling. However, the major premise of the International Style, that of planning the exterior around the interior, even though revolutionary, was quickly adopted. This idea of "form follows function" is the guiding spirit of architecture today—made possible, in reality, by the free use of a combination of old and new building methods.

With this general background of the development of domestic architecture in North America, you may wish to review Chapter 1, Architectural Styling, to obtain a more complete view of the effect of architectural history upon domestic architectural styling.

REVIEW QUESTIONS

1. What are the five structural systems? Give the civilization responsible for introducing each system.

2. Egyptian architecture is best known for what type of structures?

3. What is a mastaba?

4. Differentiate between columns and lintels. Which architectural era based its structures on columns and lintels?

5. List the three styles of Greek columns.

6. Describe the three Roman-style vaults.

7. The Gothic style focused on what type of structures?

8. What are three predominant mechanical systems in architecture?

9. Where, and by whom, did log cabins begin?

10. What is significant about an eclectic style of architecture?

ENRICHMENT QUESTIONS

1. Prepare an outline naming the major civilizations responsible for architectural development.

   a. Include the approximate dates of each.

   b. Indicate the structural systems used.

   c. Give examples of well-known buildings in each period.

2. Discuss the contribution to architectural development made by the following civilizations. Give the methods of construction of each and examples.

    a. Egyptian   c. Roman

    b. Greek   d. Gothic

3. Develop a historical time line of the major architectural eras including pictures and text.

4. Trace the development of a mechanical system from its inception to today's modern systems.

PRACTICAL EXERCISES

1. Draw or computer-produce a chart showing the five fundamental structural systems.

2. Build a model of each of the five fundamental structural systems.

3. Prepare a hand-drawn or computer-produced chart of the major civilizations responsible for architectural development.

4. Build a model of a typical residence of one or more of the following types.

    a. Egyptian

    b. Greek

    c. Roman

16

Primary Considerations

OBJECTIVES

By completing this chapter you will be able to:

- 🏛 Identify the primary considerations for a residence.
- 🏛 Describe the importance of considering an appropriate site for a residence.
- 🏛 Compare and contrast various facets of the basic structure of a residence.
- 🏛 Discuss the purpose and advantage of following building codes and zoning ordinances.

KEY TERMS

Building codes
Americans with Disabilities Act
Zoning ordinances

Talk to persons who have recently designed a home. Ask them if they would do it again, and you'll get a variety of replies. Some will tell you that this was a most satisfying and rewarding experience; others will say it was entirely frustrating. After further questioning, you'll find that those who were most successful have an important characteristic: *sensitivity.* The successful designer has the ability to be sensitive to many factors. Some of these factors are aesthetic (such as feeling the potentials of the site), some are empathic (such as understanding the desires of the client), and some are practical (such as keeping within the budget). The designer must be sensitive to many such factors throughout the entire design process. But some factors should be considered before starting to sketch a preliminary layout (Chapter 23) or even listing program requirements (Chapter 18). These are called *primary considerations* and are discussed in this chapter:

1. The site
2. Architectural styling
3. Basic structure
   Energy system
   Size
   Number of floors
   Shape of plan
   Foundation
   Roof
   Expansion
4. Building codes and zoning
5. Cost

THE SITE

To the sensitive designer, each site has an individual character which suggests the most appropriate structure and style of home. The contour of the land, shape of the plot, kind of trees, surrounding views, and type of community all give broad hints toward a satisfactory solution. For example, a *sloping* contour suggests a split-level or modern house. A *narrow* plot restricts the choice of styles and requires special planning to prevent a cramped appearance. A *wide* plot offers more freedom of planning, permitting a low, rambling design or several connected modules. Try to keep excavation to a minimum, for the less the natural contour of the site is disturbed, the less nature will try to disturb the occupants of that site (Figure 16–2).

Take advantage of existing trees, especially if they are large; it will take many years for newly planted trees to mature. Evergreen trees to the windward are excellent natural windbreaks, and deciduous trees located on the southern side of the site are natural "automatic" solar screens. They shade a building from the undesirable summer sun (thereby reducing the interior temperature by about 10°) and, when they shed their leaves in the winter, permit the desirable winter sunlight to filter through.

It has always been an architectural goal to provide a building that shields the occupant from a sometimes hostile environment. In so doing, however, the occupant does not wish to lose contact with the friendly parts of the environment. Thus exterior living space adjacent to interior living space should be provided (Figure 16–1). Also, glass window walls can face the best views and unpleasant views or noisy streets can be shielded. Be sensitive to the nature of the surrounding community and avoid building in a style that is incompatible with nearby architectural styles.

In this book, a house will be designed for Family A from beginning to end. They are not necessarily an average family, and you do not need to consider their home a model for your future designs.

EXAMPLE
The plot belonging to Family A is narrow and relatively flat with an excellent view to the rear. The views to the front and left are undesirable.
Decision: Clerestory windows or atrium.

ARCHITECTURAL STYLING

As indicated in Chapter 1, the architectural styling should be considered before preliminary sketches are drawn. Although many factors influence the choice of style, the owner's personal preference will be the outstanding consideration (Figure 16–3).

EXAMPLE
The need to eliminate all front windows prevents traditional styling.
Decision: Contemporary.

BASIC STRUCTURE

The nature of the site and the architectural styling desired provide important clues to the type of basic structure required. Now is the time to make tentative decisions on the size

Figure 16–1   A sloping site suggested this terraced decking. (Courtesy California Redwood Association.)

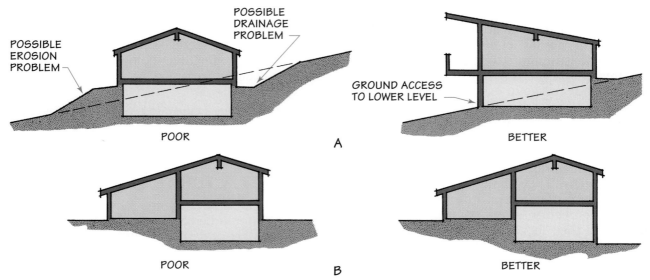

Figure 16–2 A. Choose an appropriate structure for the site. B. Choose an appropriate site for the structure.

REAL WORLD APPLICATIONS

Mean, Median, and Mode

A contractor has just purchased 12 acres of farmland for the development of a housing community. The price and size (square footage) of the houses should be typical for the general area. You decide to investigate fourteen nearby houses that have sold in the past two years. The information in the following table represents what you found.

You decide to calculate some statistics on this information. Statistics is a branch of mathematics used to understand quantities of numbers. The term average is used to describe a single number that represents a set of numbers (often called a data set). There are three types of averages used in statistics: mean, median, and mode. Determining which of these numbers is most useful depends on the type of data, and may be a matter of opinion.

Sample Housing Cost and Square Footage

House #	Cost	Sq Footage
1	$ 85,500	1,200
2	85,500	1,500
3	99,900	1,650
4	104,900	1,800
5	111,500	2,000
6	118,000	2,000
7	122,000	2,000
8	139,000	2,250
9	152,900	2,500
10	167,000	2,600
11	179,500	2,850
12	194,900	3,000
13	443,900	3,500
14	650,000	4,000

Figure 16–3 *New England colonials are a popular style choice.*

and type of structure, number of floors, and general plan shape. However, if an "alternate" energy system (such as solar heating) is to be specified, the requirements of the system may affect any of these decisions.

Energy System

A "conventional" heating system, such as an oil furnace or electric panels, will have little effect on the design of a building, but an alternate energy system often determines many of the design elements. For example, a passive solar heating system requires southern orientation to gain solar exposure (Figure 16–4). An active solar heating system having flat-plate roof panels will need a roof that is sized in multiples of the dimensions of the panels; this then determines the dimensions of the floor plan as well.

EXAMPLE
The plot belonging to Family A has an adjacent plot to the south which contains a commercial structure.
Decision: Insufficient southern exposure is available for a passive solar heating system of wall panels, but an active solar heating system would be possible if the commercial

The *mean* is the most commonly used average. It is calculated by finding the sum of all the values and then dividing this sum by the number of values. The formula for calculating the mean cost of these fourteen houses is as follows (C1 = cost of house number 1):

$$mean = \frac{C1+C2+C3+C4+C5+C6+C7+C8+C9+C10+C11+C12+C13+C14}{14}$$

The *median* is the value that divides the data set exactly in half. It is found by listing all the numbers of the set in ascending or descending order and then locating the middle number. When the number of values is even, the median is the mean of the two central numbers. The housing costs and square footage in this example are listed in ascending order. There are fourteen houses, so the median square footage can be found by performing the following calculation (S1 = square footage of house number 1):

$$median = \frac{S7 + S8}{2}$$

The *mode* is simply the most frequent number in a data set. A data set may have two modes or it may have no mode at all.

The modal housing cost and square footage for this example can be found by finding the number(s) that appears most often in each set.

Two other important terms to know when examining a data set are *outlier* and *range*. Values that are much smaller or much larger than other values in a data set are called *outliers*. They are extremes of the data set. The *range* is the difference between the smallest and largest numbers in a data set. The range of the housing costs in this example can be found using the following formula:

$$range = C1 - C14$$

1. Calculate the mean, median, mode, and range of the cost and square footage of the fourteen houses shown in the table.

2. Which of these would you use in determining the cost of the house you will build? Why?

3. Which of these would you use in determining the square footage of the house you will build? Why?

4. Are there any outliers in this data set? If so, how do these affect the mean, median, and mode?

Figure 16–4   A south-facing wall and roof windows create a passive solar house. (Courtesy VELUX-AMERICA INC.)

Figure 16–5   Even bathrooms should be carefully planned. (Courtesy Wood-Mode Cabinetry.)

structure were not too high to block southern exposure to solar collector roof panels. Or separate solar collectors could be installed in the rear yard beyond the shading of the commercial structure.

Size

The number of rooms and their special requirements are primary considerations since they determine the size of the structure. Usually a balance must be reached between the number of rooms desired and the number that can be afforded.

Careful thought should be given to the number of bedrooms. A common error on the part of young couples with no children or a single child is building a one- or two-bedroom house with the thought that they can sell or add on if another bedroom is needed. A house with fewer than three bedrooms is difficult to sell, and actually adding a room (not finishing an existing one) is more costly than including it in the orig-

inal plans, not to mention the sacrifice in exterior appearance.

Even minor rooms like bathrooms should be carefully planned to answer questions of: How many? What size? Simple or with contemporary sectioned areas? A powder room near the front entrance? These questions, and many others, must be answered before the plans are begun (Figure 16–5).

EXAMPLES

Two children, ages three and nine, are in Family A.
Decision: Three bedrooms: a master bedroom with shower and two additional bedrooms served by one main bathroom.

The A family does much entertaining (parties of four to fifty).
Decision: Provision for living area overflow to family area, basement recreation room, and patio.

Mr. A has built a 12' power boat which he stores on the property.
Decision: Oversize or double garage, provision for future shop.

Mrs. A desires more storage space than she has had in previous kitchens.
Decision: Utility room with shelves adjacent to kitchen.

Number of Floors

The number of floors is a basic decision (Figures 16–6, 16–8, 16–9, and 16–10). Should there be two floors, one and a half floors, one floor, or one of the combinations afforded by split-level designs?

The two-floor home, with sleeping areas on one floor and living-dining-cooking areas on another, usually provides more "gracious living." Areas are definitely separated, making it easier to clean (or not to clean), to entertain, and to get all the privacy desired. A special effort is needed to relieve the vertical appearance (Figure 16–7).

The one-and-a-half-story house is an attempt to capture the advantages of the two-story house and to reduce the cost by reducing the outside wall area. (Of course, portions of the bedrooms must have sloping ceilings.) Remember that the use of too many dormers will reduce any cost advantage.

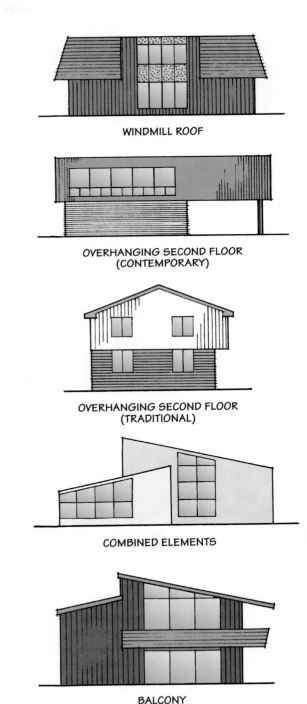

Figure 16–7 *Methods of relieving vertical appearance.*

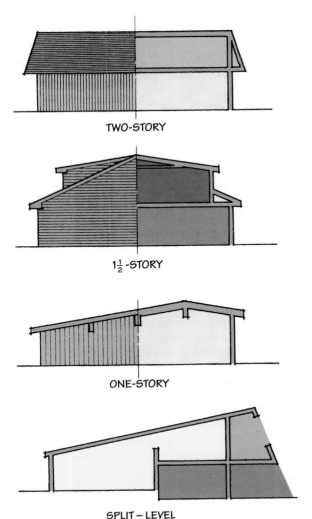

Figure 16–6 Number of floors.

Figure 16–8   A one-story residence. (See Chapter 24 for plans.) (Courtesy Mr. and Mrs. Donald W. Hamer.)

Figure 16–9   A two-story choice.

Figure 16–10   A traditional split-level design.

For those who dislike stair climbing and want a low-cost home (two bedrooms, no dining room), the one-floor house is the answer. Remember, however, that a moderately large

number of rooms (six or seven) will cost more in a one-floor plan since larger roof and foundation areas are required. In general—one floor for four or five rooms, two floors for six or more. Living room, dining room, kitchen, bedroom, den, playroom, and enclosed heated porch are each considered one room. Attached rooms such as dining ell or kitchenette are considered as half-rooms. Bath, porch, breezeway, basement, and attic are not considered rooms.

EXAMPLE

The A residence will require five and one-half rooms; an atrium is desired.
Decision: Group rooms to create atrium; all rooms on one floor for more light into atrium.

Shape of Plan

Since the shape of the floor plan affects house styling and cost, it should be given special consideration (Figure 16–11). In general, it is best to create a plan that is irregular enough to be interesting and to supply the needed wall area for lighting all rooms. But keep in mind that each extraneous corner adds to the total cost. If low cost is a primary objective, a square plan is the best choice since a square encloses a given volume with a minimum of wall area. In practice, a rectangular plan is more often used because of its appearance and planning advantages. The addition of ells (L-shaped, T-shaped, or C-shaped) is costly and should be weighed against the advantages.

EXAMPLE

To illustrate the effect of the number of floors and plan shape on the total house cost, compare the two houses in Figure 16–12. Each has identical floor area, the differences being the plan shape and the number of floors.

House A is a two-floor house with 30' × 30' = 900 sq. ft. of floor area on each floor, giving a total of 1,800 sq. ft. Its plan is square. House B is a one-floor house with 60' × 30' = 1,800 sq. ft. of floor area. Its plan is rectangular. The total area of walls, roof, and slab floor needed to enclose house A is 3,871 sq. ft. The total area needed to enclose house B is 5,583 sq ft. (See Table 16–1.) Although there are some discrepancies in any comparison such as this (house A must have an additional interior floor and heavier foundation), the increase in outside area needed to enclose this one-floor rectangular house over the two-floor square house is:

$$\frac{5,583 - 3,871}{3,871} \times 100 = 44\%$$

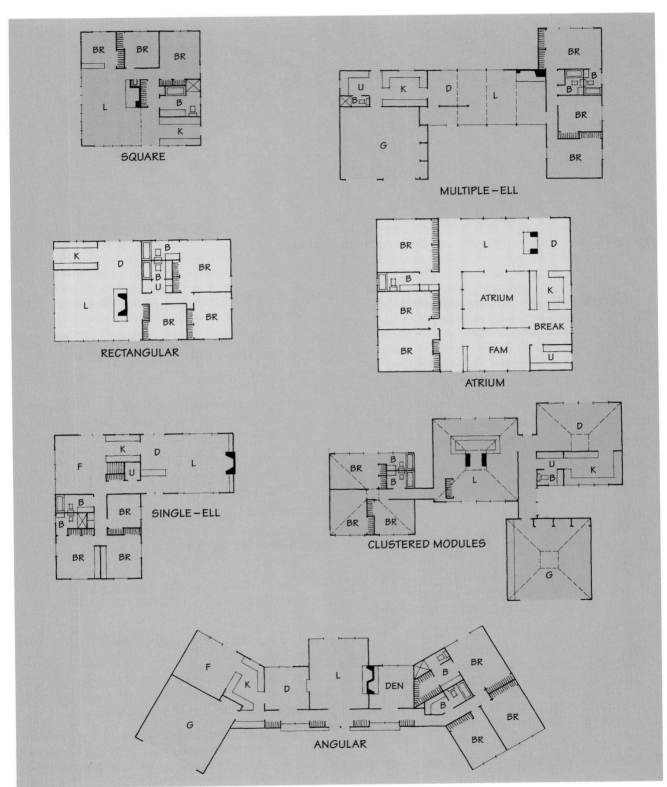

Figure 16–11 Plan shapes.

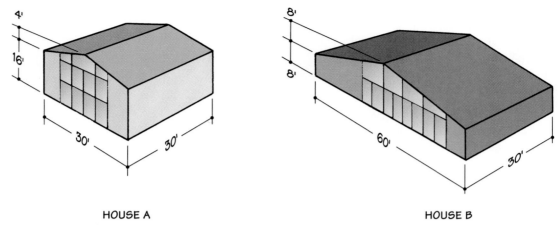

HOUSE A   HOUSE B

Figure 16–12   Identical floor area, different plan shapes.

	House A	House B
TABLE 16–1 *A Comparison of House A and House B*		
Front and rear elevations	1,080 sq. ft.	1,440 sq. ft.
End elevations	960 sq. ft.	480 sq. ft.
Roof	931 sq. ft.	1,863 sq. ft.
Slab	900 sq. ft.	1,800 sq. ft.
Total	3,871 sq. ft.	5,583 sq. ft.

Foundation

There is continuing controversy over whether or not to provide a basement in new house construction. The advantages of a basementless house are:

1. Room for the furnace, water heater, laundry, workshop, storage area, and playroom is more convenient if located on the first floor.

2. A slight saving may be realized due to the omission of floor joists and basement stairway.

3. The house can be kept just as warm and dry as if a basement were built.

4. This is the ideal construction if a ledge is encountered on the land.

5. This may be preferred in areas having a high water table (water level in the ground).

The advantages of a basement are:

1. The total cubage enclosed by a basement cannot be provided above ground at comparable cost.

2. Due to drainage problems, a basement should be provided when building on a steeply sloping plot.

3. There is little or no saving in omitting the basement from a house built in a cold climate since footings must go 4'-6" or more in depth.

4. There is little or no saving in omitting a basement from a house of two or more floors.

5. In some localities, zoning laws prohibit basementless construction.

6. Prevailing public opinion is in favor of basements. For this reason, banks and loan agencies will favor a house with a basement.

As you can see, there is no one simple answer for all conditions, because plot contour, climate, house size, zoning, and cost must be considered. The major advantage of a basementless house is elimination of stair climbing. Rooms above ground level are warmer and lighter, but unfortunately houses without basements are seldom designed with suitable space for all the functions a basement provides.

Studies have been conducted to determine actual savings in eliminating the basement. It has been found that slab construction reduces the cost of the foundation by 25 percent. But if space is added at the first-floor level for the functions of a basement, no saving is realized since additional foundation walls and roof must be provided. A large, rambling, one-floor house will produce the greatest saving if the basement is omitted. This is not true of a house with two or more floors since it requires a more substantial foundation and because in such a house, the basement takes up a smaller proportion of the total house cubage.

In your decision, various combinations should also be considered. Today many houses are built with a basement under one portion and slab construction or a crawl space under another. The split-level house is a good example of this solution. It may have the garage under the sleeping area and the basement under the living-eating area.

EXAMPLE

The A residence will include an emergency shelter and provision for a future recreation room. Also, Mr. A wants a shop for his hobby of woodworking.
Decision: Full or partial basement.

Roof

The choice of a pitched or flat roof is somewhat dependent upon house styling (Figures 16–13 through 16–15). Although the final choice will probably depend upon personal taste, both possibilities have advantages.

The advantages of a flat roof are:

1. Storage space such as is provided by an attic is more convenient if located on the ground floor.

2. A saving can be realized due to the reduction of framing and omission of the stairway.

3. A complicated floor plan is more easily covered with a flat roof.

4. A properly installed flat roof gives satisfactory service. Snow can accumulate on the roof, serving as insulation in the winter, and a *water-film roof* serves as insulation (reflector) against summer heat.

The advantages of a pitched roof are:

1. A pitched roof with attic provides cheap storage, play, and expansion area. A pitched roof with sloping ceilings provides a major design feature.

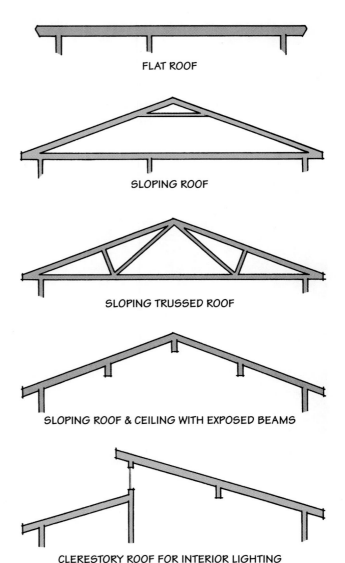

FLAT ROOF

SLOPING ROOF

SLOPING TRUSSED ROOF

SLOPING ROOF & CEILING WITH EXPOSED BEAMS

CLERESTORY ROOF FOR INTERIOR LIGHTING

Figure 16–13   Common roof types.

Figure 16–14   A basic decision: flat roof? *(Courtesy California Redwood Association.)*

Figure 16–15   Or pitched roof? *(Courtesy American Plywood Association.)*

2. Certain architectural styles demand the use of a pitched roof.

3. A pitched roof is more likely to be leak-free.

4. In some localities, zoning laws prohibit flat roof construction.

5. Prevailing public opinion is in favor of a pitched roof.

Again, there is no one answer, and many factors must influence the decision. However, if low cost is a primary objective, the flat roof will offer a saving. A flat roof will, of course, restrict the style of the house. As far as function is concerned, flat roofs have been used satisfactorily on urban buildings for centuries.

EXAMPLE

A cathedral ceiling is desired in the living room. Undesirable views suggest windows above eye level to permit light without view.
Decision: Gable or clerestory roof.

Expansion

If low original cost is an object, a house can be very carefully designed so that a room or garage can be added later without a loss to the exterior appearance either before or after the addition (Figure 16–16). The simplest forms of expansion are the expansion attic and unfinished basement. The expansion attic will cost the owner only as much as a shed dormer, an unfinished stairway, and some electrical and heating extensions, certainly an economical provision for several additional rooms.

EXAMPLE

Mr. and Mrs. A would like to provide for the possibility of a third child, increased recreational area, and a woodworking shop.
Decision: (1) One child's bedroom to be oversized for possible future use by two children. (2) A full (or partial)

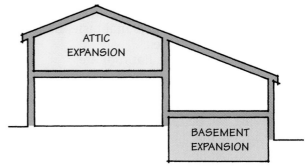

Figure 16–16   Plan for future expansion.

REAL WORLD APPLICATIONS

Conflict Resolution

A problem at work has developed between two of your fellow employees. They are working on a large project that involves the development of three basic house designs for a planned community. One insists that all houses should be one-story, which would attract senior citizens. Your other coworker asserts that the houses should be two-story condominiums, which would draw professionals and their families. The problem has negatively affected the atmosphere of the office. Your supervisor seems unaware of the situation since he is frequently out of the office. What would you do under these circumstances?

🏛 Go to your supervisor, explain the problem and suggest he address it.

🏛 Talk to the disgruntled employees at the same time and tell them they need to reach a compromise.

🏛 Discuss it with other employees to determine what they think should be done.

🏛 Talk to the disgruntled employees individually to assess their rationale for their recommended designs. Follow up by presenting a compromise to them.

🏛 Ignore it and hope it goes away.

Of the ideas listed, which would you actually carry out? Why would you choose this option? What could be the positive and negative outcomes of responding in each of these ways?

---

basement to be finished into a recreation room and shop at a later date.

BUILDING CODES AND ZONING

Building codes govern building construction while zoning ordinances govern land use. Combined, they serve to protect the health, safety, and welfare of the public.

Building Codes

Building codes are regulations which give minimum standards for building design, materials, construction, and maintenance. They are usually administered by the government of local communities. Often a local government adopts one of the model national building codes rather than writing its own unique code. The most popular model national codes are:

1. National Building Code (NBC) of the American Insurance Association

2. National Building Code of the Building Officials and Code Administrators International (BOCA)

3. Uniform Building Code (UBC) of the International Conference of Building Officials

4. Southern Standard Building Code (SBC) of the Southern Building Code Congress International

The Basic Building Code predominates in the eastern United States, the Uniform Building Code in the western states, and the Southern Standard Building Code in the southern states. In addition, there are a number of specialized codes such as the National Electric Code and the National Plumbing Code.

To ensure building code compliance, a building inspector must review architectural plans before issuing a building permit to start construction. During and after construction, the inspector visits the site to certify that plans are accurately followed before issuing an occupancy permit.

Americans with Disabilities Act

The **Americans with Disabilities Act** (ADA) is not a building code. Rather, it is a civil rights statute that requires that all public and commercial buildings must be accessible. The ADA makes it illegal to discriminate on the basis of impaired mobility, impaired sight, impaired hearing, or impaired mental condition. The ADA does not include residences or churches, but does include new and existing motels, restaurants, stores, shopping centers, gasoline stations, professional offices, banks, schools, theaters, museums, hospitals, bowling alleys, playgrounds, golf courses, and parks. Additional information and technical assistance to the public about ADA can be obtained from:

Architectural and Transportation Barriers Compliance Board
1331 F Street NW, Suite 1000
Washington, DC 20004-1111
Telephone: 800-872-2253

Zoning Ordinances

Zoning ordinances are established by local communities to coordinate land development and promote the health, safety, and welfare of the public. A zoning ordinance includes an official zoning map which establishes four basic types of land development zones:

1. Residential
2. Commercial
3. Industrial
4. Agricultural

These four categories are then further subdivided. For example, residential zones could be established as:

R1 One-family dwelling, church, school, park, or playground

R2 Any R1 use plus two-family dwelling or professional office

R3 Any R2 use plus multiple dwelling, fraternity, or sorority house

R4 Any R3 use plus rooming house or tourist home

Exceptions to zoning ordinances are permitted in special circumstances:

1. *Nonconforming use.* A structure is permitted to be used in a manner not conforming to its zone when the structure was in existence before the zoning district was established. Normal maintenance of a nonconforming structure is permitted, but it cannot be expanded.

2. *Conditional use.* After a public hearing and when found to be in the best interest of the community, facilities such as hospitals, airports, and public utilities may be permitted to be constructed in a zone normally excluding them.

3. *Variance.* Each community has a Zoning Appeals Board which may grant variances from the ordinance when there are requests for minor variations from the zoning laws or unusual circumstances (such as steeply sloping land that prevents the normal setback from a property boundary).

Local zoning ordinances may be quite restrictive as to the position of a house, its cubage in relation to surrounding dwellings, and even its style. A copy of the ordinances can be obtained from any town hall. These should be studied carefully before planning is begun.

EXAMPLE
The most important zoning limitations that apply to the location of the A residence were found to be:

30' minimum front yard setback
8' minimum side yard setback
25' maximum building height

COST

Although we list cost last, it should certainly be a major consideration, influencing nearly every other factor. To be enjoyed, a home must be within the owner's means. No one wants a house that will be impossible to keep up, and might finally be lost. The following chapter on financing relates the house cost to income.

REVIEW QUESTIONS

1. What is meant by primary considerations?

2. What should be considered when selecting a site?

3. What is meant by a split-level house?

4. What are two significant advantages of a house with a flat roof? Why did you choose these?

5. What are two significant advantages of a house with a pitched roof? Why did you choose these?

6. Should a basement or slab be specified under each of the following conditions?

    a. Low-cost housing on flat land, southern states

    b. Moderate-priced housing, New England states

    c. Known ledge 2'-6" below surface

    d. Steeply sloping land

7. Should a flat roof or pitched roof with attic be specified for:

    a. A Cape Cod house

    b. A modern house with a floor plan containing several ells

    c. A retirement cottage for a couple unable to climb stairs

ENRICHMENT QUESTIONS

1. Indicate the most suitable style of house to build on the following type of land.

    a. A wide, flat plot

    b. A narrow, flat plot

    c. A steeply sloping plot

2. A friend tells you he has purchased a small, five-and-a-half-room house. What kinds of rooms does that house probably contain?

3. List the advantages of:

    a. The two-floor house

    b. The one-and-a-half-floor house

    c. The one-floor house

4. Low-cost housing containing 1,500 sq. ft. of floor area is to be built with a flat roof and full basement. Which of the following alternatives would cost the least to construct? Which would cost the most? Why?

 a. A rectangular one-floor plan

 b. A rectangular two-floor plan

 c. An L-shaped one-floor plan

 d. An L-shaped two-floor plan

5. Would you prefer a basement or not? Why? Is it practical in the area in which you reside now?

PRACTICAL EXERCISES

1. Visit three other homes in your community and list the primary considerations for each residence according to the list below.

 a. Description of plot

 b. House style

 c. Number of rooms

 d. Number of floors

 e. Shape description

 f. Basement or no basement

 g. Type of roof

 h. Ability to expand

 i. Zoning limitations

 j. Current market value

 k. Market value at time of construction

 l. Year built

2. Complete the skeleton outline of primary considerations for your original house design:

 Type of plot:

 House style desired:

 Number of rooms needed:

 Number of floors desired:

 Plan shape desired:

 Basement or no basement:

 Flat or sloping roof:

 Expansion requirements:

 Zoning limitations:

 Tentative cost:

CHAPTER 17

Financing

OBJECTIVES

By completing this chapter you will be able to:

- 🏛 Define terminology related to mortgage calculations.
- 🏛 Differentiate between conventional and nonconventional mortgages.
- 🏛 Estimate the building cost of a residential house according to the volume and area in your locality.

KEY TERMS

Down payment
Points
Construction loan
Mortgage
Graduated payment mortgage
Adjustable rate mortgage
Renegotiated rate mortgage

Planning and building a home is an important step in the life of every family. A house is usually the largest single purchase a family will make. There are a number of well-established rules of thumb that will aid in deciding if this purchase—and how large a purchase—can be afforded. However, none of these rules can take into consideration the "sacrifice factor"—that is, that some persons with moderate incomes want their own homes so much that they are willing to make many sacrifices. The following rules and information should help with these financial decisions. In the final analysis, though, only the clients know the details of their present financial status and earning potential.

MORTGAGE CALCULATIONS

Yearly Income

The total amount invested in a house and lot should not exceed two and one-half times the yearly income.

EXAMPLE
Family A has an income of $80,000 a year. This would justify their spending $200,000 on a house and lot.

Down Payment

A **down payment** of 25 percent of the total house cost is advisable, although some lending agencies accept 20 percent or less. Of course, the larger the down payment, the smaller the mortgage interest charges.

Lending agencies accept down payments as low as 5 percent if the mortgage is insured. Mortgage insurance protects the lending agency against financial loss if a homeowner stops making the monthly mortgage payments. The additional cost for mortgage insurance is paid for by the owner.

EXAMPLE
Family A will make a down payment of 30 percent of the $200,000 total cost. This requires a down payment of $60,000, leaving $140,000 to be borrowed.

Points

In addition to the down payment, there are a number of additional expenses in obtaining a mortgage; often these fees will total about 5 percent of the total mortgage. The largest fee is the bank's service charge, a percentage of the mortgage called "**points**."

EXAMPLE
Family A is charged 4 points for their $140,000 mortgage. Therefore they must pay an additional fee of $5,600 (4% × $140,000 = $5,600).

Weekly Income

The monthly shelter expenses (mortgage payments, taxes, insurance) should not exceed the weekly income. Loan officers typically limit mortgage payments to 28% of gross income and also require that the total of all debt payments not exceed 36%.

EXAMPLE
Family A has an income of $80,000 a year, or $1,538 a week. This would justify their spending $1,538 a month on a house. Table 17–2 shows that a 15-year mortgage at 10 percent on $140,000 ($200,000 total cost minus the $60,000 down payment) would be about $1,505 a month ($1,075 for $100,000 + $430 for $40,000 = $1,505). This allows only $33 leeway for taxes and insurance. A 20-year mortgage at 10 percent would be about $1,351 a month, allowing $187 toward taxes and insurance.

Construction Loan

A mortgage can be obtained only on a finished house. During building, a **construction loan** is obtained, usually at the same interest rate as the mortgage.

REAL WORLD APPLICATIONS

Mortgage Calculations

As an architect, many clients will approach you with their "dream house" in mind. Unfortunately, many of them have no idea how much the cost will be or how much it will cost to finance the project. After the cost of the house is estimated, it would be helpful to know the total cost, including principal and interest paid to the lender. This will allow you to provide more realistic information to the client.

Computer financial programs such as Quicken, Microsoft Money, Quickbooks, and Peachtree Accounting will produce loan amortization schedules. A loan schedule will show the number of payments required, the amounts of principal and interest for each payment, and a listing of accumulated interest. Adding the principal and interest will show the required monthly payment. Complete the following scenario by using a financial program and producing a loan schedule.

Family G comes to you with ideas for their new home. You draw the plans, consult them, and then estimate the cost to be $128,000. The bank will finance the mortgage with 10 percent as a down payment and a fixed interest rate of 8.2 percent for thirty years.

1. How much money is needed for the down payment?
2. What is the monthly payment amount?
3. What is the total interest that is paid over thirty years?
4. What is the total principal that is paid over thirty years?
5. How many payments are required?

Remember that the principal amount will be the cost of the house minus the down payment. To calculate the amount of the down payment, multiply the price by the required down payment percentage:

Down Payment = Cost × Percentage

Mortgage

Nearly every home bought today is financed by means of a mortgage. Most of these mortgages provide for uniform monthly payments consisting of interest, amortization, and taxes. As time passes, the proportion of the monthly payments going toward amortization (that is, payment of the principal) increases, and the amount going toward interest decreases. Thus the early payments are nearly all interest payments, and the last payments are nearly all amortization payments. If the property tax increases, the amount of the monthly payment increases proportionately.

A good mortgage should include the following features:

1. Monthly payments, which include taxes and sometimes insurance.

2. Prepayment of part or all of the mortgage before the end of the period if the borrower wishes.

3. An open-ended clause allowing the borrower to reborrow a sum equal to the amount amortized for the purpose of expanding or finishing.

Mortgage Interest Rate   A home mortgage interest rate varies from place to place and year to year. Bank interest rates (Tables 17–1 through 17–6) have ranged from 8 to 18 percent. FHA-insured (Federal Housing Administration)

TABLE 17–2   *Monthly Payments on Mortgages at 10 Percent*

Amount of Mortgage	Length of Mortgage			
	10 yr.	15 yr.	20 yr.	25 yr.
$10,000	$132.16	$107.47	$ 96.51	$ 90.88
$20,000	$264.31	$214.93	$193.01	$181.75
$30,000	$396.46	$322.39	$289.51	$272.62
$40,000	$528.61	$429.85	$386.01	$363.49
$50,000	$660.76	$537.31	$482.52	$454.36

* Taxes and insurance are normally added to above payments.

TABLE 17–3   *Monthly Payments on Mortgages at 12 Percent*

Amount of Mortgage	Length of Mortgage			
	10 yr.	15 yr.	20 yr.	25 yr.
$10,000	$143.48	$120.02	$110.11	$105.33
$20,000	$286.95	$240.04	$220.22	$210.65
$30,000	$430.42	$360.06	$330.33	$315.97
$40,000	$573.89	$480.07	$440.44	$421.29
$50,000	$717.36	$600.09	$550.55	$526.62

* Taxes and insurance are normally added to above payments.

TABLE 17–1   *Monthly Payments on Mortgages at 8 Percent*

Amount of Mortgage	Length of Mortgage			
	10 yr.	15 yr.	20 yr.	25 yr.
$10,000	$121.33	$ 95.57	$ 83.65	$ 77.19
$20,000	$242.66	$191.14	$167.29	$154.37
$30,000	$363.99	$286.70	$250.94	$231.55
$40,000	$485.32	$382.27	$334.58	$308.73
$50,000	$606.64	$477.83	$418.23	$385.91

* Taxes and insurance are normally added to above payments.

TABLE 17–4   *Monthly Payments on Mortgages at 14 Percent*

Amount of Mortgage	Length of Mortgage			
	10 yr.	15 yr.	20 yr.	25 yr.
$10,000	$155.27	$133.18	$124.36	$120.38
$20,000	$310.54	$266.35	$248.71	$240.76
$30,000	$465.80	$399.53	$373.06	$361.13
$40,000	$621.07	$532.70	$497.41	$481.51
$50,000	$776.34	$665.88	$621.77	$601.89

* Taxes and insurance are normally added to above payments.

TABLE 17-5 *Monthly Payments on Mortgages at 16 Percent**

Amount of Mortgage	Length of Mortgage			
	10 yr.	15 yr.	20 yr.	25 yr.
$10,000	$167.52	$144.72	$139.13	$135.89
$20,000	$335.03	$289.43	$278.26	$271.78
$30,000	$502.54	$434.14	$417.38	$407.67
$40,000	$670.06	$578.85	$556.51	$543.56
$50,000	$837.57	$723.56	$695.63	$679.45

* Taxes and insurance are normally added to above payments.

TABLE 17-6 *Monthly Payments on Mortgages at 18 Percent**

Amount of Mortgage	Length of Mortgage			
	10 yr.	15 yr.	20 yr.	25 yr.
$10,000	$180.19	$161.05	$154.34	$151.75
$20,000	$360.38	$322.09	$308.67	$303.49
$30,000	$540.56	$483.13	$463.00	$455.23
$40,000	$720.75	$644.17	$617.33	$606.98
$50,000	$900.93	$805.22	$771.66	$758.72

* Taxes and insurance are normally added to above payments.

mortgages are lower, but added charges raise the total to approximately the same cost. For example, *mortgage insurance* (to protect the *bank* against loss on the mortgage) may add a 1 percent insurance premium.

Length of Mortgage In normal times, mortgage payments should be made over as short a term as possible. But in times of expected continuing inflation, long-term mortgages have the advantage of repayment with money of lessened value.

EXAMPLE

Family A calculated they would have to pay a total of $382,000 on a twenty-five-year mortgage ($140,000 amortization and $242,000 interest) at 10 percent. They decided on the fifteen-year mortgage since they would pay only $271,000 total ($140,000 amortization and $131,000 interest).

Satisfaction Piece When the last payment is made, the bank will send a document stating that the mortgage has been paid in full. This is called a *satisfaction piece*. Also, the mortgage will be returned. The satisfaction piece should be recorded; the mortgage can be burned.

Other Types of Mortgages

In addition to the conventional mortgage described above, a number of nonconventional types of mortgages are available, each having some advantages and disadvantages. The most popular types are graduated payment, adjustable rate, and renegotiated rate mortgages.

Graduated Payments

Mortgage interest rates may become so high that many people cannot afford the payments. **Graduated payment mortgages,** however, have lower payments in early years when the homeowner's income is modest, but higher payments in later years when income is expected to be greater. For example, the monthly payments for a $50,000, thirty-year, graduated payment mortgage at 10 percent would be $346.47 during the first year, but would increase to $497.41 after the fifth year (see Table 17–7). In comparison, monthly payments for a conventional 10 percent mortgage would be $454.36. Because amortization begins slowly, the graduated payment mortgage has a greater total cost.

Adjustable Rates

The interest rate on an **adjustable rate mortgage** depends on some economic index (such

TABLE 17-7 *Monthly Payments on a Graduated Payment Mortgage ($50,000 for 25 years at 10%)*

First year	$346.47
Second year	$372.46
Third year	$400.39
Fourth year	$430.42
Fifth year	$462.70
Remaining years	$497.41

REAL WORLD APPLICATIONS

Spreadsheet Basics

A new client is considering building a two-story home located in approximately 4 acres of wooded land. They found a set of plans they like in a magazine, but they wish to make changes to it. They would like to add a family room on the end and a den above it. They also wish to brick the outside, unless it would be more economical to use vinyl siding. Unfortunately, these are not the only modifications they want to make. They actually have many different ideas. Sketching only one set of plans that incorporates all proposed changes is impossible. These clients are most concerned about the cost of the house.

How would you design a spreadsheet that will detail the costs of these changes? A spreadsheet is essentially composed of three items: rows, columns, and cells. Rows are horizontal and are identified numerically while columns are vertical and labeled with letters. At the intersection of a row and a column is a cell. Cells are specified according to their row and column location. For instance, the cell at the intersection of column D and row 5 is labeled as D5. Within the cells are placed numbers, text, or formulas. Formulas are placed in a specific cell to calculate numbers in other cells and produce an answer in the cell where the formula resides. An example of a spreadsheet is shown in the first figure. The formulas used to produce the results are shown in the second figure.

Modify the first figure to include information that you select for a new plan and add a new line for painting supplies. Be sure to update and create new formulas when appropriate; do not just compute the answers with a calculator. You will find that once you learn how to do computations within a spreadsheet, it is easier and faster than using a calculator.

	A	B	C	D
1	Item	Plan 1	Plan 2	Plan 3
2	Door	$57.00	$62.00	$60.00
3	Window	$235.00	$223.00	$200.00
4	Carpet	$504.00	$356.00	$425.00
5	Electrical	$95.00	$100.00	$90.00
6				
7	Sub-total	$891.00	$741.00	$775.00
8	Tax (6%)	$53.46	$44.46	$46.50
9	TOTAL	$944.46	$785.46	$821.50
10				

	A	B	C	D
1	Item	Plan 1	Plan 2	Plan 3
2	Door	$57.00	$62.00	$60.00
3	Window	$235.00	$223.00	$200.00
4	Carpet	$504.00	$356.00	$425.00
5	Electrical	$95.00	$100.00	$90.00
6				
7	Sub-total	=SUM(B2..B5)	=SUM(C2..C5)	=SUM(D2..D5)
8	Tax (6%)	=B7*0.06	=C7*0.06	=D7*0.06
9	TOTAL	=B7+B8	=C7+C8	=D7+D8
10				

as a Treasury bill index) and may change each year. Usually the interest rate increase is limited to 2 percent per year with a maximum increase of 5 percent over the life of the mortgage. The interest rate may also decrease but only if a previous increase has occurred. Adjustable rate mortgages have the advantage of an initial interest rate that is about 1 percent lower than that of a conventional mortgage.

Renegotiated Rates

The interest rate on a **renegotiated rate mortgage** remains constant over a given period—usually five years—at which time the rate is renegotiated. An interest rate increase is usually limited to 3 percent every five years. Renegotiated rate mortgages also have interest rates that are lower than conventional mortgages.

COST ESTIMATION

To avoid building an unaffordable house, an accurate estimate of cost is needed. There are two commonly used methods of estimating house cost. One method is based on the volume, or cubage, of the building, the other on the floor area.

Cubage: $3.00–$10.00/cu. ft.
Floor area: $25.00–$50.00/sq. ft. inexpensive, unfinished
$50.00–$75.00/sq. ft. moderate
$75.00–$100.00/sq. ft. good materials and workmanship

The unit cost of a house (by cubage or floor area) will vary considerably from one location to another and even from one year to another. Also, a small house will have a greater unit cost than a large one, since some expenses (such as the heating plant) remain almost the same regardless of the size of the house.

Calculation by Cubage

Calculations for volume include all enclosed areas, such as the garage, basement, attic, dormers, chimneys, and enclosed porches. Outside dimensions are used (from the outside of walls, roof, and floor slab). Open porches and areaways are included at half volume.

Calculation by Area

All enclosed areas are included in the area calculations, but some are reduced in the following proportions:

Garage: $\frac{2}{3}$
Enclosed porch: $\frac{2}{3}$
Open porch: $\frac{1}{2}$
Unfinished basement: $\frac{1}{2}$
(finished basement: full area)
Carport: $\frac{1}{2}$

Land Cost

The total developed land cost includes the cost of the lot, street, sidewalk, water pipeline, and sanitary sewer. The total developed land cost typically ranges from 20 to 33 percent of the cost of a residence.

REDUCING COST

Many persons cut building costs by doing some of the construction themselves. Figure 17–1 shows the areas which will produce the greatest savings. The percentages on this pie chart represent percentages of total cost of a wood-framed house. Each category can be further split fifty-fifty, half of each category representing labor, half material. Several thousand dollars should be added to cover miscellaneous building equipment, landscaping, and other expenses.

Warning

Anyone who has recently built a home will tell you that they have spent 10 to 25 percent more than they originally intended. Indeed, ex-apartment dwellers will sometimes spend over 50 percent more than their estimated cost when they are furnishing and buying tools and equipment.

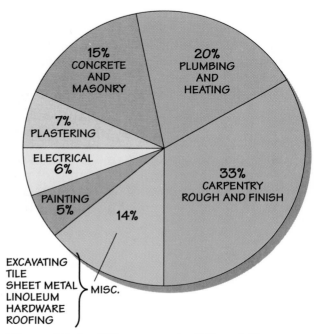

Figure 17–1 Where your building dollar goes.

REVIEW QUESTIONS

1. Define the following words.

   a. Mortgage   d. Mortgage rate

   b. Points   e. Mortgage insurance

   c. Construction loan

2. The *total cost* of a house should not exceed _____.

3. The *monthly cost* of a house should not exceed _____.

4. When obtaining a house mortgage, an advisable down payment is _____.

5. What is the difference between area cost estimation and volume cost estimation?

6. What effect does increasing the down payment on a residence have on the total interest to be paid?

ENRICHMENT QUESTIONS

1. What are the advantages and disadvantages of a 15-year mortgage and a 30-year mortgage?

2. A higher interest rate raises the payment amount because it also increases what?

3. On what are bank interest rates usually based?

4. What is an amortization schedule and how could it be useful to you as a homeowner?

5. If you needed to save money on the construction of your home, which of the items shown in Figure 17–1 would you be comfortable doing yourself? Will it be worth it in the long run? Why or why not?

6. Why do banks have to consider your credit history before giving you a mortgage?

PRACTICAL EXERCISES

1. Obtain the mortgage interest rates and an interest table from your local bank. Prepare a table of monthly mortgage payments similar to that shown in Table 17–1.

2. Mr. and Mrs. B have a yearly income of $60,000. Prepare a list of financial considerations for them as follows:

   a. Yearly income: c. House cost:

   b. Weekly income: d. House area:

3. Prepare a list of financial considerations as indicated in exercise 2 to meet the requirements of a client as assigned by your instructor.

4. Prepare a list of financial considerations for your original house design as follows:

   a. Estimated yearly income:

   b. Weekly income:

   c. House cost:

   d. Amount of mortgage:

   e. Length of mortgage:

   f. Interest rate of mortgage:

   g. Amount of down payment:

   h. Amount of monthly payments:

   i. Total interest paid:

   j. House area:

CHAPTER 18

The Program

OBJECTIVES

By completing this chapter you will be able to:

- Identify general requirements, including information about the site, orientation, rooms, and mechanical elements for a residence.
- Identify basic room requirements and room sizes for a typical residence.
- Develop a program of minimum objectives for a residence or small commercial building.

The first problem confronting an architect engaged to design a home is to find the type of house that will best suit the family's requirements and budget. To prevent items from being forgotten, a list of requirements is drawn up at the first meeting between the owner and architect. This list is called the *program*.

In studying architectural drafting, you are probably planning your own home, and thus you are taking the parts of both owner and architect. However, you will find that you should still prepare a program to help you keep your goals clearly in mind.

Your program should look somewhat like the A residence program, which follows.

GENERAL REQUIREMENTS

Size: 2,500–3,000 sq. ft.
Cost: Lot: $40,000. House: $160,000
Style: Contemporary.
Exterior finish: Vertical board and batten siding, brick or stone.

Roof: No preference.
Second floor: No.
Basement: Full or partial basement to be finished into recreation room later.

ORIENTATION

Living: West (to view).
Dining: No preference.
Kitchen: No preference.
Master bedroom: West (to view).
Other bedrooms: Not west.

ROOM REQUIREMENTS

Living area: Conversational area (fireplace with raised hearth), bridge area, and music area. Cathedral ceiling desired.
Dining area: Informal (combine with kitchen or family room).
Kitchen: U-type kitchen with baking island, built-in-wall ovens, burners, dishwasher, and garbage disposer. Allow for refrigerator-freezer and breakfast bar. Connect to garage.
Utility room: For storage of canned foods and miscellaneous items.
Family area: Include if possible without exceeding cost limitations. Provide fireplace if possible.
Bedrooms: Master bedroom with double bed, separate closets for Mr. and Mrs. A, large picture window view of Mount Nittany, and attached bathroom with shower. Separate bedrooms for two children: Mark, age nine; and Beth, age three. Beth's bedroom adjacent to master bedroom, if possible.

REAL WORLD APPLICATIONS

Client Database

Client files containing their program requirements are piling up on your desk. Papers, papers, and more papers are taking up space in your office, and retrieving specific information is becoming more and more time-consuming. You wonder what can be done to store information more efficiently.

One day while visiting the local hardware store, you notice that the cash registers have been replaced with computers. The store manager says they have found that using a database to track all their inventory and supplies has eliminated a lot of tedious paperwork. This might be an excellent way to store and retrieve your program requirements information.

A database is a computer program that stores information so that it can be easily retrieved, sorted, and printed. This tool works well for program requirements because each house design has the same general items listed on the program. You can have a complete listing of what a typical customer may desire that can be customized by entering different specifications.

Most databases contain three basic parts: the records, the fields, and the data. In your database, the records will contain information for each individual client. For instance, Family A would be one record, while families B, C, and D would be other records. The information in the records is divided into fields. The field headings could include the name, address, house size, price, number of rooms, and requirements for the kitchen, dining area, living room, bedrooms, and so on. In general, the fields would contain the program requirements described in this chapter. The information that is particular to one record and one field would be the data. For instance, the data listed for Family A's living room would be, "this room should contain a fireplace, bay window, and measure at least 12' x 20'." The data about Family B's living room would differ from the living room specifications of other clients.

When all of this information is compiled, it can be readily recalled. For example, if a client wants a 2000 sq. ft. house, they may ask to see what other clients have designed into their 2000 sq. ft. homes. This would save a tremendous amount of work because less filing, searching, and explanation is needed.

Design a sample database containing three different records. Each record should then contain twenty fields, but the data in each field should be unique to that client.

Hobby room: Use family room or basement.

Entertainment: Much entertaining (parties of 4 to 50).

Bathrooms: Main bathroom with toilet, lavatory, bathtub, and shower; master bathroom with toilet, lavatory, and stall shower; and powder room with toilet and lavatory (or use main bathroom).

Laundry: Locate on first floor. Provide for electric washer and dryer.

Storage: Two closets in master bedroom, one in other bedrooms, linen closets, large front entrance guest coat closet, broom closet, garden tools.

Basement: Unfinished. Future use will be general recreation, table tennis, snack bar and workshop. Provide outside entrance to future workshop.

Garage: One-car, attached, 12' power boat.

Porch: None.

Emergency shelter: For four persons.

Terrace: To rear of plot next to living area.

Garden: Flowers, not vegetables.

Fireplaces: Living area, family area, and recreation room.

Miscellaneous: Front entrance vestibule preferred. Atrium or clerestory required due to narrow lot and unpleasant views.

ROOM SIZES

Room sizes should be determined only after careful comparison with familiar rooms of known size.

Living area: 15' × 25'

Dining and family area: 15' × 20'

Kitchen: 12' × 15'

Master bedroom: 12' × 15'

Other bedrooms: 10' × 12', 12' × 15'

Bathrooms: Average size.

Hall width: 3' to $3\frac{1}{2}$'

Stair width: 3' to $3\frac{1}{2}$'

Closets: Large.

Garage: 14' × 21'

MECHANICAL REQUIREMENTS

Plumbing: Copper tubing, several outside bibbs.

Heating: No exposed radiators.

Air conditioning: Omit, but plan for future addition.

Electrical: Outside convenience outlets, silent switches.

Special equipment: Telephone wiring to each room. Television conduits to living room, family room, and master bedroom.

SITE

Location: West side of Outer Drive.

Size: 75' × 140' deep.

Zoning limitations: 30' minimum front yard setback. 8' minimum side yard setback, 25' maximum height.

Best view: West to Mount Nittany.

Trees: Eight birch, five sumac, one cherry (remove as few as possible).

Garden wanted: As extensive as possible.

After studying this program, it might be well to turn to the final plans (Chapter 24) and elevations (Chapter 27) to discover how the final product compares with the program. You will notice that nearly every requirement has been satisfied.

REAL WORLD | APPLICATIONS

Coordinated Effort

 For your next project your supervisor has assigned you to be on a team with another architect. The project involves establishing the program requirements for homes in a new housing development. The person you will be working with has been at the firm for twenty years, and you have been here for only two years. For this reason, you fear she may not consider your ideas and suggestions seriously. In this situation it is important that you have good interpersonal skills. You will need to build rapport with this co-worker and show respect for her ideas while at the same time presenting your own ideas as valuable. As you work together you will both need to compromise.

Look up the word *compromise* in the dictionary and list five items that you may need to compromise on during this project. Finally, count the number of times you have to compromise during the course of a typical day.

REVIEW QUESTIONS

1. Why is it important to prepare a list of room requirements for a residence?

2. Why is a program of minimum objectives important?

3. What is meant by orientation?

4. How are room sizes determined?

5. List five mechanical requirements that need to be considered.

6. Why is a description of the desired site needed when developing a program of minimum objectives?

7. Why should room sizes be determined only after careful comparison with familiar rooms of known size? How did you accomplish this?

PRACTICAL EXERCISES

1. Complete the program for your original house design using the following skeleton outline:

GENERAL REQUIREMENTS

Size: _____

Cost: _____

Style: _____

Exterior finish: _____

Roof: _____

Second floor: _____

Basement: _____

ORIENTATION

Living: _____

Dining: _____

Kitchen: _____

Master bedroom: _____

Other bedrooms: _____

ROOM REQUIREMENTS

Living room or living area: _____

Dining room or dining area: _____

Kitchen: _____

Utility room: _____

Family room: _____

Master bedroom: _____

Other bedrooms: _____

Den or study: _____

Bathrooms: _____

Laundry: _____

Storage: _____

Basement: _____

Garage: _____

Porch: _____

Terrace: _____

Fireplace: _____

Miscellaneous (special preferences or provisions for pets, hobbies, etc.): _____

_____

ROOM SIZES

Living room: _____

Dining room: _____

Kitchen: _____

Utility room: _____

Family room: _____

Master bedroom: _____

Other bedrooms: _____

Den or study: _____

Bathrooms: _____

Laundry: _____

Closets: _____

Hall width: _____

Stair width: _____

Garage: _____

MECHANICAL REQUIREMENTS

Plumbing: _____

Heating: _____

Air conditioning: _____

Electrical: _____

Special equipment: _____

SITE

Location: _____

Size: _____

Zoning limitations: _____

Best views: _____

Garden: _____

Soil type: _____

CHAPTER 19

Introduction to Room Design

OBJECTIVES

By completing this chapter you will be able to:

- 🏛 Differentiate between room sizes and shapes according to their function.

- 🏛 Use furniture templates to assist in your room designs.

There are two major considerations in the design of the floor plan of a house. First, each room must be designed so that it is pleasant, functional, and economical; and second, the rooms must be placed in the correct relationship to one another. This second consideration may be likened to working out a computerized jigsaw puzzle in which the pieces may change in shape and size, giving various solutions. Some of these solutions will be better than others, but in the design of a house, the overall plan can be no better than the design of the individual rooms.

To give some starting point, minimum and average room sizes are given in Table 19–1. Notice that the rooms are not long and narrow (which makes them too hall-like) nor square (which makes furniture placement difficult). Actually, there is no such thing as an average, or

standard, or even ideal, size or shape for a room. Design a size and shape that will best meet your requirements of function, aesthetics, and economy.

TABLE 19–1 *Minimum and Average Room Sizes*

Room	Minimum (inside size)	Average
Living room	12' × 16'	14' × 20'
Dining room	10' × 12'	12' × 13'
Dining area	7' × 9'	9' × 9'
Bedroom (master)	$11\frac{1}{2}' \times 12\frac{1}{2}'$	12' × 14'
Bedroom (other)	9' × 11'	10' × 12'
Kitchen	7' × 10'	8' × 12'
Bathroom	5' × 7'	$5\frac{1}{2}' \times 8'$
Computer room	4' × 4'	6' × 6'
Home office	6' × 8'	8' × 10'
Utility (no basement)	7' × 8'	8' × 11'
Hall width	3'	$3\frac{1}{2}'$
Closet	2' × 2'	2' × 4'
Walk-in closet	3' × 5'	4' × 6'
Porch	6' × 8'	8' × 12'
Patio	12' × 12'	16' × 20'
Garage (single)	$9\frac{1}{2}' \times 19'$	12' × 20'
Garage (double)	18' × 19'	20' × 20'
Garage door (single)	$8' \times 6\frac{1}{2}'$	9' × 7'
Garage door (double)	$15' \times 6\frac{1}{2}'$	16' × 7'

REAL WORLD APPLICATIONS

Form Follows Function

The expression "Form follows function" is often used by engineers, designers, and architects. It provides a principle to follow when creating new designs or improving existing ones. Form refers to the shape, texture, and aesthetic nature of a design. Function refers to the structural, mechanical, and practical nature of a design. This phrase should be taken literally to establish a priority of design criteria. In other words, if you have just designed a nicely *formed* dining porch that is far removed from the kitchen (lacks *function*), it is unacceptable. The room design must function first and then the form follows. For example, in a kitchen, the dishwasher should be next to the sink. The most uniquely and beautifully designed kitchen that does not meet this criterion is not suitable because it lacks practicality.

How can you determine if the floor plan of your original house design upholds this principle? Possibly the best way is to ask other people to evaluate your plans and give you some feedback. You will find that asking individuals familiar with, and unfamiliar with, architecture will yield a diverse range of responses. Keep in mind that their job is to give you feedback. Do not take their suggestions personally. Consider what was recommended, and implement the changes that you feel will significantly improve your design. Remember that the best design is the one that functions successfully under the various conditions existing during operation and is still aesthetically pleasing.

Chapter 20 will discuss the design of the kitchen. The design of a kitchen can be interesting and enjoyable, or it can be difficult and frustrating, depending upon your state of mind and personal commitment. Either way, it is usually a challenge.

Chapter 21 discusses the remaining three major house areas: the living room, dining room, and bedrooms. (The term "six-room house" usually refers to a house having a living room, dining room, kitchen, and three bedrooms). Although the living, dining, and bedrooms are considered to be major areas, you will find their design comparatively straightforward.

Chapter 22 details the principal supporting areas such as the home office and computer areas, the all-important bathrooms, laundry, storage, porch, patio, lanai, and garage.

In these chapters, the terms *rooms* and *areas* are used. A **room** (such as a living room) refers

Figure 19–1   A warm, contemporary kitchen design. (Courtesy Wood-Mode Cabinetry.)

to a space enclosed by partitions or walls. An **area** (such as a living-dining area) refers to an undivided dual-use space or to spaces separated only by furniture, or by a counter, or by a low or partial partition. A floor plan of rooms is termed a **closed plan**; a floor plan of areas is termed an **open plan**.

The desired furniture arrangement will have a considerable effect on the design of each room. Professional designers find that furniture templates or underlays greatly simplify the design process. The furniture underlay (Figure 19–5) may be used when drawing floor plans to a $\frac{1}{4}$" = 1'-0 scale. You can make a photocopy of the underlay, slip it into position under your vellum, and trace off the desired elements. Check with your instructor as well; you may be provided with a copy of the underlay

or asked to trace these symbols, and add others, rather than copying the page in your textbook.

If you are designing a house that will be occupied by a person with a major impairment, refer to Chapter 47, Accessibility, for design recommendations.

Figure 19–3 *Even utility rooms such as bathrooms deserve careful design.* (Courtesy Scholz Design, Toledo, Ohio.)

Figure 19–4 *Room design includes both inside and outside areas.* (Courtesy California Redwood Association.)

Figure 19–2 *A traditional paneled living room.* (Courtesy Wood-Mode Cabinetry.)

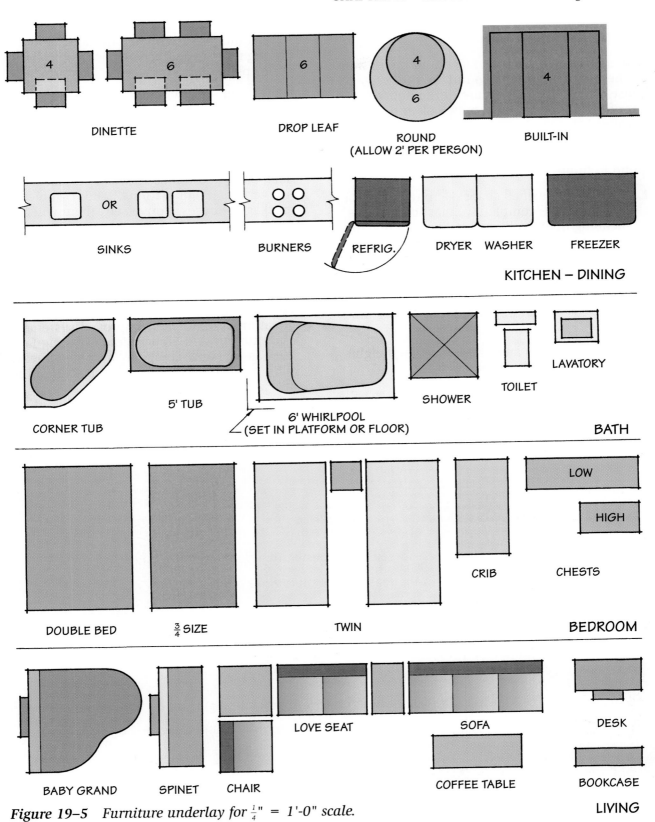

Figure 19–5 *Furniture underlay for $\frac{1}{4}" = 1'\text{-}0"$ scale.*

FUTURE TRENDS

In this chapter, we have touched on the design of the types of rooms found in the average house today. To get a picture of the house interior of the future, we need only look at the more advanced designs of contemporary architects. Using these contemporary designs as a guide, it is possible to predict the features likely to be found in future houses:

1. Rooms will be of all shapes—from circular or free-form to completely plastic surfaces, blending floor, wall, and ceiling into one uninterrupted flowing surface.

2. The trend will be away from the individual room concept toward multipurpose rooms and flexible areas that may be quickly rearranged by folding doors or retractable walls.

3. Multilevel floors and varying ceiling heights will become more common.

4. Furniture will often be an integral part of the architecture. (One example is the sunken living area surrounded by a carpeted lounge.)

5. The outdoors will be brought indoors through the greater use of glass walls and interior planting.

6. New materials will make an appearance. Wood, stone, brick, and plaster will be gradually replaced by materials manufactured by yet-to-be-discovered methods.

7. Assembly line housing will become popular—packaged rooms that can be assembled in various combinations, and completely packaged houses. Conversely, there will also be a greater demand for custom-designed homes.

8. Revolutionary appliances will appear. Many will be built-in: dust-repellent surfaces; completely concealed heating, cooling, and lighting elements; home computers and communication systems; and improved systems of food storage and preparation. Cleaning facilities as we know them may become unnecessary because of the use of disposable utensils and clothing.

9. New systems of building will be developed so that less desirable land (and water) can be used.

REVIEW QUESTIONS

1. What two items should be considered in the design of a floor plan of a house?

2. Differentiate between rooms and areas in relation to floor plans.

3. Indicate the minimum size of these architectural elements:

   a. Living room
   b. Master bedroom
   c. Bathroom
   d. Hall width
   e. Double garage

4. Using a $\frac{1}{4}$" = 1'-0 scale and the furniture underlay shown in Figure 19–5, find the size of:

   a. A typical sofa
   b. A typical coffee table
   c. Drop leaf table seating six
   d. Diameter of a round table seating six
   e. A typical refrigerator
   f. A typical washer-dryer combination

ENRICHMENT QUESTIONS

1. After conducting a thorough search of contemporary architectural periodicals, give your prediction of the type of house interior:

   a. Twenty-five years hence

   b. One hundred years hence

2. What information should an architect have about a family before designing rooms for their residence?

3. Which of the nine items listed under Future Trends do you feel will have the most impact on residential housing in the next ten years? Why?

PRACTICAL EXERCISES

1. In preparation for the layout of the rooms for your original house design, list the rooms familiar to you (your own home, neighbors' homes, and relatives' homes), their exact sizes (actually measure them), and your impression of their size (too small, satisfactory, or too large).

2. Using the information gathered in exercise 1, prepare a list of the rooms of your original house design showing the sizes you wish.

3. Using the sizes drawn up for exercise 2, sketch, draw, or computer-produce to scale individual room layouts for your original house design.

4. Design a room that incorporates one or two of the future trends indicated in this chapter.

CHAPTER 20

The Kitchen

OBJECTIVES

By completing this chapter you will be able to:

🏛 Prepare the following preliminary kitchen designs for a residence:

- Pullman
- U-shaped
- L-shaped
- Peninsula
- One-wall

KEY TERMS

Pullman kitchen	Peninsula kitchen
U-shaped kitchen	One-wall kitchen
L-shaped kitchen	

When a builder puts up a low-priced house, there are a number of ways to cut corners and reduce cost. However, you have probably noticed that even the smallest house has a carefully designed and well-equipped kitchen (Table 20–1). The reason is that builders realize prospective buyers demand an efficient and cheerful kitchen. As a matter of fact, this is the first room many will want to see. The average homemaker walks two hundred miles a year in the performance of household tasks—half of this in the kitchen. Consequently, in designing any house for yourself or others, you cannot spend too much time in perfecting the kitchen design (Figure 20–1).

A kitchen should be thought of as a group of three activity centers:

1. Storage
2. Preparation and cleaning
3. Cooking and serving

Storage

The focal point of the storage center is the refrigerator, although many cabinets for nonre-

TABLE 20–1 *Kitchen Equipment*

	Width	Depth	Height
Base cabinets		2'-0	3'-0
Wall cabinets		1'-0	2'-6"
Range, oven below	2'-6"	2'-1"	3'-0
Range, built-in	2'-8"	1'-8"	
Oven	2'-0	2'-0	2'-4"
Microwave	2'-0	1'-6"	1'-4"
Sink	2'-0	2'-0	3'-0
Sink and drainboard	3'-6"	2'-0	3'-0
Refrigerator, 7 cu. ft.	2'-1"	2'-3"	4'-6"
Refrigerator, 14 cu. ft.	2'-8"	2'-6"	5'-4"
Refrigerator, built-in	3'-0	2'-0	7'-0
Washer	2'-0	2'-3"	3'-1"
Dryer	2'-6"	2'-3"	3'-1"

Figure 20–1 *Molded trim and leaded glass accent this formal kitchen.* (Courtesy Wood-Mode Cabinetry.)

Figure 20–2 *Side-by-side refrigerator-freezer.* (Courtesy KitchenAid.)

frigerated food, dishes, and utensils must be provided. Refrigerators are usually freestanding, but built-in refrigerators are also available. To save steps, the refrigerator may be located near the delivery door, or nearest the door to the living-dining area. Also consider a freezer or refrigerator-freezer (Figure 20–2), although a freezer is often removed from the kitchen itself to another area. Pull-out storage shelves (Figure 20–3) or separate storage rooms should be considered.

Preparation and Cleaning

This center is built around the sink (Figure 20–4) and its adjoining counter space. Do you want to include an automatic dishwasher (Figure 20–5) or garbage disposal unit? Most people want their sink by a window.

Cooking and Serving

The cooking center is grouped around the range (Figure 20–7). In addition to electric coils and gas burners, range cooktops are available in

Figure 20–3 *Pull-out pantry unit.* (Courtesy Wood-Mode Cabinetry.)

Figure 20–4   *Kitchen sink with soap/lotion dispenser and a stainless steel corner sink.* (Courtesy Kohler Co.)

Figure 20–6   *Trash compactor.* (Courtesy Kitchen-Aid.)

Figure 20–5   *Automatic dishwasher.* (Courtesy KitchenAid.)

Figure 20–7   *Downdraft electric surface range.* (Courtesy KitchenAid.)

Figure 20–8   Built-in combination microwave and self-cleaning oven. (Courtesy KitchenAid.)

solid ceramic disks, magnetic induction surfaces, and tungsten halogen surfaces. A built-in oven (Figure 20–8), being less used, may be located at a more isolated position than the range. Range-oven combinations are standardized at 36" high, so counters should be designed at the same height. If a house is designed for especially tall or short persons, these major appliances may be built in at a convenient height. Choose a microwave oven if you wish to decrease cooking time. Consider an exhaust fan or hood and fan above the range cooktop.

A planning center consisting of a desk or table, computer, and telephone is included in many modern kitchens. Allow space for cookbooks, recipe index, and so forth (Figure 20–17).

As a rule, the laundry (automatic clothes washer, dryer, and ironing board) is best placed in its own utility room. If it must be placed in the kitchen, it should have its own separate area.

These activity centers may be combined in a number of ways in various-shaped rooms designed so that work progresses from right (storage) to left (serving). The often-used combinations shown in Figure 20–12 are:

1. Pullman (or corridor or two-wall)
2. U-shaped
3. L-shaped
4. Peninsula (or island)
5. One-wall

Pullman

The **pullman kitchen,** consisting of a long corridor with utilities on either side, is often used where space is at a premium. Doors may be at either end or at one end only. Although this design is somewhat factorylike, it is efficient in saving space and steps.

U-shaped

The **U-shaped kitchen** has cabinets on three walls, the sink usually in the middle, and the refrigerator and stove on opposite sides. Space for dining may be allowed near the fourth wall. This plan is adaptable to large and small rooms.

L-shaped

(See also Figure 20–9). The **L-shaped kitchen** is probably the most commonly used arrangement since it is efficient, allows for two doors without any interruption of countertop area, and may be nicely used with a breakfast table.

Peninsula

This kind of layout may be used only with large rooms. The layout of a **peninsula kitchen** includes a counter or *breakfast bar* that runs perpendicular to a wall (see also Figure 20–10). If the counter is freestanding, the kitchen is referred to as an *island* kitchen (see Figure 20–11).

One-wall

A **one-wall kitchen** layout is used when a kitchen must be fitted into a long narrow space. It is not an ideal arrangement.

CAREER PROFILES

Name: Scott A. Stancombe   *Age: 26*
Gender: Male
Occupational Title: Interior Designer/Sales
Employer: Lentz Kitchen and Bath, Indiana, PA 15701
Years of Experience: 3
Educational Background:
   Bachelor of Science, Interior Design
   Indiana University of Pennsylvania
Starting Salary:
   $15,000. Salaries are usually higher in cities and commission is usually a basis for income in smaller firms.

JOB OVERVIEW:
   My day consists of the following responsibilities: kitchen and bathroom design layouts, pricing, quotes and proposals, job site supervision, "selling the job," specifications, and drafting. One of the most significant projects in which I have been involved is designing and assisting in the selection of family room cabinetry including a furniture wall, entertainment center, and custom fireplace mantel. Designing the renovation of a kitchen by removing a wall and incorporating the kitchen within the dining room was another major project completed. Showroom layout and yearly changes of displays is another of my design activities.
   One of the greatest benefits of my position is seeing a project through to completion from an idea to a drawing to the finished job. For me, customer relations is another benefit since I enjoy working with people to solve their kitchen and bath needs. The biggest challenge I encounter on my job is ensuring that a project runs smoothly. If material is backordered it slows a project down, which can upset the customer. Obtaining material before the start of the job helps prevent dissatisfied customers.
   Finally, I would give a few words of wisdom to individuals interested in a career related to interior design. First, you need patience. Changes in layouts are common. Sometimes you think a layout is the best it could possibly be, but the customer disagrees. Then you must sell the job—remember you are a salesperson as well! Creativity is another skill individuals entering this field must possess. This means to do your own thing—do not copy!

The Work Triangle

Kitchen planners have long used the "work triangle" as a measure of kitchen efficiency. The work triangle measures the distance covered when removing food from the refrigerator to the sink, then to the range, and finally back to the refrigerator. For an efficient kitchen, the total length of the three sides of the work triangle should not exceed 20'. For example, you can calculate the work triangle of the U-shaped kitchen of Figure 20–12 by drawing and measuring lines from the front center of the refrigerator to the sink (7'), to the range

Figure 20–9 *An L-shaped kitchen.* (Courtesy Wood-Mode Cabinetry.)

Figure 20–10 *A peninsula kitchen.* (Courtesy Wood-Mode Cabinetry.)

Figure 20–11 *An island kitchen.* (Courtesy Wood-Mode Cabinetry.)

(4'), and back to the refrigerator (5'), for a total of only 16'.

Although it may seem from the previous discussion that the kitchen is greatly standardized, this is not the case. The kitchen should be original and pleasant. Avoid the monotony of the "factory" look. Consider built-in ranges, ovens, and refrigerators. Materials like brick and copper; color in appliances; and careful floor, window, and lighting design will all add to a kitchen's attractiveness. Working areas should be continuous; they may turn corners but should not be interrupted by doors. Remember that the cabinet space in an inside corner cannot be *fully* utilized.

In addition to major appliances, kitchen cabinet manufacturers offer a wide range of convenience options and specialized storage areas. A sampling is shown in Figure 20–13.

To qualify for an FHA-insured mortgage, a minimum area of 100 sq. ft. is recommended.

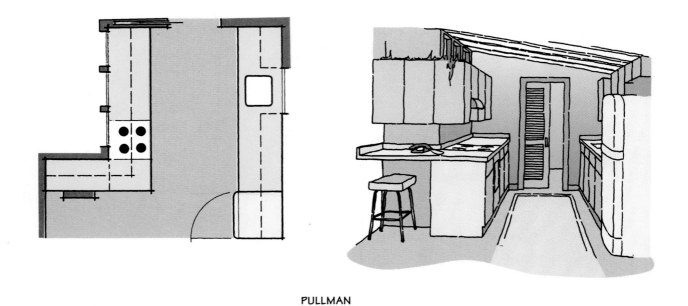

PULLMAN

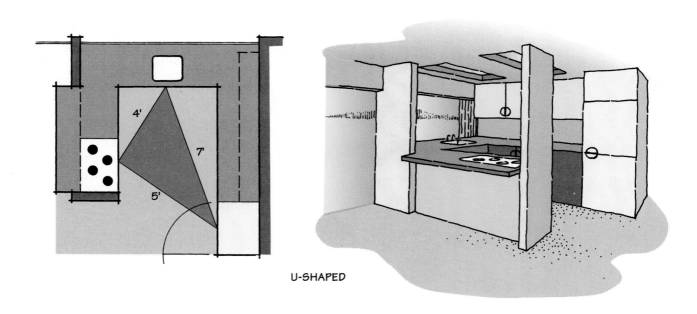

U-SHAPED

Figure 20–12   Kitchen design.

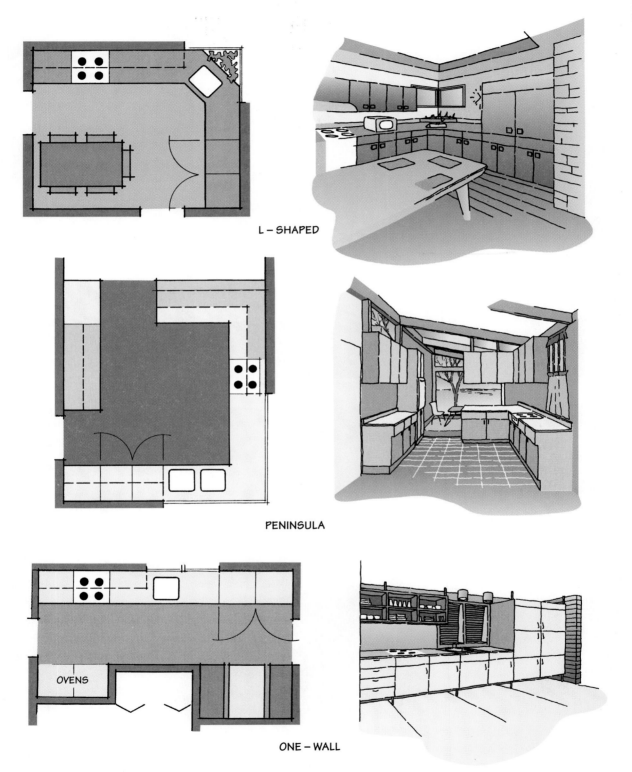

L – SHAPED

PENINSULA

OVENS

ONE – WALL

Figure 20–12   Continued.

ROLL-OUT SERVING CART

CUTLERY DIVIDER

VEGETABLE DRAWER

TRAY PARTITIONS

PULL-OUT TABLE

Figure 20–13   Kitchen convenience options and specialized storage areas. (All courtesy Wood-Mode Cabinetry.)

SPICE DRAWER INSERT

TILT-OUT WASTEBASKET

MIXER CABINET

DIVIDED DRAWERS

DOUBLE WASTEBASKET

SLICING BOARD/BREAD BOX

HOME RECYCLING CENTER

Figure 20–13 *Continued.* (*All courtesy Wood-Mode Cabinetry.*)

Figure 20–14 *A pass-through kitchen layout.*
(Courtesy Wood-Mode Cabinetry.)

Figure 20–16 *A kitchen with two counter heights for maximum convenience.* *(Courtesy Wood-Mode Cabinetry.)*

Figure 20–15 *A country kitchen design.*
(Courtesy Wood-Mode Cabinetry.)

Figure 20–17 *A kitchen's planning center.*
(Courtesy Wood-Mode Cabinetry.)

Figure 20–18   Peninsula kitchen with waffled ceiling. (Courtesy Scholz Design, Toledo, Ohio.)

Figure 20–20   Family kitchen with a beamed cathedral ceiling. (Courtesy Wood-Mode Cabinetry.)

Figure 20–19   An ultra-modern kitchen design. (Courtesy Wood-Mode Cabinetry.)

Figure 20–21   Recycled barn timbers create a unique kitchen style. (Courtesy VELUX-AMERICA, INC.)

REAL WORLD APPLICATIONS

Cooks with Special Needs

 Special design considerations may be needed for kitchens to be used by persons with physical impairments. Examples of this include such things as lower counter heights and cabinet doors that can be opened without twisting, pinching, or grasping knobs. The figure below shows a barrier-free kitchen sink with knee room under the sink and a flexible faucet head. See Chapter 47, Design for Accessibility, for more details. Other kitchen design considerations for persons with special needs include:

🏛 A cart can be used for movable storage of tabletop appliances.

🏛 For cooks with special needs, appliances such as a garbage disposal, a trash compactor, and a dishwasher are especially useful.

🏛 Pullout cutting boards are convenient because they can be removed and set on the lap.

A barrier-free kitchen sink.
(*Courtesy Kohler Co.*)

🏛 A cooktop can be used rather than a conventional stove so knee space can be provided beneath.

🏛 Built-in ovens should be installed at the most convenient height for the seated cook.

🏛 Continuous countertops allow the cook to slide heavy items rather than carry them.

🏛 A side-by-side refrigerator is most practical because this allows the cook access to both the refrigerator and freezer sections.

🏛 A spray attachment with a long hose should be installed so the cook can fill pans at the stove.

In order to become more familiar with the importance of designing for accessibility, especially in a kitchen, contact someone you know who is impaired and ask for an interview. Use the following questions as guidelines for your interview. What is the most difficult aspect of cooking for you? What types of special design features and appliances does your kitchen have? What additional changes in design or new appliances would make work in your kitchen easier? Ask them specifically if the preceding design features have been installed. Then ask if any of these would improve their kitchen accessibility.

When conducting the interview, keep the following in mind. Prepare for the interview through research and have your main questions written out. Remember to ask follow-up questions—your best information is likely to come from these. Take abbreviated notes during the interview and prepare a one-page summary of the interview.

REVIEW QUESTIONS

1. Name the three activity centers included within a kitchen. What appliances are associated with each?

2. What purpose does a planning center serve?

3. Name five types of kitchen layouts and include the advantages and disadvantages of each.

4. Name six types of specialized kitchen storage units.

5. What components of the kitchen work triangle should be considered when designing a kitchen?

ENRICHMENT QUESTIONS

1. Calculate the perimeter of the work triangle for the four remaining kitchens shown in Figure 20–12.

2. What family characteristics should be considered when designing a kitchen for a new residence?

3. How could the phrase "form follows function" be incorporated into the planning of a kitchen?

PRACTICAL EXERCISES

1. Sketch to scale, draw, or computer-produce:
   a. An 8' × 12' Pullman kitchen
   b. A 10' × 12' U-shaped kitchen
   c. A 10' × 14' island kitchen

2. Sketch to scale, draw, or computer-produce the kitchen for your original house design.

CHAPTER 21

Living, Dining, and Sleeping Areas

OBJECTIVES

By completing this chapter you will be able to:

🏛 Produce preliminary room designs for living, dining, and sleeping areas.

KEY TERMS

Circulation

Center-hall house arrangement

Dead-end room

LIVING AREA DESIGN

In addition to a living room (Figure 21–1), the living areas of a residence might include a family room (often called a recreation room, playroom, or gameroom), open porch, an open-plan living-dining area (or an open-plan living-dining-kitchen area), or a closed sunroom or porch. Many residences also offer living spaces such as open porches or verandas, lanais or loggias, decks, patios or terraces, or a breeze-way, that are not interior areas (Figure 21–2). The use of these areas will vary from the more

Figure 21–1 *Good design includes both interior and exterior living areas.* (Courtesy California Redwood Association.)

Figure 21-2 *Some exterior living area designs. (All courtesy California Redwood Association.)*

formal dining, conversation, reading, music, television, and relaxing (Figure 21–3) to the more recreational snacking, hobbies, games, dancing, and exercising. Obviously, then, you have great flexibility in living area design (Figures 21–4 and 21–5), and your final choices will depend upon your family's interests and budget goals. A family room can be designed in primary space as a separate, ground-floor room, but it is often positioned in a secondary area such as a basement, porch, or large breezeway.

The Living Room

Many people feel that since guests are entertained in the living room, it should be as large and gracious a room as can be afforded. The

REAL WORLD APPLICATIONS

Listening Skills

A meeting has been scheduled with a client to discuss their plans and ideas for their living room. They need someone to listen to all of their ideas and then help them decide on the better options. You plan to take your apprentice with you to this meeting.

Realizing how important it will be for you and your apprentice to listen carefully to your clients' ideas, you briefly discuss listening skills with your apprentice. You explain that to be a good listener one must focus attention on the speaker. Make eye contact with the speaker and do not allow other thoughts to crowd into your mind. While listening, do not draw a floor plan of the living room in your mind or evaluate the client's ideas. If this happens you are not open to all the suggestions the client is making and you may miss an important point.

Another important listening skill to discuss with your apprentice is the importance of body language. Sitting with your arms crossed suggests to the speaker you are not open to their perspective. Therefore, an open body posture is recommended. Leaning toward the speaker during emphatic points also indicates true openness and attentiveness.

Listening also requires some notetaking, but, in a meeting such as this, notetaking should be minimal so as not to make the speaker uncomfortable. Responding appropriately to the speaker's comments also demonstrates attentiveness. Responses should be brief and reflect what the speaker has stated. For instance, if the client is discussing the possibility of a skylight in the living room, the listener may respond, "We've often specified skylights in our most successful designs." In this way the listener responds without taking over the conversation and encourages the speaker to expand on that thought.

In groups of three, practice your listening skills. You may use the above situation if you would like. One person should be the architect, one the apprentice, and one the client. Both the architect and apprentice will listen but the architect will also observe the apprentice's listening skills. The architect will then provide feedback to the apprentice. The architect should consider eye contact, attentiveness, body language, notetaking, and feedback responses. The architect should question the apprentice to determine if listening and understanding truly occurred.

Figure 21–3 *A living area designed as a home entertainment center. (Courtesy Scholz Design, Toledo, Ohio.)*

Figure 21–5 *A living room with an Asian flair. (Courtesy Wood-Mode Cabinetry.)*

Figure 21–4 *An interior decorated to obtain a classic flair of old Europe. (Courtesy Pella Corporation.)*

Figure 21–6 *A spectacular residence as built in Dallas, Texas. (Courtesy Pella Corporation.)*

minimum area recommended by the U.S. Office of Housing is 160 sq. ft. The shape of the living area is often rectangular, the length one-third greater than the width to avoid a long, hall-like feeling or a square, box-like shape. Design factors such as dramatic window arrangements (Figures 21–6, 21–7, and 21–8), a beamed, cathedral ceiling, a sunken floor (Figure 21–9), or a sunken conversation pit do much to achieve the desired graciousness.

Two factors should be given careful consideration in the design of the living area: circulation and furniture grouping.

Figure 21–7 *An entrance view of the living area for the residence pictured in Figure 21–6.* (Courtesy Pella Corporation.)

Figure 21–9 *A sunken living room.* (Courtesy American Plywood Association.)

Figure 21–8 *Clean, basic lines accent a contemporary home with modern flair.* (Courtesy Pella Corporation.)

Circulation

Movement within a room and from one room to another is called **circulation**. Most persons feel the **center-hall house arrangement** is ideal because circulation to all rooms is easily accomplished without using the living room as a hallway or vestibule. When a room is not used as a throughway, it is called a **dead-end room**. There is much merit in dead-end rooms, but remember that they are not as interesting as a room that "goes somewhere." At any rate, a living room should never have to be used for trips between bedroom and bath.

Furniture Grouping

The major furniture grouping should be arranged so that the circle of chairs falls within a 12' diameter. A living room that has a fireplace as the focal point of this major furniture grouping is at a disadvantage when the fireplace is not in use. Rather than place furniture facing a black hole, a picture window or television set might be better choices. Consider also any special living room areas for music, reading, study, or cards.

Figure 21–10 shows a living room designed for a contemporary executive mansion. The client required an area that could be used for large, formal receptions as well as small, intimate parties. The design requirements included provision for indoor and outdoor gatherings in the day or evening. Notice that this design met these various requirements by the creation of areas that could be used individually or combined. The room itself, for example, is separated

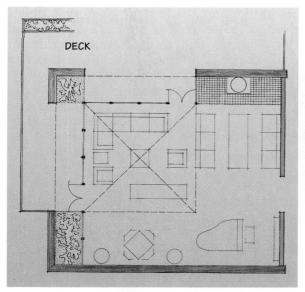

Figure 21–10   *Living room design.*

into four major furniture groupings: (1) the major formal conversational grouping, emphasized by the ceiling design; (2) a secondary conversational grouping in front of the fireplace; (3) a music area; (4) an area for card playing. For large gatherings, the outside deck could be added by opening the French doors at the two locations. To provide contrast with the long distance view through the window walls, plantings were specified immediately outside these windows. These plantings, backed by brick piers, could be illuminated in various ways depending upon the mood desired.

Undoubtedly you will want to design a much smaller living room than this example. Please notice, though, that the plan by itself does not tell the entire story. An interior perspective of the room was needed to verify that the design requirements were met satisfactorily.

The Entryway

Although an economical house may be designed with an entrance door opening directly into the living room, the addition of an entryway (a separate space between the entrance door and interior rooms) presents an inviting and gracious introduction to a residence (Figures 21–11 and 21–12). Rather than walking directly into the living room, an entryway (also called entrance hall, foyer, or vestibule) offers a transitional

Figure 21–11   *A rear entryway from a porch deck.* *(Courtesy VELUX-AMERICA, INC.)*

space from the outdoors to the building interior. Guests are greeted, and outdoor garments can be removed and stored in the "front hall closet".

Figure 21–12   A dramatic contemporary entryway. (Courtesy Pella Corporation.)

Current design trends for larger houses produce entryways that are two-story and even two and one-half stories high, though at the sacrifice of energy efficiency and ease of window cleaning and operation. When designing an entryway, consider weather protection by means of a front entrance overhang or a recessed entrance. Also consider entrance sidelights or other means to identify your visitors.

DINING AREA DESIGN

Periodically, the dining room loses and gains popularity. It can be dropped from new house plans to reduce costs, but a separate dining room is sometimes wanted.

The size of a dining room depends entirely upon the size of the table and amount of accessory furniture (Figure 21–13). Typical dining room furniture includes a table with expandable leaves, two armchairs, two or four side chairs, china cabinet (also called highboy or hutch), buffet (also called low boy or sideboard), and hot-plate serving cart. A good compromise between the dining room and breakfast nook (Figure 21–14) is a living-dining area or kitchen-

CAREER PROFILES

Name: Anita Soliday, AIA   Age: 35
Gender: Female
Occupational Title: Registered Architect
Employer: Macallister Group, Plymouth Meeting, PA
Years of Experience: 10
Educational Background:
    Bachelor of Science in Architecture
    Pennsylvania State University
Starting Salary:
    Low $20,000s. Salary increases quickly with experience.

JOB OVERVIEW:
    My first step every day is to review my calendar and plan the day. Included in my daily routine are the following tasks: talking to clients, CAD, sketching and design work, organizing those who work for me, visiting project sites, and planning project coordination. Of these, co-

Figure 21–13   *A symmetrical dining room design. (Courtesy Wood-Mode Cabinetry.)*

Figure 21–14   *A sunny breakfast room. (Courtesy Pella Corporation.)*

ordinating projects for specific clients, structural calculations, and billing clients are my primary responsibilities. Larger firms have both project managers and project architects. I do both. I have found that everything you learn in school is expanded on the job. In addition to the schooling at Penn State University, I have had to learn to deal with clients, develop my scope of work, coordinate proposals and bids, and use a computer to produce architectural drawings and organize billings.

All the projects I have worked on have been in residential architecture. Projects have involved private custom homes, homes in developments, and residential additions. The total cost of these projects range from $10,000 to $2,000,000. For me, the most enjoyable projects have been additions because I enjoy helping people solve their home problems and meet their housing needs. Meeting people's needs is one of the major benefits of my job. Ensuring quality building, both structurally and architecturally, also provides satisfaction.

There are also challenges on my job. One of my greatest challenges is explaining to builders and clients the value of architectural services and justifying the expense. I am constantly battling with builders, trying to convince them that they should employ an architect rather than a draftsperson, who will do the work at a lower cost.

One tip I would give to students pursuing a career in architecture is that you always need to be learning so you are informed about new information and technology in the industry. Remember, architecture is a lot more than just design. All design ideas need to be adapted to the real world without losing the creative design. This challenge is at the heart of architecture.

dining area. The living-dining combination is economical and tends to increase the apparent size of the living room (Figures 21–15, 21–16, and 21–17). Kitchen eating space may also be provided if desired. A dining porch (Figure 21–18), located adjacent to the kitchen, can be substituted for a dining room.

The dining room shown in Figure 21–19 was planned for a small city lot that allowed no

Figure 21–15   *Folding doors can be used to separate living and dining areas.* (Courtesy Pella Corporation.)

Figure 21–17   *A solar dining room.* (Courtesy VELUX-AMERICA, INC.)

Figure 21–16   *A combined dining room and library.* (Courtesy Wood-Mode Cabinetry.)

Figure 21–18   *An informal dining bay.* (Courtesy Pella Corporation.)

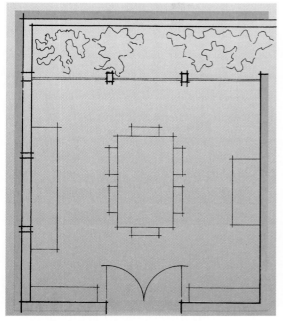

Figure 21–19   Dining room design.

pleasant views. The windows were arranged high on the walls to block the view without preventing natural lighting. On the rear wall, the window was combined with an enclosed courtyard designed to be landscaped. The double-entry doors, centered on the wall opposite the courtyard, provide a dramatic entrance.

BEDROOM DESIGN

The importance of the number of bedrooms in a house is emphasized when houses are described as two-bedroom, three-bedroom, or four-bedroom homes. Most houses are built with three bedrooms: a master bedroom and a second bedroom for one or two girls and a third for one or two boys. When children become adults and leave home, the second bedroom is often used as a guest bedroom, and the third bedroom as an office or den. A bedroom for a baby is termed a nursery.

To qualify for a FHA-insured mortgage, a minimum area of 80 sq. ft. is recommended for a single bedroom, and 120 sq. ft. for a master bedroom. As a general rule of thumb, bedrooms are often designed with the length 2' longer than the width. The rooms should be large enough that the beds can be freestanding (that

is, with only the head of the bed touching the wall), to make them easier to make (Figures 21–20, 21–21, and 21–22). In a small bedroom, combine areas used for dressing, circulation,

Figure 21–20   A contemporary bedroom with mirrored wall. (Courtesy California Redwood Association.)

and closet access into one larger area rather than three small ones.

Small, high windows allow greater freedom of furniture arrangement and provide privacy. However, windows should not be so small and high that escape in case of fire is impossible.

The bedroom shown in Figure 21–24 has been planned to include an element of symmetry. The designer has located all the major items of furniture so that there will be enough room for each. This is particularly important when unusually large or irregularly shaped furniture is to be used (Figure 21–25).

Each bedroom must contain a clothes closet. FHA recommends a minimum size of 2' by 4' for a man and 2' by 6' for a woman. Closets that are carefully designed to provide room efficiently for all sorts of clothes are becoming popular. These *storage walls* may have drawers, replacing bureaus and chests (Figures 21–26, 21–27, and 21–28). Table 21-1 gives sizes which might be of help in closet design.

Figure 21–22   *An ultra-modern bedroom design.* (Courtesy Pella Corporation.)

Figure 21–21   *An attic or above-garage bedroom.* (Courtesy VELUX-AMERICA, INC.)

Figure 21–23   *Bay window seats are popular in bedrooms and other areas.* (Courtesy Pella Corporation.)

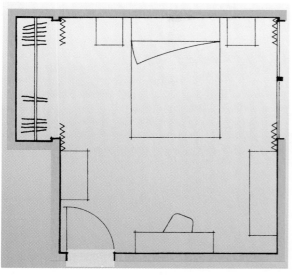

Figure 21–24   Bedroom design.

Figure 21–25   A modern canopy bed in a striking contemporary bedroom. (Courtesy California Redwood Association.)

Figure 21–26   A walk-in closet with dust-free storage. (Courtesy Scholz Design, Toledo, Ohio.)

Figure 21–27   A contemporary bedroom design with built-in storage. (Courtesy Wood-Mode Cabinetry.)

Figure 21–28   Consider toy shelves in a child's bedroom. (Courtesy VELUX-AMERICA, INC.)

TABLE 21–1   *Closets*

	Width	Depth	Height
Clothes closet	4'-0	2'-0	
Walk-in clothes closet	6'-0	4'-0	
Linen closet	2'-0	1'-6"	1'-0 between shelves
Suits, trousers, jackets, shirts, skirts		2'-0	3'-9"
Dresses, overcoats		2'-0	5'-3"
Evening gowns		2'-0	6'-0
Children, 6–12 years		1'-8"	3'-9"
Children, 3–5 years		1'-4"	2'-6"

REVIEW QUESTIONS

1. What proportions should be used when designing a living room?
2. What is furniture grouping?

3. List two advantages and disadvantages of a two-story entryway.

4. Identify four items to consider when deciding upon the placement of the kitchen on a floor plan.

5. Define room circulation.

6. What features should be offered in a typical bedroom?

7. What are some typical sizes for a closet in a master bedroom?

8. What are the minimum FHA-insured mortgage bedroom sizes?

9. List the furniture contained in an average:

   a. Living room
   b. Dining room

   c. Master bedroom
   d. Child's bedroom

ENRICHMENT QUESTIONS

1. Characterize and differentiate between living areas typically found in a residence.

2. Why should furniture groupings be considered when designing a floor plan?

3. What are some advantages and disadvantages of a dining-living room combination as opposed to a separate dining room and living room?

4. What should be considered when locating bedrooms in a residence?

5. Should the FHA be able to require minimum room sizes? Why or why not?

PRACTICAL EXERCISES

1. Sketch, draw, or computer-produce individual room layouts for the following rooms. Show the placement of major furniture.

   a. 15' × 22' living room with fireplace.

   b. 12' × 14' dining room with alcove for a serving table.

   c. 14' × 18' family room with fireplace and pass-through counter from the kitchen.

   d. 12' × 14' master bedroom with "his" and "hers" closets.

2. Sketch, draw, or computer-produce the living, dining, and sleeping areas for your original house design.

CHAPTER 22

Service Areas

OBJECTIVES

By completing this chapter you will be able to:

🏛 Identify the service areas found in a residence.

🏛 Prepare preliminary room designs for the service areas of a residence.

KEY TERMS

Home office	Porch
Computer area	Patio
Bathroom	Lanai
Laundry	Garage
Storage	

Having designed the major areas of kitchen, living room, dining room, and bedrooms, let's consider the principal supporting areas, including a home office, computer area, bathrooms, laundry, storage areas, porch, patio, lanai, and garage. As you design individual rooms, remember that these areas will have to be fitted into an integrated plan. Therefore, consider alternate acceptable sizes and shapes, and don't be too disappointed when you find that trade-offs will finally be required.

HOME OFFICE DESIGN

A separate room for a **home office** is often necessary for professionals such as architects, engineers, physicians, attorneys, bankers, teachers, and authors. The size of a home office can vary greatly, often equaling the size of a bedroom or even a master bedroom. Office-computer areas and office-libraries (Figure 22–2) are popular combinations. The home office is often called a *study* or *den*.

COMPUTER AREA DESIGN

Home computer equipment can be included in a contemporary office design (Figure 22–1) or a traditional office design (Figure 22–3). Consider the family room, kitchen planning center, and

Figure 22–1   A dramatic computer office design. (Courtesy Wood-Mode Cabinetry.)

Figure 22–2 *An office-library combination.* (Courtesy Wood-Mode Cabinetry.)

Figure 22–3 *A graceful and reserved computer center designed after an Old English office.* (Courtesy Scholz Design, Toledo, Ohio.)

Figure 22–4 *A computer office fitted into an expansion attic.* (Courtesy Wood-Mode Cabinetry.)

even basement and attic locations as sites for a **computer area** (Figure 22–4). When space is at a premium, computer equipment can also be set up in temporary pull-out tables (Figure 22–5), and battery-operated laptop computers, of course, can function at any convenient indoor or outdoor location.

BATHROOM DESIGN

A **bathroom** may be designed to be as small and compact or as large and compartmented as you wish. The trend is toward a compartmented room—keeping the toilet and tub or shower in separate, enclosed areas.

REAL WORLD   APPLICATIONS

Brainstorming

Your architectural team has been assigned the task of designing luxurious bathrooms for the honeymoon suites of a new five-star hotel being built in your city. Room size, floor plan, accessibility, ventilation, type and size of bath or shower, windows, vanity size and shape, mirrors, lighting, and type of flooring and wall covering will all need to be considered. There are many options for each design element. At first, the task may seem overwhelming, but if you begin with an important first step in problem solving—generating ideas—the job will quickly begin to come into perspective.

It is decided that a brainstorming session will be held. Since there are so many design elements to consider, the first session will focus on the type of bath and general layout of the bathroom. Once decisions are made in these areas other design elements will follow.

This session is designed to collect all ideas without limiting the possibilities. All team members are challenged to think freely and be open to all suggestions no matter how seemingly outrageous. Criticism and evaluation during the brainstorming process is strictly forbidden! One person's ideas stimulate the creativity of others. Each member is encouraged to be spontaneous, imaginative, and to use sketching, doodling, or any other techniques that increase creativity. Attention

should always be given to the person presenting the idea. Others are then invited to build upon these ideas. Time limits will be placed on each problem topic. In this case there will be a time limit for collecting ideas on type of bath and another time limit for general room layout. So that no ideas are forgotten and so all can be revisited, one person is designated to take notes. Finally, for this to flow smoothly, team members are encouraged to conduct a preliminary investigation before the meeting so everyone is informed, thereby allowing many alternative solutions to be generated.

The meeting goes very well. Although some of the solutions suggested will not be feasible, several satisfactory solutions are generated. The next step will be for the team to agree on one solution. This can be accomplished by eliminating any impractical or undesirable ideas and, thus, narrowing the number of possibilities. The final choice in design can be decided by having team members vote for their preferred design or simply by narrowing solutions to one.

In teams of three or four, brainstorm design solutions for the honeymoon suite bathrooms. You may choose any specific design element on which to focus. Be sure to follow the brainstorming guidelines delineated above and have one member of your team take notes.

Figure 22–5   *A pull-out computer desk for dual use of space.* (Courtesy Wood-Mode Cabinetry.)

Figure 22–6   *A luxurious and colorful bathroom design.* (Courtesy Kohler Co.)

For economy in plumbing, keep the bathrooms near the kitchen. If possible, avoid specifying a window behind a bathroom fixture. A window behind a tub or toilet is awkward, and one behind a lavatory interferes with the mirror.

The economy bathroom shown in Figure 22–8a has been provided with a wall-hung toilet and lavatory. These will give the effect, visually, of enlarging the room. In addition, the floor will be easier to clean. A one-piece plastic tub and shower unit eliminates the need for wall tile.

The compartmented bathroom shown in Figure 22–8b has been provided with two lavatories (for a larger family), a built-in tiled vanity, a corner whirlpool bath, and a stall toilet with swinging doors for privacy.

Bathrooms must have adequate ventilation, either by a window or an exhaust fan. For safety reasons, bathroom convenience outlets should include ground fault interrupters (GFIs), and electrical switches should not be placed within reach of the bathtub.

Design for Accessibility

If you are designing a bathroom for a building inhabited by a person with physical impairments, specify ADA-approved fixtures such as the wheelchair-accessible shower shown in Figure 22–7. This shower measures $40\frac{1}{2}$" × $40\frac{1}{2}$" × 80" and has a folding transfer seat, a hand-held

Figure 22–7   *Barrier-free shower.* (Courtesy Florestone Products.)

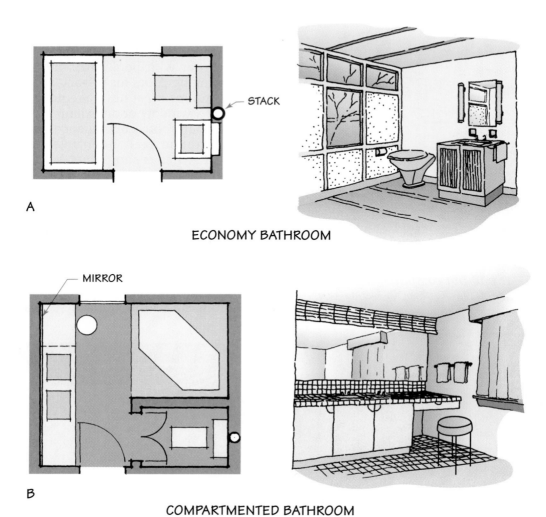

A

ECONOMY BATHROOM

B

COMPARTMENTED BATHROOM

Figure 22–8   Bathroom design.

TABLE 22-1   *Bathroom Equipment*

	Width	Depth	Height	Minimum clearance between:	
Lavatory (washstand)	1'-10"	1'-7"	2'-9"	Front of toilet and wall	1'-6"
Built-in lavatory	2'-0 min.	1'-10"	2'-9"	Side of toilet and wall	6"
Bathtub	5', 5'-6", 6'	2'-6"	1'-4"	Front of lavatory and wall	2'-0
Whirlpool	5', 6', 7'	3', 4', 5'	1'-6"	Front of lavatory and lower fixture	1'-6"
Shower	3'-0	3'-0	7'-0	Edge of tub and wall	2'-0
Toilet	1'-8"	2'-4"	2'-4"	Edge of tub and other fixture	1'-6"
Medicine cabinet	1'-2"	4"	1'-6"		

shower head, and grab bars. The manufacturer is Florestone Products, 2851 Falcon Drive, Madera, CA 93637.

LAUNDRY DESIGN

The homemakers of today spend less time in the laundry than did their predecessors. However, more time is spent on laundry chores than on any other work with the exception of food

preparation. The **laundry**, then, should be carefully planned. There is a sequence of laundry work just as there is a sequence of kitchen work. Proper planning should provide for each of the following steps:

1. Collection of laundry (hampers, laundry chute)

2. Treating of spots (a sink for pretreating spots and stains)

REAL WORLD APPLICATIONS

Water Quality

A new client has just purchased 4 acres of wooded land on which they plan to build a new home. Often the city water supply is not available to a new residence in rural areas. A water well is necessary, but there is some concern about the quality of the water supply since there are some coal mines nearby. A water quality test is important for new home construction because additional changes may need to be made to plumbing and electrical plans to accommodate extra equipment. A typical water test will commonly check for the following items:

🏛 **pH.** The pH measures the hydrogen ion concentration. In this case, if the pH of the groundwater were low (a value of 6 or less), it would indicate contamination from mine water. The normal pH of groundwater is between 6.5 and 8.0.

🏛 **Hardness.** The hardness of the water measures the amount of dissolved minerals, mostly calcium salts. Hard water can be a problem because soap scum can easily form, scale can build up in pipes, and the water may have an unpleasant taste. Water with a hardness greater than 100 mg/L of calcium can

be considered hard. Water hardness can be treated with a water softener.

🏛 **Iron.** Iron contamination with levels above 0.5 mg/L would indicate contamination from mining. Iron is not toxic but will stain ceramic tubs and sinks. A commercial water treatment and filtration system can remove iron and other dissolved minerals.

🏛 **Fecal Coliform.** Bacteria from animal waste in drinking water is a serious problem. Sources can include animals getting into the well or spring area and farm runoff. Chlorination corrects this problem.

To gain more insight into water quality and testing, take a sample of your water to an environmental laboratory. For a nominal fee they will conduct these and other tests to determine the quality of your water. What additional tests were performed? Are there any tests that are recommended specifically for your local area? What recommendations, if any, were made for improving your water quality?

Contributed by P. J. Palko

3. Washing and drying (the core of the laundry)

4. Finishing (a table or counter for folding and ironing; may be away from the area where the washing is done)

A storage area for soap and supplies must be specified also.

There are a number of possible locations for a laundry (Figure 22–17).

Bathroom

The most efficient location for a laundry is near the bedrooms and bath, where most soiled clothes originate. Additional advantages of a bathroom location are economy of plumbing and suitable interior finish. A disadvantage to be considered is the sacrifice of privacy. A combination bathroom and laundry should be a good-sized room that is well vented to remove excess moisture. Figure 22–17 shows a possible arrangement. The hamper is filled from the hall and emptied from the laundry. The shower doubles for drip-drying clothes.

Kitchen

The kitchen is the most obvious laundry location since placement here simplifies plumbing connections. However, a well-planned kitchen-laundry combination provides for separate areas or rooms so that clothes and cooking are not competing for the same space. Notice that the M residence has a small utility room containing the laundry adjacent to the kitchen.

Family Room

The family room is an ideal laundry location for the large family; the homemaker can watch the children while doing the laundry chores. Although some might object to the appearance and noise of laundry equipment here, this can be overcome by the use of sound-reducing folding, sliding, or accordion partitions.

Entry

A rear hall or area may be the perfect solution in some cases, especially if a rear mud room, which already requires plumbing connections, is specified.

Figure 22–9 *Pedestal lavatory with integral countertop.* *(Courtesy Kohler Co.)*

Figure 22–10 *"His" and "hers" double vanity.* *(Courtesy Wood-Mode Cabinetry.)*

Figure 22–11 *A corner shower. (Courtesy Kohler Co.)*

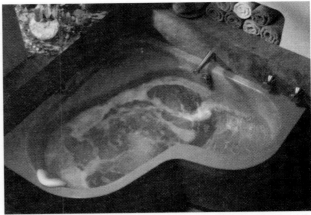

Figure 22–14 *A corner whirlpool for singles or doubles. (Courtesy Kohler Co.)*

Figure 22–12 *An 8-jet hydro-massage shower.*
(Courtesy Kohler Co.)

Figure 22–15 *A palladian window helps brighten this bathing alcove. (Courtesy Andersen Corporation.)*

Figure 22–13 *A steeping bath or whirlpool.*
(Courtesy Kohler Co.)

Figure 22–16 *Bathroom with pillowed whirlpool, bidet, toilet, and double lavatories. (Courtesy Kohler Co.)*

BATHROOM LAUNDRY

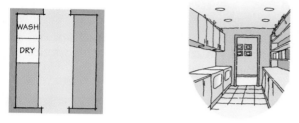

KITCHEN LAUNDRY

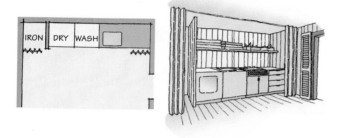

FAMILY ROOM LAUNDRY

Figure 22–17   Laundry design.

Basement

Basement locations are often desired, since the "free" basement area is used instead of prime first-floor area. Also, a washer that accidentally overflows will cause less damage in the basement. The major disadvantage of the basement location is the additional stair climbing required.

STORAGE DESIGN

Anyone will tell you that no house can have too much **storage** room. As a minimum, a well-designed house will have one closet per bedroom, two closets in the master bedroom, a linen closet, a utility closet, coat closets, and provision for exterior storage. The most economical way to build storage space into a resi-

TABLE 22–2 *Closet Sizes*	
Each bedroom closet	2' × 4' or 6'
Master bedroom closets	two 2' × 4' or 6'
Linen closet	1'-6" to 2' × 2'
Utility closet	2' × 4'
Guest closet	2' × 2'
Family coat closet	2' × 4'
Exterior storage	2' × 10'

dence is by use of a *closet wall:* a double wall containing closets that serve rooms on opposite sides. Closets should be designed 2' deep and wide enough to serve their function.

A small house should have approximately the sizes of closets shown in the accompanying Table 22–2.

In addition, it is advisable to plan for storage walls for special activities: a home computer and printer, video recording equipment, a card table and chairs, games, hobbies, toys, musical instruments, shop tools, and seldom-used equipment.

PORCH DESIGN

If a **porch** is desired, it is usually located at the rear of the house for maximum privacy. Also, it should be conveniently near the kitchen if outdoor meals are to be served. Consider screening the porch for comfort in the summer and glazing it for winter use. A roofed, partially enclosed porch is termed a "veranda." An open, unroofed porch is termed a "deck." A porch with extensive windows or skylights is termed a "sunroom" or "Florida room" (Figures 22–18 and 22–19).

PATIO DESIGN

The Spanish term **patio** or the Italian term "terrace" are used to describe a courtyard, usually paved, and often adjacent to a residence and used as an outdoor living-dining (and even kitchen) area. A patio deck (floor) may be constructed of pressure-treated lumber, brick, flagstone, or concrete (Figures 22–20 and 22–22). The size of patios can vary greatly, from a minimum size of 12' × 12' to a more typical size of 16' × 20'.

Figure 22–18 *Some sunroom designs.* (Courtesy Andersen Corporation.)

Figure 22–19 *The Florida room.* (Courtesy Andersen Corporation.)

Figure 22–20 *A wood-decked patio.* (Courtesy Pella Corporation.)

A central, unroofed patio is called an "atrium." Plans for the atrium A residence are presented throughout this text.

LANAI DESIGN

The Hawaiian term **lanai** or the Italian term "loggia" are used to describe a roofed outside

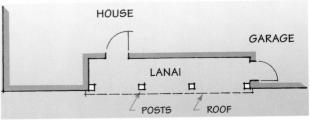

Figure 22–21 *A typical lanai.* (Courtesy Pella Corporation.)

hallway with one or more open sides. The lanai shown in Figure 22–21, for example, offers a covered passageway from the front entrance to the garage entrance, and at a construction cost less than an interior hall.

GARAGE DESIGN

A **garage** is usually attached to the house in some manner to provide a sheltered entry. This may be done by an attached first-floor garage, a breezeway, or a garage built into the basement. A basement garage will fit well in a split-level house; it is economical and offers heat as well as shelter. An attached garage or breezeway fits well in a small house design since it lengthens the house exterior. The breezeway may double as an open or closed porch. When completely enclosed, this space may be more efficiently used for living.

Consider a carport if you do not wish an enclosed garage. The carport is economical, but it does not offer the same protection from snow and rain. The deciding factor between garage and carport is usually climate. Whatever method is used to house the automobile, remember to include a sizable storage wall for automotive equipment, gardening tools, and toys.

Figure 22–22   *A luxurious pool patio.* (*Courtesy Pella Corporation.*)

REVIEW QUESTIONS

1. What equipment should be considered when designing a home office or computer area?

2. What fixtures are required in a full bathroom?

3. Name the four activity centers included in a laundry room. What appliances are associated with each?

4. Identify the minimum sizes for four different types of closets.

5. Differentiate between a veranda and a deck.

6. What is the typical size for a patio?

7. What are the advantages and disadvantages of a carport compared to a garage?

8. Indicate the alternate names for:
   a. Porch
   b. Sunroom
   c. Patio
   d. Lanai

ENRICHMENT QUESTIONS

1. How many bathrooms should be specified for the following houses? How did you arrive at that number?
   a. Economy one-bedroom summer house
   b. Moderate-priced three-bedroom ranch house
   c. Higher-priced four-bedroom, two-story colonial house

2. Why should the location of a laundry be carefully considered?

3. What should be considered when placing the bathrooms, garage, and family rooms with regard to the overall floor plan design?

PRACTICAL EXERCISES

1. Sketch, draw, or computer-produce to scale room layouts for the following rooms. Show furniture placement.
   a. 8' × 10' home office
   b. 6' × 6' computer area
   c. 6' × 8' bathroom
   d. 7' × 8' laundry area
   e. 20' × 20' double garage

2. Sketch, draw, or computer-produce to scale the service areas for your original house design.

CHAPTER 23

Preliminary Layout

OBJECTIVES

By completing this chapter you will be able to:

- Design a house using methods of:
  - Prototype
  - Templates
  - Interior planning
  - Overall planning
  - Computer-aided planning
- Prepare thumbnail sketches, preliminary sketches, and finished sketches for your original house design.
- Administer an effective critical plan analysis for a house design.

KEY TERMS

Orientation	Wind orientation
Solar orientation	Circulation
Topographical orientation	Efficiency
View orientation	Prototype

The preparation of a satisfactory preliminary layout is undoubtedly the most difficult but yet most satisfying phase of architectural design. As mentioned in Chapter 22, the placing of all the rooms in the correct relationship to each other may be likened to completing an unusual jigsaw puzzle composed of flexible pieces which must fit together in a workable pattern. This puzzle is further complicated by some considerations in addition to those of the individual room designs and their relationships:

1. The program requirements
2. Orientation
3. Circulation
4. Efficiency
5. Elevations

Each of these will be considered in turn.

THE PROGRAM REQUIREMENTS

The designer must meet *all* the requirements set up in the program. The program, you recall, reflects the decisions previously made on the primary considerations of type of plot and energy system, styling, number of rooms and floors, shape of plan, inclusion of basement or attic, expansion, zoning, and cost (Figure 23–1).

ORIENTATION

The term **orientation** refers to the compass location of the various rooms of the house to make best use of sun, topography, breezes, and views.

Solar Orientation

Solar orientation refers to orientation designed to take full advantage of exposure to the sun at all times. Major living areas should face south for winter solar heat, and the minor areas

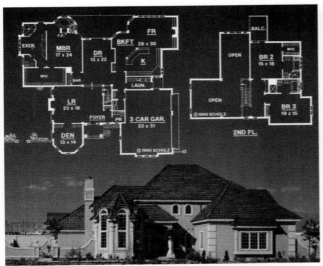 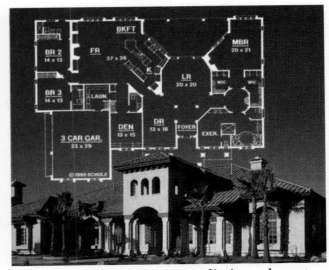

Figure 23–1 *A floor plan and front elevation are major considerations in any preliminary layout.*
(Courtesy Scholz Design, Toledo, Ohio.)

should face north. The breakfast room should face east to get the morning sun. Children's bedrooms should *not* face west if the occupants are expected to go to bed early.

Topographical Orientation

Topographical orientation refers to orientation designed to take full advantage of the land contour. For example, a house on a sharply sloping plot might be designed so that lower-level rooms would open directly to the low portion of the plot, and the upper-level rooms would open to the high portion of the plot. In general, it is good to keep excavation to a minimum, because the less the natural contour of the site is disturbed, the less Nature will try to disturb the occupants of that site.

View Orientation

View orientation refers to orientation designed to take advantage of the surrounding scenery. Major rooms such as the living room and master bedroom should be oriented to obtain the best possible view of the interesting features of the surrounding countryside. Minor rooms such as the utility room, laundry, and garage can face the less desirable views. In fact, some city dwellings have been built with no windows at all in the outside walls—all windows open upon landscaped interior courts.

Wind Orientation

Wind orientation refers to orientation designed to take full advantage of wind patterns at different times of the year. Major living areas should be located to take advantage of prevailing summer breezes, and the minor areas may serve to block off the winter winds.

Designing for Orientation

The best method to design for optimum orientation is to sketch the plot showing the direction of the sun, slope of land, wind, and best views. Figure 23–2 shows an orientation sketch for the plot of the A residence. A study of this sketch and the program requirements will indicate the best location of many of the rooms.

CIRCULATION

The designer must carefully consider the patterns of traffic among rooms to ensure convenient and, at the same time, efficient circulation. **Circulation** areas are halls, stairs, and the lanes of travel through rooms. Successful circulation means that there are convenient pathways between the rooms or areas that have the most connecting traffic. Usually this is accomplished by planning these rooms adjacent to each other with a common doorway. When halls are used, they should be kept as short as possible; when rooms are

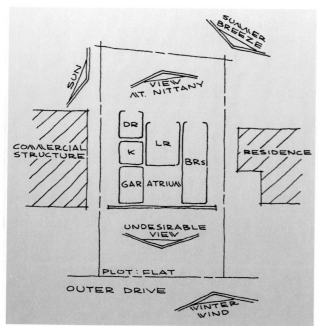

Figure 23–2   Orientation sketch for the A residence. (Courtesy Mr. and Mrs. Donald W. Hamer.)

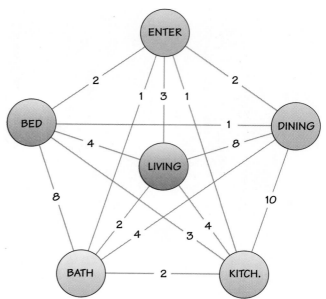

Figure 23–3   House circulation guide. The numbers indicate the average number of daily trips between rooms.

used for circulation, they should be planned so that the most direct route is across a corner or along one side of the room. With this arrangement, traffic will be less likely to disturb those using the room.

Probably the best way to ensure satisfactory circulation is to group all the individual rooms into three main zones:

1. Living (living room, front entrance foyer, powder room)
2. Sleeping (bedrooms and bathrooms)
3. Dining (dining areas, kitchen, service door)

Rooms of the same zone must be planned to have excellent circulation. For example: (1) The living room should be adjacent to the entrance foyer. (2) It should be easy to reach a bathroom from any bedroom. (3) The kitchen and dining room should not be separated by a hall.

In addition, the most often used rooms of different zones should be planned to have good circulation. The dining areas should be easily reached from the living area, for instance. Figure 23–3 shows the relative circulation between

the areas of an average home. The numbers in the lines connecting the rooms are the number of trips made each day between those rooms by the average occupant. Compare circulation between *entrance* and *living* (three trips) and *entrance* and *kitchen* (one trip). This shows that it is more important to have the living room near the front entrance.

Design for Accessibility

If you are designing a building that will be inhabited by persons with physical impairments (such as impaired mobility, sight, or hearing), special design adjustments are necessary. For example, all hallways must be at least 3'-0 wide to accommodate wheelchairs. See Chapter 47, Design for Accessibility, for further detail. Remember the ADA requirement that all public and commercial buildings should be accessible.

EFFICIENCY

Halls serve a very definite purpose, but should be kept at a minimum. They should not occupy more than 10 percent of the total area. The efficiency, or *tare*, of a house shows the percentage of area actually occupied by rooms. As you know, the first-floor area may be taken up by stairs, heating unit, halls, walls, closets, and vestibules. Although these are all necessary items, they must be kept to a minimum in the efficient—or low-cost—house. Seventy-five percent is considered excellent efficiency. Under 70 percent is considered poor efficiency.

ELEVATIONS

The plans of the house should not be developed too far before elevations and framing are considered. Actually, the good architect will not completely finish one drawing at a time, but will jump from one to another until all the major plans are finished simultaneously.

Accessibility

If you are designing an accessible building, use entrance ramps in place of steps. See Chapter 47 for more detail.

PLANNING PROCEDURE

Designers do not all have the same approach to obtaining a satisfactory preliminary layout. Five methods are widely used:

1. Prototype
2. Templates
3. Interior planning
4. Overall planning
5. Computer-aided planning

Let's look at each of these methods from the simple to the sophisticated. Then you can decide on the method you will want to use.

Prototype

The method of planning recommended for inexperienced persons is the **prototype**. A prototype is an example that you wish to copy fully or partially. The prototype may be an actual house you have seen and admired or a set of plans that appears to meet your requirements.

REAL WORLD 🚪 APPLICATIONS

Calculating Percentages

 A percentage is a share or proportion of a whole that is based on 100 units. The units could be millimeters, inches, square feet, pounds, or any other unit of measure. Follow these steps to calculate a percentage:

A living room is 300 sq. ft., of which 90 sq. ft. will be taken up by a sunken conversation pit. What percentage of the living room is occupied by the pit?

1. Set up the formula: $\frac{part}{whole} \times 100 = $ %

2. Fill in the appropriate numbers: $\frac{90}{300} \times 100 = $ %

3. Divide the fraction: $90 \div 300 = 0.3$

4. Multiply the result by 100: $0.3 \times 100 = 30$

5. Explain the result in words: "The conversation pit represents 30 percent of the total living-room area."

Try this problem yourself.

🏛 A small residence is 1850 sq. ft. There are two bedrooms measuring 9' × 11' and 12' × 14'. What percentage of the total square footage is occupied by the bedrooms?

Plans may be found in a variety of sources: popular home magazines, books on residential design, and newspapers. These sources publish houses that have been designed by architects, and they are able to furnish a complete set of drawings and specifications at comparatively low cost. Figure 23–4 shows three typical layouts from booklets published by Weyerhaeuser. Some other sources are listed at the end of this chapter. Building material suppliers also have sets of "stock" plans that may be borrowed or purchased.

Probably you will not be fortunate enough to find a good plan that meets all the requirements of your program, but you may find a plan that will be satisfactory after some modifications and alterations are made. If major alterations are necessary (such as moving the location of a stairway), you will find that you will need to redesign the entire plan.

Excellent results have been obtained using the prototype method, but remember that this is not original designing; it is merely copying someone else's design!

Templates

A second method of planning is by use of templates: first a set of furniture templates to obtain the individual room design, and then a set of individual room templates to obtain the preliminary layout.

A furniture template should be made for each existing or proposed piece of furniture. The templates may be cut from graph paper or cardboard using a reasonable scale. The furniture underlay in Chapter 19 was drawn to a scale of $\frac{1}{4}$" = 1', which has proved to be a satisfactory working scale. Obviously much measuring will be saved if $\frac{1}{4}$" grid paper is used.

The furniture templates can be moved to various positions until a satisfactory grouping is obtained. Then draw the outline of the furniture templates together with lines showing

Figure 23–4   Source books provide many prototypes. (Courtesy Weyerhaeuser Company.)

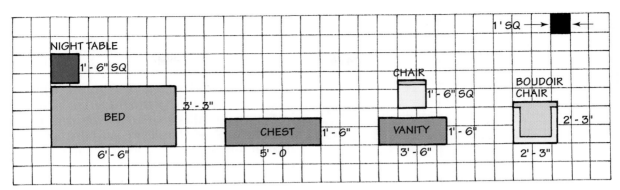

Figure 23–5 *Furniture templates.*

the size and shape of the entire room. Remember to allow for circulation areas, doors, windows, and closets.

EXAMPLE

To design one of the bedrooms for the A residence, templates (Figure 23–5) are made for the principal pieces of furniture: the bed, night table, chest, vanity and its chair, and boudoir chair. These templates are shown in Figure 23–6 arranged in two groupings. When you study Figure 23–6, you will notice an inherent disadvantage in the template method of design: The furniture will have to be rearranged if the doors and windows must be placed at another location. And any change in furniture arrangement usually means a change in room dimensions also.

After the dimensions of each room are determined by use of furniture templates, room templates may be made and used in a similar manner to determine alternate floor plan layouts. However, you must remember to leave spaces between the room templates—small 6" spaces for the wall thickness and large 3' or 4' spaces for closets, stairs, and halls.

EXAMPLE

To design the first-floor layout for the A residence, templates are made for the living area, dining area, kitchen, garage, and bedrooms as shown in Figure 23–7. Since the sizes of the atrium and lavatories are not considered to be critical, no templates are made for them. These areas are kept in mind, however, just as are halls, closets, walls, and a stairway. The first trial proved unsatisfactory for the reasons noted in Figure 23–8. A later scheme shows improvement, but the last scheme shown in Figure 23–8 meets all established requirements. It is important that a record be made of all acceptable schemes so that they may be compared to arrive at the very best solution.

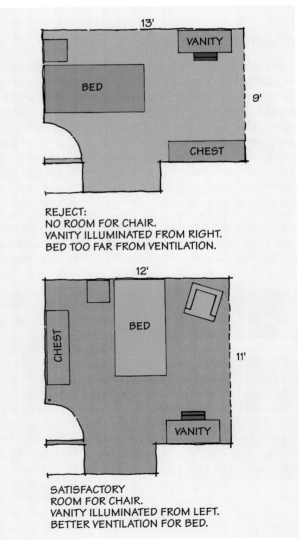

Figure 23–6 *Room layout using method of templates.*

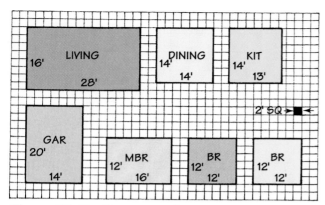

Figure 23–7   Room templates.

Interior Planning

A widely used system of architectural planning is called interior planning. Neither prototype drawings nor templates are used, but rather a series of sketches is made to develop the house plan. The architect first shows only basic concepts by use of *thumbnail* sketches, and then works up through a series of larger and more detailed sketches until a satisfactory finished sketch is obtained. The order of procedure is shown by an example from the A residence.

Thumbnail Sketches   Figure 23–9 shows several sketches made to show the general location of various areas. These are purposely drawn

very small, or *thumbnail*, so that there is no possibility of getting bogged down with detail at this early stage. Thus the designer is forced to consider only the essential elements.

Preliminary Sketches   Using the thumbnail sketch as a guide, a preliminary sketch as shown in Figure 23–10 is prepared. This will normally be done to $\frac{1}{8}$" = 1' scale on graph paper, as many beautiful ideas are seen to be obviously impossible when sketched to scale. Main attention is directed to room layout, circulation, and orientation, with no attention yet given to details such as door swings and exact window locations. The dwelling front will face the bottom of the sheet regardless of compass direction. Compass direction is indicated, however. Notice that the walls have been *pochéd* or darkened so that the plan will be easier to read. This is standard practice in preliminary and finished sketching.

Finished Sketches   The finished sketch may be a carefully drawn freehand sketch or a drawing done with drafting instruments. The scale is usually $\frac{1}{4}$" = 1' for a small residence and $\frac{1}{8}$" = 1' for larger buildings. Rough outside dimensions and approximate inside room dimensions are indicated.

Figure 23–11 shows a finished sketch of the A residence made using the preliminary sketch as a model. Notice that all the requirements of the program have now been met and that con-

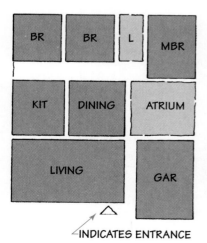

REJECT:
DESIRE ENTRANCE THROUGH ATRIUM.
KITCHEN TOO FAR FROM GARAGE.
VIEW IS ONLY FROM BEDROOMS.

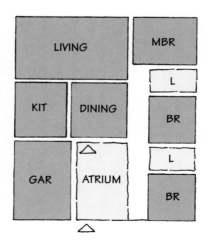

BETTER:
KITCHEN IS NEXT TO GARAGE & VIEW
IS FROM LIVING AREA, BUT ENTRANCE
IS NOW THROUGH DINING AREA.

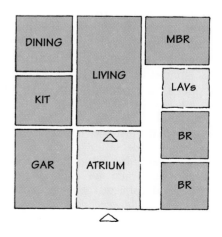

SATISFACTORY:
PROVISION FOR ALL REQUIREMENTS.

Figure 23–8   House layout using method of templates.

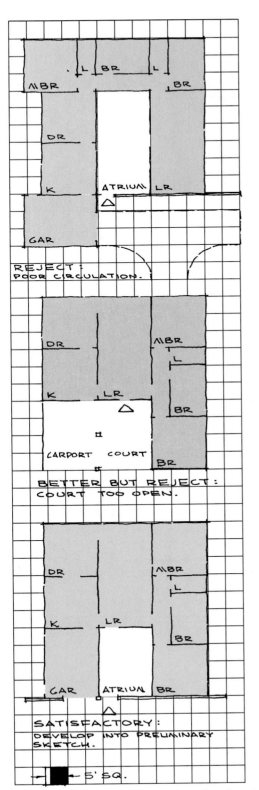

Figure 23–9 *Thumbnail sketches for the A residence.* (Courtesy Mr. and Mrs. Donald W. Hamer.)

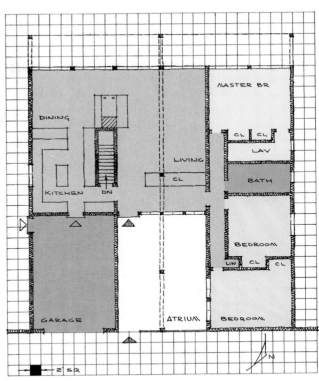

Figure 23–10 *Preliminary sketch of the A residence.* (Courtesy Mr. and Mrs. Donald W. Hamer.)

siderably more detail has been added. Some additional minor changes will be made and details added when the finished sketch is eventually redrawn as the final plan.

Overall Planning

Experienced designers and architects use the concept of overall planning since it is likely to produce the best results. The actual procedure for overall planning is identical to the procedure for interior planning except that other elements in addition to the floor plan are considered. Thus preliminary elevations are sketched along with preliminary floor plan, plot plan, and details.

The choice of house styling may affect the type of planning. A ranch house or modern house might be planned from inside-out (interior planning), but a symmetrical, traditional structure would be planned somewhat from outside-in (overall planning). For example, Figure 23–12 shows two alternate front elevations of the A residence which were sketched when the plan was developed. Notice that the side elevations were not sketched, allowing the

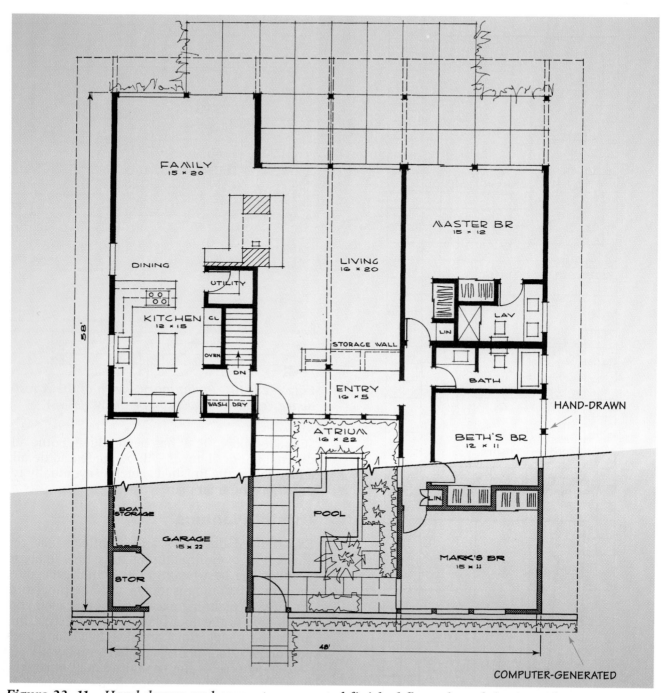

Figure 23–11 *Hand-drawn and computer-generated finished floor plan of the A residence. Compare with the thumbnail sketches and preliminary sketches, and with the finished drawings.* (Courtesy Mr. and Mrs. Donald W. Hamer.)

windows and doors to fall at random. The front elevation, then, represents overall planning; the side elevations represent interior planning only.

Computer-Aided Planning

Your computer graphics system can be used as an architectural planning tool to accomplish any

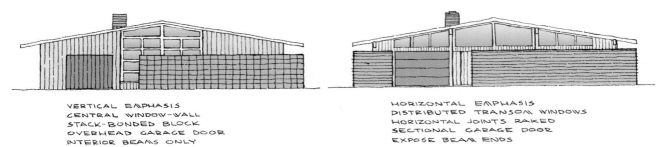

VERTICAL EMPHASIS
CENTRAL WINDOW-WALL
STACK-BONDED BLOCK
OVERHEAD GARAGE DOOR
INTERIOR BEAMS ONLY

HORIZONTAL EMPHASIS
DISTRIBUTED TRANSOM WINDOWS
HORIZONTAL JOINTS RAKED
SECTIONAL GARAGE DOOR
EXPOSE BEAM ENDS

Figure 23–12   *Preliminary front elevation sketches of the A residence.* (*Courtesy Mr. and Mrs. Donald W. Hamer.*)

of the preceding four planning methods efficiently. As shown in Chapter 6 (Computer-Aided Drafting) and Chapter 7 (Architectural Design with CAD), your preliminary ideas can be entered by puck or stylus on a digitizer tablet, displayed on your CRT screen, stored on diskette, and printed or plotted quickly and accurately. For example, a prototype layout can be entered on your screen and then modified and compared until a satisfactory plan is obtained. Or furniture and room templates can be entered and then moved or resized. Interior or exterior planning proceeds by advancing from thumbnail to preliminary to finished CRT displays. A series of finished hard-copies can then be compared to obtain the optimum solution.

PLAN ANALYSIS

The ability to critically analyze a plan—whether designed by you or by someone else—is a talent you must develop. The entire design process is really a series of these analyses, but the most comprehensive analysis should be made of the finished sketch, since it will serve as the basis for the working drawings.

Procedure

Each planning item that has been discussed should be studied for possible flaws or omissions:

1. *Program.* Each program requirement must be satisfied. Do not trust to memory; go through the program item by item.

2. *Orientation.* Check each room for the orientation of sun, wind, and views desired. Has best use been made of topography?

3. *Circulation.* Do not trust a visual check. Rather, take a pencil and actually trace the path of common circulation patterns. They should be simple and uncomplicated.

EXAMPLE

A typical early-morning pattern would be:

bedroom → bathroom → bedroom → kitchen → dining area → kitchen → coat closet → garage

4. *Efficiency.* Calculate hall area percentage (should be less than 10 percent) and house efficiency (should be greater than 70 percent).

EXAMPLE

Check hall area percentage and house efficiency of the A residence (see Figure 23–11).
Approximate total area: 2,200 sq. ft.
Approximate hall area: 150 sq. ft.
Approximate room area: 1,600 sq. ft.

Hall area percentage $= \dfrac{150}{2,200} \times 100 = 7\%$ O.K.

House efficiency (tare) $= \dfrac{1,600}{2,200} \times 100 = 73\%$ O.K.

5. *Elevations.* Carefully consider how the floor plan will coordinate with the exterior elevations. Also consider details and construction. The framing method may affect the maximum room size.

6. *Utilities.* Has space been allowed for utilities such as plumbing walls, heating, and chimney? Are the kitchen and bathroom plumbing adjacent for economy? Are bearing partitions on different floors directly over one another?

7. *Individual room design.* Study each room separately to see if it meets its requirements of function, aesthetics, and economy. Compare the proposed room sizes with your list of familiar rooms and their sizes. Are the rooms too commonplace and ordinary? Or have you planned extreme features which will be costly and eventually embarrassing? Figure 23–13 shows some common planning errors.

REFERENCE PLAN SOURCES

Better Homes and Gardens, 1716 Locust Street, Des Moines, IA 50336

Brick Institute of America, 11490 Commercial Park Drive, Suite 300, Reston, VA 22091

The Garlinghouse Company, 34 Industrial Park Place, Middletown, CT 06457

Good Housekeeping, 959 Eighth Avenue, New York, NY 10019

Homes for Living, Inc., 107-40 Queens Boulevard, Forest Hills, NY 11375

Home Planners, Inc., 3275 W. Ina Road, Tucson, AZ 85741

National Forest Products Association, 1250 Connecticut Avenue NW, Washington, DC 20036

National Home Planning Service, 37 Mountain Avenue, Springfield, NJ 07801

Portland Cement Association, Old Orchard Road, Skokie, IL 60077-4321

Sunset, The Magazine of Western Living, Lane Publishing Co., 80 Willow Road, Menlo Park, CA 94025-3691.

REAL WORLD APPLICATIONS

Presentation Outline

Five contractors are interested in buying your house plans. They have already seen the plans but want to hear from you regarding the benefits of your design. All five parties will be coming to a meeting where you will present the rationale behind your design and its advantages. Plan to focus on the preliminary layout, for you have taken great care to place rooms in a functional pattern. The program requirements are of little concern as all five contractors have the same basic requirements, which your plans meet. The focus of your presentation will be on orientation, circulation, efficiency, and elevations.

As a handout for the contractors, prepare a topic outline that they can follow as you make the presentation. A topic outline is designed to provide the reader with a general overview of the information. Short phrases are used rather than complete sentences or paragraphs. Most topics will be supported by at least two subtopics. An example of a topic outline is given below.

I. Topic

I. Interior Design of Plan "A"

A. Subtopic
  1. Supporting Detail
     a. Specific example
     b. Specific example

  2. Supporting Detail
     a. Specific example

     b. Specific example

B. Subtopic

A. Orientation
  1. Topographical Orientation
     a. Sloping front yard
     b. Level, wooded back yard

  2. View Orientation
     a. Master bedroom overlooks valley

     b. Formal and informal entrances

B. Circulation of traffic flow

Outline the remainder of your presentation using a topic outline format. Following this outline will not only put your audience at ease because they know what to expect, but you are now more prepared. Finally, provide enough spacing between topics and subtopics for the contractors to write notes.

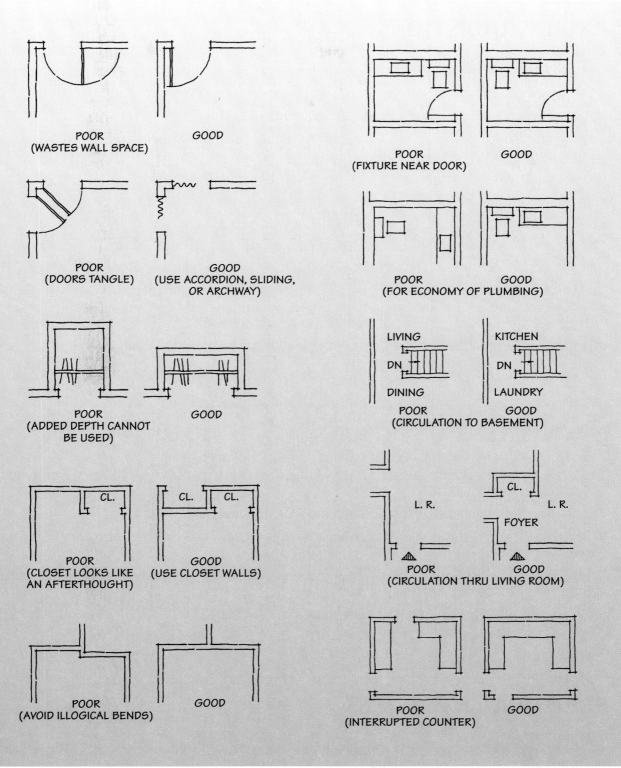

Figure 23–13 *Correction of common planning errors.*

REVIEW QUESTIONS

1. In architecture, what is meant by orientation?
2. Name four different types of architectural orientation.
3. What is meant by circulation when studying room layout?
4. Indicate which sets of rooms should and should not be planned adjacent to each other in residential design:
   a. Kitchen and dining room
   b. Kitchen and bedrooms
   c. Bath and bedrooms
   d. Bath and dining room
   e. Living and dining room
5. What special considerations should be given to room layout for individuals with physical impairments?
6. Halls should not exceed what percentage of the total area of the house?
7. What is the main advantage of planning the kitchen, bathrooms, and laundry near each other?
8. List the seven items that should be considered when analyzing the final plan.

ENRICHMENT QUESTIONS

1. How does the information determined by the primary considerations relate to preliminary layout?
2. Why is determining the orientation of a building important?
3. In what ways will the kind of family (e.g., young couple, three children, or elderly) influence the room layout?
4. Purchase a booklet that contains prototype houses. What are some advantages and disadvantages of using these plans?
5. What advantages does CAD have over hand-sketched ideas?

PRACTICAL EXERCISES

1. Hand-sketch or computer-sketch a preliminary layout of the A residence making any improvements you feel are desirable. Scale: $\frac{1}{8}$" = 1'-0.

2. Using the method of templates, design a residence having the following rooms: 13' × 19' living area, 9' × 9' dining area, 10' × 12' kitchen, 11' × 13' master bedroom, 9' × 11' bedroom, small bathroom, and a full basement.

3. Using the thumbnail sketches shown in Figure 23–14 as a guide, draw or computer-produce the following preliminary or finished sketches for a low-cost passive solar house.

   a. First floor plan

   b. Balcony floor plan

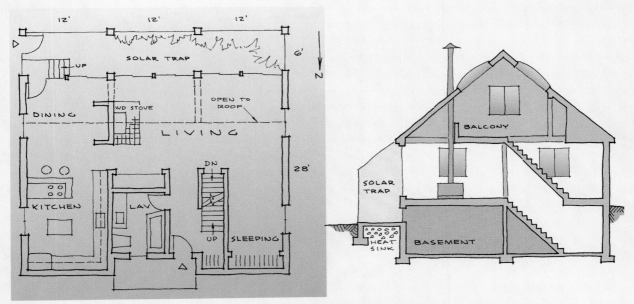

Figure 23–14 *Thumbnail sketches of a passive solar home.*

4. Complete the preliminary layout for your original house design using the planning procedure assigned by your instructor.

   a. Prototype

   b. Templates

   c. Interior planning

   d. Exterior planning

   e. Computer-aided planning

5. Make a critical plan analysis for your original house design by listing and explaining its strong and weak features.

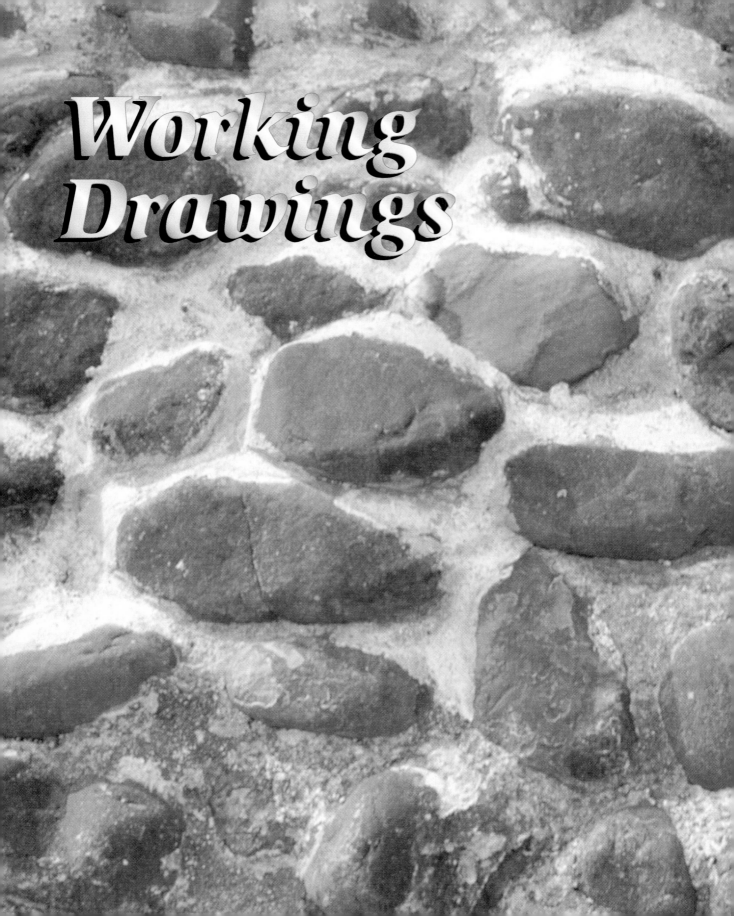

Working Drawings

SECTION IV

CHAPTER 24

Floor Plans

OBJECTIVES

By completing this chapter you will be able to:

- 🏛 Make a list of sheets included in a complete set of working drawings.
- 🏛 Prepare the foundation plan and floor plans for a residence or small commercial building, including:
  - ◼ Wall layout
  - ◼ Wall completion
  - ◼ Plan details
  - ◼ Built-in equipment
  - ◼ Dimensioning
  - ◼ Lettering
  - ◼ Checking
- 🏛 Prepare the foundation plan and floor plans for your original building design.

KEY TERMS

Working drawings
Specifications

WORKING DRAWINGS

The finished drawings made by the architect and used by the contractor are called **working drawings**. The working drawings, together with the **specifications** and the general conditions, form the legal contract between the owner and contractor. Since the working drawings are a major portion of the contract documents, they should be very carefully drawn.

A complete set of working drawings includes architectural, structural, electrical, and mechanical sheets in this order:

1. Title page and index (a perspective is often included)
2. Plot plan
3. Foundation plan
4. First-floor plan
5. Second-floor plan
6. Elevations
7. Sections
8. Typical details
9. Schedules
S1. Floor framing plan
S2. Roof framing plan
S3. Column schedule
S4. Structural details
E1. Electrical requirements
H1. Heating and air conditioning
P1. Plumbing
V1. Ventilation

Usually all of the working drawings are drawn to the same scale ($\frac{1}{8}$" = 1' or $\frac{1}{4}$" = 1'), with the exception of details, which are drawn to a larger scale, and the plot plan, which is drawn to an engineer's scale.

Before starting finished drawings, most drafters prefer to prepare *mock-ups* of each sheet. This organizes the set of working drawings so that related information fits on the same or adjacent sheets, and ensures that no drawing will have to be redrawn due to its poor placement on a sheet. Mock-ups are merely sketched forms (usually rectangles) in proportion to the finished drawings. They should allow room for all expected dimensions and notes. Mock-ups are also called *layouts, dummies,* or *cartoons.*

FLOOR PLANS

Of all the different kinds of working drawings, the floor plan is the most important since it includes the greatest amount of information. The floor plan is the first drawing started by the designer, but it may be the last finished because

Figure 24–1 *These dramatic designs represent only four of an infinite variety of floor plans and elevations.* (*Courtesy Andersen Windows, Inc., American Plywood Association, California Redwood Association, Andersen Windows, Inc.*)

the designer will transfer attention to the sections, elevations (Figure 24–1), and details required to complete the floor plan design.

A floor plan is actually a sectional drawing obtained by passing an imaginary cutting plane through the walls about 4' above the floor (midway between floor and ceiling). The cutting plane may be offset to a higher or lower level so that it cuts through all desired features (such as a high strip window). In the case of a split-level house, the cutting plane must be considerably offset.

If the finished sketch has been carefully made, the floor plan can be drawn without much trouble. Notice in the accompanying floor plan of the A residence (Figure 24–3), the similarity with the finished sketch of Figure 23–11. Of course, if the designer feels a sketch can be improved upon, it is done.

The steps used in drawing a floor plan are illustrated in Figure 24–2. A portion of the first-floor plan of the A residence (Figure 24–3) is used as an example.

FIRST-FLOOR PLAN

Step 1: Wall Layout

Lay out the exterior and interior walls very lightly on tracing vellum using a hard, sharp pencil. A scale of $\frac{1}{4}$" = 1' should be used for a residence, $\frac{1}{8}$" = 1' for a larger structure. Always indicate the scale in the title block or on the drawing. To save time measuring the thicknesses of exterior and interior walls, a wall guide may be made by marking the wall thicknesses on a strip of paper. The wall sizes shown in Table 24–1 may be used.

Step 2: Wall Completion

Still using a hard, sharp pencil, locate the windows and doors on the wall layout.

Windows   The final placement of windows is determined, keeping fenestration, compass direction, pleasantness of view, and amount of light and air required in mind. Window selection depends on the style, design, and appearance of the building. See Chapter 19 for more detail.

The width of the windows given on the plan is that of the sash opening. Remember

TABLE 24–1   *Wall Sizes*

	Thickness
Wood-frame walls	
Exterior walls with 2 × 4 studs	6"
Exterior walls with 2 × 6 studs	8"
Interior partitions	5"
Brick-veneered walls	10"
Brick walls	
With two courses of brick	8"
With two courses of brick and air space	10"
With three courses of brick	12"
Concrete block walls	
Light	8"
Medium	10"
Heavy	12"

to allow room for the surrounding framing in close conditions.

Doors   The final door placement should take into consideration the door swing and furniture placement. Plan for unbroken wall spaces where they are needed, making sure no unnecessary doors are specified. Remember that every swinging door takes up valuable floor space (the area through which it swings) and two wall areas (the area containing the door and the area the door swings against) (Table 24–2). Do not discount sliding

TABLE 24–2   *Door Sizes*

Interior doors:	2'-6" × 6'-8" × $1\frac{3}{8}$"
Allow for approximately 2'-9" rough opening	
Front door:	3'-0 × 6'-8" × $1\frac{3}{4}$"
Allow for approximately 3'-3" rough opening	
Rear door:	2'-8" × 6'-8" × $1\frac{3}{4}$"
Allow for approximately 2'-11" rough opening	

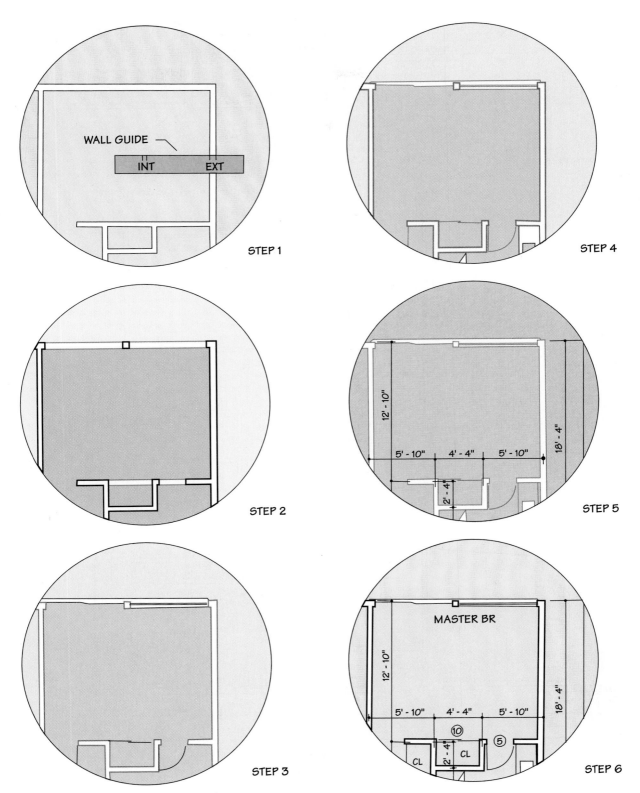

Figure 24–2   *Preparing a floor plan.*

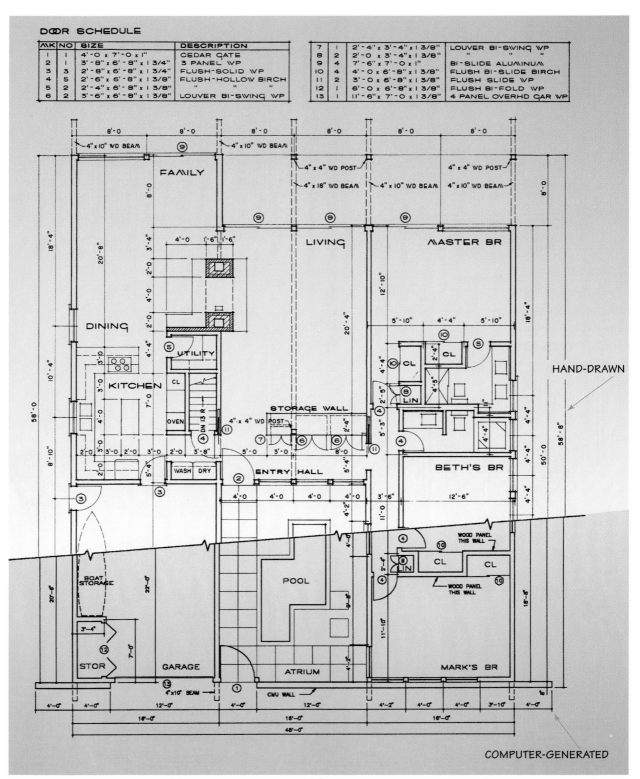

DOOR SCHEDULE

MK	NO	SIZE	DESCRIPTION
1	1	4'-0 x 7'-0 x 1"	CEDAR GATE
2	1	3'-8" x 6'-8" x 1 3/4"	3 PANEL WP
3	3	2'-8" x 6'-8" x 1 3/4"	FLUSH-SOLID WP
4	5	2'-6" x 6'-8" x 1 3/8"	FLUSH-HOLLOW BIRCH
5	2	2'-4" x 6'-8" x 1 3/8"	" " "
6	2	3'-6" x 6'-8" x 1 3/8"	LOUVER BI-SWING WP

7	1	2'-4" x 3'-4" x 1 3/8"	LOUVER BI-SWING WP
8	2	2'-0 x 3'-4" x 1 3/8"	" "
9	4	7'-6" x 7'-0 x 1"	BI-SLIDE ALUMINUM
10	4	4'-0 x 6'-8" x 1 3/8"	FLUSH BI-SLIDE BIRCH
11	2	3'-0 x 6'-8" x 1 3/8"	FLUSH SLIDE WP
12	1	6'-0 x 6'-8" x 1 3/8"	FLUSH BI-FOLD WP
13	1	11'-6" x 7'-0 x 1 3/8"	4 PANEL OVERHD GAR WP

Figure 24–3 Hand-drawn and computer-generated floor plan of the A residence. *(Courtesy Mr. and Mrs. Donald W. Hamer.)*

and accordion doors for closets, and especially for little-used openings.

Doors opening into rooms from a hall should swing into the room against the wall. Use sliding, folding, or accordion doors if space is at a premium, but remember that these should not be specified for often-used bedroom and bath doors since they require more time and energy to operate. As a rule, swinging doors should be used on all but closets.

Occasionally it is better to use archways in place of doors for the sake of economy, appearance, and spaciousness. They should be shown on the plan by two dashed lines.

After all windows and doors have been located, the wall lines are darkened using a soft

REAL WORLD   APPLICATIONS

Facts, Judgments, and Inferences

You have finally completed a very complicated set of residential plans. However, changes may need to be made based on the client's review of your work. The client made a number of comments during your meeting which could be very confusing if you cannot separate their facts, judgments, and inferences. A fact is a statement that is based solely on truth. "The dining area is separated from the kitchen by a counter" is an example of a fact. When the client states a fact you may want to clarify whether or not this is satisfactory to him or her. Otherwise, there is no way to know if any changes should be made.

A judgment expresses one's opinions about something, someone, or a situation. "I do not like the counter that separates the dining area from the kitchen" is the client's opinion about the size or placement of the counter. Judgments can be positive or negative and are important because you are attempting to satisfy the client. It is clear from this evaluation of the counter that you may need to change your plans. It is important and only fair to you, however, that you discover why the client does not like this counter. One way to identify judgment statements is to listen for terms of evaluation. These include right, wrong, good, bad, like, and dislike.

Finally, an inference draws a conclusion from facts or premises. "The counter will cause too much traffic congestion between the kitchen and dining area" is an inference. The client is not simply stating a fact or making a judgment but drawing a conclusion about traffic congestion based on the fact that a counter separates the two areas. You may be able to provide further information that will eliminate the concern about traffic congestion, or you may decide the inference is logical and change your plans.

Decide if these statements are facts, judgments, or inferences and identify a follow-up question you would ask the client to better analyze the statement.

1. The laundry room is too far from the bedrooms.

2. The master bedroom is 175 square feet.

3. Many of the rooms are not of adequate size.

4. There will not be enough room for two beds in the children's room.

5. The stairs leading to the basement are too steep for children.

6. I am right-handed and the dishwasher is on the left side of the sink in the kitchen.

7. It will be difficult to carry laundry down the stairway.

pencil. The wood frame wall convention is shown in the illustration of Step 2 in Figure 24–2.

Step 3: Details

Using a sharp, but black, line weight, add the floor plan details.

Windows Show the sill and glass as indicated in Chapter 28.

Doors Show the doors and their swings. Show the sill on exterior doors.

Stairs A preliminary cross-sectional layout must be first drawn to determine the total run and the number and size of treads and risers. This layout is saved and used later as the basis for the finished stair details. Draw about half of the full stair run and letter a small "UP" at the stair foot, and "DN" (down) at the stair head. Indicate the number of risers (for example: UP-14R). See Chapter 31 for more detail.

Fireplace Show the overall width and depth of brickwork, location of basement flue, ash drop, and the outline of the fireplace opening and hearth. Cross sectioning is used to indicate the brickwork. See Chapter 32 for more detail.

Step 4: Equipment

Show all built-in equipment, such as bathroom fixtures (bathtub, toilet, lavatory, and medicine cabinet), kitchen fixtures (cabinets, sink, built-in wall ovens, countertop burners, and built-in refrigerator or freezer), and closet fixtures (shelves and clothesrod). The location of a movable stove or refrigerator is shown even if it is not included in the contract. Notice that the wall cabinets are shown as hidden lines. A hidden line on an architectural plan may refer to a feature (cabinet, archway, or beam) *above* the level of the imaginary cutting plane. The locations of lighting and heating devices are included only when they will not make the floor plan crowded. Occasionally all furniture placement is shown.

Step 5: Dimensioning

Dimension all walls and partitions using dimension lines spaced approximately $\frac{3}{8}$" apart. The dimension lines are continuous with the $\frac{3}{32}$"-high figures lettered above them. The dimension figures should be in feet and inches (to the nearest $\frac{1}{16}$"), and should read from the bottom or

right-hand side of the drawing. In frame construction, the dimensions are given from the outside faces of the studs in exterior walls and to the center lines of interior partitions. Cumulative dimensions should lie in one line if possible, but try to avoid dimensioning over other features. Also remember to allow some open spaces for lettering the room names later on. Dots or small circles may be used in place of arrowheads. Check carefully that no *necessary* dimension is omitted, but some *minor* dimensions can be omitted for the sake of clarity. For example, the width of some closets may have to be scaled from the drawing. This is accepted practice since the exact location of these partitions is not critical. See Chapter 13.

Step 6: Lettering

Letter room names in the center or lower left-hand corner of each room without lettering over other features. Room lettering should be approximately $\frac{3}{16}$" high. Special explanatory notes on materials and construction are added using $\frac{3}{32}$"-high lettering. See Appendix D for the proper spelling of words often misspelled on architectural drawings.

A $\frac{1}{8}$"-high schedule mark (a number for doors and a letter for windows) is assigned to each door or window of the same size and type. The door numbers are placed within $\frac{1}{4}$"-diameter circles in a convenient location near the doors. The door schedule is added as shown in Chapter 30. Window marks and the window schedule may be included also, although they usually appear on the elevations.

Any additional information necessary for proper construction should be added at this point. Include such details as concrete porches, window areaways, hose bibbs, and electrical fixtures if they do not appear elsewhere.

Step 7: Checking

In a professional drafting room, checkers have a most responsible position. They must certify that they have checked and approved every line and dimension on the drawings. If done correctly, this procedure should guarantee that nothing has been omitted and that nothing is in error.

In a classroom, however, there is no professional checker, and you must check your own work. This is very difficult to do properly since

REAL WORLD  APPLICATIONS

Resource Allocation

Prospective homeowners, designers, and builders are constantly making decisions concerning the appropriate allocation of available financial resources. Typically, choices are made between requirements and needs compared to desires and luxuries. A requirement is usually chosen over a desire, but a choice between two desires is often a more difficult decision. For example, a fireplace is not a necessity, but some families would place it higher on a priority list than a formal dining area or a whirlpool bath. When allocating resources, a bit of flexibility can be obtained by postponing some construction. For example, an open carport can be designed so that it could be converted into an enclosed garage at a later time. Other delayed projects could include interior carpeting, painting, and wall papering, exterior landscaping, finishing of recreation and storage rooms, or the addition of an enclosed porch.

Practice allocating your available resources wisely while designing the following residence.

Assume you have been hired by a contractor to design a moderately priced, 2,000 sq. ft. three-bedroom home. The contractor plans to build a development of similar homes on $\frac{1}{2}$-acre lots. The site that was chosen is located in the northeast part of the country and is gently sloping. Decisions regarding the total number of rooms, room layout, shape design, types of rooms, and basement options are up to you. The cost of materials should not exceed $30,000. These homes should be able to be constructed in a period of two months. Considering the time deadline and material, the house should be rather simple.

Design a basic floor plan layout with these specifications that can be built for less than $100,000 total. This figure includes the lot cost, tap-in fees for water, gas and sewage, labor costs, cost of materials, and your fee.

---

you will, in all likelihood, repeat any drawing errors in your checking. If possible, it is a good idea to trade the responsibility of checking with another student. If this is not possible, you should check the drawing only after it has been put aside for several days to assure a fresh outlook.

SECOND-FLOOR PLAN

After the first-floor plan is completed, related plans such as those for the basement and second floor are started by tracing common features from the first-floor plan. The exterior walls may usually be traced, together with bearing partitions, plumbing walls, stairways, and chimney location. Principal walls should be over one another from foundation to roof insofar as possible, to ensure a stiffer frame and facilitate

TABLE 24–3   *Foundation Wall Sizes*

	Thickness
Under wood-frame house	
Concrete block	8", with pilasters 16' oc
Poured concrete	10"
Under veneered construction	
Concrete block	10"
Poured concrete	10"
Under solid masonry construction	
Concrete block	12"
Poured concrete	12"

plumbing and heating installations. Also, the fenestration may require that first- and second-floor windows be directly over one another.

BASEMENT PLAN

The basement plan or foundation plan (Figure 24-4) is begun by tracing common features from the first-floor plan: the outside lines of exterior walls, the stairway, and the chimney location. The foundation wall sizes shown in Table 24-3 may be used.

A hidden line is used to indicate the size of foundation footings and column footings.

Dimensioning practice is somewhat different from that used for the floor plans, in that dimensions will run from the faces of all masonry walls or partitions rather than from their center lines. Also, the openings for windows and doors in masonry walls are dimensioned to faces rather than to center lines.

Some special considerations in designing basements must be remembered:

1. The size and location of columns and girders.
2. The size, spacing, and direction of joists (double-headed arrow indicates joist direction).
3. Stiffening pilasters, needed on long, straight runs of wall.
4. The furnace, located near chimney flue (within 10').
5. Provision for fuel storage, near the driveway but removed from the furnace (beyond 10').
6. The water heater.
7. The water meter, located near water line entrance.
8. Floor drains.
9. Electrical entrance panel, located near entrance pole.
10. Labeling of unexcavated areas, crawl spaces, and concrete floors, together with any additional helpful information. For example, the thickness of a basement floor is usually noted on the basement plan. Specify 3" minimum concrete for a basement floor and 4" minimum concrete for a garage floor or finished slab floor.

Notice that the basement of the A residence has been designed for the future construction of a recreation room, game room, shop, and lavatory. This allows better planning in locating necessary utilities.

COMPUTER-AIDED FLOOR PLANS

After a satisfactory preliminary floor plan has been entered into your computer graphics system, enlarge the plan to an appropriate finished scale and add refined window, door, stair, fireplace, and built-in details. Complete the dimensioning and lettering. Take advantage of available diskette menus of standard built-in symbols such as kitchen and bathroom equipment, or custom-design your own symbols if you prefer. When satisfied with a finished soft-copy floor plan, use your printer or plotter to produce a hard-copy, but retain the plan on diskette storage for future use or revision. For example, floor plans will be recalled later to serve as the basis (first layer) for roof framing, electrical, plumbing, and heating plans.

Refer to Chapter 7 for operating details on providing computer-generated floor plans.

REVIEW QUESTIONS

1. List the drawings normally included in a set of residential working drawings.

2. Hand-sketch or computer-sketch a section of each of the following walls, showing the actual size of materials to illustrate the total wall thickness normally used on a plan:

   a. Exterior wood-frame wall

   b. Interior wood-frame partition

   c. Exterior brick-veneered wall

   (Continued)

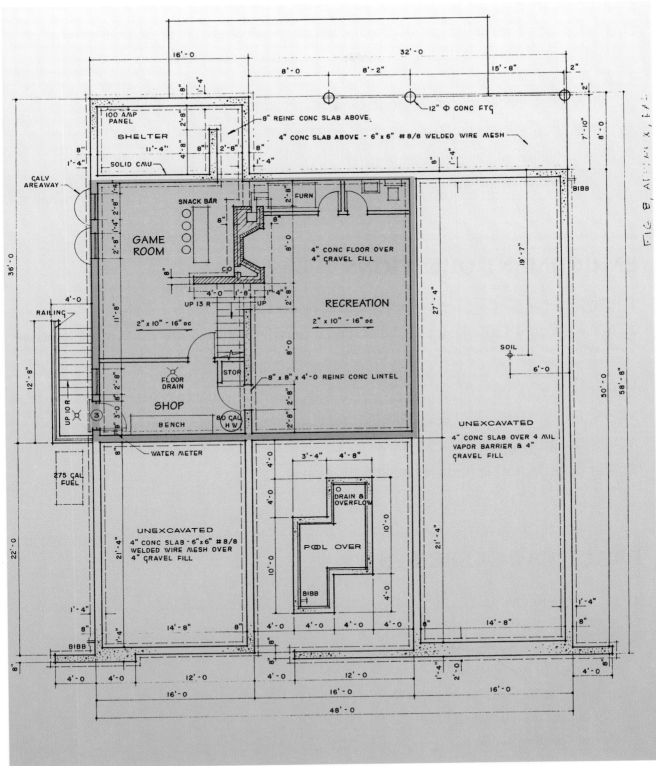

Figure 24–4   *Basement plan of the A residence. (Future construction in scarlet)* (*Courtesy Mr. and Mrs. Donald W. Hamer.*)

3. List the most commonly used size (width, height, and thickness) of the following doors:
   a. Front entrance door
   b. Rear entrance door
   c. Interior doors

4. What does a hidden line on an architectural plan represent? Give three examples.

5. When drawing a second-floor plan, what features may be traced from the first floor plan?

6. When drawing a basement plan, what features may be traced from the first-floor plan?

7. What is the purpose of a mock-up?

ENRICHMENT QUESTIONS

1. Why do other drawings need to be considered before the floor plan is completed?

2. In the layout of a basement plan, what special considerations must be made in the location of the:
   a. Furnace
   b. Fuel storage
   c. Water meter
   d. Electrical entrance panel

3. Why is it necessary to include different types of working drawing sheets?

4. Describe the steps used to draw a floor plan.

5. Why do walls need to be drawn at different widths?

PRACTICAL EXERCISES

1. Draw or computer-produce a title page and index for your set of working drawings to include:
   a. Title information (name of project, location, name of designer, date)
   b. Index listing number and title of each sheet
   c. Legend of symbols used
   d. List of abbreviations used
   e. Modular dimensioning note
   f. Interior perspective
   g. Exterior perspective

2. Using a scale of $\frac{1}{4}$" = 1', draw or computer-produce the following plans of the A residence:

   a. First-floor plan

   b. Basement plan

3. Draw or computer-produce a plan showing your concept of a dwelling to be built under one of the following conditions:

   a. Flexible plan (must be used equally well by families with one to four children)

   b. Low-cost, mass-produced

   c. Unpleasant view in all directions

   d. Physically handicapped couple

4. Draw or computer-produce a plan showing your concept of a dwelling of the future to be built:

   a. In a desert area

   b. On extremely mountainous terrain

   c. Entirely over water

   d. Completely underground

   e. Occupying a minimum amount of land

5. Draw or computer-produce the plans of your original house design:

   a. First-floor plan

   b. Second-floor plan

   c. Basement plan

CHAPTER

25

The Plot Plan

OBJECTIVES

By completing this chapter you will be able to:

🏛 Prepare the plot plan for a residence or small commercial building, including:

- ■ Property lines
- ■ Zoning requirements
- ■ Location of building
- ■ Location of utilities
- ■ Location of trees

🏛 Prepare the plot plan for your original building design.

KEY TERMS

Property lines
Contour lines
Zoning

The plot plan shows the location of the house on the plot together with information on terraces, walks, driveways, contours, elevations, and utilities. A roof plan or landscaping plan may be included.

The steps used in drawing a plot plan are illustrated in Figure 25–1. We will use the plot plan of the A residence (Figure 25–2) as an example.

STEP 1: PROPERTY LINES

Lay out and dimension the **property lines** using a medium-weight center line. The information is

obtained from the site survey (Figure 33–1, page 397). The plot plan is usually drawn to an engineer's scale (such as 1" = 20') rather than to an architect's scale (such as $\frac{1}{16}$" = 1'). Show the "north" arrow.

Bearings are given to show the compass direction of the property lines. Notice in Figure 25–2 that the opposite and parallel property lines have opposite bearings (such as S 65° 10′W and its opposite N 65° 10′E), because the surveyor started at one corner and surveyed clockwise around the perimeter until reaching the starting corner again.

STEP 2: CONTOUR LINES

A **contour line** is an imaginary line representing a constant elevation on the lot. The vertical distance between adjacent contour lines is called the *contour interval* and is usually 1' for a residential lot.

The easiest way to learn to read contour lines is to imagine that you are constructing a model of the plot. For example, a model of the small section of sloping land shown in Step 2 of Figure 25–1 could be built by cutting out cardboard sections shaped to the contour lines.

Contour lines are drawn freehand and dimensioned by indicating their elevation above sea level or some other datum plane such as a nearby street or house. This information is also obtained from the site survey. On the plot plan of the A residence (Figure 25–2) notice that the contour lines indicate gently sloping ground, with the lowest point being the south-west corner. The elevations of any permanent markers at the plot corners should also be transferred from the site survey.

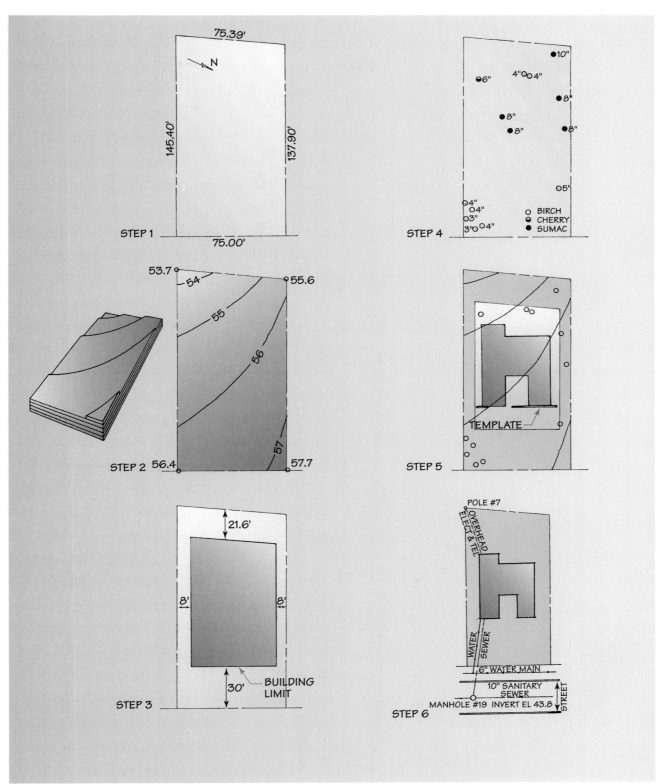

Figure 25–1 *Preparing a plot plan.*

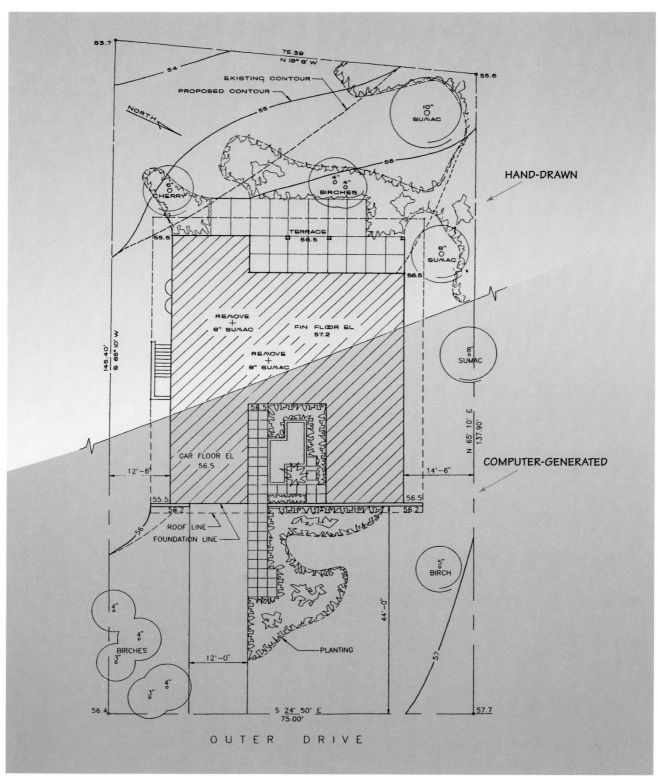

Figure 25–2   *Hand-drawn and computer-generated plot plan of the A residence.* *(Courtesy Mr. and Mrs. Donald W. Hamer.)*

STEP 3: ZONING

The first-known **zoning** law was passed in ancient Rome to prevent industries from locating too near the central forum of the city. This established the principle that private property can be restricted in favor of the general welfare. The first comprehensive zoning ordinance in the United States was passed in New York City in 1916 as a consequence of a tragic fire at the

CAREER PROFILES

Name: Gregory C. Parker   *Age:* 44
Gender: Male
Occupational Title: Professional Land Surveyor
Employer: Self-employed, Parker Surveying, Indiana, PA 15701
Years of Experience: 22
Educational/Training Background:
       2 years Indiana University of Pennsylvania
       1 year Allegheny County Community College
       1 year Penn State University
       Served required 8-year apprenticeship prior to taking registration exam in 1982.
Starting Salary:
       Low Level Entry Assistant—$7.00 to $8.00/hour
       Experienced Assistant—$9.00 to $11.00/hour
       Party Chief—$12.00 to $13.00/hour
       Professional Land Surveyor—$15.00 to $16.00/hour

JOB OVERVIEW:
    I arrive at work early to prepare for the day ahead. At this time I prepare data for the field crews. This includes: deed plots, topographic and aerial map overlays, any data from previous surveys, and job descriptions for specific projects. I also make sure clients are scheduled as necessary. When the field crews arrive, most days are spent conducting field surveys. Field surveys include visual inspection for physical evidence of possession. This includes locating monuments from previous surveys, and setting stones, fences, ridges, and timber lines. Field measurements are taken to locate the assigned tract and any necessary adjacent tracts. On a typical day, I correlate these with geometry in deeds to evaluate similarities or inconsistencies. This means deciding what is correct and what is not.
    Additional responsibilities include estimating costs and time frames, presenting finished plans to clients, attorneys, realtors, or municipal planning boards, handling complaints, and scheduling future projects.
    The best aspect of my job is variety. No day is the same as we do many types of surveys. I enjoy meeting people, and (most days) working outdoors is a benefit.
    Creating public awareness of what a surveyor actually does is a challenge. The realities of daily life as a surveyor are difficult to convey. The skill required and the liability accepted is commensurate with legal or engineering professions. You must either enjoy surveying or choose a different career.

Triangle Shirtwaist factory. Over one hundred dressmakers, mostly young girls, died in that building, which was higher than the reach of firefighting equipment, had no sprinklers, and had an uncompleted fire escape.

Most communities now have zoning ordinances that restrict the size and location of buildings to prevent crowding and encourage the most appropriate use of the land. It is imperative that the designer be familiar with, and adhere to, all zoning regulations and building codes. Several regulations of one community's zoning ordinance are given in Table 25–1. Using these regulations, the following calculations are made for the A residence:

Check on lot area: $75 \times 141.6 = 10,620$ sq. ft. O.K. (greater than required 10,000 sq. ft.)
Check on lot width: 75' O.K. (equal to required minimum of 75')
Minimum front yard depth: 30'
Minimum side yard width: 8'
Minimum rear yard depth: $0.20(137.9 - 30) = 21.6'$ O.K. (greater than required 15')
Check on dwelling height: 14' O.K. (less than 26' maximum)

Using light construction lines, show the front, side, and rear yard limits calculated above so that you know exactly where you are permitted to place your house.

STEP 4: TREES

The lot of the A residence contains birch, sumac, and cherry trees. These are plotted so that the house and driveway may be located without removing many of them. The type and diameter of each tree should be noted on the plot plan to aid in identification. No tree should remain within 5' of the foundation because the excavation will disturb the roots so much that it will eventually die.

STEP 5: LOCATION OF HOUSE

The house may now be located in the position that will satisfy all requirements of solar orientation, wind orientation, topography, landscaping, zoning, and utilities. A paper template in the shape of the basement plan is useful since it can be shifted to various locations. When the fi-

TABLE 25–1 *Sample of Zoning Regulations*

Article V—Residence District

Section 501. Each lot in this district shall comply with the following minimum requirements:

501-1)	Lot area:	
	One-family dwelling	10,000 sq. ft.
	Two-family dwelling	12,000 sq. ft.
501-2)	Lot width:	
	One-family dwelling	75'
	Two-family dwelling	100'
501-3)	Front yard depth:	
	Dwelling	30'
	Nondwelling	40'
501-4)	Side yard width:	
	Dwelling and accessory building	8'
	Nondwelling	20'
501-5)	Rear yard depth:	

The rear yard depth shall be at least 20 percent of the depth of the lot measured from the front building line to the nearest point of rear lot line but in no case shall this be less than 15'.

Section 502. The maximum height of structures in this district shall be:

502-1)	Dwellings	26' (not exceeding two stories)
502-2)	Accessory building	16' (not exceeding one story)
502-3)	Nondwelling	40'

nal position is determined, draw in the plan with heavy lines using the outside basement dimensions. The plan may be sectioned if desired. Hidden lines are used to indicate roof overhang. On structures with fairly involved roof intersections, a roof plan is shown in place of a section to indicate ridges, valleys, and hips.

How to Become a Surveyor

Four-year surveying technology programs are recommended for persons having the goal of achieving professional registration as a surveyor. Two-year surveying technology programs produce a graduate with the title of surveying technician. Baccalaureate surveying technology programs provide instruction in the principal subdivisions of surveying:

- 🏛 Cadastral surveying
- 🏛 Geodetic surveying
- 🏛 Photogrammetry
- 🏛 Remote sensing
- 🏛 Geographic information systems

- 🏛 Precise positioning systems
- 🏛 Land development
- 🏛 Legal aspects
- 🏛 Professional practice
- 🏛 Ethics

Associate surveying technology programs provide technical instruction in:

- 🏛 Plane surveying
- 🏛 Curves and earthwork
- 🏛 Computer-assisted mapping
- 🏛 Construction surveying applications

- 🏛 Cadastral surveying
- 🏛 Geodetic surveying
- 🏛 Photogrammetry
- 🏛 Legal aspects

All surveyors must be proficient in mathematics (especially trigonometry) for they often depend upon mathematical calculations to complete their surveys.

The Accreditation Board for Engineering and Technology (ABET) is the official accrediting agency for both associate and baccalaureate surveying programs. Currently, ABET has accredited five universities or technical institutes offering associate surveying majors and eight universities or technical institutes offering baccalaureate surveying programs. For additional information contact ABET at:

111 Marketplace Suite 1050
Baltimore, Maryland 21202
Phone: 410-347-7700
Internet address:
http://www.abet.ba.md.us

To become registered as a surveyor, most states require four years of practical experience under the supervision of a registered surveyor following graduation from an institution offering an ABET-accredited baccalaureate surveying major, and six years of experience following graduation from an institution offering an ABET-accredited associate surveying major. Also, a comprehensive written examination must be completed. Without a degree from an accredited institution, most states accept eight years of experience.

REAL WORLD *APPLICATIONS*

Calculating Acreage

1/4"=1'-0 The plot of land on which a house is located, whether it be a zoned lot in a town or a tract of land in the countryside, is measured in feet and inches. Once the size of the plot is determined, these measurements can then be converted into acres. Many clients want to know the number of acres of land they own. An acre is a measurement of area equal to 43,560 square feet. To calculate a plot of land's acreage, follow these steps:

First, measure the plot:

For this example, use a plot measuring 150' × 250'. Then, calculate the square footage:

Square footage = 150' × 250' = 37,500 sq. ft.

Now divide the area of the lot (37,500 sq. ft.) by the number of square feet in one acre (43,560).

Acreage = 37,500 sq. ft. ÷ 43,560 sq. ft. = 0.86 acres

What is the total acreage for the following plots of land?

   a. 60' × 150'

   b. 520' × 125'

   c. 5280' × 5280'

Show the elevation of the finished first floor and complete all details such as walks, drives, street names, notes, and scale.

STEP 6: UTILITIES

Existing utilities—such as electric, telephone, sewer, water, and gas lines—are often noted or actually located since they may affect the house location. For example, some of the factors to be considered in the location of the branch drain from the house to the street manhole connection are:

1. A minimum downward grade of $\frac{1}{4}$"/ft. must be maintained to the invert elevation (the bottom inside) of the sewer line.

2. The branch line must remain below the frost line to prevent damage from freezing.

3. The line must be straight, since changes in direction or grade, which collect sediment, should occur only at manholes.

4. If there are trees, cast-iron drains with poured-lead joints must be used instead of vitreous tile to prevent the roots growing into and blocking the drain.

STEP 7: ADJUSTMENT OF CONTOUR LINES

When the existing land contour is not satisfactory, it must be adjusted by cutting away or filling in. This is indicated on the plot plan by showing the proposed position of the contour lines with a solid line; the existing contours are still indicated, but as broken lines. Figure 25–2 shows how all of these steps are finished into the plot plan. It was decided to show the utilities on later mechanical plans to prevent overcrowding of the plot plan.

The ultimate goal of a plot plan is to show how the building will integrate with its environment. Successful results are shown in Figures 25–3 through 25–12.

Figure 25–3 *A multi-level contemporary building designed to fit a sloping site.* (Courtesy American Plywood Association.)

Figure 25–4 *Raised decking overlooking natural landscaping.* (Courtesy American Plywood Association.)

Figure 25–5 *For privacy, an enclosed patio.* (Courtesy California Redwood Association.)

Figure 25–6 *A shaded pool for a southern climate.* (Courtesy California Redwood Association.)

Figure 25–7 *A contemporary backyard pavilion.*

Figure 25–8 *Rear terrace of the Z residence.*

Figure 25–9   *Even a toolshed deserves careful design.*

Figure 25–10   *Landscaping can convert any difficult plot into an asset.*

CAREER PROFILES

What Does a Surveyor Do?

A *land* surveyor uses a transit to locate the boundaries and elevations of land plots and buildings. Accomplished surveyors can also produce *engineering* surveys, *topographic* surveys, and *geodetic* surveys. Engineering surveys are required for construction of large structures, roads, bridges, and dams. Topographic surveys and aerial surveys are needed to create maps. Geodetic surveys are required to create small-scale maps in which the earth's curvature is taken into account.

Some surveyors adopt job titles that identify their specialty, such as a highway, railway, mine, oil well, pipeline, or hydrographic surveyor. Surveyors may specialize in fields such as wetlands preservation, location of hazardous waste sites, tectonic plate movements, or mapping areas ranging from the Amazon River to the polar ice cap. Surveyors may choose public or private practice. Thus, as a career path, surveying offers numerous options.

More information about the nature of the work, working conditions, training, job outlook, and salary of a surveyor can be found in the Occupational Outlook Handbook Home Page. You can obtain this resource at your library or access it on the Internet at:
http://www.bls.gov/ocohome.htm
Information on state, national, and international professional organizations for land surveyors can be found on the Internet at:
http://www.gsys.com/gpsnet/society.html

Figure 25–11   *Landscaping as a major element of architectural design.* (Courtesy California Redwood Association.)

Figure 25–12   *Multi-level decking.* (Courtesy California Redwood Association.)

COMPUTER-AIDED PLOT PLANS

The same procedure used to obtain computer-aided floor plans can be used to obtain a computer-aided plot plan. However, the finished plot plan should be designed to an engineer's scale rather than an architect's scale. Your CAD software will probably have features that allow you to sketch contour lines and plot landscaping features.

Compare the computer-generated and hand-drawn portions of the A residence plot plan in Figure 25–2.

REVIEW QUESTIONS

1. Distinguish between a contour line and a contour interval.

2. List the utilities to be considered when drawing a plot plan.

3. What does a plot plan show?

4. What does it mean for a lot to be zoned?

5. Why are zoning ordinances created?

6. Why must trees be located on a plot plan?

ENRICHMENT QUESTIONS

1. Why is it necessary to be familiar with the local zoning ordinance when drawing a plot plan?

2. List the bearings of the remaining property lines of a rectangular lot that has a front bearing of N 72° E and a right side bearing of N 18° W.

3. Describe the steps used to draw a plot plan.

4. Why are zoning regulations called ordinances and not laws?

5. What is a frost line?

PRACTICAL EXERCISES

1. Draw or computer-produce the plot plan of the A residence.

2. Draw or computer-produce the plot plan for your original house design.

3. Visit the local courthouse and locate various local zoning ordinances for the city in which you live.

CHAPTER 26

Roofs

OBJECTIVES

By completing this chapter you will be able to:

- 🏛 Identify and design roof types (gable, hip, gambrel, mansard, flat, shed, clerestory, butterfly, folded, parasol, free-form, A-frame, and warped).

- 🏛 Identify the type of architecture (traditional or contemporary) each roof is used with.

- 🏛 Identify and describe dormer types.

- 🏛 Describe different types of materials used for roofing.

- 🏛 Define rise and run as they relate to roof pitch.

- 🏛 Explain how guttering and flashing relate to roof construction.

KEY TERMS

Gable roof	Clerestory roof
Hip roof	A-frame roof
Gambrel roof	Roof pitch
Mansard roof	Rise
Dormer	Run
Flat roof	Flashing
Shed roof	Guttering

The type of roof specified for a house is a very important factor in exterior design. It also affects the interior. In a traditional house, the type of roof is pretty well determined by the house style. When the building is in a contemporary style, however, a wide variety of roof types may be used. The shape of the plan may also affect the roof choice. Whereas a rectangular plan could be roofed in nearly any manner, a rambling plan would be most economically covered by a flat roof. A hip roof, in such a case, would require much additional cutting and fitting.

The various types of roofs may be classified very broadly into two categories: roofs used mostly on traditional houses (Figure 26–1) and roofs used mostly on contemporary houses (Figure 26–8). Let us look at each in turn.

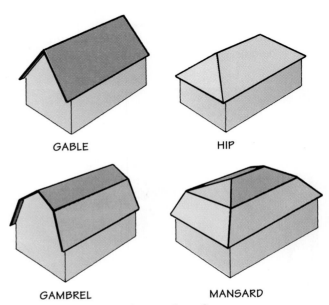

Figure 26–1   Traditional roofs.

TRADITIONAL ROOF TYPES

Gable

To be perfectly correct, the gable is the triangular portion of the end of the house, and this type of roof is a gabled roof. The **gable roof** is the most common form of roof since it is easy to construct, is pitched for drainage, and is universally accepted. This roof should be pitched high enough that louvers can be installed to allow warm air and moisture to escape. This roof may also be used in contemporary design (Figures 26–2 and 26–3).

Hip

A **hip roof** (Figure 26–4) is more difficult to construct than a gable, but is still used with a low

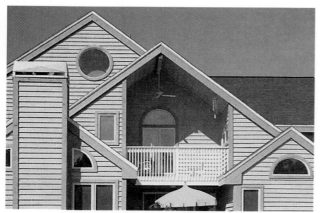

Figure 26–2 *A gable roof with horizontal siding.* (Courtesy Pella Corporation.)

Figure 26–3 *A gable roof with shingle walls.* (Courtesy Pella Corporation.)

Figure 26–4 *A hip-roofed residence.* (Courtesy Pella Corporation.)

pitch on ranch houses. However, every hip or valley increases chances for leakage.

Gambrel

The **gambrel roof** is used on Dutch Colonial designs to increase headroom on the second floor. Since the framing is complicated, it is not widely used today.

Mansard

This type was named after the French architect who originated it. The **Mansard roof** is not built today but it was extensively used on French-styled houses at one time.

DORMER TYPES

Gable

A **dormer** (Figure 26–5) is used to let light into an otherwise dark area. The designer must weigh the expense of dormers against the expense of an additional story.

Shed

A shed dormer (Figure 26–6) will give nearly all the advantages of a full story, without disturbing the one-floor look of the house. This dormer

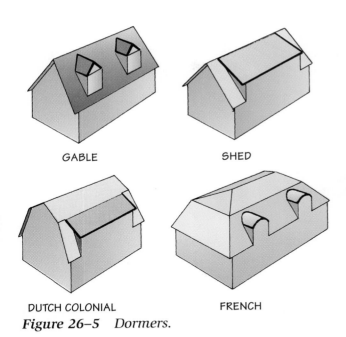

GABLE · SHED · DUTCH COLONIAL · FRENCH

Figure 26–5   *Dormers.*

Figure 26–7   *Gable dormers in a dustpan.*
(Courtesy Pella Corporation.)

French

Rounded dormer tops indicate French influence.

CONTEMPORARY ROOF TYPES

Flat

The **flat roof** (Figures 26–9 and 26–10) is the most common type of commercial roof. Most roofs that appear flat actually have a slight slope of $\frac{1}{4}$"/ft. to $\frac{1}{2}$"/ft. for drainage. Roofing is laid in layers (called *plies*) of tar and gravel or with membranes of plastic or other flexible material.

Some roofs are designed as perfectly flat, thus allowing water to remain on them. This is called a *water-film roof* and the water serves as insulation from cold in winter and insulation from heat (by reflection and evaporation) in summer. Other roofs are designed to slope, but may become deflected so that *ponding* (puddles) occur. In either case, there is a great possibility of roof leaks. Standing water encourages growth of vegetation, producing roots that force their way through the roof seams, and fungi growth that deteriorates organic roofing materials. Also water that penetrates the roofing plies will freeze and further delaminate the roof.

Figure 26–6   *A shed dormer.* *(Courtesy Pella Corporation.)*

type is sometimes called a *dustpan* (Figure 26–7).

Dutch Colonial

A Dutch Colonial dormer might be termed a shed dormer on a gambrel roof. The second-floor wall may be set back from the first-floor wall.

Shed

The **shed roof** is ideally suited to the solar house if the high wall faces south. It will also give interesting interior effects if the beams

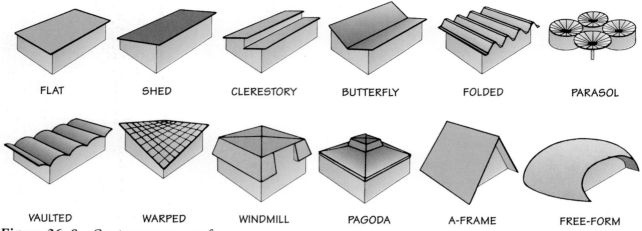

Figure 26–8   *Contemporary roofs.*

Figure 26–9   *A flat roof with vertical siding.* (*Courtesy California Redwood Association.*)

Figure 26–10   *A flat roof with horizontal siding.* (*Courtesy California Redwood Association.*)

are left exposed. Beams measuring 4" × 10" spaced 4' oc might be used. This roof takes standard roofing material. Contemporary roofs are often unique and do not fit into standard categories. The roof shown in Figure 26–11 is an example.

Clerestory

This roof solves the problem of introducing light into the center of a house. The **clerestory roof** (Figure 26–11) may be used with a sawtooth roof as shown in Figure 26–8 or with other roof types.

Butterfly

The pitch and length of each side of the butterfly roof need not be equal. This roof "opens up" the

Figure 26–11 *A clerestory roof. (Courtesy California Redwood Association.)*

house, providing plenty of light and air. Drains may be at the end of the valley or in the middle— running down through the center of the house.

Folded

Some roofs are so new that some imagination must be used to find descriptive names. As the name implies, the folded or pleated roof looks as though it were folded from a sheet of paper, and is quite popular in office and motel design. Roofing material may be exterior-grade plywood or metal.

Parasol

The parasol roof has become popular since the success of Frank Lloyd Wright's Johnson's Wax Building. Round and square variations are possible. Often the material used is reinforced concrete.

Free-form

The shape of free-form roofs may depend upon the method of construction. Urethane foam can be sprayed on a *knit jersey* material stretched over a pipe frame. It has proved to be strong, weather-resistant, and self-insulating, but some foamed plastics are dangerously flammable and toxic.

A-frame

Originally specified for low-cost summer or winter cabins, the **A-frame roof** has been adapted to larger structures such as churches. The clas-

sic A-frame roof is a gable roof that reaches the ground on two sides.

Warped

Beginnings have been made in the development of warped surfaces for roofs. In most cases, this warped surface is a hyperbolic paraboloid, that is, a surface generated by a line moving so that its ends are in contact with two skew lines. This produces a superior roof due to the high resistance to bending. Warped roofs have been constructed of molded plywood, reinforced concrete, and sprayed plastic.

Windmill

The windmill roof is often used on two-story town houses to lend interest to an otherwise simple design. This roof looks best when wood-shingled.

Pagoda

The pagoda roof provides an interesting interior, as well as exterior, design. The Oriental flavor should be extended to other elements of the building. Connected clusters of pagoda-roofed rooms have been successfully designed.

Vaults and Domes

Vaults and domes have staged a comeback from the Byzantine days. More often used on commercial than residential buildings, they limit considerably the possible shapes of the floor plan.

A *geodesic dome* is framed of members nearly equal in length which are joined to form triangular patterns. The triangles are then joined to form pyramids, giving a double-faced structure of great strength (Figure 26–12). The framing members are usually straight rather than curved. Consequently, the roof shape is a polyhedron rather than a true spherical dome.

ROOFING MATERIALS

Although several new roofing materials and methods have been mentioned, the large majority of residential roofs are constructed of built-up tar and gravel or membrane roofing when the roof is flat and of shingle when the roof slopes more than 3 in./ft.

Figure 26–12 Close-up of the geodesic shaping of the Princeton University gymnasium. (Courtesy American Iron and Steel Institute.)

Built-up Roofing

Built-up tar and gravel roofing is used on flat and slightly sloping roofs. It is constructed of alternate layers of roofing felt and mopped-on hot tar or asphalt. Three to five layers (called *plies*) are used and topped by crushed gravel or marble chips imbedded in the tar. Roofing contractors will bond a three-ply economy roof for ten years and a five-ply roof for twenty years.

Membrane Roofing

Membrane roofing is rapidly replacing built-up roofing for use on flat and slightly sloping roofs. Easy to apply and repair, membrane roofing is lighter in weight and a better heat reflector than built-up roofing. Most membranes are also available in various colors. The membrane is a thin layer (about $\frac{1}{8}$" thick) of plastic, rubber, bitumen, or aluminum. Available in rolls or large sheets, the membrane is applied over mopped-on adhesive or may be loose-laid and covered with ballast of crushed gravel or marble chips. Joints are heatsealed or spliced using a contact adhesive pressure-sealed by roller. Membranes also can be applied in fluid form by spray, brush, or squeegee. The membrane then hardens through chemical curing or evaporation of a solvent in the mixture. The manufacturer's bond for defective material is usually for fifteen years, but the roofer's guarantee for proper installation is usually only for two to five years.

Shingles

Fiberglass shingles, consisting of a fiberglass mat saturated with asphalt, are now commonly used in house construction because they are lightweight and their cost is moderate (Figure 26–13). Asphalt shingles, consisting of a cellu-

Figure 26–13 A gable roof of fiber glass shingles. (Courtesy California Redwood Association.)

lose mat saturated with asphalt, and organic shingles, consisting of a felt mat saturated with asphalt, are both still available. Wood shingles and shakes (Figure 26–14) are often of cedar or redwood and present a handsome appearance. Unfortunately the fire hazard is great, but imitation shakes of aluminum are available.

Roofing shingles are sold by *squares*. A square is the amount of shingling needed to roof 100 sq. ft.

Slate and Tile

Slate (Figure 26–15) and tile (Figure 26–16) are occasionally used for roofing materials. Both are

Figure 26–16   *A tile roof. (Courtesy Pella Corporation.)*

relatively heavy and expensive, however. Imitation Spanish tile of colored galvanized steel panels are available.

Metal Roofing

Ribbed or corrugated metal roofs of "tin" (galvanized steel), terne-plated steel, copper, and aluminum are used on commercial and residential buildings. Designers can choose steel roofing with baked enamel finishes in a variety of colors (Figure 26–17).

Figure 26–14   *A cedar-shingled roof. (Courtesy California Redwood Association.)*

Figure 26–15   *A Tudor residence with slate roof.*

Figure 26–17   *A metal-roofed contemporary residence.*

ARCHINET http://www.

Roof Planning

One of the most important parts of any house is the roof. Not only does a roof protect the house from the weather, it also significantly adds or detracts from the overall beauty of the house. If the roof is built poorly, the rest of the house will suffer.

Use the Internet to locate a homepage called The Sound Home Resource. The address is:

http://www.soundhome.com

This homepage provides, among many other topics, comparative information about roofs. It includes comparisons of manufacturers' warranties, life expectancy, relative cost, special maintenance requirements, and fire rating comments.

⛫ What material is the most expensive?

⛫ What special maintenance do cedar shakes require?

⛫ What roofing material is good for heavy weather conditions?

⛫ Besides slate, what other material has a life expectancy of over 50 years?

Plastic Foam

Sprayed urethane foam roofs offer exciting possibilities for new architectural concepts. All free-form shapes are possible, construction is economical, and it has three times the insulating value of common insulating materials. These roofs are formed by spraying on large balloons, inflatable forms, or fabric stretched over lightweight tubing. Spraying is continued until a thickness of 5" is obtained.

Urethane foam is an organic polymer formed by chemical reaction between two liquids (isocyanate and polyol). These liquids are pumped from separate tanks to a mixing spray gun where the mixture begins to foam immediately, expanding to thirty times its original volume. Within a few minutes, the foam has hardened. It will adhere to most surfaces. Final density is about 3 lb./cu. ft. A serious disadvantage is that some forms of urethane foam are very flammable. Consequently, the foam should be covered by a fireproof material rather than installed exposed. Sprayed urethane foam should always be protected against any possible source of combustion.

ROOF PITCH

Some **roof pitch** terms are shown in Figure 26–18. The terms **rise** and **run** are used in two ways:

1. To describe the rise per unit of run. For example, the roof shown in Figure 26–18 has a rise of 4" in a run of 1'.

2. To describe the *actual* dimensions of a roof. For example, the roof in Figure 26–18 has a rise of 6', run of 18', and span of 36'.

The roof pitch is also described in two ways:

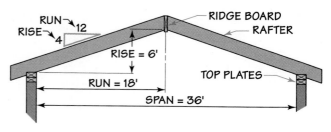

Figure 26–18   Roof pitch terms.

REAL WORLD APPLICATIONS

Angle Review

 A roof is not satisfactory if it is designed or constructed incorrectly. For instance, the minimum pitch (angle) of a roof is determined by the type of roofing material specified, as well as the expected snowfall. A roof designer needs a basic working knowledge of angles to avoid errors of roof angle design.

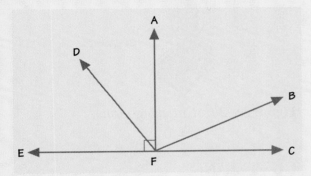

Angles are identified with three capital letters, which refer to points on a plane, while a line is identified by its two ending points. For example, angle BFC is formed at the intersection of lines BF and CF at the vertex, which is F.

A right angle measures exactly 90 degrees. It is denoted by a small corner box at the intersection of the lines that form the 90-degree angle. The figure shows the right angle AFE.

Acute and obtuse angles also refer to their measurement. Acute angles are greater than 0 degrees but less than 90 degrees. Obtuse angles, on the other hand, are greater than 90 degrees but less than 180 degrees. Angle BFC in the figure is an acute angle and angle BFE is an obtuse angle.

Two angles whose sum is 90 degrees are called complementary angles, whereas two angles whose sum equals 180 degrees are called supplementary angles. In the figure, angles BFC and AFB are complementary to one another. Angles DFC and DFE are a set of supplementary angles.

Answer these questions about types of angles using the figure:

1. Identify two acute and two obtuse angles.

2. Identify a set of complementary angles.

3. Identify a set of supplementary angles.

4. When designing a roof, how could complementary and supplementary angles be useful?

1. By a pitch triangle on the elevations showing the rise in whole inches per 12" of run. For example, the 4-12 pitch triangle shown in Figure 26–19 indicates a rise of 4" for a run of 12".

2. By a fraction whose numerator is the rise and denominator is the span. For example, a roof with a rise of 6' and a span of 36' has a $\frac{6}{36}$ pitch, which would be reduced and called a $\frac{1}{6}$ pitch. See Figure 26–19.

In general, the steeper roofs are found in areas where there are heavy snowfalls, since snow is naturally shed from a steep roof.

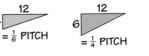

Figure 26–19 Alternate method of specifying pitch.

Special steps must be taken with low-pitched, shingled roofs to prevent wind from lifting the tabs and allowing water to seep under them. Several solutions are possible, as indicated in Table 26–1:

1. Use interlocking tab shingles.

2. Use self-sealing shingles that have factory-applied adhesive on the underside.

3. Cement each tab with a spot of quick-setting adhesive during installation.

4. Use heavyweight shingles (300#, 15" × 36"), which are stiffer than regular shingles (210#, 12" × 36") and are sized for triple overlap.

FLASHING

Thin sheets of soft metal are used to prevent leakage at critical points on roofs and walls. These sheets are called **flashing** and are usually of lead, zinc, copper, or aluminum. Areas that must be flashed are:

> Intersections of roof with chimney, soil pipe, and dormer (see Figure 32–13, page 386).
> Roof valleys

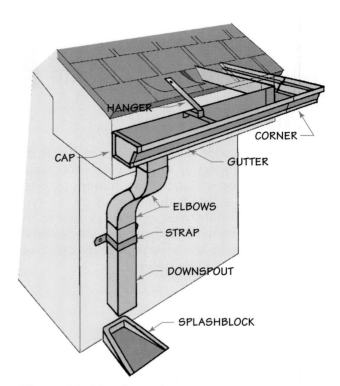

Figure 26–20 Guttering terms.

GUTTERING

Gutters are used when the soil is likely to be eroded by rain dripping from the roof, or when roof overhangs are less than 12" (in a one-story structure) or 24" (in a two-story structure). Downspouts conduct the water down the wall to a storm sewer (*not* a sanitary sewer), dry well, or splashblock (a concrete pad placed to prevent soil erosion) as shown in Figure 26–20. When gutters are omitted, a diverter is used to protect the entrances to the house from rainwater. **Guttering** is made of galvanized iron, copper, aluminum, zinc alloy, or wood, or it may be built into the roof as shown in Figure 33–29 (page 415). Seamless guttering is formed from 200'-long rolls of aluminum.

SKYLIGHTS

Plastic domes are often used to light the interior of industrial buildings. Recently, skylights and

---

TABLE 26–1 *Recommended Roof Pitch*	
Type of Roofing	**Recommended Roof Pitch**
Built-up	Under 3-12
Fiber glass shingles	
Interlocking*	2-12 or more
Self-sealing*	2-12 or more
Cemented*	2-12 or more
Heavyweight*	2-12 or more
Heavyweight	3-12 or more
Regular	4-12 or more
Wood shingles	4-12 or more
Slate shingles	6-12 or more

* Double layer of roofing for underlay must be used.

roof windows have been used in residences to obtain specific effects. Both fixed and ventilation types are available in a variety of shapes and sizes (see Figures 26–21 through 26–24).

Figure 26–21 *Skylights aligned over casement windows. (Courtesy VELUX-AMERICA, INC.)*

Figure 26–23 *Roof windows can convert unused attic space into usable rooms. (Courtesy VELUX-AMERICA, INC.)*

Figure 26–22 *Contemporary skylights blended with a traditional roof. (Courtesy VELUX-AMERICA, INC.)*

Figure 26–24 *Roof windows in a hip roof. (Courtesy VELUX-AMERICA, INC.)*

CONSTRUCTION

Roof framing is described in Chapter 33. Tables for the selection of rafters are given in Chapter 34. Figure 26–25 shows some common terms used in roof construction.

Framing Plan

When unusual or difficult roof construction is necessary, a roof framing plan is included in the working drawings. Figure 26–26 shows the roof framing plan for the roof shown in Figure 26–25. Notice that a single heavy line is used to indicate each rafter. Floor framing plans and ceiling joist framing plans are also included when necessary.

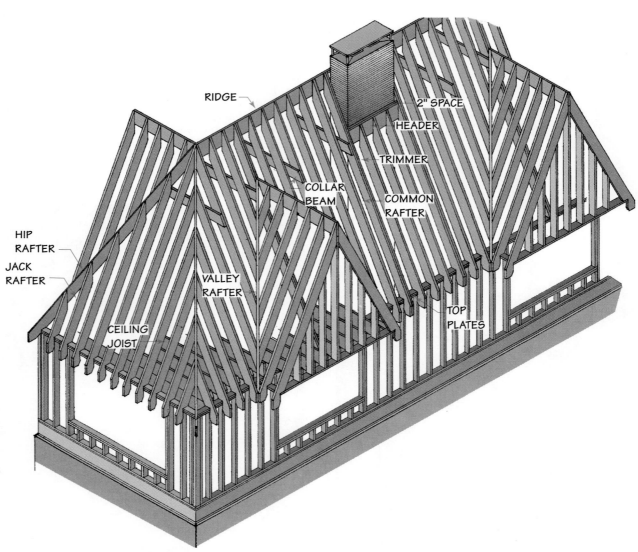

Figure 26–25 *Roof framing terms.*

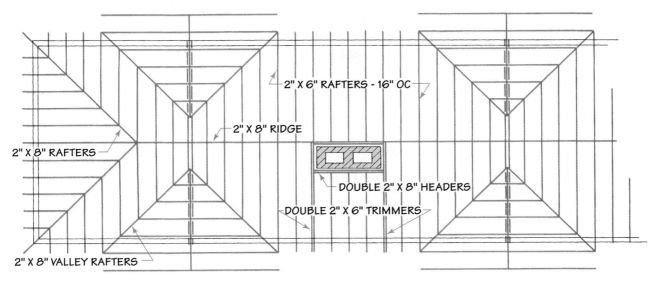

2" X 6" RAFTERS - 16" OC

2" X 8" RIDGE

2" X 8" RAFTERS

DOUBLE 2" X 8" HEADERS

DOUBLE 2" X 6" TRIMMERS

2" X 8" VALLEY RAFTERS

Figure 26–26   Roof framing plan.

REVIEW QUESTIONS

1. List and pencil-sketch or computer-sketch:
   a. Four types of traditional roofs
   c. Four types of dormer windows
   b. Twelve types of contemporary roofs

2. What is meant by a dormer?

3. Sketch and label the pitch triangle for a $\frac{1}{4}$ and $\frac{3}{8}$ pitch roof.

4. Give two methods of indicating the pitch of a roof having an 8' rise in a 32' span.

5. What is the purpose of flashing? Give several examples of where flashing is used.

6. Describe in words or by sketches the following terms:
   a. Common rafter
   f. Collar beam
   j. Run
   b. Jack rafter
   g. Header
   k. Rise
   c. Hip rafter
   h. Trimmer
   l. Roofing square
   d. Valley rafter
   i. Span
   m. Roofing ply
   e. Ridge board

7. List four methods to prevent wind from lifting roof shingles.

8. Distinguish between built-up roofing, membrane roofing, and shingle roofing.

ENRICHMENT QUESTIONS

1. What advantages and disadvantages do dormer-type windows offer?

2. Should a roof sloping 3 in./ft. be covered by ordinary fiber glass shingles? Why or why not?

3. Why is the location of a chimney important when considering roof design?

4. Name three types of roof shingles and give the advantages of each.

5. Compare and contrast the cost, durability, and longevity of four different types of roof coverings. For calculation purposes, assume the roof is a total of 850 sq. ft.

PRACTICAL EXERCISES

1. Draw or computer-produce the roof framing plan for the A residence.

2. Draw or computer-produce the roof framing plan for your original house design.

3. Design a type of roof that is entirely different from those now commonly used.

4. Draw or computer-produce a classroom display showing:
   a. Traditional roofs
   b. Contemporary roofs
   c. Dormer windows
   d. Roof construction terminology

27 Elevations and Sections

OBJECTIVES

By completing this chapter you will be able to:

- Draw interior and exterior elevations of a residence or a small commercial building.
- Utilize sketching techniques when preparing an elevation.
- Describe in words and demonstrate with sketches various techniques (harmony, symmetry, fenestration, shadows, and so on) used to enhance the appearance of a building.
- Draw sections of different walls in a building.

KEY TERMS

Interior elevations	Areawall
Proportion	Structural section
Fenestration	Wall section
Footing	Detail sections

An architectural elevation is a view of a building containing a height dimension. When elevations show the inside of a building, they are called **interior elevations**; when they show the outside, they are called simply *elevations*.

INTERIOR ELEVATIONS

Interior elevations are included in a set of working drawings only when there is some special interior construction to be illustrated. This is quite often the case in kitchen design. Figure 27–2 shows the interior elevations of the kitchen of the A residence. Notice that the arrangement of the elevations is in relation to the floor plan—as though the four walls had fallen backward. To prevent this awkward appearance, the interior elevations may be removed to an upright position and even placed on a separate drawing sheet. The relation of each elevation to the plan is then shown by sight arrows as indicated in Figure 27–3. A right arrow shows the drawing number on which the detail appears. Sight arrows $\frac{1}{4}$, $\frac{2}{4}$, etc., are interpreted as detail #1 on drawing #4, detail #2 on drawing #4, and so on.

ELEVATIONS

The exterior elevations are as necessary to the satisfactory appearance of a building as the floor plan is to its satisfactory functioning (Figure 27–1). Normally the elevations of the four sides of a building are sufficient to describe it.

Figure 27–1 *A front elevation provides the first indication of the character of a building.*
(Courtesy Andersen Windows, Inc.)

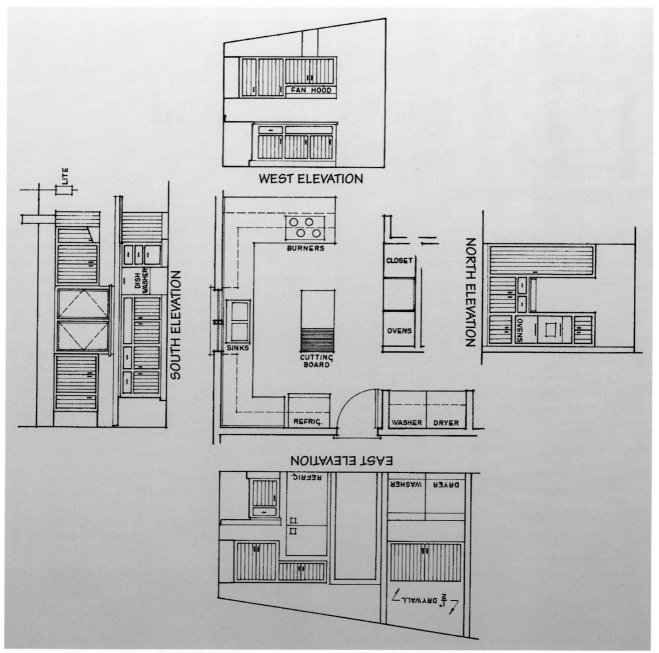

Figure 27-2 Interior elevations of the A residence kitchen. (*Courtesy Mr. and Mrs. Donald W. Hamer.*)

In some cases, however, more than the four elevations are needed. For example, a structure built around an open court would require additional exterior elevations to illustrate the building as seen from the court.

ELEVATION DESIGN

The procedure used in the design of elevations is similar to the procedure used in the design of floor plans.

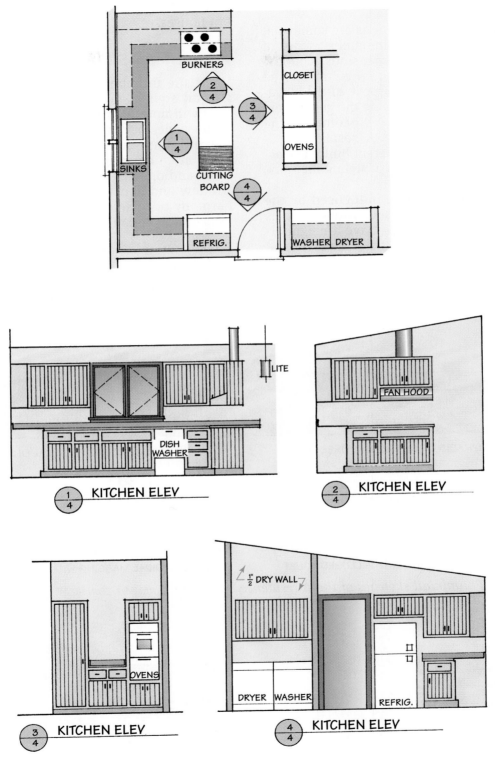

Figure 27–3 Interior elevations, preferred layout.

Thumbnail Sketches

Thumbnail sketches of elevations are somewhat simpler to draw than the thumbnail sketches of a floor plan since overall dimensions, window locations, and door locations may be transferred from the plan. The prime purpose of the elevation thumbnail is to help decide the general exterior styling. In Figure 27–4, compare the thumbnail sketches for the A residence. Although they appear to be different buildings, actually they are different ways of styling the same building.

The final choice may be influenced by the character of the neighborhood. Although it is not necessary or desirable to have a neighborhood composed entirely of the same style and size houses, the surroundings will influence the appropriateness of a particular design to some extent. For example, it would not be appropriate to place a charming Cape Cod house in the midst of a section of highly experimental modern houses.

Preliminary Sketches

Using the thumbnail sketch as a guide to the general styling, prepare a preliminary sketch. This will normally be done to $\frac{1}{8}$" = 1' scale on graph paper because the location and proportion of features are very important at this stage. Main attention is given to the proportions of walls and openings and the fenestration of the window and door openings. Also consider the harmony of materials and features and the effects of orientation and shadows.

Proportion   Most people can look at a finished building and decide if it is well proportioned (Figure 27–5). However, designing a building so that it will have good proportion requires talent and training (Figure 27–6). In general, the

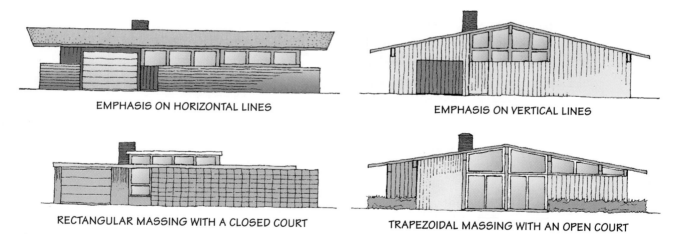

EMPHASIS ON HORIZONTAL LINES

EMPHASIS ON VERTICAL LINES

RECTANGULAR MASSING WITH A CLOSED COURT

TRAPEZOIDAL MASSING WITH AN OPEN COURT

Figure 27–4 *Thumbnail sketches of the A residence.* (Courtesy Mr. and Mrs. Donald W. Hamer.)

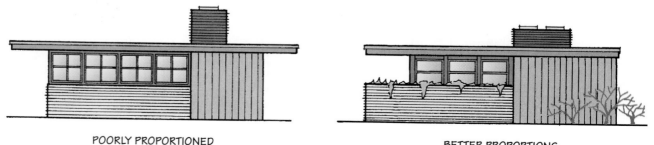

POORLY PROPORTIONED

BETTER PROPORTIONS

Figure 27–5 *Effect of proper proportions.*

term **proportion** deals with the size and shape of areas and their relation to one another. Some of the most important rules of good proportion are:

1. It is often advisable to avoid an isolated square area or multiples of squares since a rectangular area that cannot be visually divided into squares is usually more interesting and pleasing (Figure 27–6). However, see the clustered modular plan in Chapter 16 for one example of a successful design using repeated squares.

2. Balance areas so that they do not appear unstable. Thus the lightest-appearing material or area should be above the heaviest or darkest (Figure 27–7).

3. Areas should be either completely symmetrical or obviously unsymmetrical (Figure 27–8).

4. One leading area should dominate the entire design, with the other areas subordinate (Figure 27–9).

5. Repetition of elements may be used to advantage or disadvantage depending upon the circumstances. The repeated window arrangement shown in Figure 27–10 is superior, but the repeated drawer arrangement in Figure 27–11 is not.

Fenestration   The term **fenestration** refers to the arrangement of windows (and doors) in a wall. Some of the rules for satisfactory fenestration are:

1. Arrange windows symmetrically in a symmetrical elevation (Figure 27–12), but off center in an unsymmetrical elevation (Figure 27–13).

2. Line up windows on different floors. This is important for both aesthetic and structural reasons (Figure 27–14).

3. Do not use a variety of types and sizes of windows (Figure 27–15).

4. Arrange windows in groupings when possible (Figure 27–16).

Harmony   All features should harmonize to present a uniform elevation. Materials also should be selected to harmonize with one another. In general, it is wise to use no more than two different types of materials on one building (Figure 27–24).

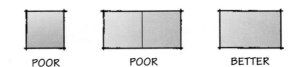

Figure 27–6   *Shape.*

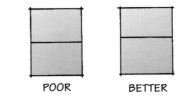

Figure 27–7   *Stability.*

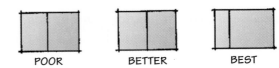

Figure 27–8   *Symmetry.*

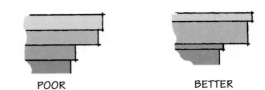

Figure 27–9   *Domination.*

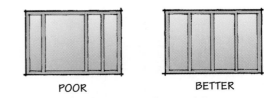

Figure 27–10   *Horizontal repetition.*

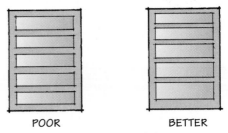

Figure 27–11   *Vertical repetition.*

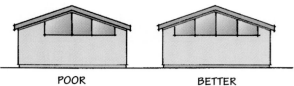

Figure 27–12   Symmetrical fenestration.

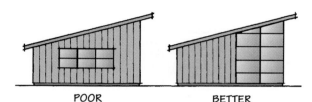

Figure 27–13   Unsymmetrical fenestration.

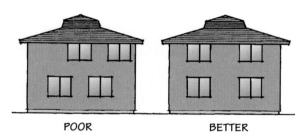

Figure 27–14   Vertical alignment of windows.

VERY POOR

Figure 27–15   Disorderly fenestration.

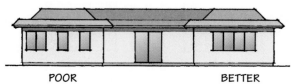

POOR   BETTER

Figure 27–16   Orderly fenestration.

Shadows   Consider the effect of solar orientation upon each elevation. A very simple elevation may become more interesting when designed to take full advantage of shadows. Figure 27–25 shows a house designed with a second floor slightly overhanging the first floor. The simple shadow lines are made more interesting by a variation in the surface casting the shadow (Figure 27–26), or by varying the surface receiving the shadow (Figure 27–27).

Finished Sketches

The finished sketch may be a carefully drawn free-hand sketch or a drawing done with drafting instruments. If the preliminary sketch was very carefully drawn, the designer may omit the finished sketch and proceed directly to the finished elevation drawing. The scale of both is usually $\frac{1}{4}$" = 1' for a small residence and $\frac{1}{8}$" = 1' for larger buildings. Occasionally one or two major elevations are drawn to the large $\frac{1}{4}$" = 1' scale, and the less important elevations are drawn to the smaller $\frac{1}{8}$" = 1' scale. Since these are fairly large-scale drawings, the exact size of all features must be considered together with their correct representation.

Window and Door Sizes   As a general rule, it is a good idea to align the tops of all exterior doors and windows. This simplifies construction by allowing the builder to use one size of header for all normal wall openings and, incidentally, simplifies the drawing of the elevations. A front door will usually be 3'-0 × 6'-8" (actual size of the door), and a rear door will be 2'-8" × 6'-8". A single garage door averages 9'-0 × 7'-0 high, and a double garage door 16'-0 × 7'-0 high. Window sizes

Figure 27–17   Horizontal massing on a hillside plot. (Courtesy California Redwood Association.)

Figure 27–18   *Block and cylindrical massing in a contemporary structure.* (Courtesy American Plywood Association.)

Figure 27–21   *An open, but shaded, patio.* (Courtesy California Redwood Association.)

Figure 27–19   *A whirlpool and firepit in outdoor decking.* (Courtesy California Redwood Association.)

Figure 27–22   *Even a wood fence deserves careful design.* (Courtesy California Redwood Association.)

Figure 27–20   *Fenestration as a major design element.* (Courtesy California Redwood Association.)

Figure 27–23   *Horizontal massing is the principal design element in this office building.* (Courtesy American Plywood Association.)

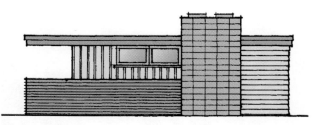

Figure 27–24   *Lack of harmony—too many materials used.*

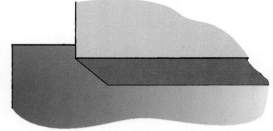

Figure 27–25   *Shadow too simple.*

CAREER PROFILES

Name: Paul M. Churchill   *Age:* 55
Gender: Male
Occupational Title: Builder
Employer: Self-employed, Churchill Construction Company, Indiana, PA 15701
Years of Experience: 30
Educational Background: Three years of college—physics major and mathematics minor
Starting Salary: $6.00 per hour, unemployment compensation, workman's compensation, FICA, company medical plan

JOB OVERVIEW:

As supervisor of the building project, my typical day involves planning and coordinating work assignments and seeing that materials and equipment are on hand. Coordination of the project with the owner is another daily task. The continuing success of a small business is a big responsibility. Meeting with prospective clients, making suggestions, idea input, preparing estimates, insuring timely project completion, billing, and servicing customer complaints are all key responsibilities.

Some of the major projects I have completed during my career include:

🏛 Warehouse office complex for an automotive parts supplier

🏛 Maintenance and office for Indiana Borough in Indiana, Pennsylvania

🏛 Church camp youth meeting building in Stoneboro, Pennsylvania

I receive great enjoyment in seeing a project to successful completion. Another benefit of my job is knowing clients are pleased with our work. The greatest challenge is to produce quality craftsmanship in an area of very stiff competition due to a high unemployment rate. To survive and prosper, we are constantly watching for new and better ways of doing our job.

Important things for students to consider in their pursuit of a career is to develop their skills in mathematics and communications (speaking, writing, and vocabulary skills). Also, it is important to retain your childhood curiosity about how things are constructed and function. This means becoming a lifelong student, continually learning new things so your job remains interesting and enjoyable.

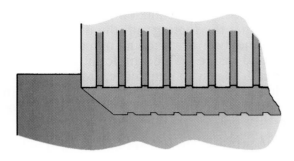

Figure 27–26   Shadow made more interesting by use of boards and battens.

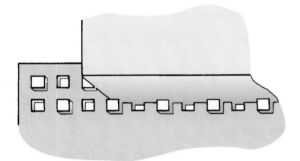

Figure 27–27   Shadow made more interesting by use of extended and recessed concrete blocks.

must be chosen from manufacturers' catalogs, which offer a great variety of sizes and types. Windows fall into the following general types:

Fixed
Double-hung (slides vertically)
Sliding (slides horizontally)
Awning (hinged at top and swings outward)
Hopper (hinged at bottom and swings inward)
Casement (hinged at side and usually swings outward)
Pivoted (hinged at center, half swings outward and half inward)
Jalousie (many individually hinged panes)

Only one or two types of windows should be specified for a house, although their sizes can be varied to suit the need.

Window and Door Representation   Since most architectural features are too complicated to draw in detail, certain simplifications and conventions have been established to lessen the

work of the drafter. For example, windows and doors are shown considerably simplified. Figure 27–28 shows how a door and window would appear if drawn completely. Fortunately, this type of representation is never used. Figure 27–29 shows standard representations—with lines only for the opening, the trim, and panels. Notice that even the doorknob should be omitted.

A further simplification is often used: one window on an elevation is detailed as in Figure 27–29 and all other similar windows are merely outlined by a rectangle. The hidden lines in Figure 27–30 indicate a hinged window. The hinge is located where the hidden lines meet.

Material Representation   Like windows and doors, materials are also represented by drawing only a few lines. Brick, for example, is indicated by several horizontal lines spaced about 3" apart (to the proper scale) rather than showing each brick and mortar joint. When bricks are laid on edge for windowsills or window and door heads,

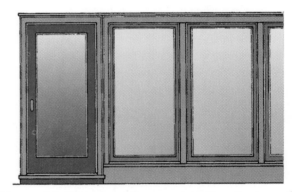

Figure 27–28   Actual representation (not used).

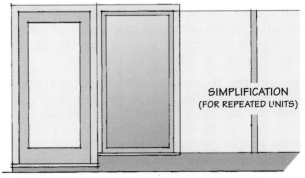

Figure 27–29   Standard representation.

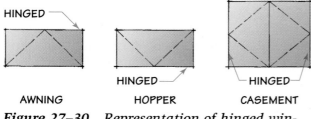

Figure 27–30 *Representation of hinged windows.*

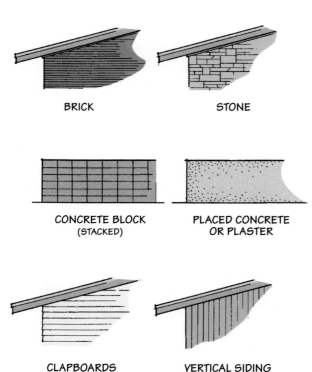

Figure 27–31 *Representation of exterior materials.*

they should be shown also. Figure 27–31 shows the usual representation of brick, stone, concrete block, placed (poured) concrete, clapboards, and vertical siding. Roofs may be left blank.

Footings and Areawalls   Hidden lines are used to indicate the location of footings, below-grade windows, and their areawalls. An areawall is a retaining wall that holds the earth back from a below-grade opening. Common materials used for areawalls are concrete, masonry, and corrugated sheets of galvanized iron. Take particular note of the hidden foundation and footing lines. Notice that the footing and the outside wall of the foundation *are* shown, but the basement floor and inside wall of the foundation are *not* shown. It is easy to remember which lines to include in an elevation view: just imagine that the ground has been removed and show only those lines that you would then see.

Labeling Views   Two methods are used to label elevation views:

1. Front elevation
   Rear elevation
   Right-end elevation
   Left-end elevation

2. North elevation
   East elevation
   South elevation
   West elevation

In the second method, the north elevation is the elevation that faces generally northward, but it does not have to face exactly north. When an interior elevation is designated as a north elevation, this means the *outside* of the wall faces north, and the *inside* of the wall faces south.

Dimensioning   Elevation dimensions are limited to vertical dimensions since horizontal dimensions have already been shown on the plan. Show the depth of the footing below grade, the finished-floor-to-finished-floor heights (or finished-floor-to-finished-ceiling height for the topmost story), the roof height, and the height of the chimney above the roof. Check the local building code for inside room heights (finished floor to finished ceiling). The following minimum allowable heights are typical.

Basement:	6'-2" (clear of all low beams, ducts, or pipes)
First floor:	7'-10"
Second floor:	7'-4"
Garage:	8'-0

Of course, when you give finished-floor-to-finished-floor heights, you must allow for the thickness of the floor (finished flooring, sub-flooring, joists, and ceiling). This will amount to about 1 ft.; the exact amount can be calculated by adding the actual sizes of these members as indicated on the typical section. Chimneys should extend at least 2' above a nearby ridge line. The window schedule may be placed in any convenient location.

Changes   Remember that it is quite probable that some changes and additions in the elevations will be necessary after all the other drawings are completed.

Elevation Drawing

After the elevations have been designed by use of thumbnail, preliminary, and finished sketches, the final drawing may be started. The steps used in drawing elevations are illustrated as follows. We will use the rear elevation of the A residence as an example.

Step 1: Layout   Lay out the elevation very lightly on tracing vellum using a hard, sharp, pencil (Figure 27–32). Using dividers or a scale, transfer horizontal dimensions from the floor plan and vertical dimensions from the sectional drawing. A scale of $\frac{1}{4}$" = 1' or $\frac{1}{8}$" = 1' is used and indicated in the title block or near the drawing. If the plan and section are drawn to the same scale as the required elevation, they may be taped in position and dimensions projected directly using a triangle and T square. (If the plan and section have not been drawn yet, they should be drawn in rough form at this point. Save these rough drawings since they will help finish the final plan and section.)

Windows and doors are located horizontally by projecting from the plan; they are located vertically by projecting from the window and door details or simply by aligning the tops of the windows with the tops of the doors.

Step 2: Details   The elevation details to be included will vary depending upon the style of the house. In the case of the A residence, the following details are added (Figure 27–33):

1. Roof fascia
2. Roof beams
3. Chimney, saddle, and flashing
4. Window representation (if you wish, only one window is detailed)
5. Grade lines
6. Footings
7. Material representation
8. Darkened building outline

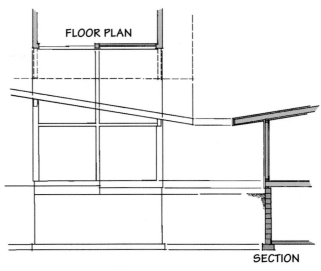

Figure 27–32   Step 1: Layout.

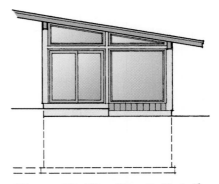

Figure 27–33   Step 2: Details.

Step 3: Dimensioning   Elevation dimensions and notes are added (Figures 27–34 through 27–38):

1. Height of roof (in this example, the roof beam height determines the roof height)
2. Depth of footings
3. Height of other features such as masonry wall and chimney
4. Height of windows (in this example, the windows fit directly under the roof and roof beams)
5. Roof slope indication

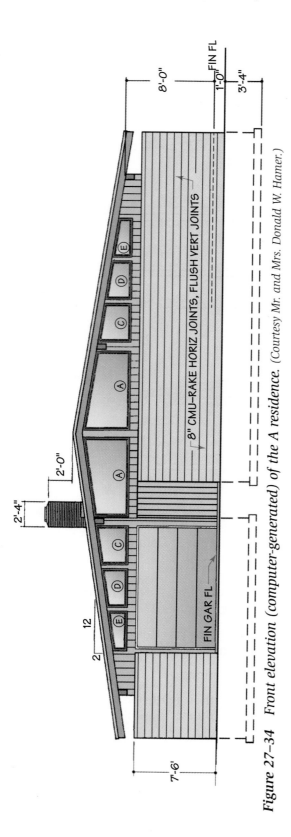

Figure 27–34 Front elevation (computer-generated) of the A residence. (Courtesy Mr. and Mrs. Donald W. Hamer.)

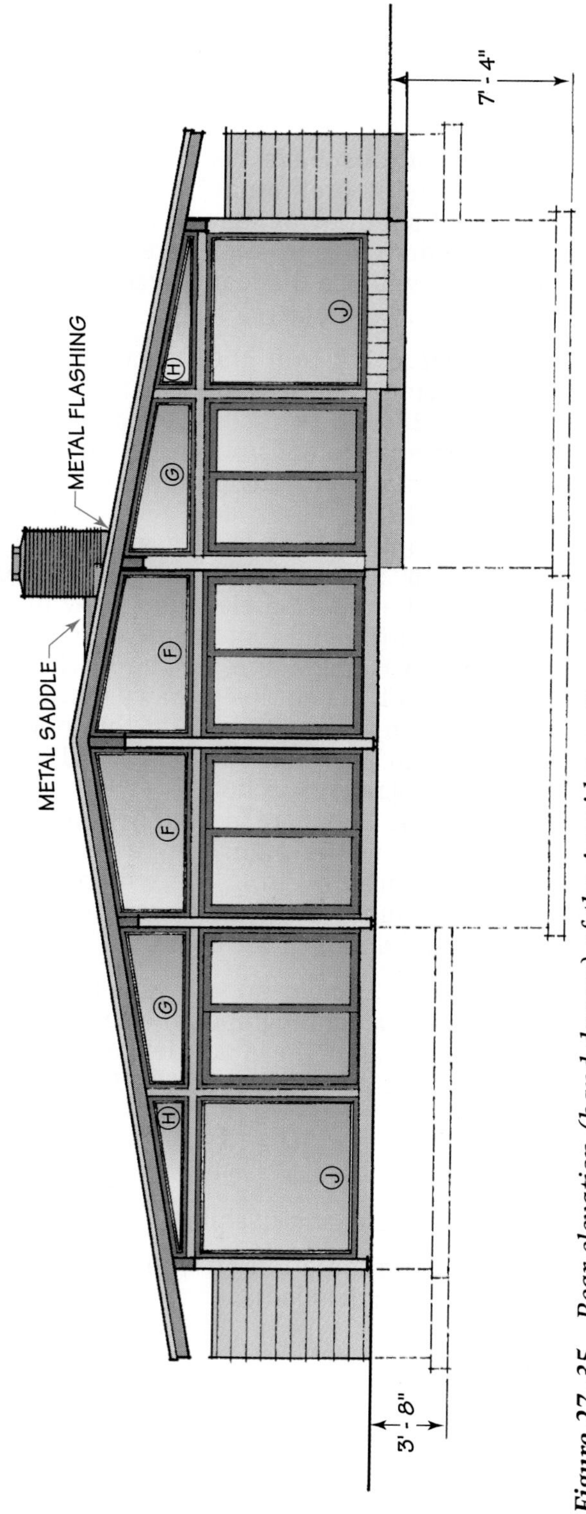

Figure 27–35 Rear elevation (hand-drawn) of the A residence. (Courtesy Mr. and Mrs. Donald W. Hamer.)

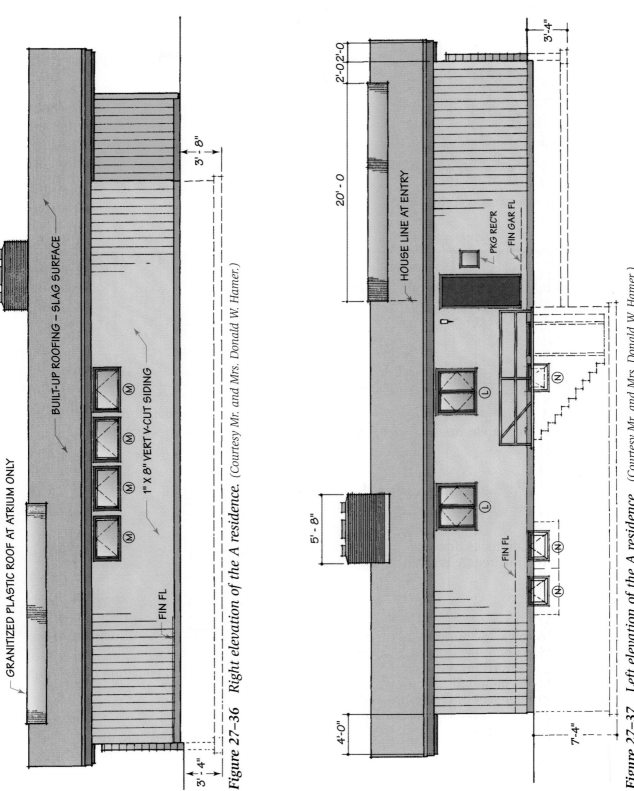

Figure 27–36 *Right elevation of the A residence. (Courtesy Mr. and Mrs. Donald W. Hamer.)*

Figure 27–37 *Left elevation of the A residence. (Courtesy Mr. and Mrs. Donald W. Hamer.)*

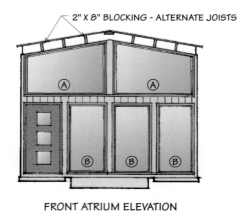

FRONT ATRIUM ELEVATION

MK	NO	SIZE	DESCRIPTION			
A	2	90" X 42" X 57"	1/4" FIXED	PLATE	GLASS	
B	3	44" X 80"	"	"	"	"
C	2	44" X 34" X 41"	"	"	"	"
D	2	44" X 26" X 33"	"	"	"	"
E	2	44" X 18" X 25"	"	"	"	"
F	2	90" X 38" X 53"	"	"	"	"
G	2	90" X 22" X 37"	"	"	"	"
H	2	90" X 6" X 21"	"	"	"	"
J	2	90" X 84"	"	"	"	"
K	2	AP421	ANDERSEN AWNING-FIXED			
L	2	C235	" CASEMENT			
M	4	A41	" AWNING			
N	3	2820	" BASEMENT			

WINDOW SCHEDULE

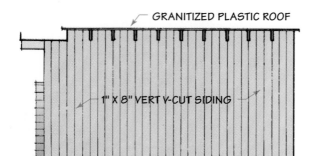

LEFT ATRIUM ELEVATION

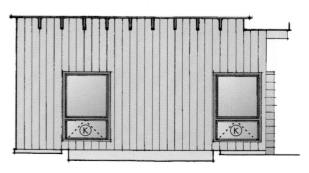

RIGHT ATRIUM ELEVATION

Figure 27–38   *Atrium elevations of the A residence.* (Courtesy Mr. and Mrs. Donald W. Hamer.)

6. Window schedules

7. Titles and notes indicating materials and special details

Compare the elevations as drawn with the photographs of the finished A residence, Figures 27–39 to 27–41.

SECTIONS

The designer shows the entire building construction by means of a few drawings called *sections.* He lays out the sections in much the same order that the workmen will use in actual construction. A complete set of construction drawings would contain one or more of each of the following types of sections:

Structural Section

A **structural section** shows the entire building construction, as shown in Figure 27–42. A $\frac{1}{4}$" =

Figure 27–39   Front elevation of the A residence. (Courtesy Mr. and Mrs. Donald W. Hamer.)

1' scale is often used. Figure 27–43 shows a structural section for the A residence. This would be useful in planning for structural strength and rigidity, determining the length of members, and specifying sizes.

Figure 27–40   Rear elevation of the A residence. (Courtesy Mr. and Mrs. Donald W. Hamer.)

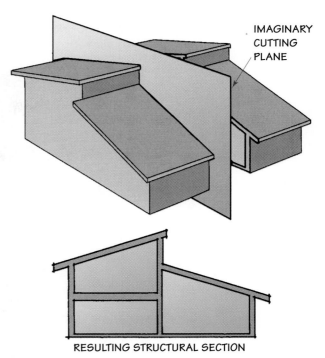

RESULTING STRUCTURAL SECTION

IMAGINARY CUTTING PLANE

Figure 27–42   Imaginary cutting plane used to obtain a structural section.

Wall Section

A **wall section** shows the construction of a typical wall to a larger scale than the structural section ($1\frac{1}{2}$" = 1' is often used). Figure 27–44 shows a wall section for the A residence. Notice

Figure 27–41   Atrium of the A residence. (Courtesy Mr. and Mrs. Donald W. Hamer.)

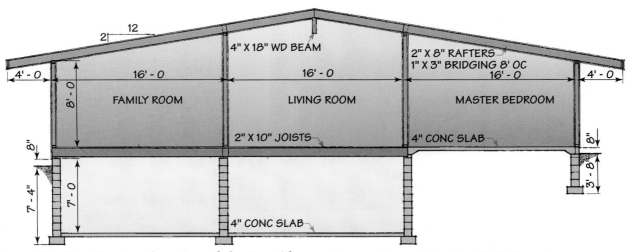

12
2

4" X 18" WD BEAM

2" X 8" RAFTERS
1" X 3" BRIDGING 8' OC

4' - 0   16' - 0   16' - 0   16' - 0   4' - 0

8' - 0

FAMILY ROOM   LIVING ROOM   MASTER BEDROOM

2" X 10" JOISTS   4" CONC SLAB

8"

8"   3' - 8"

7' - 4"   7' - 0

4" CONC SLAB

Figure 27–43   Structural section of the A residence. (Courtesy Mr. and Mrs. Donald W. Hamer.)

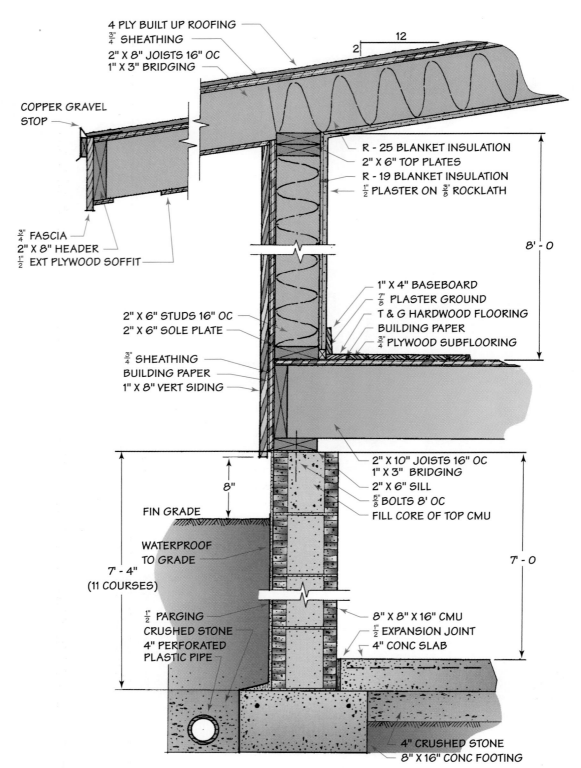

Figure 27–44   *Typical section of the A residence.* (*Courtesy Mr. and Mrs. Donald W. Hamer.*)

that floor-to-ceiling heights are shown, together with sizes and material specifications for all rough and finished members.

Detail Section

Any deviations from the typical wall sections may be shown in **detail sections**. Also any special or unusual construction must be detailed, as shown in Figure 27–45. These sections may be drawn to a large scale (up to full size).

COMPUTER-GENERATED ELEVATIONS

Architectural elevations (exterior elevations, interior elevations, and sections) can be entered into your computer graphics system by puck or stylus on a digitizer tablet and displayed on your monitor. Start with thumbnail elevations and sections. Proceed to preliminary drawings, finished drawings, and working drawings. When completely satisfied with the result, use your printer or plotter to produce hard copies,

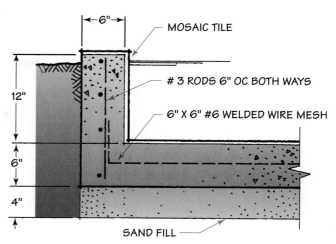

Figure 27–45 *Atrium pool detail for the A residence. (Courtesy Mr. and Mrs. Donald W. Hamer.)*

but always store your CAD work on diskette for the inevitable future revisions.

Compare the computer-generated front elevation of the A residence (Figure 27–34) with the hand-drawn rear elevation (Figure 27–35).

REVIEW QUESTIONS

1. Describe the purpose of, and give the probable scale of, each of the following:
   a. Thumbnail elevation sketches
   b. Preliminary elevation sketches
   c. Finished elevation sketches
   d. Finished elevation drawings

2. How many exterior elevations are needed to describe a building having:
   a. A rectangular floor plan
   b. An L-shaped floor plan
   c. A U-shaped floor plan
   d. An S-shaped floor plan

3. Give the most commonly used size for the following:
   a. Front door
   b. Rear door
   c. Single garage door
   d. Double garage door

4. State two methods of simplifying the representations of doors and windows.

5. How is the location of window hinges indicated on elevation drawings?

6. Differentiate between an interior and an exterior elevation.

7. Indicate information on wall sections:
   a. Not found on other kinds of architectural plans
   b. Repeated on other kinds of architectural plans
8. What is the rule for determining which below-grade lines should be shown by hidden lines and which below-grade lines should be entirely omitted?

ENRICHMENT QUESTIONS

1. Explain how each of the following words is associated with the rules of good proportion:
   a. Squares
   b. Stability
   c. Symmetry
   d. Domination
   e. Repetition
2. Explain how each of the following words is associated with the rules of fenestration:
   a. Symmetry
   b. Alignment
   c. Variety
   d. Grouping
3. Pencil-sketch or computer-sketch and name three types of sectional drawings. What scale might be used for each?
4. What purposes do sections serve?
5. If an interior elevation of a room is called a *south elevation*, in which direction does the inside wall face?

PRACTICAL EXERCISES

1. Using a scale of $\frac{1}{4}$" = 1', draw or computer-produce the following elevations of the A residence:
   a. Front elevation
   b. Rear elevation
   c. Right-end elevation
   d. Left-end elevation
   e. Atrium elevation
2. Draw or computer-produce the elevations for your original house design.
3. Draw or computer-produce the typical wall section of the A residence.
4. Draw or computer-produce the structural section for your original house design.
5. Draw or computer-produce the typical wall section for your original house design.
6. Draw or computer-produce the detail sections for your original house design.

Windows

OBJECTIVES

By completing this chapter and studying Appendix B you will be able to:

- 🏛 Differentiate between casement, awning, hopper, projected, sliding, double-hung, and fixed windows.

- 🏛 Explain how most windows are installed.

- 🏛 Describe how to specify windows with the aid of sizing tables.

- 🏛 Demonstrate how to detail windows with the aid of tracing details.

- 🏛 Utilize window specifications to select and appropriately draw the architectural window symbols.

KEY TERMS

Casement window	Pivoted window
Awning window	Jalousie window
Hopper window	Fixed window
Projected window	Mullion
Sliding window	Muntin
Double-hung window	

The architectural drafter should be thoroughly familiar with *all* the building components to be specified and detailed. However, the window is probably the most important single component in the design of a successful building. The proper selection and placement of windows is necessary for aesthetic as well as functional reasons. In addition to providing light and air, windows can change the interior of a room by providing framed views or window walls, and can change the exterior of a building by the fenestration (Figure 28–1). The energy costs of a building may be dramatically affected by the size and orientation of windows. For example, heating costs in cold climates are increased by the heat lost through windows on north-facing

Figure 28–1 *Successful designs created by attractive fenestration.* (Photos courtesy Pella Corporation.)

walls, and cooling costs in warm climates are increased by the heat gained through windows on south-facing walls.

TYPES

There are many different types of windows now on the market. Some of the most commonly used are:

> Casement
> Awning
> Hopper
> Projected
> Sliding
> Double-hung
> Pivoted
> Jalousie
> Fixed

These windows can be obtained in many metals, wood, clad wood (wood with vinyl or metal sheathing), and extruded vinyl in nearly any size ranging from small lavatory windows to entire window walls. Let's look at each in turn.

Casement

A **casement window** (Figure 28–2) is hinged at the side and usually swings outward so that the inside drapes are not disturbed. Screens, then, are hung on the inside. When more than two casement windows are installed side by side, it is often the practice to specify fixed-sash for the middle windows, and a hinged-sash at each end.

When the sash is hinged at the top, it is called an **awning window**; when the sash is hinged at the bottom, it is called a **hopper window**. To prevent rain from entering the open windows, awning windows swing outward and hopper windows swing inward.

Projected

A **projected window** (Figure 28–3) is somewhat different from a casement window in that some form of linkage other than the hinge is used. A projected window will swing open and slide at the same time. Projected windows may also be classified as casement, awning, or hopper according to the direction of swing. Metal projected windows are commonly used on commercial buildings, whereas wooden casement (nonprojected) windows are commonly used on residences.

Sliding

Sliding or gliding **windows** (Figure 28–4) are designed to run on horizontal tracks, either in sliding pairs as shown or as a single sliding window paired with a fixed sash. Most sliding windows contain sash that can be removed from the frame for easy cleaning. The screen is installed on the outside of the window.

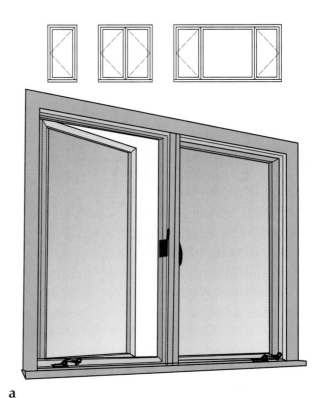

a

b

Figure 28–2 Casement windows. *(Part **a** Courtesy Andersen Windows, Inc.; part **b** Courtesy Pella Corporation.)*

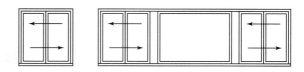

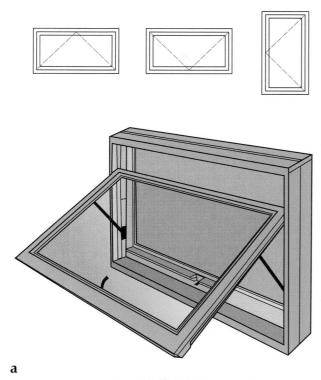

a

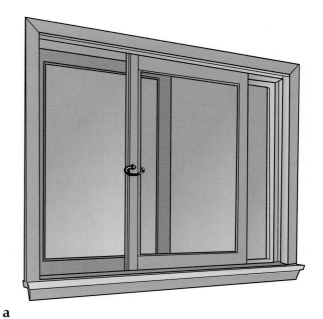

a

b

Figure 28–3 *Projected windows.* (Part **a** Courtesy Andersen Windows, Inc.; part **b** Courtesy Pella Corporation.)

b

Figure 28–4 *Sliding windows.* (Part **a** Courtesy Andersen Windows, Inc.; part **b** Courtesy Pella Corporation. Architect: The Troyer Group.)

Double-hung

The **double-hung window** (Figure 28–5) is usually specified for Colonial-type houses. This window contains two sash which slide in vertical tracks. A spring-balance arrangement is used to counterbalance the weight of the sash and hold them in any desired position. The sash are easily removed for window cleaning. Double-hung windows may be obtained with self-storing screens and storm windows.

Pivoted

The **pivoted window** revolves on two pivots—one at the center of the top of the sash, and the other at the center of the bottom of the sash. Not often specified for houses, the pivoted window is common in taller buildings because of the ease of cleaning.

Jalousie

The **jalousie window** is a series of small awning panes, all operated together. It has not become very popular because the view through it is interrupted by the many intersections.

Fixed

When views and light are desired without ventilation (as is often true of air-conditioned buildings), **fixed windows** are specified. Large fixed windows are sometimes called *picture windows*. Some fixed windows are designed so that they can be opened only by window washers for cleaning. Fixed windows are stocked in both rectangular and trapezoidal shapes (Appendix B), or they can be custom-shaped (Figure 28–6).

Basement Windows

The most often specified basement window is a reversible awning-hopper window (Figure 28–7). The sash is designed so that it can be easily removed from the frame and installed to swing up (awning) or down (hopper). In both cases the swing is toward the inside. Since a basement window is installed at the inner side of the foundation wall, there is no danger of rain entering. The screen is installed on the outside of the frame.

a

b

Figure 28–5 ***Double-hung windows.*** (*Courtesy Andersen Windows, Inc.*)

Figure 28–6 *Custom-shaped fixed windows.*
(Courtesy Pella Corporation.)

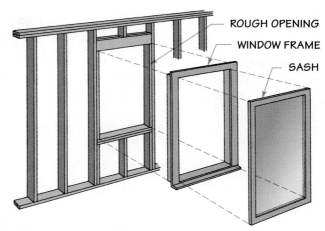

Figure 28–8 *Window installation.*

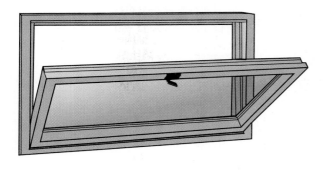

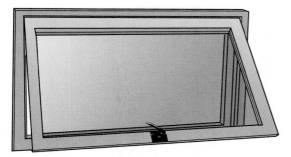

Figure 28–7 *Basement windows.* *(Courtesy Andersen Windows, Inc.)*

INSTALLATION

Before studying specifications and the detailing of windows, it is necessary to understand how a window is installed in a building wall (Figure 28–8). The glass and its immediate framing members are called the *sash*. Except in fixed windows, the sash is designed to be opened for ventilation or entirely removed for easy cleaning. The sash is surrounded by the *window frame*, which is permanently fastened to the rough wall (studs). The window frame has an L-shaped cross section, the outer portion of which is called the *blind stop*. The blind stop helps to position the frame properly in the rough opening. Stock windows are obtained with the sash already installed in the frame so that the entire window unit can be set into the rough opening. The rough opening is constructed several inches larger than the window frame to allow for leveling the window. After the window is in place, exterior and interior trim is used to close the cracks between the frame and the rough opening.

TERMINOLOGY

Many special terms are used to describe the various parts of a window. Figure 28–9 shows a cutaway pictorial of a double-hung window and the corresponding sectional details. The terms *head, jamb, rail,* and *sill* indicate that the sectional cuts were taken through the head (the upper horizontal members), jamb (side vertical members), meeting rail (middle horizontal members), and sill (lower horizontal members).

Sash

The horizontal members of the upper sash are called *top rail* and *meeting rail.* The lower sash

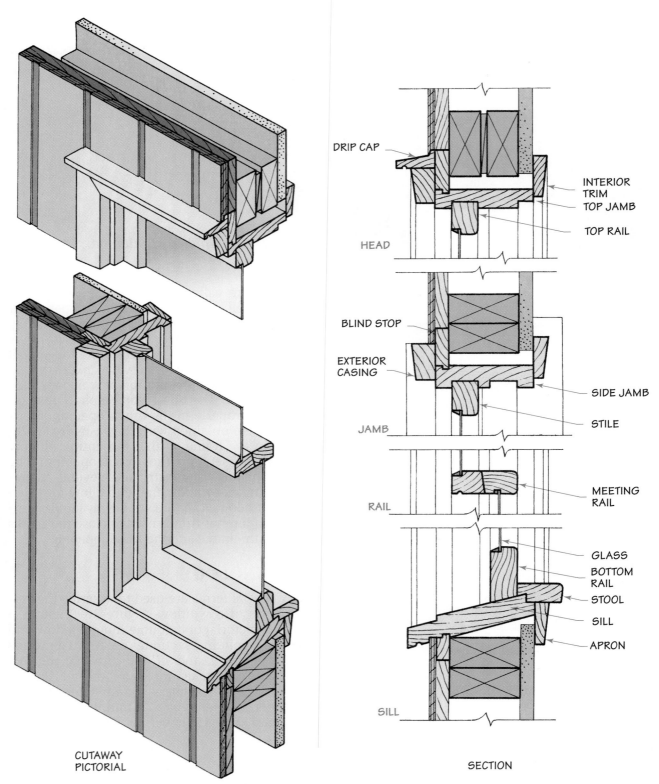

Figure 28–9 *Double-hung window in a wood-frame wall.*

horizontal members are called *meeting rail* and *bottom rail*. All vertical members are called *stiles*.

Window Frame

The members of the window frame are called *top jamb, side jamb, sill,* and *blind stop* (or *windbreaker*).

Interior Trim

The interior trim and *apron* cover the crack between the window frame and the interior finished wall.

Exterior Casing

The exterior casing (which may be called *trim*) also covers wall cracks. In addition, it serves as the frame around the *storm sash* or *screens*.

Drip Cap

The drip cap prevents water from seeping into the window head. The *drip groove* on the underside prevents water from seeping inward underneath the drip cap. *Flashing* can be used in place of the wooden drip cap.

Double Glazing

A second glass pane may be installed on the sash, creating a dead-air space to provide insulation and prevent condensation on the inside pane. Also, the outside pane can be manufactured with a micro-thin transparent coating, called *low-E glass*, which reflects outside heat in summer and inside heat in winter. Manufacturers also provide windows with a low-conductance gas (argon) between panes to increase insulating value.

Mullions and Muntins

Shown in Figure 28–10, **mullions** are members (usually vertical) that separate adjacent windows; **muntins** are smaller members used to subdivide large glass areas. Many manufacturers offer removable muntins so that the windows may be subdivided to any taste. Also, they may be removed for easy cleaning.

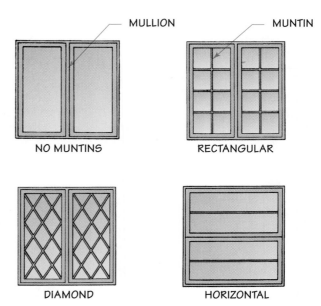

Figure 28–10   Removable muntins.

Figure 28–11   Muntins can be specified to create a unique style. (Courtesy Pella Corporation.)

Figure 28–11 shows some custom designed muntins.

SPECIFYING WINDOWS

The type of window to be specified for a building will be determined by the building style and the client's desire. Once the style is decided, the size must be determined from the many sizes available in each style. As an example, let us study the Andersen casement windows. Eight window heights are obtainable, each height recommended for a different condition. Figure 28–12 shows how different window sizes will fit into different types of rooms.

Manufacturers do not offer a wide variety of window widths since wide windows are obtained by specifying a number of individual units side by side. Andersen casements may be

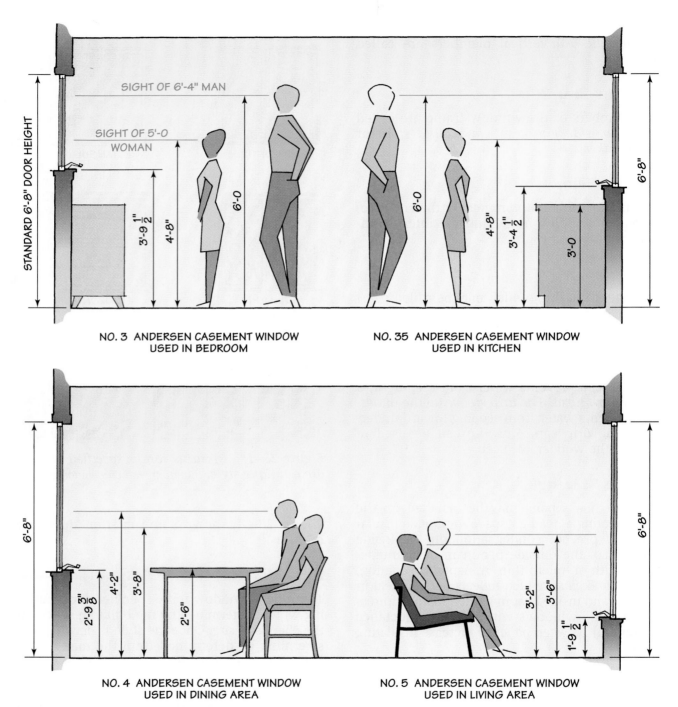

NO. 3 ANDERSEN CASEMENT WINDOW
USED IN BEDROOM

NO. 35 ANDERSEN CASEMENT WINDOW
USED IN KITCHEN

NO. 4 ANDERSEN CASEMENT WINDOW
USED IN DINING AREA

NO. 5 ANDERSEN CASEMENT WINDOW
USED IN LIVING AREA

Figure 28–12   Selection of window height.

obtained in four widths, CR (reduced casement), CN, C, and CW (wide casement).

Each manufacturer has its own set of window sizes and catalog numbers. To simplify window specifications, tables showing the size and numbers may be obtained from the manufacturers (Table 28-1). Tables for Andersen casement, sliding, double-hung, and basement

TABLE 28–1　*Andersen Casement Windows*

Andersen Catalog Number	Rough Opening Height	Recommended-for:
2	$2'-0\frac{5}{8}''$	Extreme privacy
3	$3'-0\frac{1}{2}''$	Lavatory, Bedroom
35	$3'-5\frac{3}{8}''$	Kitchen
4	$4'-0\frac{1}{2}''$	Dining area
45	$4'-5\frac{3}{8}''$	Family room
5	$5'-0\frac{3}{8}''$	Living area
55	$5'-5\frac{3}{8}''$	Sunroom
6	$6'-0\frac{3}{8}''$	Solar wall

Andersen Catalog Number	Rough Opening Width	Preferred for:
CR	$1'-5\frac{1}{2}''$	Emphasis on vertical lines
CN	$1'-9''$	General applications
C	$2'-0\frac{5}{8}''$	General applications
CW	$2'-4\frac{7}{8}''$	Wider uninterrupted view

computer workstation, and desired window details can be transferred to the designer's CAD plans. Also, manufacturers provide window descriptions in the form of (X, Y, Z) coordinates of the corners and edges of the windows. This greatly simplifies the task of adding these details to pictorial views. Although the manufacturer attempts to supply tracing details for nearly all possible types of construction, some modification is often necessary to fit the window into a particular type of wall. Tracing details for Andersen casement, sliding, double-hung, and basement windows are shown in Appendix B.

WINDOW ARRANGEMENTS

Stock windows may be stacked horizontally or vertically to obtain countless distinctive arrangements. Figures 28–13 through 28–18 show various combinations of Andersen, Pella, and Velux windows.

Figure 28–13　*A charming oval fixed window.*
(*Courtesy Andersen Windows, Inc.*)

windows and a table for Pella fixed trapezoidal windows (much used in contemporary design) are shown in Appendix B.

DETAILING

In addition to supplying size tables for the convenience of architectural drafters, manufacturers also supply *tracing details* on paper and computer diskette. The paper details may be slipped under tracing vellum and all applicable details copied. Or they may be digitized. The diskettes are inserted in the disk drive of a

REAL WORLD APPLICATIONS

Window Condensation

Have you ever wondered why a pair of eyeglasses steams up when the wearer enters a warm house in the winter? This happens for the same reason that moisture appears on some windows when it becomes cold outside—condensation.

Under normal conditions, air contains some water vapor. The amount of water vapor present is often expressed as the relative humidity, which is given as a percentage that indicates how full of moisture the air is. Air that is "full" is called saturated and has a relative humidity of 100 percent. Air temperature is one of the major factors affecting the relative humidity because warm air is able to hold more water vapor than cold air. All else being equal, as air is cooled, its relative humidity increases. If the relative humidity reaches 100 percent, water vapor will begin to condense onto any nearby surfaces. As warm room air passes over cold window panes (or eyeglass lenses), a thin layer of air is cooled. If the relative humidity in this layer reaches 100 percent, water will condense from the air onto the glass. This condensation can create problems in a home ranging from damaged sills to deterioration of wall studs that are saturated with water.

To combat condensation, newly designed windows have incorporated additional panes. The advantage of this is that gases trapped between the panes increase the window's insulating value, slowing the movement of heat from the inside of the window to the outside. This not only results in an energy savings, but also helps avoid condensation by keeping the inside window pane warm enough to avoid cooling air to the point of saturation. When designing a house, however, always remember to consider the climate so as to avoid specifying unnecessarily costly windows. How can you determine what type of windows are appropriate for your original house design?

Home Energy Magazine has developed an Internet home page that can provide some answers. Their address is:
http://beijing.dis.anl.gov/homeenergy/index.html

Once you are there, find the section labeled "Links for Consumers" and click on "Windows." Scroll through the article until you find Figure 2, which illustrates Conditions for Window Condensation. Briefly describe what this line graph is illustrating. If the outside air temperature is 30 degrees Fahrenheit, at what relative humidity will condensation form on single-pane, double-pane, and triple-pane windows? Why do these three types of windows have a larger humidity range when the temperature is –30 degrees and a smaller humidity range when the temperature is 60 degrees?

Once you have answered these questions, spend some time investigating these topics about windows:

🏛 Insulating value

🏛 Recommendations for selecting window U-factors

🏛 Window orientation and solar control

🏛 Ultraviolet protection

🏛 Ventilation and airtightness

🏛 Window energy glossary

🏛 The difference a U-factor can make

🏛 Window labels

Contributed by C. A. Pepper

Figure 28–14 *Skylights in a second floor bathroom. (Courtesy VELUX-AMERICA, Inc.)*

Figure 28–16 *Casement and fixed windows frame a full-sized bay. (Courtesy Pella Corporation.)*

Figure 28–15 *An arched double-hung window with adjacent windows form a magnificent Palladian window. (Courtesy Andersen Windows, Inc.)*

Figure 28–17 *Awning roof windows create a sunny, yet private, retreat. (Courtesy VELUX-AMERICA, Inc.)*

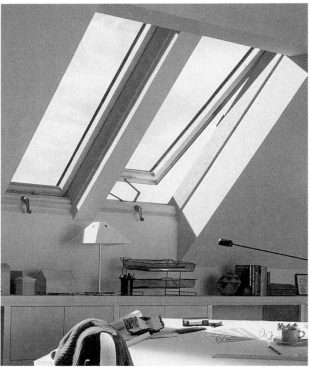

Figure 28–18 *Tilting roof windows permit the sash to pivot inside for easy cleaning. (Courtesy Andersen Windows, Inc.)*

REAL WORLD   APPLICATIONS

Windows in the Year 2050

The use of windows in house design and window styles have changed throughout time and will continue to change. Creativity is an important component of architecture. To develop this creativity, your challenge is to write an essay describing how windows might be used in a house of your design in the year 2050. Include style of windows, any specific purposes they may have, energy efficiency needs, and special features.

Before writing, think quietly and dismiss any preconceived ideas you have about the current use of windows in house design. The point of this activity is to be creative, not necessarily practical. Once you have conceived an innovative idea, then practicality can be taken into consideration. For this exercise, all ideas, however seemingly impractical, are welcome! So put on your creative cap and design a window for the year 2050.

REVIEW QUESTIONS

1. List the different types of windows and indicate how each type is opened.

2. Define the following window terms:

   a. Sash

   b. Window frame

   c. Rough opening

   d. Blind stop

   e. Apron

   f. Casing

   g. Stool

   h. Removable muntin

   i. Clad window

   j. Low-E glass

3. What is the difference between these window terms?

   a. Head, jamb, and sill

   b. Top rail, bottom rail, and meeting rail

   c. Storm sash and double glazing

   d. Drip cap and drip groove

   e. Mullion and muntin

   f. Blind stop and windbreaker

   g. Awning window and hopper window

   h. Awning window and jalousie window

   i. Fixed window and picture window

   j. Pivoted window and projected window

4. Where in a house would Andersen CR and CW windows be used?

5. What is the rough opening for an Andersen G65 gliding window?

6. What is the difference between Andersen windows 3062 and 3462?

7. What is the rough opening height for an Andersen window typically used in the kitchen?

8. What is a tracing detail?

ENRICHMENT QUESTIONS

1. What are advantages and disadvantages of using tracing details called up from a computer diskette compared to those obtained from a window catalog?

2. Prepare a cost analysis that compares the following windows that are manufactured by different companies:

   a. Double-hung

   b. Casement

   c. Awning

   d. Pivoted

3. Prepare a report about various types of glass used for windows and the gases that are infused between them.

PRACTICAL EXERCISES

1. Draw or computer-produce a classroom illustration showing:

   a. Types of windows   b. Window terminology

2. Design a type of window entirely different from those commonly used. Draw or computer-produce as appropriate.

3. Complete the design of the windows for your original building design.

Doors

OBJECTIVES

By completing this chapter and Appendix C you will be able to:

- 🏛 Describe the following principal types of doors:
  - ■ Hinged
  - ■ Sliding
  - ■ Folding
  - ■ Accordion
  - ■ Kalamein
  - ■ Garage
- 🏛 Explain how most doors are installed.
- 🏛 Describe how to specify doors with the aid of manufacturers' catalogs.
- 🏛 Demonstrate how to detail doors with the aid of tracing details.
- 🏛 Utilize door specifications to select and appropriately draw the architectural door symbols.

KEY TERMS

Hinged door	Kalamein door
Sliding door	Entranceway
Folding door	Sidelight
Accordion door	

Doors, like windows, are important components in the design of a successful building. The main entrance door is particularly important, since it will be the first detail experienced by visitors. The other doors should also be carefully chosen; a building can be no better than its details.

TYPES

Doors are available in a wide range of types and materials. Residential doors most commonly used are:

> Hinged
> Sliding
> Folding
> Accordion

These doors may be obtained in single and double units. Wood is usually used, but metal and glass doors are also popular.

Hinged

Hinged doors (Figure 29–1) may be flush, paneled, or louvered. The flush door is most popular due to its perfectly clean lines and low cost.

Figure 29–1   Hinged doors.

It may be either solid-core or hollow-core. The solid-core flush door is constructed of solid wood covered with wood veneer and is preferred for exterior doors. The hollow-core flush door has an interior of honeycombed wood strips also covered by veneer. The most popular veneers are mahogany and birch.

Paneled doors (Figure 29–2) consist of Ponderosa pine members framing wood or glass panels. The horizontal framing members are called *top rail* and *bottom rail.* The midheight rail is called a *lock rail.* The vertical framing members are called *stiles.*

Louvered doors are constructed like paneled doors, but with louvers replacing the panels. They are often used as closet doors to permit circulation of air.

Hinged doors may be installed as a *double* unit (two doors, one hung on the right jamb, the other on the left) to allow a larger and more dramatic passageway. Hinged double doors with glass panels are called *French doors* (Figure 29–7). A *Dutch door,* on the other hand, is a single door that has been cut in halves so that the top half can be opened for light and air without opening the bottom half. Simulated Dutch doors which open like ordinary doors are specified to give the appearance of a Dutch door without the function. A door hung on special hinges which permit it to swing in both directions is called a *swinging door* and is often used between the kitchen and dining area to permit operation by a simple push.

Hinged interior doors are usually installed at room corners so that they swing into rooms with a 90-degree swing against the adjacent wall. Exterior doors on public buildings are installed with panic hardware to swing outward, but exterior doors on homes are installed to swing inward, thus allowing storm and screen doors (that are more weather-resistant) to be installed on the outside.

Sliding

Sliding doors (Figure 29–3) are used to save the floor space that is required for hinged doors. They are especially useful in small rooms. Sliding doors are hung from a metal track screwed into the door frame head. A single sliding door slides into a pocket built into the wall. Double sliding doors are usually installed so that one door slides in front of the other. This has the disadvantage of opening up only one half of the doorway space at a time. Exterior sliding doors of glass serve the double purpose of a doorway and window wall.

Folding

A **folding door** (Figure 29–4) is partially a hinged and partially a sliding door. Two leaves are hinged together, one being also hinged to the door jamb. The other leaf has a single hanger sliding in a track. Although some floor space is required for the folding door, it has the advantage of completely opening up the doorway space. Single (total of two leaves) and double (total of four leaves) units are available (Figure 29–5).

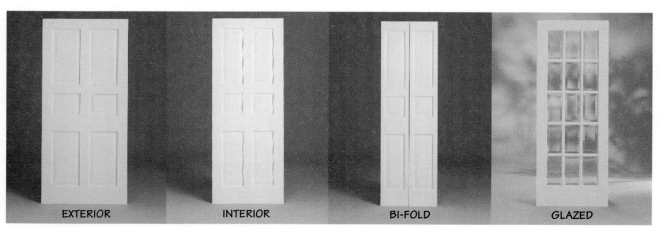

EXTERIOR   INTERIOR   BI-FOLD   GLAZED

Figure 29–2   Some commonly specified panel doors. (Courtesy Morgan Manufacturing.)

Figure 29–3   *Sliding doors.*

Figure 29–4   *Folding doors.*

Accordion

A door that operates on the principle of the folding door, but contains many narrow leaves, is called an **accordion door** (Figure 29–8). These leaves may be made of hinged wood or a flexible plastic material. Accordion or folding partitions are used to provide an entire movable wall between rooms. The folds of the retracted partition may be left exposed or hidden by a wall pocket (Figure 29–6).

MATERIALS

In addition to wood doors, metal-clad wood doors (called **kalamein doors**) or hollow metal doors are used for fireproofing and strength. Metal door frames (called *door bucks*) are also available. Even all-glass doors are used. Bronze, aluminum, and glass doors are commonly used for public buildings, but wood remains the popular choice for residential construction. Figure 29–9 shows how a door frame

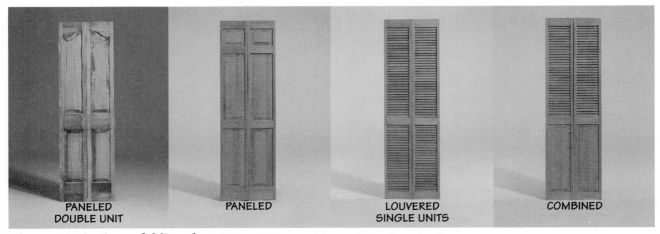

PANELED
DOUBLE UNIT   PANELED   LOUVERED
SINGLE UNITS   COMBINED

Figure 29–5   *Some folding door types.* *(Courtesy Morgan Manufacturing.)*

Figure 29–6 *Accordion doors offer maximum flexibility and efficiency.* (Courtesy Pella Corporation.)

Figure 29–7 *French doors coordinated with windows form a major design feature.* (Courtesy Pella Corporation.)

Figure 29–8 *Accordion doors.*

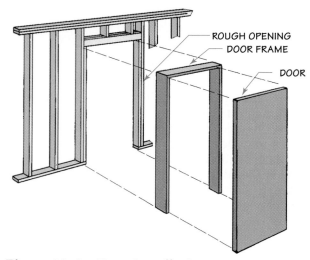

Figure 29–9 *Door installation.*

fits into the rough opening in the same way as a window. The door, in turn, fits into the door frame.

TERMINOLOGY

Figure 29–10 shows a cutaway pictorial of a hinged exterior door, and the corresponding sectional details. Figure 29–11 shows the pictorial and details of a sliding interior door. Notice that a saddle is used to weatherproof the exterior door, but such protection is not necessary for the interior door.

Architectural Engineering Technologist

Looking for a career that is more hands-on yet interesting and challenging? Architectural engineering technology may be for you. Baccalaureate engineering technology programs are similar to baccalaureate engineering programs in many respects. The principal difference is that the baccalaureate engineering technology programs prepare students for basic engineering design, production, and construction rather than for positions that require extensive knowledge of science and mathematical theory. Many baccalaureate engineering graduates continue on for a master's degree or a doctorate.

Baccalaureate architectural engineering technology programs, unlike architectural engineering programs, are four-year technical institute or university programs. Most baccalaureate engineering technology institutions accept transfer students from pre-engineering or associate engineering technology programs, giving full credit for courses completed. Such programs are often called 2 + 2 programs. Typical junior and senior technical courses in an architectural engineering technology major include:

- 🏛 Architectural presentation
- 🏛 Architectural design
- 🏛 Architectural materials and methods of construction
- 🏛 Timber and steel design
- 🏛 Reinforced concrete design

- 🏛 Lighting and electrical distribution
- 🏛 Plumbing and fire protection
- 🏛 Heating, ventilating, and air conditioning
- 🏛 Estimating
- 🏛 Construction management

Several other baccalaureate technology programs supply specialized graduates to architectural offices: graduates from programs such as civil engineering technology, electrical engineering technology, and mechanical engineering technology.

Most state professional engineering registration boards award four years of credit for baccalaureate engineering technology graduation from an Accreditation Board for Engineering and Technology (ABET)-accredited institution (there are over 100 in the United States), and they award two years of credit for associate engineering technology graduation from an accredited institution (over 100 in the United States). However, many graduate engineering technologists prefer to apply for certification from the National Institute for Certification in Engineering Technologies (NICET), 1420 King Street, Alexandria, Virginia 22314-2794. Phone: 703-684-2835.

NICET offers two levels of technologist certification: *Associate Engineering Technologist* and *Certified Engineering Technologist*. The requirement for the first level is graduation from a baccalaureate engineering technology major at an ABET-accredited technical institute or university. Requirements for the second level are five years of technical experience acquired after the baccalaureate technology degree was received, plus two endorsements by qualified persons having direct knowledge of the candidate's technical experience.

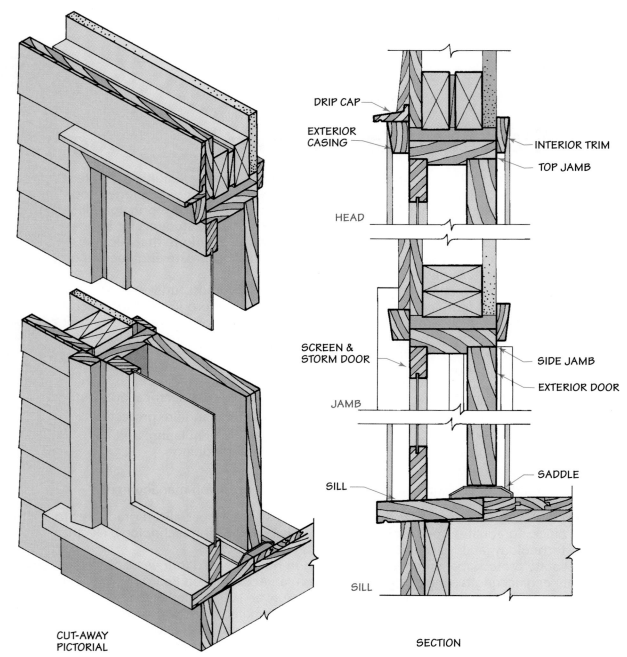

Figure 29–10   *An exterior door in a wood-frame wall.*

SPECIFYING DOORS

The type of door to be specified for each location will be determined by functional and aesthetic considerations. Manufacturers' catalogs should be consulted for available styles and sizes. In general, exterior doors are $1\frac{3}{4}$" thick and interior doors are $1\frac{3}{8}$" thick. Widths from 2'-0 to 3'-0 in even inches are available, although some manufacturers offer doors as narrow as 1'-6" and as wide as 4'-0. Residential doors are obtainable in 6'-6" to 7'-0 heights in even inches. The most popular height is 6'-8". See Appendix C.

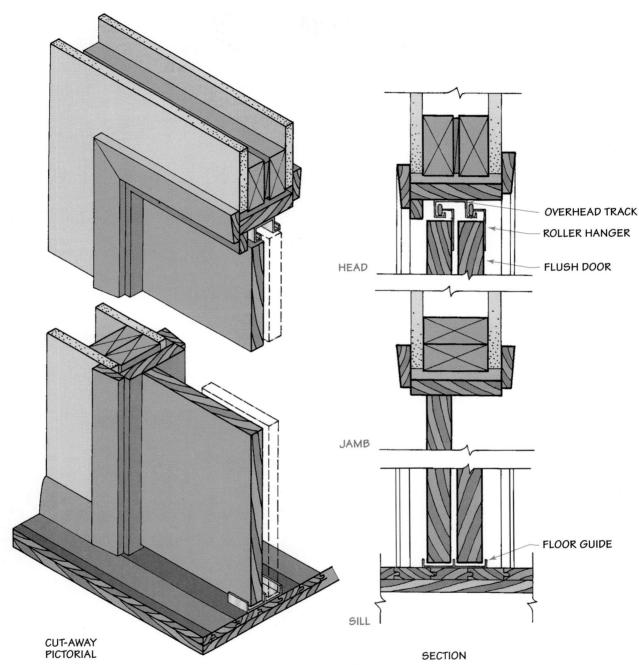

CUT-AWAY PICTORIAL

- OVERHEAD TRACK
- ROLLER HANGER
- FLUSH DOOR

HEAD

JAMB

FLOOR GUIDE

SILL

SECTION

Figure 29–11   A double sliding door in an interior wall.

Interior doors:	$2'\text{-}6'' \times 6'\text{-}8'' \times 1\frac{3}{8}''$
Front entrance door:	$3'\text{-}0 \ \times 6'\text{-}8'' \times 1\frac{3}{4}''$
Other entrance doors:	$2'\text{-}8'' \times 6'\text{-}8'' \times 1\frac{3}{4}''$

The rough opening must be considerably larger than the door sizes. The frame for a $3'\text{-}0 \times 6'\text{-}8''$ door, for instance, would require a $3'\text{-}2\frac{3}{4}'' \times 6'\text{-}11''$ (top of subfloor to bottom of header) rough opening.

Design for Accessibility

If you are specifying doors for a residence that will be inhabited by a person with impaired mobility, all open doorways must offer a clear passageway of 2'-8" minimum to accommodate wheelchairs. Always remember that the Americans with Disabilities Act (ADA) requires that all public and commercial buildings must be accessible. This includes buildings such as motels, restaurants, shopping centers, professional offices, banks, schools, theaters, and hospitals. Refer to Chapter 16, Primary Considerations, for more detailed ADA information.

ENTRANCES

The **entranceway** provides the first close introduction to a building and therefore should be attractive and representative as well as functional (Figure 29–12). A completely designed entrance may be obtained from stock. Details and sizes for a Morgan entrance are shown in Appendix C. **Sidelights** are obtainable as separate units in a variety of sizes, colors, and patterns. A selection of contemporary entranceways is shown in Figures 29–13 through 29–16.

HARDWARE

The hardware for doors is specified by indicating the desired manufacturer and catalog number in a hardware schedule. Some of the many hardware items are:

    Door butts (hinges)
    Lock sets
    Doorstops
    Door checks (for public buildings)
    Cabinet hinges, handles, and catches

Figure 29–13 *A semicircular transom above double entrance doors. (Courtesy Pella Corporation.)*

Figure 29–14 *Double entrance doors flanked by monumental casements provide a stylish entrance. (Courtesy Pella Corporation.)*

Figure 29–12 *Traditional entranceways. (Courtesy Morgan Manufacturing.)*

Figure 29–15   *A compact, yet distinctive entrance. (Courtesy Pella Corporation.)*

Figure 29–16   *A distinguished classical entranceway. (Courtesy Pella Corporation.)*

REAL WORLD APPLICATIONS

Finding the Perfect Front Door

As part of a job, you have been asked by your clients to compare and contrast three types of doors in relation to their cost, dimensions, and special features. The door will be used on the front of their house, contain windows on each side not to exceed twenty inches each, and will be centered on a large front porch.

Choose any three types of doors and write a report for your clients that delineates the similarities and differences between these types of doors. Your clients should be able to use this report in their own cost-benefit analysis to decide what type of door will best suit their needs and yet remain within their budget. Remember, you are not trying to sell your clients any specific type of door; you are simply providing them with the facts they will need to make an informed decision.

Use a word processor to prepare the report and ensure it is free from spelling and grammatical errors.

GARAGE DOORS

Most residential garage doors are of the overhead type. An overhead door is composed of several hinged sections that roll up to the ceiling on tracks. Adjustable springs are used to counterbalance the weight of the door. Between 3" and $13\frac{1}{2}$" of headroom (depending on the hardware type and door size) is required above the bottom of the header. Garage doors may be operated by means of remote-controlled motors.

Residential garage doors are usually stocked in 6'-6" and 7'-0 heights $\times 1\frac{3}{8}$" thick. Common widths for single doors are 8'-0 and 9'-0; double doors are 16'-0 wide. See Figure 29–17 and Appendix C for more detail.

In addition to wood, garage doors may be obtained in aluminum and fiberglass in the same range of sizes. Fiberglass doors are also available in 18'-0 widths.

CABINETS

A wide range of cabinets for the kitchen, laundry, and bathroom are available from stock. They include storage, china, corner, all-purpose, and ironing board cabinets.

DETAILING

In addition to supplying size tables for the convenience of architectural drafters, manufacturers also provide *tracing details* on paper and computer diskette for all types of doors and cabinets. The paper details may be slipped under tracing vellum and all applicable details copied. Or they may be digitized. The diskettes are inserted in the disk drive of a computer workstation, and desired door details can be transferred to the designer's CAD plans. In both instances, some modification is often necessary to fit a door into a particular type of wall. Also, manufacturers provide door descriptions in the form of (X, Y, Z) coordinates of the corners and edges of the doors. This greatly simplifies the task of adding these details to pictorial views.

RESIDENTIAL GARAGE DOOR
AVAILABLE FROM 8'-0" X 6'-6" TO 18' X 7'

COMMERCIAL GARAGE DOOR
AVAILABLE TO 20' X 16'

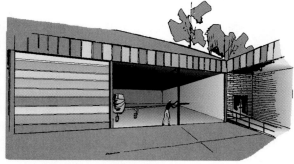

INDUSTRIAL GARAGE DOOR
AVAILABLE TO 24' X 20'

Figure 29–17   *Garage door sizes.* (*Courtesy Frantz Manufacturing Company.*)

Figure 29–18 *Interior view of Figure 29–19.*
(Courtesy Pella Corporation.)

Figure 29–19 *A commercial building with out-swinging glass doors. (Courtesy Pella Corporation.)*

REVIEW QUESTIONS

1. What is the difference between a hinged, sliding, folding and accordion door?

2. What are kalamein doors used for?

3. What is the difference between the following door terms:
   a. Solid-core and hollow-core door
   b. Flush, paneled, and louvered door
   c. Top rail, bottom rail, side rail, and lock rail
   d. French door and Dutch door
   e. Door butt and door buck
   f. Doorstop and door check
   g. Door rail and stile

4. What is meant by a door's rough opening size?

5. Give the common sizes for the following doors:
   a. Interior door
   b. Front entrance door
   c. Rear entrance door
   d. Single garage door

6. ADA is the abbreviation for _____.

7. What is the minimum door width for a wheelchair?

ENRICHMENT QUESTIONS

1. Compare the cost of a hollow-core bi-fold, louvered bi-fold, sliding, and accordion door when used for the same size closet.

2. Describe how an accordion door functions.

3. What is the jamb opening of M-2FD-518 if it is $1'-2\frac{11}{16}"$ wide? Use Appendix C.

4. What is the door width of M-4FD-1088 if the jamb opening is 5'-0? Use Appendix C.

5. What are the door dimensions (thickness, height, and width) of garage door #6326?

6. What do you think will be the major difference between garage doors #6319 and #5919?

PRACTICAL EXERCISES

1. Design a simple room that includes the two doors below. The specifications can be found in Appendix C.

   a. Door #1: 4416 (height = 6'-8")

   b. Door #2: M-4FD-512 (height = 6'-8")

2. Design a type of door entirely different from those commonly used. Draw or computer-produce as appropriate.

3. Complete the design of the doors for your original building design.

CHAPTER 30

Schedules

OBJECTIVES

By completing this chapter and studying Appendix E you will be able to:

🏛 Prepare architectural schedules on working drawings and in written specifications.

KEY TERMS

Schedules

A building is composed of a tremendous number of parts. In fact, if all of these parts were indicated on the plans, the plans would become so crowded that they would not be readable. Therefore the designer includes much of this information in **schedules** on the working drawings or in the written *specifications* (see Chapter 50 for a discussion of specifications writing).

DOOR AND WINDOW SCHEDULES

Figures 30–1 and 30–2 show minimum layouts for door and window schedules. Although this information may be included on the drawings in the form of notes, it is usually considered better practice to use schedules and keep the actual plans and elevations uncluttered. Of course, a reference mark or symbol must be placed upon each door in the plan and each window in the elevations. These marks are repeated in the door and window schedules, the schedules then giving all the necessary sizes and information. A numeral is used for doors and most other scheduled items, but a letter is used for windows. The usual place for the door schedule is near the

DOOR SCHEDULE

MK	NO.	SIZE	ROUGH OPENING	DESCRIPTION	REMARKS
1	1	3'-0 X 6'-8" X 1 3/4"	3'-2 3/4" X 6'-11"	14 PANEL WP, 4 LTS	ALUMINUM HDW
2	1	2'-8" X 6'-8" X 1 3/4"	2'-10 3/4" X 6'-11"	FLUSH WP, 1 LT	
3	2	2'-8" X 6'-8" X 1 3/4"	2'-10 3/4" X 6'-11"	2 PANEL WP, 3 LTS	
4	10	2'-6" X 6'-8" X 1 3/8"	2'-8 3/4" X 6'-11"	FLUSH BIRCH	
5					
6					

Figure 30–1   Sample Door Schedule.

WINDOW SCHEDULE

MK	NO.	SIZE	ROUGH OPENING	DESCRIPTION	REMARKS
A	1	G65	6'-0 1/2" X 5'-0 1/2"	ANDERSEN GLIDING	DOUBLE GLAZED
B	1	C24-2	8'-0 5/8" X 4'-0 1/2"	" CASEMENT	" "
C	3	C23	4'-0 1/2" X 3'-0 1/2"	" "	" "
D					
E					

Figure 30–2   Sample Window Schedule.

plan, and the window schedule should be placed near the elevations. Use a different mark for different sizes or types of doors and windows, but use the same mark for similar doors and windows. Such marks should be enclosed in circles, which are drawn about $\frac{1}{4}$" in diameter.

ADDITIONAL SCHEDULES

In addition to doors and windows, other materials may be specified by the use of schedules. Figures 30–3 through 30–7 show the outlines of some other commonly used schedules. The plans for a large, well-detailed building might contain many other types of schedules in addition to those illustrated.

Some of the customs and rules of writing schedules follow.

COLUMN AND BEAM SCHEDULE

MK	NO.	DESCRIPTION	LENGTH	REMARKS
1	4	3 1/2" STEEL PIPE COLUMN W/PLATES	7'-0	
2	1	S 7 X 15.3 FLOOR GIRDER	42'-0	
3				
4				

Figure 30–3   Sample Column and Beam Schedule.

LINTEL SCHEDULE

MK	NO.	DESCRIPTION	LENGTH	REMARKS
1	6	L 5 X 3 1/2" X 3/8"	4'-8"	
2				
3				

Figure 30–4   Sample Lintel Schedule.

FOOTING SCHEDULE

MK	A	B	C	ELEV	REINF	REMARKS
1	8"	1'-4"	8"	93.4'	2 #4	
2	10"	1'-6"	8"	93.4'	2 #4	
3	8"	1'-4"	8"	97.4'	NONE	
4						

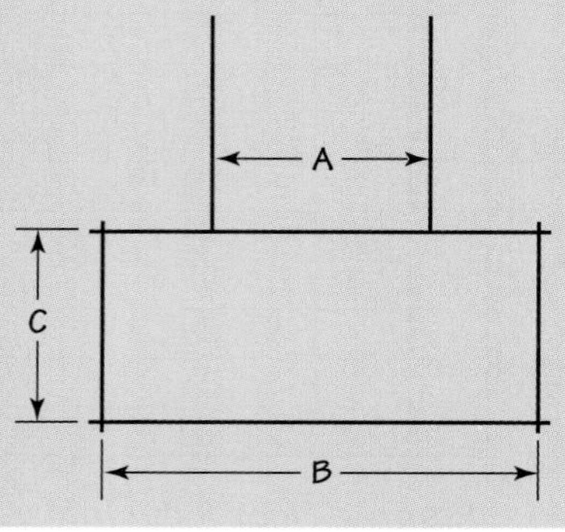

Figure 30–5   Sample Footing Schedule.

REAL WORLD  APPLICATIONS

Maintaining a Log

How did you design that roof for the addition to the Smith home? What pitch did you use? How many squares of shingles were needed? What type of shingles were ordered? How did it look when it was done? Why are you asking yourself all these questions? It is because seven years ago you designed a spectacular roof for the Smith family and you were just handed a very similar job. Have you ever solved a particular problem once before, but been unable to recall how you did it? Unfortunately, this happens all too often. However, most of the time it could be avoided simply by maintaining a professional log.

A log is a written chronological ledger of your activities and projects. For instance, the pages that were used for the Smith project would include a title page, daily summary pages, and a final report. A title page would contain the starting date, your name, the title of the project, and clientele information. The following pages would include a daily summary of your work on this project. An entry for June 5, 1990 might read: "9:30 a.m.: Met with the Smiths regarding roof design." On this page notes from this meeting would be recorded. These summary pages may also include sample sketches and research notes. A final report and photographs of the completed work will conclude the Smith project entry.

A log is also useful in sharing your work with your teammates and supervisors. As you accept more responsibility and more complex assignments, you need to keep a careful record of decisions you make so that you can have them reviewed and approved by the appropriate architect or engineer. If you are in the habit of maintaining a log, these records will be easily available in a professional format.

As you proceed through this course, you may wish to maintain a log of the progress you make on your original house design. Be sure to include preliminary sketches and ideas as well as discussions with your classmates and instructor.

1. Rather than take the time to letter the same words many times over, use the ditto mark (as in Figure 30–2), or the note "DO" (short for *ditto*, as in Figures 30–6 and 30–7), or a vertical arrow (as in Figure 30–1).

2. Abbreviations are often used to reduce the size of the schedule. Standard abbreviations should be used and listed in a table of abbreviations on the drawings. The abbreviations used on Figures 30–1 through 30–7 are:

CL	Closet
ELEV	Elevation
FIN	Finish
S	American Standard I beam
LAV	Lavatory
LT	Light (window glass)
MK	Mark
NO.	Number
REINF	Reinforce
T & G	Tongue-and-groove
WP	White pine
W/	With
L	Angle
#	Pounds or number

For a more extensive list of commonly used abbreviations see Appendix E.

FINISH SCHEDULE

MK	ROOM	FLOOR	FIN	WALL	FIN	CEILING	FIN	TRIM	FIN
1	LIVING ROOM	T&G OAK	VARN.	SHEET ROCK	PAINT	ACOUST. TILE	—	PINE	PAINT
2	LAV	ASPHALT TILE	—	CERAMIC TILE	—	SAND PLASTER	—	DO	DO
2A	LAV CL	DO	—	SAND PLASTER	—	DO	—	DO	DO
3									
4									

Figure 30–6 *Sample Finish Schedule.*

ELECTRICAL SCHEDULE

LOCATION	SYMBOL	NO.	WATT	DESIGNATION	EXAMPLE
GARAGE	⊖ / ○A	2 / 2	100 / 100	DUPLEX OUTLET / CEILING MOUNT	PASS & SEYMOUR #41
LIVING ROOM	⊖ / ○B / ○C	6 / 1 / 12	100 / 100 / 60	DUPLEX OUTLET / WALL MOUNT / WALL VALANCE	GENERAL #1606 / PASS & SEYMOUR #41
KITCHEN	⊖ / ○D / ○E	4 / 1 / 1	100 / 100 / 100	DUPLEX OUTLET / FLUSH CEILING / DO	HOLOPHANE #RL-732 / DO #RL-796
BEDROOM					

Figure 30–7 *Sample Electrical Schedule.*

3. When spaces on the schedule do not apply to a material, they may be left blank (as in the Figure 30–3 remarks column) or filled with a strike line (as in the Figure 30–6 ceiling finish column).

4. The desired manufacturer of a product may be specified (as in Figure 30–2, Andersen windows) or merely given as an example of an acceptable product (as in Figure 30–7, Pass & Seymour electric fixtures).

COMPUTER-PRODUCED SCHEDULES

Computer-generated and plotter-produced architectural schedules for the Z residence are shown in Figures 30–8 and 30–9. Letters and numerals are added to a schedule by typing on a computer keyboard. The drafter must select the lettering style (font), starting point, lettering direction, and lettering height, width, and slant. When symbols are also included, they can be selected

DOOR SCHEDULE

MK	NO.	SIZE	DESCRIPTION
1	1	3'-0" X 6'-8" X 1 3/4"	14 PANEL WP, 4 LTS
2	1	2'-8" X 6'-8" X 1 3/4"	FLUSH WP, 1 LT
3	2	2'-8" X 6'-8" X 1 3/4"	2 PANEL WP, 3 LTS
4	1	2'-8" X 6'-8" X 1 3/4"	FLUSH BIRCH
5	1	2'-6" X 6'-8" X 1 3/8"	" " DOUBLE SWINGING
6	10	2'-6" X 6'-8" X 1 3/8"	" "
7	1	2'-4" X 6'-8" X 1 3/8"	" "
8	2	2'-2" X 6'-8" X 1 3/8"	" "
9	3	1'-6" X 6'-8" X 1 3/8"	" "
10	2	5'-4" X 6'-8" X 1 3/8"	" " DOUBLE SWINGING
11	1	6'-0" X 6'-8" X 1 3/8"	LOUVERED DOUBLE FOLDING, WP
12	2	3'-0" X 6'-8" X 1 3/8"	" SINGLE " "
13	1	9'-0" X 6'-8" X 1 3/8"	18 PANEL WP, 6 LTS, OVERHD GAR

Figure 30–8 *Computer-produced door schedule for the Z residence.* (*Courtesy Mark D. Weidhaas, Snyder Associates, and Autodesk, Inc.*)

WINDOW SCHEDULE

MK	NO.	SIZE	DESCRIPTION	REMARKS
A	1	CX26	ANDERSEN CASEMENT	DOUBLE GLAZED
B	1	C15/35/15	" "	" "
C	3	C24-2	" "	" "
D	7	C24	" "	" "
E	1	C335	" "	" "
F	1	C235	" "	" "
G	3	C135	" "	" "
H	2	C33	" "	" "
J	1	C23	" "	" "
K	1	CR135-3	" "	" "
L	6	2820	" BASEMENT	" "

Figure 30–9   Computer-produced window schedule for the Z residence. (Courtesy Mark D. Weidhaas, Snyder Associates, and Autodesk, Inc.)

from standard menus of architectural, structural, mechanical, and electrical symbols. Or symbols can be custom-designed to meet special needs.

Refer to Chapter 7 for operating details on producing the elements of architectural schedules.

REVIEW QUESTIONS

1. What is a schedule?

2. List several building components that may be specified by means of schedules.

3. When using schedules, what mark is used to indicate windows and what mark is used for doors?

4. What does "DO" or a vertical arrow mean when written in a schedule?

5. Identify the words that correspond to these abbreviations. For some you may need to refer to Appendix E.

   a. ₵

   b. T & G

   c. W/

   d. #

   e. OC

   f. LGTH

   g. FD

   h. S4S

   i. WP

ENRICHMENT QUESTIONS

1. What advantages do schedules have in preference to notes or dimensions?

2. Why is it important to use standard symbols when preparing schedules?

3. What purpose does leaving blank lines in a schedule serve?

PRACTICAL EXERCISES

1. Draw or computer-produce the window and door schedules for Figure 14–8 (page 164). Select appropriate windows and doors from Appendix B and C or a manufacturer's catalog.

2. Draw or computer-produce the schedules for your original house design.

CHAPTER 31

Stairways

OBJECTIVES

By completing this chapter you will be able to:

- 🏛 Define various terms related to stairway design.
- 🏛 Explain the construction of a stairway.
- 🏛 Design a stairway and complete its details.
- 🏛 Distinguish between types of stairways.

KEY TERMS

Winder	Stringer
Step	Handrails
Tread	Guardrails
Headroom	Treillage
Nosing	

An important consideration in a house having more than one level is the stairway design. Often the stairway is specifically planned for its architectural effect and it can be a major design element in both traditional and contemporary house styles. In addition, the stairway must be carefully planned to perform its function of vertical circulation conveniently. Since the stairway is usually associated with the halls, it may be the key to circulation throughout the entire house.

TYPES

See Figure 31–1. Straight-run stairs take up the least amount of floor area and are the simplest to construct. However, some designs require a stair that turns or is shorter in length. In these cases, the designer specifies a U-type or L-type stair with a platform at the turn. The platform has the safety feature of breaking up a long run of stairs and providing a place to pause and rest. When

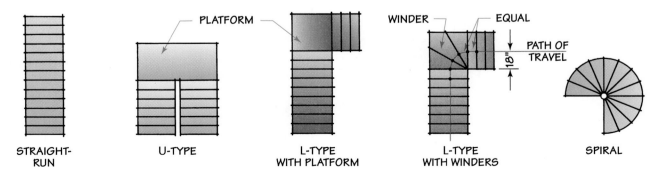

Figure 31–1   Stair types.

space is restricted, diagonal steps called **winders** are used in place of the platform. Winders are designed so that the same tread depth is maintained at the normal path of travel: 18" from the inside corner. Since the tread depth is reduced inside the normal path of travel, it is obvious that winders are dangerous and should be used only as a last resort. Also avoid single steps to sunken rooms. There is less likelihood of a person tripping on two or more steps. Spiral stairs may be obtained in packaged units and will satisfy unique design requirements.

TERMS

The terms generally used in stairway design are illustrated in Figure 31–2.

> **Step** or *riser:* The vertical distance from one tread top to another.
>
> **Tread** or *run:* The horizontal distance from the face of one riser to the next. Notice that there will always be one fewer tread than risers.
>
> *Total rise:* The vertical distance from one finished floor to the next. This is a basic measurement in stair planning.
>
> *Total run:* The horizontal distance of the entire stairway.
>
> **Headroom**: The vertical distance from the outside edge of the step to the ceiling above.
>
> **Nosing**: The projection of the tread beyond the riser. It should be about $1\frac{1}{8}$", as shown in Figures 31–3, 31–4, and 31–5 and is not

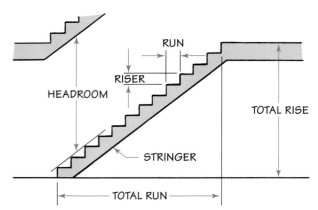

Figure 31–2 *Stairway terminology.*

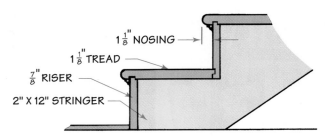

Figure 31–3 *Closed-riser stair.*

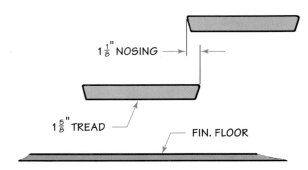

Figure 31–4 *Open-riser stair.*

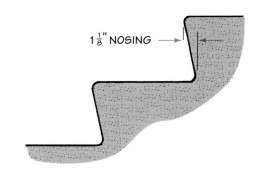

Figure 31–5 *Concrete stair.*

considered as part of the stair tread when laying out the stairway.

DESIGN

All stairways do not have the same slope, or pitch. They may vary from a 5-16 pitch (that is, a 5" rise with a 16" run) to an 8-9 pitch (an 8" rise with a 9" run). A pitch less than the 5-16 limit will require a ramp; a pitch greater than 8-9 will require a stepladder or rung ladder. Figure 31–6

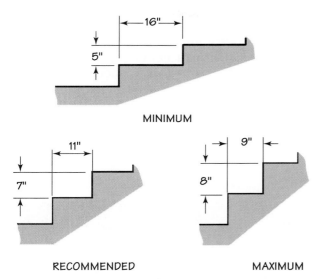

MINIMUM

RECOMMENDED   MAXIMUM

Figure 31–6   Stair pitch.

shows the minimum pitch of 5-16, which may be used for outside or monumental stairs; the recommended pitch for most stairways of 7-11; and the maximum possible pitch of 8-9. It is assumed that no one would use such steep stairs without holding the handrails. An 8-9 pitch is only for basement stairs, but lower pitches are safer.

In Figure 31–6, notice that as the run dimension decreases, the rise dimension increases so that the overall distance from step to step remains about the same. This fact has given us this general rule:

Rule 1   rise + run = 17 to 18

For example, 7 + 11 = 18. Also 8 + 9 = 17. Two slightly more sophisticated rules may be used for additional checks:

Rule 2   2 × rise + run = 24 to 25

The 7-11 pitch would be satisfactory according to this rule since 2 × 7 + 11 equals 25.

Rule 3   rise × run = 72 to 77

The 7-11 pitch used in the two preceding examples equals 7 × 11 = 77, which is acceptable. None of the preceding rules should be used to check for monumental stairs.

Allow 7'-0 minimum headroom on a stairway, but 7'-6" is preferred. A 6'-6" headroom may be used for basement stairs. The most comfortable handrail height is 34" on a stair and 38" on a landing, as shown in Figure 31–7. The minimum comfortable stair width is 3'-0, but 3'-6" or 4'-0 is better for moving furniture.

Warning

In stair design, as in *all* architectural design, it is vitally important that the designer be

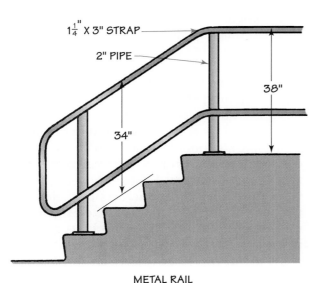

METAL RAIL

Figure 31–7   Contemporary handrails.

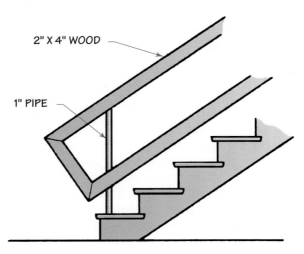

WOOD RAIL

thoroughly familiar with an updated version of relevant building codes. For example, in 1996 the BOCA code, which had required $8\frac{1}{4}$" maximum riser and 9" minimum tread for one- and two-family dwellings, changed to $7\frac{3}{4}$" maximum riser and 10" minimum tread in order to increase stair safety.

Design for Accessibility

If you are designing a residence for a person with physical impairments (such as impaired mobility, sight, or hearing), special design adjustments are needed. For example, for use of a wheelchair, a ranch home would be preferred

ARCHINET http://www.

Stairway Design

 Stairways are an important part of architectural design and can be the centerpiece of a room or home. Unique stairway design can greatly enhance the overall effect of an architectural project. Stairway design changes should be considered in many remodeling projects. For example, removing a wall from an enclosed stairway opens the entire floor plan of a home. What other changes can be made to an existing stairway to change the design effect?

When considering stairway design, it is important for you to know stairway terminology. First, you should be able to distinguish between straight-run, U-type, L-type, winder, open-riser, closed-riser, circular, and spiral stairs. Using the text and the glossary of stairway terms you will find on Century Stair Company's homepage (**http://www. business1.com/century**) write a paragraph explaining the differences between these types of stairways and give an example of a house style (such as those listed in Chapter 1) for which each stairway design might be most appropriate. Conversely, provide an example of a house style for which these stairway designs would be *in*appropriate.

This glossary of terms provides additional technical stairway vocabulary not found in a common dictionary. Such terms include bullnose tread, glue block, gooseneck, and horse cut. These terms may prove beneficial to you as you consider stairway design. Study this glossary and use it as a resource.

To be successful in stairway design, it is important to become aware of many possible stairway designs as well as creative railing options. The Internet is one place to investigate and view some unique designs and products. Two companies that provide pictures of some of their stairway designs are Century Stair Company and the Taney Corporation (**http:// www.taney.com**). The Taney Corporation even provides CAD drawings of their stairways including straight, circular, and spiral designs. You can download these onto your personal computer and use them in drawing proposals. Designer's Choice, which has been creating copper, bronze, and stainless steel architectural features for twenty years, provides descriptions and pictures of their stair railing designs (**http://www.designers-choice.com/stairs.htm**). This company may have what you need for the clients who want a stairway that makes an impact on everyone entering their home.

over a split-level home. Other alternatives are ramps, wheelchair lifts, and elevators. See Chapter 47, Design for Accessibility, for further detail. Remember that all public and commercial buildings should be accessible.

LAYOUT

Tables like Table 31–1 have been devised to simplify some of the stair calculations. A stairway is best laid out using the following method:

Step 1. Draw two horizontal lines representing the finished floors to a scale of $\frac{1}{2}$" = 1'-0. In Figure 31–8, the distance between finished floors (total rise) is 9'-0.

Step 2. Using Table 31–1, we shall have 15 risers of 7.20" each and 14 treads (treads = risers − 1) of $9\frac{3}{4}$" each, with a total run of 11'-$4\frac{1}{2}$". Lay out the total run to scale.

Step 3. Since there are 15 risers, divide the total rise into 15 equal parts using your scale at an angle so that the zero mark is on the other finished floor.

Step 4. Since there are 14 treads, divide the total run into 14 equal parts in the same manner using the zero mark and the 7" mark.

Step 5. Darken in the outline of the stairway, adding details like those shown in Figure 31–17.

A similar procedure is used for U-type, L-type, and circular stairs.

CONSTRUCTION

Figures 31–10 through 31–13 show different types of stairs. A 2" × 12" member would be

TABLE 31–1 *Stair Dimensions*

Total Rise	Number of Risers	Riser	Tread	Total Run
8'-0	13	7.38"	$9\frac{1}{2}$"	9'-6"
8'-6"	14	7.29"	$9\frac{3}{4}$"	10'-$6\frac{3}{4}$"
9'-0	15	7.20"	$9\frac{3}{4}$"	11'-$4\frac{1}{2}$"
9'-6"	16	7.13"	10"	12'-6"
10'-0	17	7.06"	10"	13'-4"

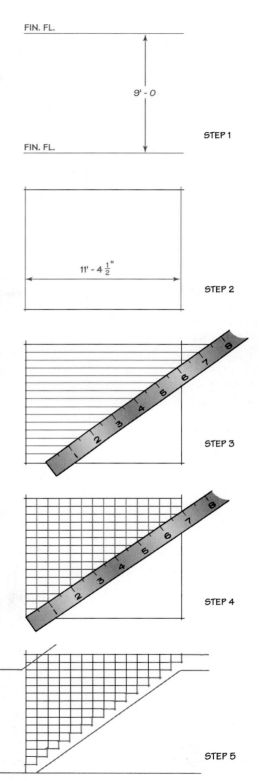

Figure 31–8 *Stair layout (new steps shown in slate blue).*

Figure 31–9 *Under-stair storage. (Courtesy Wood-Mode Cabinetry.)*

Figure 31–11 *A cantilevered stairway. (Courtesy California Redwood Association.)*

Figure 31–10 *A contemporary open stairway with treads suspended from ceiling by tension rods. (Courtesy California Redwood Association.)*

Figure 31–12 *Handsomely massed exterior steps. (Courtesy California Redwood Association.)*

Figure 31–13 *Exquisite curved stairway in traditional styling.* (Courtesy Scholz Design, Toledo, Ohio.)

Figure 31–15 *Rail caps are available in a variety of colors.* (Courtesy Julius Blum & Co., Inc.)

used for the open **stringer** in Figure 31–14. Incidentally, the triangular pieces cut from this 2" × 12" are often nailed to a 2" × 4"—serving as a middle stringer for extra support. Closed stringers are slightly different from open stringers in that no triangular pieces are cut from the stringer (Figure 31–14). Rather, $\frac{1}{2}$" grooves are routed to receive the treads and risers, which are wedged and glued in place. This completely conceals their ends. The tongue-and-groove construction between riser and tread is often omitted for economy. Preformed metal stair guides can be used to simplify stair construction. Prefabricated stairways are becoming increasingly popular.

Stock components are often used for stairs and railings in commercial buildings (Figure 31–15). In addition to the metal systems shown in Figures 31–19 and 31–20, a $\frac{1}{2}$" tempered-glass railing is available (Figures 31–18 and 31–21).

STAIR DETAILS

A complete set of stair details includes a section or elevation together with a plan view of each stairway. Details of tread construction and handrail construction may be included. Some of these drawings may be incorporated with other plans. The stair plan, for example, may be satisfactorily shown on the floor plan. Notice in Figure 31–17 that arrows with the notation "UP" and "DN" are used to show stair direction. The number of risers is also included. Always use capital letters for "UP" and "DN" notations since, when viewed from the opposite direction, a lowercase "up" looks like "dn" and a lower case "dn" looks like "up." When there are both "up" and "down" stairs over one another, they are separated by a break line.

Figures 31–16 and 31–17 show details of the stairs for the A residence and the Z residence.

RAILINGS

Most building codes distinguish between *handrails* and *guardrails*. **Handrails** are used for *support* (usually with stairs) while **guardrails** prevent *accidental falls* (usually from balconies). Handrail heights are usually between 34" and 38" (Figure 31–7).

Guardrails for balconies and retaining walls require special attention to safety considerations.

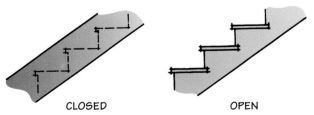

CLOSED   OPEN

Figure 31–14 *Stringer types.*

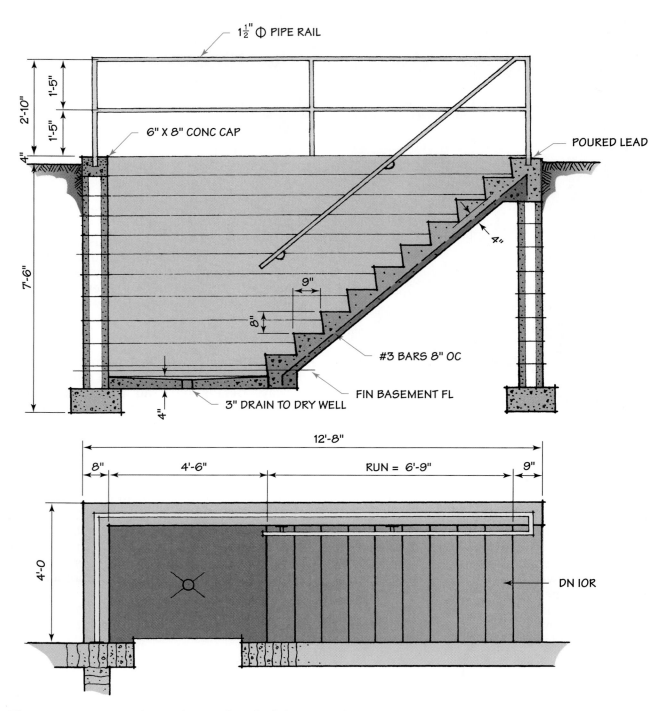

Figure 31–16   *Exterior stairway detail of the A residence.*

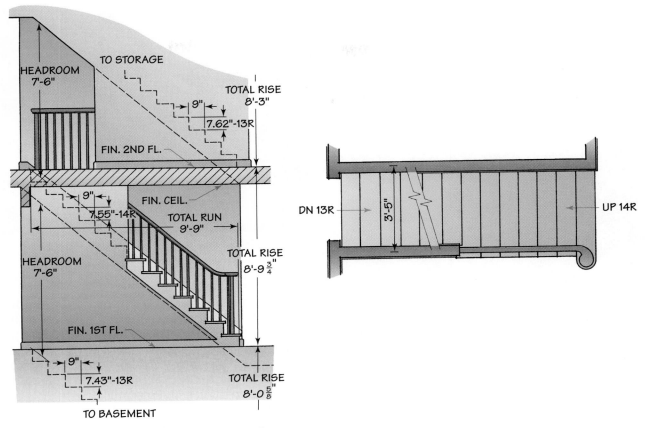

Figure 31–17 *Stair detail of the Z residence.*

Figure 31–18 *Glass and metal railing. (Courtesy Julius Blum & Co., Inc.)*

Figure 31–19 *Handrail and wooden guardrail. (Courtesy Julius Blum & Co., Inc.)*

Figure 31–22 A handsome rail design. (Courtesy California Redwood Association.)

Figure 31–20 Handrail and metal guardrail. (Courtesy Julius Blum & Co., Inc.)

Figure 31–21 Another glass and metal railing design. (Courtesy Julius Blum & Co., Inc.)

The national standard requires a minimum height of 42" because that height is above the center of gravity of even a tall person. Railing bars should be less than 4" apart to prevent children from squeezing through. Most standards presently require that railings be strong enough to withstand at least 200 pounds of impact pressure, but much greater strength than that is advisable. Often codes require that a guardrail have a handrail included as well.

Handrails for stairs and balconies must be small enough to grasp easily. BOCA requires that handrails be between 4" and $6\frac{1}{4}$" in circumference.

TREILLAGE

Treillage is used for building facing, partitions, room dividers, privacy fences, or concealment of unsightly elements. (Pronounced *trail-lige* in French, with the accent on the first syllable. The last syllable is pronounced like the last syllable in the word *pillage*.) Stock components are available in many patterns (Figure 31–23).

COMPUTER-AIDED DETAILS

Stair, railing, and treillage sections and details can be entered into your computer graphics system by puck or stylus on a digitizer tablet and displayed on your monitor. Start with preliminary layouts and proceed to finished draw-

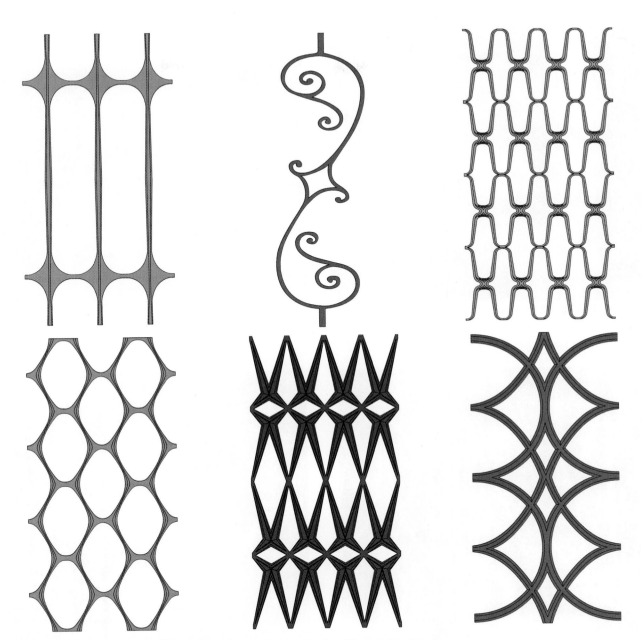

Figure 31–23 *Some treillage patterns.* *(Courtesy Julius Blum & Co., Inc.)*

ings and working drawings. When satisfied with the result, use your printer or plotter to produce hard copies, but also store your work on diskette. Refer to Chapter 7 for operating information on producing and correcting stair details (such as dimensioning, lettering, sectioning, and symbols).

Numerous specialized software programs are available to help design stairways. Soft-desk's *Auto-Architect*® software, for example, produces straight-run, U-type, L-type, winding, and spiral stairs in two or three dimensions, including handrails (Figure 31–26). Both English and metric dimensioning are available.

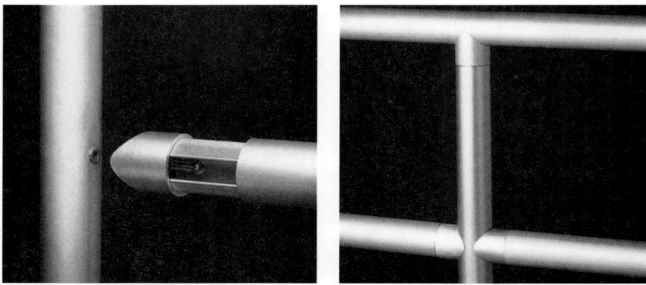

Figure 31–24 *Mechanical fasteners secure continuous metal post handrails.* (*Courtesy Julius Blum & Co., Inc.*)

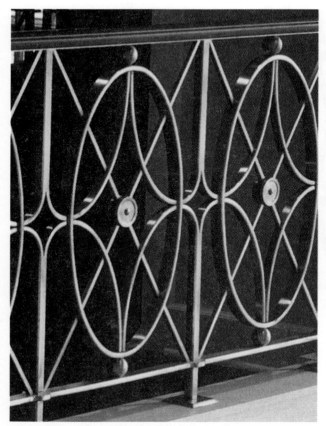

Figure 31–25 *Handrail and treillage.* (*Courtesy Julius Blum & Co., Inc.*)

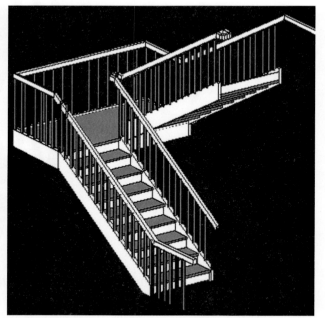

Figure 31–26 *Computer-produced pictorial of a U-shaped stairway.* (*Courtesy Softdesk, Inc.*)

REVIEW QUESTIONS

1. What is the difference between:

   a. Rise and total rise

   b. Run and total run

   c. 8-9 and 9-8 pitch

   d. Open stringer and closed stringer

   e. Open riser and closed riser

2. Show how to divide any line into three equal parts using the following. Do not use trial-and-error solutions.

   a. Your scale

   b. Your dividers

3. Distinguish between five types of stairways by listing their advantages and disadvantages.

4. Define the following terms:

   a. Riser

   b. Run

   c. Total rise

   d. Total run

   e. Headroom

   f. Nosing

5. How is a stringer used in stairway construction?

6. How is stair direction indicated in a floor plan?

7. Differentiate between a handrail and guardrail according to their intended function.

ENRICHMENT QUESTIONS

1. Using each of the three rules of rise and run proportions, figure the tread dimensions for:

   a. A 7" rise

   b. A $7\frac{1}{2}$" rise

   c. An 8" rise

2. How many risers and treads should be specified for a total rise of:

   a. 8'-9"

   b. 4'-9" (split-level house)

   c. 1'-9" (sunken room)

3. After a thorough library search, prepare an illustrated paper on the history of the development of the various methods of vertical circulation in buildings. Include your estimate of the status fifty years hence.

4. Draw or computer-produce a classroom illustration showing:
   a. Stairway terminology
   b. Stairway construction
   c. Recommend stairway dimensions for tread, rise, width, headroom, and handrail heights
5. Explain how the location of a stairway can impact traffic flow in a residence.

PRACTICAL EXERCISES

1. Draw or computer-produce the plan and sectional elevation of:
   a. Interior stairway of the A residence
   b. Exterior stairway of the A residence
   c. Main stairway of the Z residence
2. Design a means of vertical circulation that is entirely different from those commonly used.
3. Draw or computer-produce the stairway details for your original house design.

32 Fireplaces and Chimneys

OBJECTIVES

By completing this chapter you will be able to:

- 🏛 Describe the influences that styling has on a fireplace.
- 🏛 Define terminology related to a fireplace and a chimney.
- 🏛 Identify and explain the purpose of the parts of a typical fireplace and chimney.
- 🏛 Use appropriate charts to design an elevation and plan view of a fireplace and chimney detail.

KEY TERMS

Hearth	Smoke chamber
Mantel	Damper
Ash dump	Flue
Clean-out door	Fireplace liner
Firebrick	

Figure 32–1 A fireplace and expansive windows in a contemporary living area. (Courtesy California Redwood Association.)

The fireplace, although no longer a necessity as the major heat source, is considered by many to be a "must" luxury. A blazing fire or glowing embers on a cold winter day create a mood of cheerfulness and comfort that cannot be achieved by a concealed heating system. The fireplace is usually the major element of interior design in a living room or family room (Figures 32–1 through 32–5). In addition, it is occasionally specified for the family-type kitchen or master bedroom (Figure 32–6). Careful planning

and thoughtful design are always required to obtain proper styling, the best location in a room, and coordination with the heating plant and the other fireplaces in the house.

STYLING

Several types of fireplaces are illustrated in Figure 32–7. Notice that the designer may work with a great variety of sizes and styles of fireplace openings, hearths, mantels, and materials. Fireplace openings may be single-faced (the basic type), double-faced (with faces on adjacent

Figure 32–2   *A classical raised-hearth fireplace with built-in cabinetry.* (Courtesy Wood-Mode Cabinetry.)

Figure 32–3   *A mirrored fireplace as the centerpiece of this luxurious, yet comfortable, living room.* (Courtesy Scholz Design, Toledo, Ohio.)

Figure 32–4   *The fireplace as the centerpiece of a high-classic decor.* (Courtesy Pella Corporation.)

Figure 32–5   *A southwestern-styled family room with fireplace.* (Courtesy Scholz Design, Toledo, Ohio.)

Figure 32–6   *Master bedroom with free-standing fireplace.* (Courtesy Scholz Design, Toledo, Ohio.)

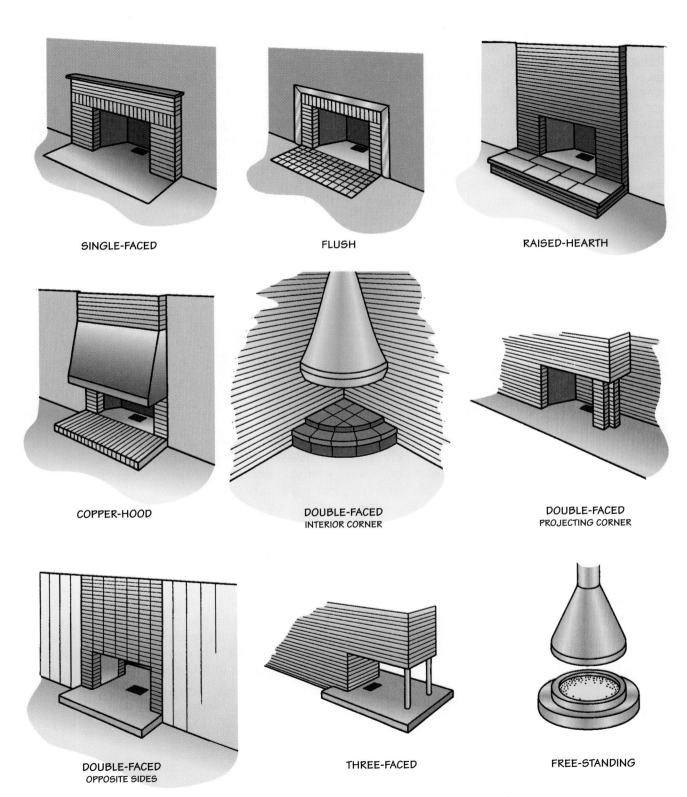

SINGLE-FACED

FLUSH

RAISED-HEARTH

COPPER-HOOD

DOUBLE-FACED
INTERIOR CORNER

DOUBLE-FACED
PROJECTING CORNER

DOUBLE-FACED
OPPOSITE SIDES

THREE-FACED

FREE-STANDING

Figure 32–7   Fireplace types.

or opposite sides), three-faced (serving as a peninsula partition between two areas), or even freestanding (in the center of an area). Multi-faced fireplaces are associated with contemporary design, but corner fireplaces have been in use for many years. The double-faced (opposite sides) and three-faced fireplaces are often used as room dividers, but they are not particularly useful for heating since the open design reduces the reflection of heat. Also, large flues must be used to obtain an adequate chimney draft. Cross-drafts may cause smoke to enter the room unless glass fire screens are added.

Hearths may be flush with the floor or raised to any desired height. The back hearth serves as the base for the fire; the front hearth protects a combustible floor from sparks. The front hearth and edges of the fireplace opening may be surfaced with an ornamental material such as a ceramic tile, which is highly heat resistant. **Mantels** may be of various designs (Figure 32–8) or omitted entirely. Brick, block, stone, tile, and metal are used for fireplace construction. When the fireplace is used as the primary element in the decorative scheme, log bins, shelves, or cabinets are often included.

FUEL

The usual fuel is wood: softwood for kindling and logs of hickory, birch, beech, ash, oak, or maple. Andirons or some form of grating may

REAL WORLD APPLICATIONS

Fireplace History

Did you ever wonder why Native Americans built teepees in a conical or pyramid shape? At least one reason is that this shape was a natural chimney. When a fire was built in the middle of the teepee, smoke went out the opening at the peak. Uses and designs of fireplaces have changed through time. At one point, fireplaces served as the main source of heating and cooking in homes. Today it is questioned whether fireplaces should even be constructed since they lack energy efficiency.

One means of broadening your perspective in the area of fireplace design is to study the use and design of fireplaces during a different time period. Choose a specific culture in history and research the use and design of fireplaces in this era. You may choose Colonial America, Native Americans, Victorian England, the Roman Empire, or any other historical era or culture. In fact, you may decide to choose a different culture of the modern era. Refer to Chapter 15 for an overview of more architectural eras. Whatever you decide, begin your research at your school library's card catalog. Enjoy this project as you discover unique uses and designs of fireplaces and enhance your understanding of how fireplaces and chimneys function.

Finally, in addition to your historical or cultural research of fireplaces, access the following Internet sites to view custom-designed modern fireplaces. These include wood-burning fireplaces and stoves, gas logs and appliances, and pellet stoves. These design ideas will help expand your creativity as you design your original house.

🏛 Hot Designs—**http://www.hotdesigns.com./index.html**

🏛 Temco—**http://www.hearth.com/temco/three.html**

🏛 Rhodes Masonry—**http://www.rhodesmasonry.com**

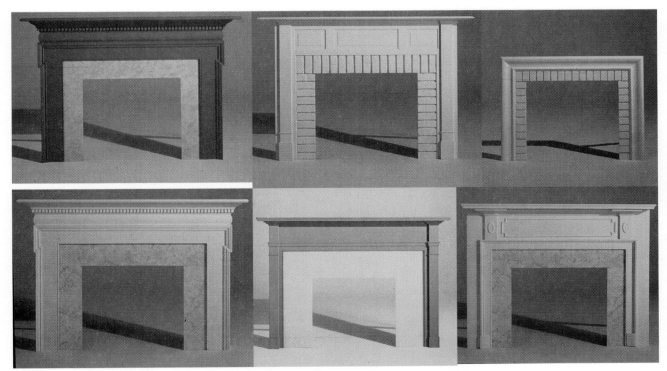

Figure 32–8 *Some traditional fireplace mantels. (Courtesy Morgan Manufacturing.)*

be an integral part of the fireplace design. A suspended grill might be considered for indoor charcoal grilling.

LOCATION

A fireplace may be located on a wall, in a corner, on a projecting corner, or freestanding. When the fireplace is to be located on a wall, the designer must choose between interior and exterior walls. Most chimneys in older houses were located in the center of the house; here they drew better and the heat could not escape directly outdoors. Newer houses tend to have exterior-wall chimneys to save floor space. The exterior chimney can be a distinct feature of the exterior design; however, more flashing and finish brick is required. Remember that a frame structure is weakened by having a masonry fireplace and chimney on an exterior wall; a masonry structure is stiffened.

Split-level houses pose a special problem because the chimney should not emerge at a location close to higher-elevation roofs. The chimney will not draw properly unless it is extended at least 2' higher than any portion of the roof located within 10' (see Figure 32–9).

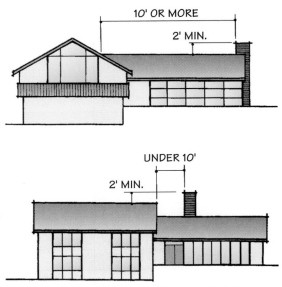

Figure 32–9 *Minimum chimney height.*

SIZE

Figures 32–10 and 32–11 give minimum and recommended dimensions for single- and multifaced fireplace designs; some of these dimensions will vary according to the size category as shown in Table 32–2. The size category will depend upon the room size and the emphasis to be placed upon the fireplace. For example, the *medium* fireplace might be specified for a living room of 300 sq ft.; the *medium-large* for a living room of 350 sq. ft. Notice that the dimensions in Figure 32–10 are nearly all multiples of 4". To reduce the amount of brick trimming and waste, it is important to establish a modular system such as that shown in Figure 32–12. Here a $2\frac{1}{6}" \times 3\frac{1}{2}" \times 7\frac{1}{2}"$ modular brick is laid up with $\frac{1}{2}"$ joints, resulting in a 4" module. Table 32–3 may be consulted for a quick reference of modular sizes of brick, tile,

and block. If you are using common brick, Norman brick, Roman brick, or some other tile or block size, it is a good idea to make up a similar table. The actual sizes of these bricks are given in Table 32–1.

TABLE 32–1 *Masonry Sizes*

Modular brick	$2\frac{1}{6}" \times 3\frac{1}{2}" \times 7\frac{1}{2}"$
Common brick	$2\frac{1}{4}" \times 3\frac{3}{4}" \times 8"$
Norman brick	$2\frac{1}{6}" \times 3\frac{1}{2}" \times 11\frac{1}{2}"$
Roman brick	$1\frac{1}{2}" \times 3\frac{1}{2}" \times 11\frac{1}{2}"$
Fire brick	$2\frac{1}{2}" \times 4\frac{1}{2}" \times 9"$
Tile	$4\frac{5}{6}" \times 7\frac{1}{2}" \times 11\frac{1}{2}"$
Block	$7\frac{5}{8}" \times 7\frac{5}{8}" \times 15\frac{5}{8}"$

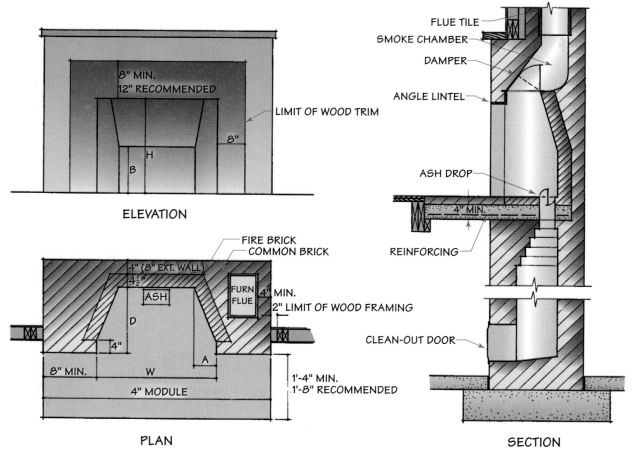

Figure 32–10 Fireplace dimensions.

TABLE 32–2   *Sizes for Fireplace Design*

Size Category	Width of Opening (W)	Height of Opening (H)	Depth of Opening (D)	A*	B*	Nominal Flue Size
Single-faced						
Very small	2'-0	2'-0	1'-5"	6'	1'-0	8" × 12"
Small	2'-8"	2'-3"	1'-8"	$6\frac{1}{2}$"	1'-2"	12" × 12"
Medium (most common)	3'-0	2'-5"	1'-8"	$6\frac{1}{2}$"	1'-2"	12" × 12"
Medium large	3'-4"	2'-5"	1'-8"	$6\frac{1}{2}$"	1'-2"	12" × 16"
Large	4'-0	2'-8"	2'-0	9"	1'-4"	16" × 16"
Very large	5'-0	3'-1"	2'-0	9"	1'-4"	16" × 20"
Double-faced, corner						
Small	2'-8"	2'-3"	1'-8"		1'-2"	12" × 16"
Medium	3'-0	2'-5"	1'-8"		1'-2"	16" × 16"
Medium large	3'-4"	2'-5"	1'-8"		1'-2"	16" × 16"
Large	4'-0	2'-5"	2'-0		1'-2"	16" × 16"
Double-faced, opposite sides						
Small	2'-8"	2'-5"	3'-0			16" × 16"
Medium	3'-0	2'-5"	3'-0			16" × 20"
Medium large	3'-4"	2'-5"	3'-0			16" × 20"
Large	4'-0	2'-8"	3'-0			20" × 20"
Three-faced						
Small	3'-4"	2'-3"	3'-0			20" × 20"
Medium	3'-8"	2'-3"	3'-0			20" × 20"
Medium large	4'-0	2'-3"	3'-0			20" × 20"
Large	4'-8"	2'-3"	3'-0			20" × 24"

*See Figure 32–10 or 32–11

CONSTRUCTION

A chimney is a complete structure in its own right—unsupported by any wooden member of the house framing. Recall how often you have seen a house that has burned to the ground, while the chimney remained standing. It is equally improper to use the chimney as a support for girders, joists, or rafters since a wooden member framing into a chimney may eventually settle and crack it. Actually, no framing lumber should come closer than 2" from the chimney due to the fire hazard. The 2" space should be filled with noncombustible

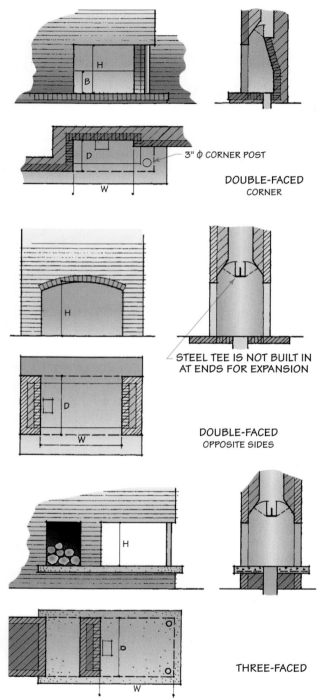

DOUBLE-FACED
CORNER

3" φ CORNER POST

STEEL TEE IS NOT BUILT IN
AT ENDS FOR EXPANSION

DOUBLE-FACED
OPPOSITE SIDES

THREE-FACED

Figure 32–11   Multifaced fireplace designs.

material to act as a fire stop. Subflooring, flooring, and roof sheathing may come within $\frac{3}{4}$" of the chimney.

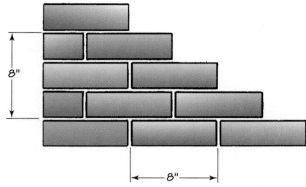

Figure 32–12   Common brick modular dimensions.

Figure 32–13 shows the type of overlapping flashing used at the roof. This allows movement between the chimney and roof due to settling without damage to the flashing.

The construction of a fireplace and chimney should be entrusted only to experienced workers since improper construction can cause a fire hazard or a smoking fireplace. Due to its great weight, the chimney should have a sizable footing. Each fireplace is fitted with an **ash dump** (5" × 8" is a common stock size) to the ash pit which is fitted in turn with a **clean-out door** (12" × 12" is often used). The hearth is supported by a 4"-thick concrete slab reinforced with $\frac{3}{8}$"-diameter bars spaced 6" oc both ways. The opening to the fireplace is spanned by a steel angle lintel using 4" × $3\frac{1}{2}$" × $\frac{5}{16}$" stock. The back and sides of the fireplace are **firebrick** laid up in fire-clay mortar, which is more heat-resistant than ordinary brick and mortar. The sides are sloped to direct the heat toward the room and the smoke to the **smoke chamber**. A metal **damper** set below the smoke chamber is used to control the draft. The damper and the base of the smoke chamber

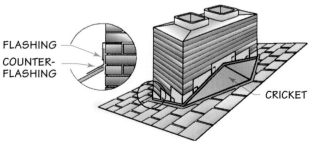

FLASHING
COUNTER-
FLASHING

CRICKET

Figure 32–13   Chimney flashing.

TABLE 32–3 *Masonry Sizes (3 brick + 3 joints = 8")*

Modular Brick	Tile	Block	Size	Modular Brick	Tile	Block	Size
1			$2\frac{2}{3}$"	51		17	11'-4"
2	1		$5\frac{1}{3}$"	52	26		$11'\text{-}6\frac{2}{3}$"
3		1	8"	53			$11'\text{-}9\frac{1}{3}$"
4	2		$10\frac{2}{3}$"	54	27	18	12'-0
5			$1'\text{-}1\frac{1}{3}$"	55			$12'\text{-}2\frac{2}{3}$"
6	3	2	1'-4"	56	28		$12'\text{-}5\frac{1}{3}$"
7			$1'\text{-}6\frac{2}{3}$"	57		19	12'-8"
8	4		$1'\text{-}9\frac{1}{3}$"	58	29		$12'\text{-}10\frac{2}{3}$"
9		3	2'-0	59			$13'\text{-}1\frac{1}{3}$"
10	5		$2'\text{-}2\frac{2}{3}$"	60	30	20	13'-4"
11			$2'\text{-}5\frac{1}{3}$"	61			$13'\text{-}6\frac{2}{3}$"
12	6	4	2'-8"	62	31		$13'\text{-}9\frac{1}{3}$"
13			$2'\text{-}10\frac{2}{3}$"	63		21	14'-0
14	7		$3'\text{-}1\frac{1}{3}$"	64	32		$14'\text{-}2\frac{2}{3}$"
15		5	3'-4"	65			$14'\text{-}5\frac{1}{3}$"
16	8		$3'\text{-}6\frac{2}{3}$"	66	33	22	14'-8"
17			$3'\text{-}9\frac{1}{3}$"	67			$14'\text{-}10\frac{2}{3}$"
18	9	6	4'-0	68	34		$15'\text{-}1\frac{1}{3}$"
19			$4'\text{-}2\frac{2}{3}$"	69		23	15'-4"
20	10		$4'\text{-}5\frac{1}{3}$"	70	35		$15'\text{-}6\frac{2}{3}$"
21		7	4'-8"	71			$15'\text{-}9\frac{1}{3}$"
22	11		$4'\text{-}10\frac{2}{3}$"	72	36	24	16'-0
23			$5'\text{-}1\frac{1}{3}$"	73			$16'\text{-}2\frac{2}{3}$"
24	12	8	5'-4"	74	37		$16'\text{-}5\frac{1}{3}$"
25			$5'\text{-}6\frac{2}{3}$"	75		25	16'-8"
26	13		$5'\text{-}9\frac{1}{3}$"	76	38		$16'\text{-}10\frac{2}{3}$"
27		9	6'-0	77			$17'\text{-}1\frac{1}{3}$"
28	14		$6'\text{-}2\frac{2}{3}$"	78	39	26	17'-4"
29			$6'\text{-}5\frac{1}{3}$"	79			$17'\text{-}6\frac{2}{3}$"
30	15	10	6'-8"	80	40		$17'\text{-}9\frac{1}{3}$"
31			$6'\text{-}10\frac{2}{3}$"	81		27	18'-0
32	16		$7'\text{-}1\frac{1}{3}$"	82	41		$18'\text{-}2\frac{2}{3}$"
33		11	7'-4"	83			$18'\text{-}5\frac{1}{3}$"
34	17		$7'\text{-}6\frac{2}{3}$"	84	42	28	18'-8"
35			$7'\text{-}9\frac{1}{3}$"	85			$18'\text{-}10\frac{2}{3}$"
36	18	12	8'-0	86	43		$19'\text{-}1\frac{1}{3}$"
37			$8'\text{-}2\frac{2}{3}$"	87		29	19'-4"
38	19		$8'\text{-}5\frac{1}{3}$"	88	44		$19'\text{-}6\frac{2}{3}$"
39		13	8'-8"	89			$19'\text{-}9\frac{1}{3}$"
40	20		$8'\text{-}10\frac{2}{3}$"	90	45	30	20'-0
41			$9'\text{-}1\frac{1}{3}$"	91			$20'\text{-}2\frac{2}{3}$"
42	21	14	9'-4"	92	46		$20'\text{-}5\frac{1}{3}$"
43			$9'\text{-}6\frac{2}{3}$"	93		31	20'-8"
44	22		$9'\text{-}9\frac{1}{3}$"	94	47		$20'\text{-}10\frac{2}{3}$"
45		15	10'-0	95			$21'\text{-}1\frac{1}{3}$"
46	23		$10'\text{-}2\frac{2}{3}$"	96	48	32	21'-4"
47			$10'\text{-}5\frac{1}{3}$"	97			$21'\text{-}6\frac{2}{3}$"
48	24	16	10'-8"	98	49		$21'\text{-}9\frac{1}{3}$"
49			$10'\text{-}10\frac{2}{3}$"	99		33	22'-0
50	25		$11'\text{-}1\frac{1}{3}$"	100	50		$22'\text{-}2\frac{2}{3}$"

Fireplace Options

Clients are interested in having a fireplace in their living room but are open to other options, such as a wood-burning stove, gas log, ventless gas stove, pellet stove, or other heating appliances. They would like you to investigate these alternatives and compare and contrast the energy efficiency, cost, maintenance, and aesthetic design of each. You are to draw the living room as it might look with each heating appliance. Drawing a room or home in several different ways is part of an architect's normal routine as clients often want to see how different options affect the final design. Research for clients is not often required but in this case you feel the research can be used for future jobs as well.

Your research can be done traditionally in the library or you can use the Internet. If you decide to use the Internet, your search will begin on one of the search engines such as Lycos, Infoseek, or Yahoo!. Here, type in your search request. Fireplace, ventless gas stove, wood-burning stove, pellet stove, or heating

appliance are possible searches. A number of findings will be listed from which you can choose those most relevant to your research.

After researching these heating options, present your findings in a professional, easy-to-understand manner. You may choose to create a chart or table on which the headings could be energy efficiency, cost, maintenance, aesthetic design, and any other pertinent aspect. An alternative way to create your chart might be to use advantages and disadvantages as the headings. To be truly innovative, create a bar graph or pie chart showing the cost of each alternative. This would allow the client to easily see which option is least or, alternatively, most expensive. Energy efficiency may also be charted or graphed. Photos could be used to show aesthetic appeal. The point is to be creative in communicating your findings. This information packet will be used again and again so it must be attractive and effective in comparing and contrasting these various heating options.

form a smoke shelf, which is important in preventing backdraft. The smoke chamber is corbeled into the flue itself, which conducts the smoke and waste gas safely outside.

Tile Flue

The **flue** is often constructed of rectangular terra-cotta tiles surrounded by 4" or 8" masonry. For proper draft, the area of the flue should not be less than $\frac{1}{10}$ the area of the fireplace opening. When computing the flue size of multifaced fireplaces, the area of *all*

faces must be included. The flue tile sizes given in Table 32–2 meet this requirement. A sharp chimney capping, usually obtained by extending the flue tile 4" above the masonry, also improves draft. Remember to extend the chimney 2' higher than any portion of a roof located within 10'. A tall chimney has a naturally better draft than a short chimney.

Each fireplace or furnace must have a separate flue to prevent the interference of drafts, although these flues may all be combined within a common chimney. This is usually stated, "a flue for every fire." For example, an average home may have two flues: one for the oil furnace and another for the living room fireplace. Both the flues should be set side by side in one chimney separated by 4" of brick (called a *wythe*). Even a house with no fireplace must have a chimney if it is heated by an oil, coal, or gas-fired furnace. For the flue size, see the manufacturer's specifications; 12" × 12" is often used.

A flue size should always be approved by an architectural engineer, because an improperly sized flue (either too small or too large) can create problems such as carbon monoxide release or inadequate draft.

Prefabricated Flue

The prefabricated, nonmasonry flue and chimney shown in Figure 32–14 consist of insulated and fireproof flue sections and a metal housing

placed above the roof to simulate a masonry chimney. This costs considerably less than a masonry chimney and is often used for oil and gas furnace flues.

Fireplace Liner

Since a skilled mason is needed for the proper construction of a masonry fireplace, metal **fireplace liners** are often used (Figure 32–15). These liners consist of the fireplace sides and back, damper, smoke shelf, and smoke chamber all in one prefabricated unit and provide a form for the mason to work to. They also contain a duct system that draws in the room air through inlet registers, warms it, and then discharges it back to the room through outlet registers. This increases the heating capacity of the fireplace.

Metal Fireplaces

Fireplaces constructed entirely of metal have been used for many years, and faithful simulations of these early models are still manufactured. A contemporary type made of sheet steel is shown in Figure 32–16.

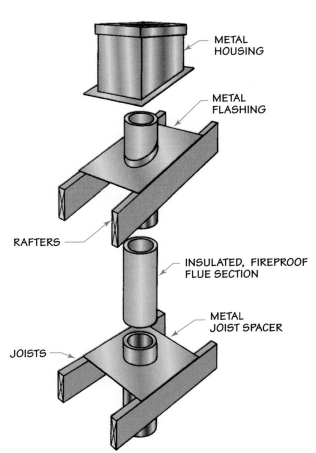

METAL HOUSING

METAL FLASHING

RAFTERS

INSULATED, FIREPROOF FLUE SECTION

METAL JOIST SPACER

JOISTS

Figure 32–14 Prefabricated flue.

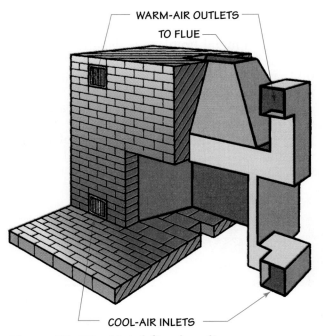

WARM-AIR OUTLETS

TO FLUE

COOL-AIR INLETS

Figure 32–15 Metal fireplace liner.

Figure 32–16 *Free-standing wood stove. (Courtesy Blaze King Royal Heir.)*

Since a metal fireplace is usually used with a metal chimney, the cost is quite low. In addition, a metal fireplace heats up faster and gives more heat than a masonry fireplace.

Metal fireplaces are sold in a variety of sizes, colors, and coatings. Some are made of porcelain enamel steel in a choice of colors. Others use black-painted sheet steel. Others come with a factory prime coat over heavy-gauge sheet steel—leaving the final choice of color to the user. Steel fireplaces can be hung from a wall or ceiling, stand on a platform on their own legs, or be recessed into the floor.

FIREPLACE AND CHIMNEY DETAILS

The architect shows the fireplace and chimney design by means of detail drawings and specifications. The detail drawings usually consist of a front elevation showing the design of the fireplace opening and trim, and sectional plans showing dimensions of the fireplace and chimney together with flue placement. A side vertical section may be included. A scale of $\frac{1}{2}$" = 1'-0 is often used. Some dimensions, such as the height of the chimney above the roof, are shown in the house elevation views.

Figure 32–17 shows the detail drawings of the fireplaces for the A residence. Notice that dimensions are based upon the 4" module to reduce the amount of brick cutting.

Computer-aided Details

Fireplace and chimney sections and details can be entered into your computer graphics system by puck or stylus on a digitizer tablet and displayed on your monitor. Start with preliminary layouts and proceed to finished drawings and working drawings. When satisfied with the result, use your printer or plotter to produce hard copies, but also store your work on diskette. Refer to Chapter 7 for operating information on producing and correcting fireplace and chimney details (such as dimensioning, lettering, sectioning, and symbols).

REVIEW QUESTIONS

1. Describe what is meant by the styling of a fireplace.
2. Why should a chimney be extended higher than any nearby construction? *(Continued)*

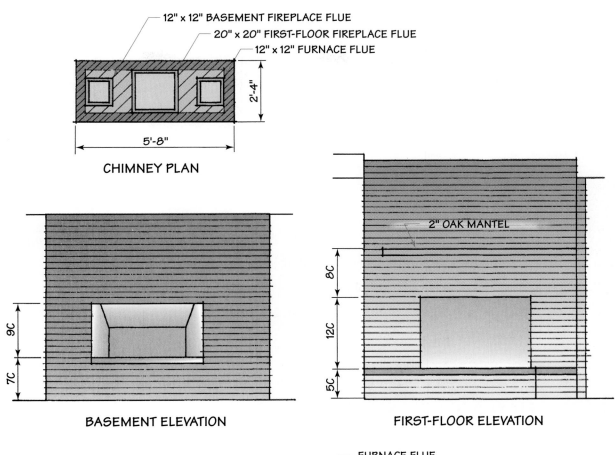

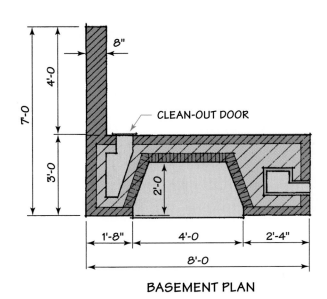

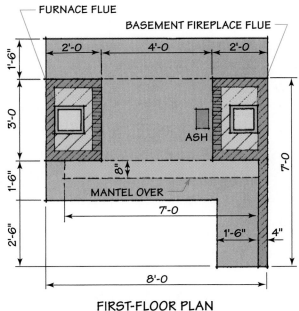

Figure 32–17   *Fireplace details of the A residence.*

3. What is the actual size of:
    a. Common brick
    b. Modular brick
    c. Fire brick
    d. Roman brick
    e. Norman brick

4. Why is it considered poor practice to frame a girder into a chimney?

5. What is the purpose of:
    a. Ash drop
    b. Clean-out door
    c. Front hearth
    d. Damper
    e. Smoke shelf
    f. Fire brick
    g. Flue
    h. Steel angle lintel over fireplace opening

6. How many flues are required for an oil-heated house with two fireplaces?

7. What are the advantages of a fireplace liner?

8. What are the advantages of multifaced fireplaces?

9. What is meant by a prefabricated flue?

ENRICHMENT QUESTIONS

1. Compare the advantages and disadvantages of a chimney located on an interior wall with a chimney located on an exterior wall.

2. Why do hardwoods (ash, oak, hickory, beech) provide more heat than softwoods (pine, poplar, fir)?

3. Describe in words what is being shown in Figure 32–9.

4. Determine the following measurements:
    a. The flue size for a large single-faced fireplace
    b. The height, width, and depth of a medium double-faced opposing side fireplace
    c. The height of ten courses of block
    d. The height of 68 courses of modular block

5. Why is a fireplace commonly referred to in modular sizes?

PRACTICAL EXERCISES

1. Research and explain the development of the fireplace during the past two hundred years.

2. Prepare a report about some other types of fireplaces not mentioned in the chapter.

3. Draw or computer-produce fireplace details for the A residence. Include fireplaces in the basement and on the first-floor.

4. Draw or computer-produce a fireplace detail for your house design.

5. Draw or computer-produce a fireplace that incorporates seven features discussed in this chapter.

   a. Location

   b. Size

   c. Styling

   d. Fireplace opening

   e. Material

   f. Hearth

   g. Accessories (log bin, book shelves, record cabinet, tempered glass doors, outside air intake)

Light
Construction

SECTION V

CHAPTER

33

Light Framing

OBJECTIVES

By completing this chapter you will be able to:

- 🏛 Explain the procedure used to construct a footing and foundation.
- 🏛 Explain how girders and joists are used to construct a floor.
- 🏛 Describe how walls and roofs are constructed.
- 🏛 Differentiate between platform, balloon, plank and beam, and metal framing.
- 🏛 Identify methods of exterior and interior finishing.

KEY TERMS

Platform frame	Bridging
Staking	Subflooring
Excavation	Finished floor
Footing	Stud
Drainage	Sheathing
Foundation	Rafter
Sill	Open-web joist
Header	Cornice
Girder	Balloon framing
Column	Plank-and-beam
Joist	framing

It is extremely important that you, as a student of architectural drafting, be thoroughly familiar with standard construction practices. An architectural drafter attempting to design a building without knowledge of its construction would be like a contractor trying to build without being able to read blueprints. In either case, chaos would result.

Although there are many different types of acceptable construction in use today, we shall study in detail the most commonly used type: the **platform frame**. This will be done step by step, starting with the site survey and excavation, and working up to the roof and finish materials—as though we were actually building. Later we will look at other kinds of construction to see how they differ from the platform frame.

SURVEY

Before starting construction, the house lot should be surveyed by a registered surveyor who will accurately locate the property lines. The property corners are marked with 30"-long galvanized iron pipes driven almost completely into the ground. The surveyor may also stake out the corners of the house, checking the local building ordinance for requirements on minimum setback from the road, and minimum side and rear yards. The ordinance may state that the front of the building must align with adjacent buildings.

As previously mentioned, if there are trees on the site, the house should be placed to save a maximum number. As a rule, all trees within 5' of the proposed building are cut down, since the excavation will disturb their roots so much that they will die.

The surveyor will also establish correct elevations, usually using the adjoining road or house as references. All of this information is placed on a survey map which will be the basis of the plot plan. Figure 33–1 shows the survey map for the A residence.

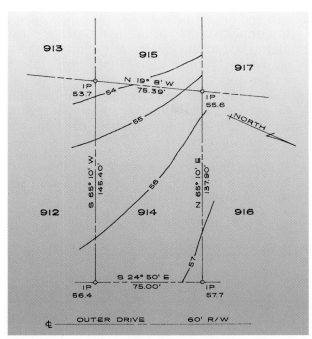

Figure 33-1   Site survey for the A residence.

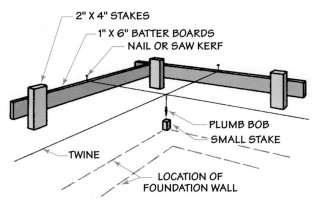

Figure 33-2   Staking.

STAKING

Although **staking** out the house foundation may be done by the surveyor, this is usually done by the building contractor who locates the future outside corners of the foundation and marks these points with tacks in small stakes. Then, since these stakes will be disturbed by the excavation, larger 2"×4" stakes are driven 4' beyond the foundation lines, three at each corner. Figure 33-2 shows 1"×6" batter boards nailed to these stakes so that their tops are of the same elevation. Using a plumb bob, the contractor stretches stout twine across the batter boards directly above the corner tacks. Saw kerfs (cuts) or nails are located on the batter boards where the twine touches to establish a more permanent record of the foundation lines (Figure 33-3). Of course, it is particularly important to check for squared corners. This may be done with surveying instruments, by measuring diagonals, or by using the principle of the 3-4-5 triangle (Figure 33-4).

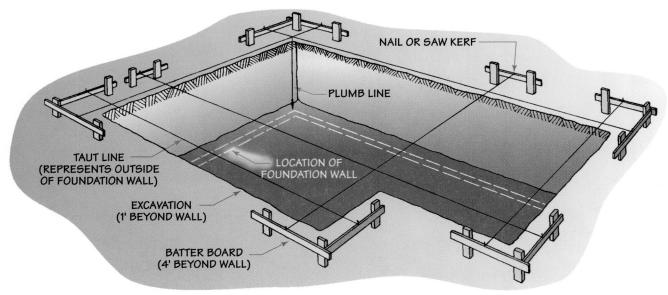

Figure 33-3   Excavation.

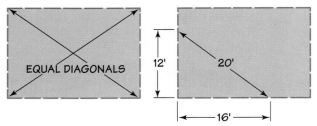

Figure 33–4 Methods of squaring corners.

EXCAVATION

The **excavation** is usually done by power equipment. First, about 1' of topsoil is removed and stored at the side of the lot to be used later in the finish grading. Then the excavation itself is made, the depth depending upon these factors:

1. On sloping land, the foundation must extend above the highest perimeter point of finished grade by:

   a. 8" in wood-frame construction to protect the wood from rotting due to moisture.

   b. 2" in brick construction to protect the first brick course from constant exposure to moisture which may eventually work into the joints.

2. The footing must extend below the lowest perimeter point of finished grade by the prevailing frost-line depth. This is necessary to prevent upheaval when the ground freezes. Figure 33–5 may be consulted for a general indication of frost-line depth, but a more accurate depth for a particular area may be obtained from local architects and builders.

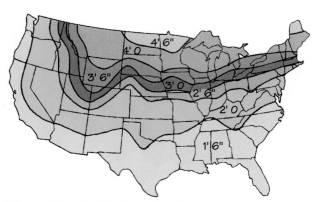

Figure 33–5 Footing depth.

3. When a full foundation is to be built, allow a minimum of 6'-9" from the top of the basement floor to the bottom of the floor joist as shown in Figure 33–6. Remember that a girder under the joist will reduce the headroom. The minimum comfortable ceiling height for a habitable room is considered to be 7'-6"; 8'-0" more often is used for the main living areas.

4. When a crawl space (for inspection and repair) is to be built, allow a minimum of 18" to the bottom of the joists. Specify 2'-6"

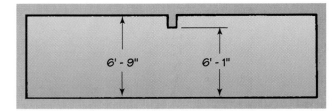

MINIMUM BASEMENT HEIGHTS

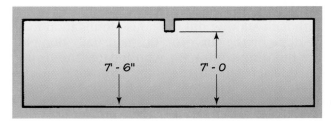

MINIMUM HABITABLE ROOM HEIGHTS

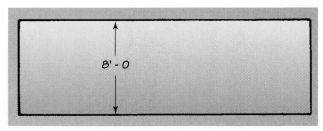

RECOMMENDED ROOM HEIGHT

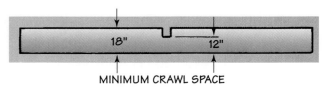

MINIMUM CRAWL SPACE

Figure 33–6 Minimum design heights.

for a more comfortable working height. Prevent water from accumulating in the crawl space by:

a. Locating the crawl space $1\frac{1}{2}$" above the outside finish grade.

b. Providing a special drain to a lower elevation or storm sewer.

c. Relying on local soil conditions which may be such that water will naturally drain from the crawl space.

5. The excavation should extend down to *unfilled* ground. Because it is so important that a good bearing surface be provided, the trench for the footings should be dug shortly before pouring the concrete to prevent a possible softening of the bearing ground by exposure to rain and air.

FOOTING

Footings increase the bearing surface of the house upon the ground so that there will be less settling. The footings should be of concrete poured on undisturbed land. Average residential construction on firm land calls for footings twice as wide as the foundation wall, from 16" to 24". The depth of the footing should equal the wall thickness, ranging from 8" to 12". Side forms may be omitted if the ground permits sharply cut trenches. Reinforcing steel is used when the footing spans pipe trenches. As mentioned previously, the frost

REAL WORLD   APPLICATIONS

Frost Upheaval

Frost upheaval, commonly known as frost heave, is a natural phenomenon associated with freezing water in the soil above the frost line. As water freezes between the soil molecules and within the pores, it expands and may increase in volume by as much as 10 percent. This expansion can create an increase in pressure of several hundred pounds per square inch. When a foundation is constructed above the frost line and the underlying soil is allowed to freeze, the expansion and increase in pressure may result in the foundation being lifted or "heaved." As the soil thaws it becomes loose and unconsolidated. The foundation may not return to its original position. The result is differential settlement, often contributing to structural cracks and uneven window and door sills. For this reason, foundations should be constructed below the frost line. Outdoor utilities susceptible to freezing should also be constructed below the frost line. To help alleviate the effects of frost, retaining walls, bridge abutments, and foundation walls are often backfilled with a granular material such as gravel. The backfill may be used in conjunction with a drain, most commonly perforated pipe. By reducing or eliminating the water in the backfill, the effects of frost are also reduced.

The force of expanding ice associated with frost action in a confined space, such as fissures or cracks in rock or structures, acts like a wedge to split the material apart. This technique was often used in early rock quarries. This mechanical erosion affects small and large objects alike, from roadways to mountains. The depth to which frost will penetrate depends upon climate and geographical area.

Contributed by M. J. Shanshala III, P. E., B.S.C.E.

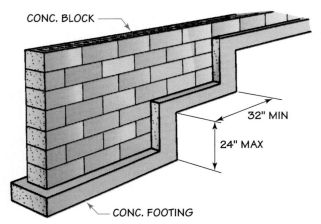

Figure 33–7   Stepped footing.

line determines the minimum depth of footing excavation, this varying from 1'-6" in Florida to 4'-6" in Maine.

The bottom of footings should always be horizontal, never inclined. Thus on sloping land, *stepped footings* such as shown in Figure 33–7 are used. The horizontal portion of a step footing should not be less than 32"; the vertical portion should not exceed 24". To reduce cutting when building the foundation wall, these dimensions should be in modular block units. The horizontal and vertical portions of the stepped footing should be of equal thickness, and both portions should be poured at the same time.

Footings are also required for chimneys and columns. Since column footings must support as much as one-quarter of the total weight of the house, they are stepped out even farther (usually to 24" square or 30" square). By the use of the tables in Chapter 34, it is possible to calculate the required sizes of foundation, chimney, and column footings.

DRAINAGE

To provide **drainage** around the foundation and ensure a dry basement, 4" perforated pipes are laid in the foundation excavation or footing excavation. Two types of drain pipe are available:

1. A rigid, perforated plastic pipe, 4" diameter in 10' lengths joined by plastic fittings.

2. A flexible, perforated plastic pipe, 4" diameter in 250' rolls.

Ground fill above either type encourages drainage to the drain pipes. The drain line should slope slightly ($\frac{1}{16}$"/ft. minimum) to a catch basin, dry well, or sewer.

FOUNDATION

Foundations (Figure 33–8) may be constructed to provide a basement or crawl space. When a concrete floor is poured at ground level, it is called *slab* construction. It may be floating or perimeter. The floating slab requires reinforcing since it is meant to "float" as an integral unit on the ground. Although this construction has been used in cold climates, it is best suited to areas where frost penetration is no problem. The perimeter foundation, on the other hand, provides a complete foundation wall for the protection of the slab from frost. The rigid insulation reduces heat loss from the house.

Materials

The two most common foundation materials are poured concrete (8"–12") and concrete blocks (8", 10", or 12").

Poured Concrete   The poured concrete foundation is usually considered superior because it is more likely to be waterproof and termite-proof. A 1-2$\frac{1}{2}$-5 concrete mix is often used. This means 1 part cement, 2$\frac{1}{2}$ parts fine aggregate (sand), and 5 parts coarse aggregate (gravel or crushed stone). Poured concrete walls are sometimes battered (sloped) from 12" thickness at the bottom to 8" thickness at the top. This is to prevent any adhesion between the walls and clay ground due to freezing and to guard the wall from being lifted by frost action. The outside faces of the foundation and footing are mopped with hot tar or asphalt for additional protection against water. For this purpose, emulsified or hot tar (pitch) is superior to asphalt, since asphalt in continual contact with moisture may eventually disintegrate.

Concrete Block   A $\frac{1}{2}$" layer of cement plaster (called *parging*) is applied to the outside block wall and covered with hot tar or asphalt waterproofing. It is good practice to fill the cores of the top course of concrete blocks with concrete to prevent passage of water or termites.

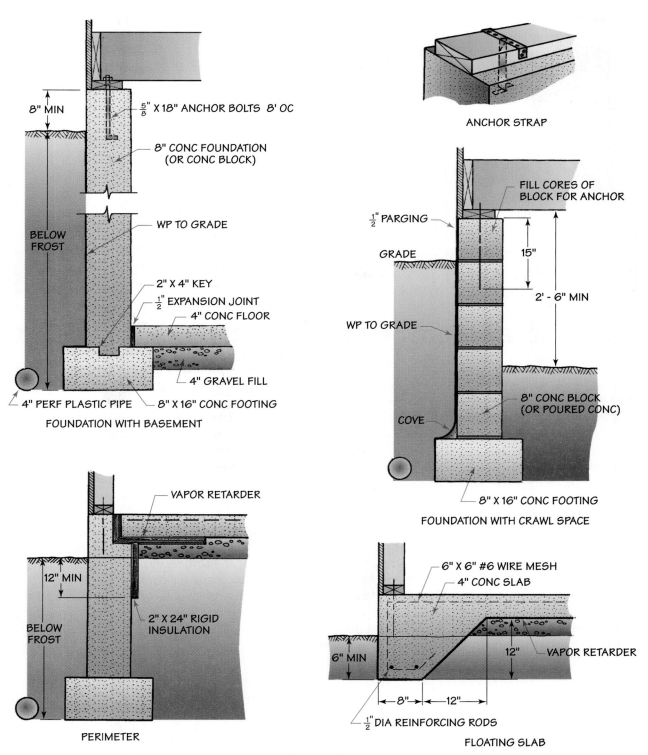

Figure 33–8 *Foundations.*

Long stretches of wall are often stiffened with 8" × 16" pilasters every 16' as shown in Figure 33–9. This is particularly important when using walls of only 8" block.

Foundation Height

Remember that the foundation must extend above the highest perimeter point of finished grade by 8" in wood-frame construction to protect lumber from ground moisture. When a brick-veneer construction is used, the foundation should extend at least 2" above the finished grade.

Reinforcing

Concrete has excellent strength in *compression*, but is weak in *tension.* Therefore when any portion of a concrete member is expected to be subjected to tension, steel rods or steel wire mesh are cast in that portion to resist the tension.

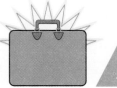

CAREER PROFILES

Civil, Electrical, and Mechanical Engineers

The personnel of a small architectural office might consist of the registered architect who founded the office (called a principal), a not-yet-registered architect-in-training, a registered architectural engineer, one or more architectural drafters, and a receptionist/secretary/specifications writer. Rather than one architectural engineer, a larger office might employ three specialized engineers or engineering technologists; a civil engineer for structural design in steel and concrete, an electrical engineer for electrical and illumination design, and a mechanical engineer for heating ventilation and air conditioning (usually called HVAC) and plumbing design.

According to the *Occupational Outlook Handbook*, the job outlook for all three engineering disciplines is favorable and employment is expected to grow about as fast as average through 2005. In 1994, 40 percent of civil engineering jobs were in firms that provide engineering consulting services, primarily developing designs for new construction projects. Most of the jobs of electrical engineers, the largest branch of engineering, were also in engineering and business consulting firms. Although six out of ten mechanical engineering jobs were in manufacturing, a significant minority held jobs in business and engineering consulting services.

Civil, electrical, and mechanical engineers are usually baccalaureate graduates of Accreditation Board of Engineering and Technology (ABET)-accredited engineering or engineering technology programs. The engineering programs are nominally four-year degrees, but due to the large number of credits required, about half of engineering students graduate in four and one-half or more years. Some typical technical courses for a civil engineering major are:

- Structural analysis
- Steel structures design
- Concrete structures design
- Foundation engineering
- Transportation engineering
- Highway engineering
- Geotechnical engineering
- Environmental engineering
- Fluid mechanics
- Hydraulic engineering
- Water and wastewater systems
- Hazardous waste management

This is called *reinforcing steel.* In light construction, reinforcing is used in concrete bond beams, concrete lintels, and occasionally in concrete slabs and footings. This reinforcing should be placed near the *bottom* of bond beams, lintels, column footings, and suspended slabs to best resist the tension there. It is common building practice to also place reinforcing near the bottom of wall footings and on-grade slabs upon the assumption that these members are similar to bond beams. However, many engineers now specify that reinforcing rods in wall footings be

near the *top* to better prevent cracking which could then extend up into the wall above. Reinforcing wire mesh in slabs poured on grade is also specified near the *top* of the slab to better control cracking which would affect the exposed floor surface. See Figure 33–10.

SILL

The sill is a 2" × 6" plank resting directly on top of the foundation wall. Notice in Figure 33–11 that the sill is set back about 1" from the outside

Some typical technical courses for an electrical engineering major are:
- 🏛 Signals and circuits
- 🏛 Electronic devices and circuits
- 🏛 Electronic circuit design
- 🏛 Electromagnetics
- 🏛 Analog and digital interfacing
- 🏛 Optical fiber communications
- 🏛 UHF and microwave engineering
- 🏛 Linear network analysis
- 🏛 Linear and nonlinear holography
- 🏛 Applications of lasers and masers

Some typical technical courses for a mechanical engineering major are:
- 🏛 Thermodynamics
- 🏛 Fluid flow
- 🏛 Compressible flow
- 🏛 Turbomachinery
- 🏛 Machine dynamics
- 🏛 Internal combustion engines
- 🏛 Mechanical design
- 🏛 Dynamics of mechanical systems
- 🏛 Mechanical engineering measurements
- 🏛 Automatic control systems
- 🏛 Industrial robotics
- 🏛 Air pollution control systems

If you are interested in a civil, electrical, or mechanical engineering career, obtain a catalog and application from a university having ABET-accredited majors. A listing of such universities (over 300 in the United States) can be obtained from ABET, 111 Marketplace Suite 1050, Baltimore, Maryland 21202. Phone: 410-347-7700. Additional information about ABET can be obtained through their Internet homepage at:
http://www.abet.ba.md.us/
Currently, thirteen universities in the United States offer ABET-accredited majors in architectural engineering. A list can be obtained from ABET.

To achieve engineering registration, the civil, electrical, or mechanical engineer must graduate from an ABET-accredited university, obtain four years of practical experience under the supervision of a registered engineer in the same field, and successfully complete a comprehensive written examination. Without a degree, most states accept eight years of practical experience as a substitute.

For more information on engineers' salaries, working conditions, nature of work, training, and job outlook you can use the *Occupational Outlook Handbook.* You can access it on-line at:
http://www.bls.gov/ocohome.htm
From this page go to Professional Specialty Occupations and then to Engineers.

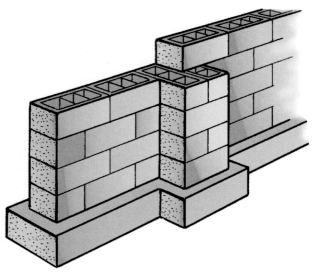

Figure 33–9   *Pilaster construction.*

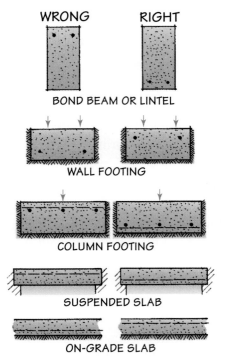

Figure 33–10   *Placement of reinforcing steel in light construction.*

wall so that the sheathing, which is nailed to the sill, will be flush with the outside foundation wall. Some builders allow for irregularities in the face of the foundation wall by setting the sill

flush so that the sheathing projects beyond the outside of the foundation wall. This is illustrated in Figure 14–11 (page 167). The sill should be fastened by $\frac{1}{2}$", $\frac{5}{8}$", or $\frac{3}{4}$" bolts spaced 8' apart. These extend 6" into a poured concrete foundation and 15" into a concrete block foundation. Holes are drilled into the sill, a bed of mortar (called *grout*) is spread on the foundation, and the sill is tapped into a level position. The nuts and washers are tightened by hand. Several days later they may be wrench-tightened. The grout provides a level bed for the sill and makes an airtight joint.

Other acceptable connectors are $\frac{1}{16}$" × $1\frac{1}{2}$" × 15" metal anchor straps spaced 4' apart and perforated for spiking to the sill. Some manufacturers are Panel Clip, Simpson Strong-Tie, and TECO.

An alternate to grout is preferred by some builders: a 1" × 6" strip of fiberglass insulation between the foundation and sill.

HEADER

Headers and joists are the same size. The header is spiked upright to the top outside edge of the sill. Where a basement window or door breaks the foundation wall, it is good practice to let the header, rather than the sill, act as the spanning member. This is best accomplished by a ledger strip spiked to the header and extending at least 6" beyond the opening, as shown in Figure 33–12. The joists are cut to rest on the ledger and are also spiked to the header. A steel angle lintel may also be used.

GIRDER

The dimensions of most houses are so great that joists cannot span the foundation walls. In that case, a wood **girder** (that is, several 2"-thick members spiked together) or steel beam is used, as shown in Figure 33–13. Notice that the girder is also too long to span the foundation walls and must be supported by wood or steel **columns** as in Figure 33–14. Steel pipe columns (usually referred to by the trade name *Lally columns*) are capped with a steel plate to increase the bearing surface with the wood girder. A $3\frac{1}{2}$"-diameter column is large enough for ordinary requirements.

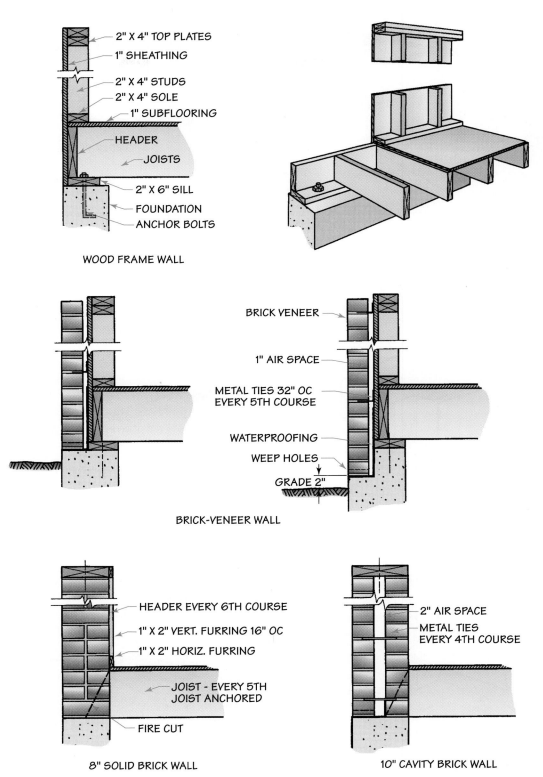

2" X 4" TOP PLATES
1" SHEATHING
2" X 4" STUDS
2" X 4" SOLE
1" SUBFLOORING
HEADER
JOISTS
2" X 6" SILL
FOUNDATION
ANCHOR BOLTS

WOOD FRAME WALL

BRICK VENEER
1" AIR SPACE
METAL TIES 32" OC
EVERY 5TH COURSE
WATERPROOFING
WEEP HOLES
GRADE 2"

BRICK-VENEER WALL

HEADER EVERY 6TH COURSE
1" X 2" VERT. FURRING 16" OC
1" X 2" HORIZ. FURRING
JOIST - EVERY 5TH
JOIST ANCHORED
FIRE CUT

8" SOLID BRICK WALL

2" AIR SPACE
METAL TIES
EVERY 4TH COURSE

10" CAVITY BRICK WALL

Figure 33–11   *Walls.*

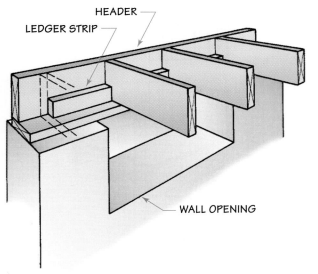

Figure 33–12   *Use of a ledger strip.*

Figure 33–13   *Use of a girder.*

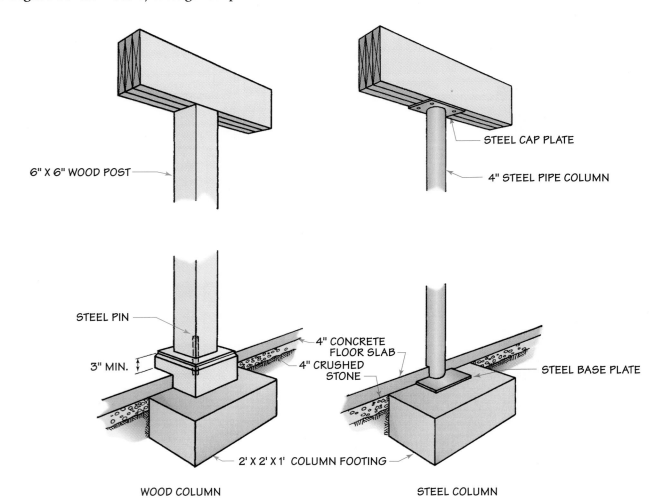

Figure 33–14   *Use of columns.*

The girder is framed into the foundation wall as shown in Figure 33–16 so that it bears a minimum of 4". Incidentally, this is a good rule to remember: *always provide the greatest possible bearing surface between two members.* In case of wood, the length of bearing surface should not be less than 4" (see Figure 33–15) for safest construction. However, to save head-room, **joists** are occasionally framed level with the girder using iron stirrups, ledger strips, or framing anchors. Joists may also be set "level" with a steel beam. (Actually, such joists are installed with their upper edges 1" above the steel beam to allow for wood shrinkage and to prevent a bulge in the floor above.)

The wall pocket for a wood girder should be large enough to allow a minimum of $\frac{1}{2}$" air space at the sides and ends of the girder. This allows moisture to escape and reduces the possibility of decay.

Sizes of wood girders and steel beams may be calculated as shown in Chapter 34.

FLOOR JOISTS

Because so much material is made in 4' lengths (4' × 8' plywood, plasterboard, rigid insulation, 4' rocklath lengths, and so forth), it is desirable that floor joists, wall and partition studs, and rafters be spaced either 12", 16", or 24" oc (all

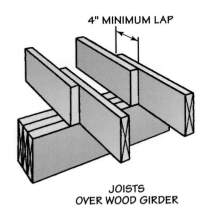

4" MINIMUM LAP

JOISTS
OVER WOOD GIRDER

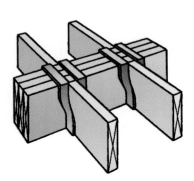

USING IRON STIRRUPS

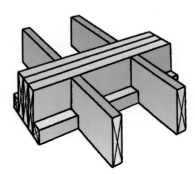

USING LEDGER STRIPS
JOISTS LEVEL WITH WOOD GIRDER

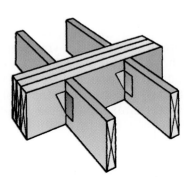

USING FRAMING ANCHORS

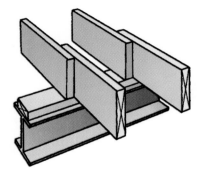

JOISTS
OVER STEEL BEAM

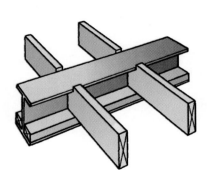

RESTING ON WOOD NAILERS

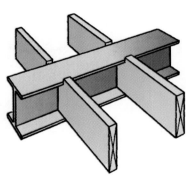

RESTING ON BEAM

JOISTS LEVEL WITH STEEL BEAM

Figure 33–15 *Methods of framing joists.*

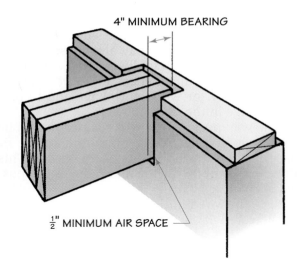

Figure 33–16   Girder pocket.

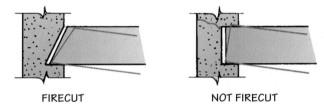

Figure 33–18   Use of firecutting.

even divisions of 4') to avoid cutting. Since 24" oc is usually too weak and 12" oc is wasteful, 16" oc is normally used. The joist sizes are determined by the tables in Chapter 34. Joist spans are often 14'–16'.

Joists and headers are doubled around all openings (such as stairwells and chimneys) as shown in Figure 33–17. When a partition runs parallel to a joist, its entire weight must be supported by one (or two) joists. Since this weight might cause excessive bending, such a joist is also stiffened by doubling. When partitions run

at right angles to joists, no extra support is necessary. Joists may be spaced 12" oc instead of the usual 16" oc under bathrooms and occasionally under kitchens to allow for weakening caused by pipes being set into the floor.

When joists frame into masonry walls as in Figure 33–11, their ends should be firecut to prevent the walls being pushed outward if the joists should sag. Firecutting also helps prevent cracks in the masonry wall due to the joists settling (see Figure 33–18).

BRIDGING

Bridging (Figure 33–19) is used to keep the joists vertical and in alignment, and to distribute a concentrated load on more than one joist. Solid wood blocking, 1" × 3" wood bridging, or metal straps may be used. Rows of bridging should be spaced a maximum of 7' apart. Since the subflooring has a tendency to align the joist

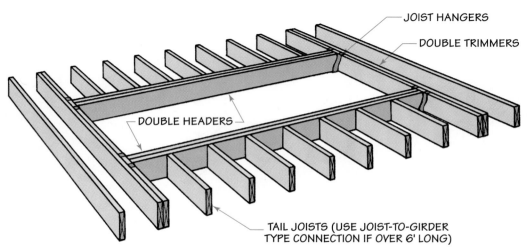

Figure 33–17   Stairwell framing.

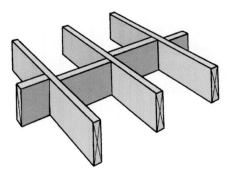

SOLID BLOCKING

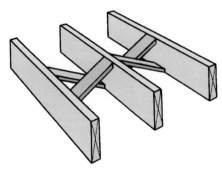

1" X 3" WOOD BRIDGING

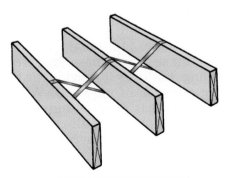

METAL STRAP BRIDGING

Figure 33–19   Bridging.

tops, the lower end of the wood and strap bridging is not nailed until the subflooring is laid.

SUBFLOORING AND FLOORING

Subflooring is a wood floor of $\frac{5}{8}$" plywood or 1" boards laid over joists to serve as a base for the finished floor. The **finished floor** is usually of tongue-and-grooved hardwood; oak, maple, and birch are used. When plywood or boards laid diagonally are used for subflooring, the fin-

ished flooring is laid parallel to the long dimension of the room; when boards laid perpendicular to the joists are used for subflooring, the finished flooring must be laid perpendicular to the subflooring regardless of the room proportions. Building paper is laid between subflooring and flooring as a protection against air and moisture.

STUDS

A *wall* means an exterior wall; a *partition* means an interior wall. Partitions may be either bearing or curtain (nonbearing). A wall or partition consists of vertical members spaced 16" oc called **studs**, a lower horizontal sole plate, and doubled top plates, as shown in Figure 33–11. All of these members may be 2" × 4" lumber. When a 4" cast-iron soil stack is used, however, the wall is made of 2" × 6" lumber to conceal it. In cold climates, studs of 2" × 6" lumber spaced 24" oc are used to increase the wall thickness so that more insulation can be installed in the walls. Some communities have codes that require insulative sheathing whenever 2" × 4" studs are used. Often an entire wall, including sole and top plates, is assembled horizontally on the subflooring and then raised and braced in position while the sole is spiked to the subflooring. This method avoids toenailing the stud to the sole. Sheathing serves as an additional tie between wall, header, and sill.

WINDOW AND DOOR OPENINGS

The horizontal framing member above a window or door opening is called a *header*, and the horizontal framing member below the window is called the *rough sill* (Figure 33–20). All members framing an opening should be doubled for greater strength and to provide a nailing surface for trim. The headers are laid with the long edge vertical to provide greater strength. They must be shimmed, however, to increase their 3" (2 × $1\frac{1}{2}$") thickness to $3\frac{1}{2}$". The size of the headers ranges from doubled 2" × 4" members up to doubled 2" × 12" members, depending upon the span and superimposed load. Table 34–12 in Chapter 34 may be used to determine the required size of headers.

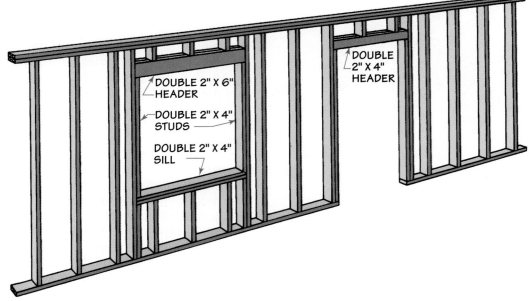

Figure 33–20   Rough framing of windows and doors.

CORNER POSTS

Corner posts must provide surfaces for nailing the sheathing at the corner and the lath at both interior walls. Two methods of accomplishing this are illustrated in Figure 33–21.

CANTILEVER FRAMING

Figure 33–22 shows the method of framing for cantilevered construction such as the second-floor overhang of a garrison house. The length of the lookouts should be at least three times the length of the overhang.

SHEATHING

Sheathing is nailed to the exterior of the studs in a manner similar to that used for the subflooring. Common sheathing materials are:

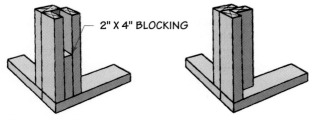

Figure 33–21   Methods of framing corner post.

1. $\frac{5}{16}$" minimum × 4' × 8' plywood

2. $\frac{1}{2}$" minimum × 4' × 8' composition board. Composition board has the advantage of providing some additional insulation but does not make a good base for exterior finish nailing, nor does it provide the diagonal bracing strength of plywood. When composition board is used as outside sheathing, it should be asphalt-coated to prevent disintegration and serve as a moisture barrier.

3. 1" rigid insulation. When rigid insulation is used for sheathing, plywood is substituted at corners to stiffen the wall, or corner braces are let into the studs.

Plywood should be used for roof sheathing.

BUILDING PAPER AND HOUSEWRAP

Building paper is asphalt-saturated felt or paper used between subflooring and finished flooring, sheathing and finished wall covering, and roof boards and roof covering. It prevents wind and water from entering the building between cracks, while still allowing water vapor to escape.

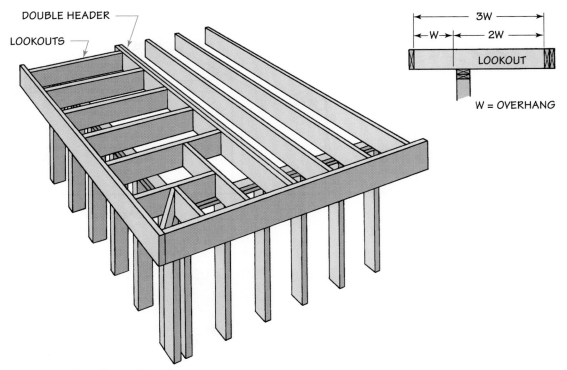

Figure 33–22   *Cantilever framing.*

Housewrap is a very strong plastic sheet, such as Dupont's Tyvek, often used in place of building paper. Housewrap is more resistant to wind and water infiltration, yet it is breathable. Also, housewrap is available in large-sized rolls (up to 10' wide). But building paper, being less slippery than housewrap, is always preferred for roof underlayment.

RAFTERS

Rafter size and spacing may be determined by Table 34–10. A 16" or 24" spacing is often used. The upper end of the rafter is spiked to a 1"- or 2"-thick ridge board, the depth of which is not less than the end cut of the rafter. The lower end of the rafter is cut to obtain a full bearing on the top plate. This cut is called a *bird's mouth* or *crow's foot* (see Figure 33–23).

ROOF THRUST

A sloping roof exerts not only a *downward* thrust on the exterior walls, but also an *outward* thrust which tends to push the exterior walls

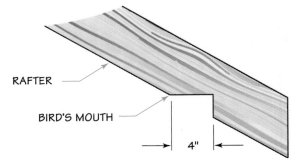

Figure 33–23   *Bird's mouth.*

apart as shown in Figure 33–24. The result of this outward thrust may be prevented by the following measures.

1. Run the ceiling joists parallel to the rafters together with 1" × 6" or 2" × 4" collar beams spaced 4' oc.

2. Support the rafters at the ridge by a bearing partition or beam.

3. Roof trusses or trussed rafters may be used for large spans without bearing partitions,

ROOF LOAD

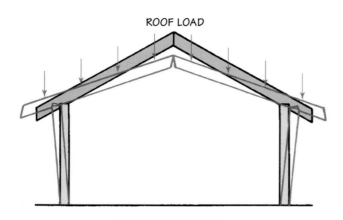

UNSTABLE FRAMING

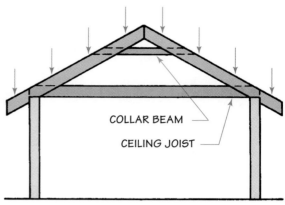

COLLAR BEAM

CEILING JOIST

STABLE FRAMING

Figure 33–24   Framing for outward thrust.

Figure 33–25   Erecting rafters. (Courtesy APA—the Engineered Wood Association.)

Figure 33–26   Solid blocking between rafters. (Courtesy APA—the Engineered Wood Association.)

allowing great freedom in room planning. (A trussed rafter is a truss spaced close enough to adjacent trusses that purlins are unnecessary.) Notice, though, that a truss greatly reduces the usefulness of the attic space. Because a truss is composed of a number of small spans, the members need not be heavy. A typical trussed rafter is shown in Figure 33–27. The Timber Engineering Company of Washington, D.C., publishes a reference book that is very helpful to designers planning to specify trusses or trussed rafters.

FLAT ROOFS

Flat roofs may be laid level to hold water on the roof (called a *water-film roof*) or, more commonly, sloped slightly to prevent water from collecting. The roof joists rest directly on the top plates and serve a double purpose, as roof rafters and ceiling joists. When a wide overhang is desired, the roof joists are framed for cantilever framing, as shown in Figure 33–22. Although wood roof joists are used to frame flat roofs in residential construction, steel **open-web joists** (also called *bar joists*, Figure 33–28) are normally used in commercial construction since they can span 48' or more. Because shingles cannot be used for a flat roof covering, a *built-up* roof finish is used. A built-up roof is constructed by laying down successive layers of roofing felt and tar or asphalt topped with roll roofing, gravel, or marble chips.

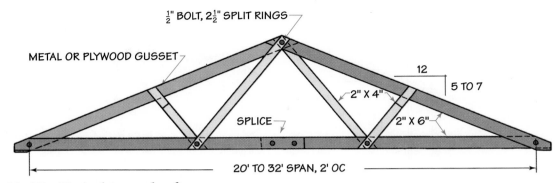

Figure 33–27   *Typical trussed rafter.*

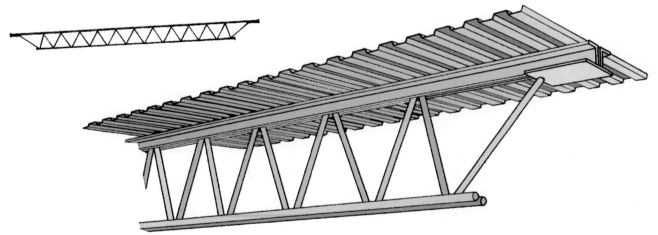

Figure 33–28   *Open-web joist.*

CORNICES

Figure 33–29 shows **cornice** construction over a frame wall, brick wall, and brick-veneer wall. Also, the methods of framing various overhangs are shown, ranging from wide overhangs to a flush cornice. The raised cornice is used to provide an additional foot of headroom in the attic space or, in the case of wide overhangs, to provide more clearance above the windows beneath the cornice. Since roof gutters are often unsightly, they may be built into the roof as shown. Obviously, built-in gutters must be very carefully flashed.

EXTERIOR FINISH

Exterior wall finish (Figure 33–30) covers the sheathing and building paper. Since the choice of finish will greatly influence the final appearance and upkeep of a house, the materials should be carefully selected.

Plastic and metal siding, installed over sheathing, is available in horizontal clapboard and vertical V-groove styles. Plastic siding, usually vinyl, and metal siding, usually aluminum, is manufactured in a variety of colors and textures. Advantages include durability and low maintenance.

Wood siding is usually of red cedar, cypress, or California redwood because these materials have superior weather resistance. Corners may be mitered for a neat appearance, but wood or metal corners are more durable (see Figure 33–31). The style of the house will also influence the corner treatment.

Board-and-batten wood siding is relatively inexpensive and presents an attractive finish.

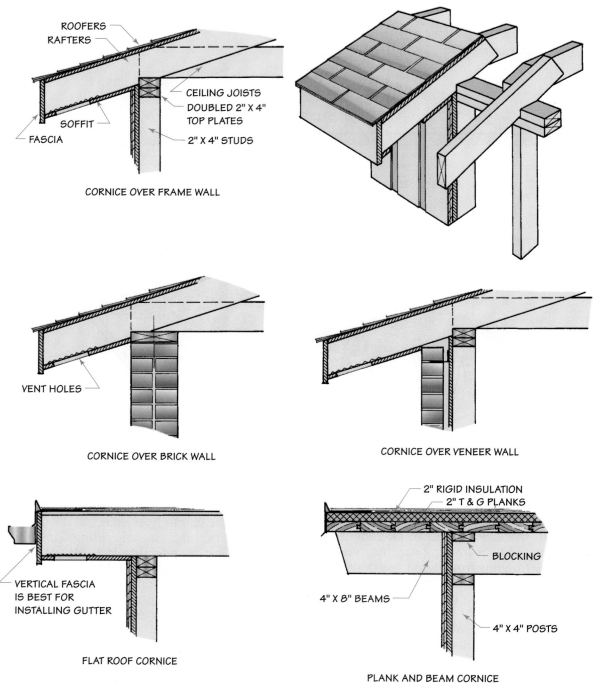

ROOFERS
RAFTERS
CEILING JOISTS
DOUBLED 2" X 4" TOP PLATES
2" X 4" STUDS
SOFFIT
FASCIA

CORNICE OVER FRAME WALL

VENT HOLES

CORNICE OVER BRICK WALL

CORNICE OVER VENEER WALL

VERTICAL FASCIA IS BEST FOR INSTALLING GUTTER

FLAT ROOF CORNICE

2" RIGID INSULATION
2" T & G PLANKS
BLOCKING
4" X 8" BEAMS
4" X 4" POSTS

PLANK AND BEAM CORNICE

Figure 33–29 Cornices.

Wood shingles are also of red cedar, cypress, or redwood. Hand-split shakes may be used for a special effect. Shingles are often left unpainted and unstained to obtain a delightfully weath-ered finish. Red cedar and California redwood weather to a dark gray color; cypress weathers to a light gray with a silver sheen. Various types of composition siding—hardboard, fiberboard,

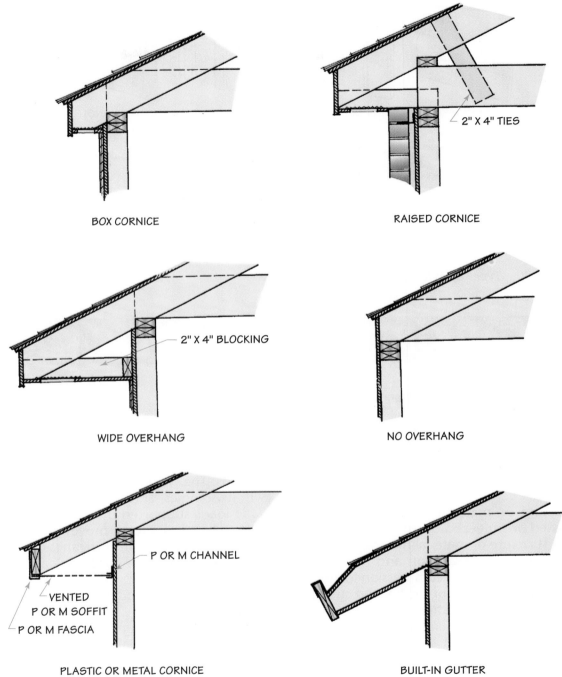

BOX CORNICE

RAISED CORNICE

2" X 4" TIES

WIDE OVERHANG

2" X 4" BLOCKING

NO OVERHANG

PLASTIC OR METAL CORNICE

P OR M CHANNEL

VENTED
P OR M SOFFIT

P OR M FASCIA

BUILT-IN GUTTER

Figure 33–29 *(continued)   More cornices.*

asbestos, asphalt, and so forth, in imitation of wood, brick, or stone—have certain advantages. But they must be carefully specified so that they do not cheapen the appearance of the building.

Brick and stone finishes are durable, require very little upkeep, and present a fine appearance. Types of brick bonds are shown in Figure 33–32. The word *bond* used in reference to masonry has

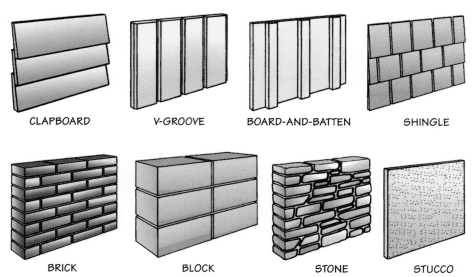

CLAPBOARD   V-GROOVE   BOARD-AND-BATTEN   SHINGLE

BRICK   BLOCK   STONE   STUCCO

Figure 33–30   Types of exterior finish.

REAL WORLD APPLICATIONS

Forces on a Structure

 When designing any structure it is important to consider its *compressive* and *tensile* strength. These strengths refer to how well a structure holds up under compression and tension. When a structure or material is compressed, it is pressed or squeezed. Tension refers to how far a structure or material may be stretched. Lightly step on a balloon and notice how your weight acts as a compression force. You will see that the rest of the balloon stretches as far as its tensile strength allows until it breaks. As in this example, both forces act on most structures at the same time. Therefore, structures must be designed to balance these two forces.

When a single large beam cannot support the forces exerted on it, a *truss* is often used. In architecture, the truss is a specially designed triangular brace made up of smaller members used to balance these forces. Forces exerted on a structure are called *loads.* Static loads are forces that are constant and do not move. The weight of the structure and objects it supports are static loads. The weight of a roof truss, the roof with its shingles, and snow are examples of static loads supported by a truss. Dynamic loads change as a force moves around. Severe winds, including hurricanes or tornadoes, are examples of dynamic loads a roof truss must withstand.

Make a simple model of a roof truss and determine the amount of weight it can support by hanging a weight from its peak. Were you able to balance tension and compression forces? How? When designing for the load it is important to understand how certain types of structures are used to support loads. Using a dictionary or other reference, define the following structures and explain how each is specially designed to support a load: post and lintel, flying buttress, space frame, cable roof, and rigid frame.

Figure 33–31   *Siding corner construction.*

MITERED   CORNER BOARD   METAL CORNERS

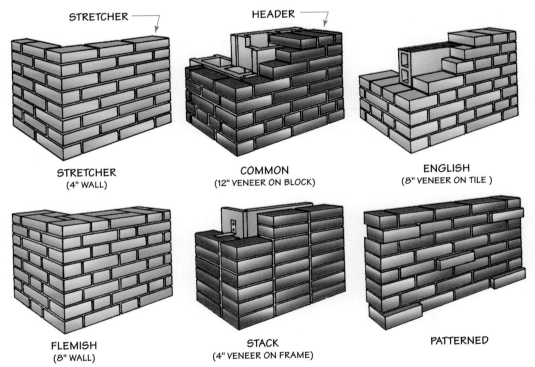

STRETCHER   HEADER

STRETCHER
(4" WALL)

COMMON
(12" VENEER ON BLOCK)

ENGLISH
(8" VENEER ON TILE)

FLEMISH
(8" WALL)

STACK
(4" VENEER ON FRAME)

PATTERNED

Figure 33–32   *Brick bonds.*

several meanings. *Mortar bond* is the adhesion of the mortar to the brick or block units. *Structural bond* is the method of overlapping the masonry units so that the entire wall is a single structural member. *Pattern bond* is the decorative pattern formed by the use of various units in different combinations. The pattern may result from the type of structural bond specified (as an 8" solid brick Flemish bond wall using full brick laid as stretchers and headers), or it may be purely decorative (as a 4" brick-veneer Flemish bond wall using full brick and half-brick).

The *stretcher* or *running bond* is the most popular bond. Since no headers are used, this is often used in single-width walls (veneer and cavity) with metal ties. The *common bond* is a variation of the running bond with a course of headers every fifth, sixth, or seventh course to tie the face wall to the backing masonry. The *English bond* is laid with alternate courses of headers and stretchers, and the *Flemish bond* is laid with stretchers and headers alternating in each course. The *stack bond* is a popular contemporary pattern. Because of the alignment of all vertical joints, reinforcing is needed in the horizontal joints. A masonry wall may be varied by diamond, basket weave, herringbone, and other patterns. Also, brick can be recessed or

Figure 33–33   Hip roof framing. (Courtesy APA— the Engineered Wood Association.)

Figure 33–34   Installing wood siding. (Courtesy APA—the Engineered Wood Association.)

projected for special shadow effects. Decorative variations are endless.

Concrete block lends itself well to contemporary designs. Special effects may be obtained by the bond or by the block itself, which can be specified in many textured and sculptured surfaces.

Stucco is a cement plaster which may be used on exterior walls for special effects.

Combinations of Exterior Finishes

As mentioned in Chapter 27, it is not wise to mix many different kinds of exterior finishes on one house, as this will result in a confusing appearance. Normally no more than two finishes should be used, and the lighter finish should be above the heavier finish. (Notice that the M and the Z residences have wood siding stories over masonry stories.)

INTERIOR FINISH

Interior walls are often of plaster or dry-wall finish. Ceilings may be of the same finish or of ceiling tile.

Plaster

Plaster finishing is considered superior to drywall finish, but it has the following disadvantages:

1. It is more likely to crack.

2. Wet plaster requires many days to dry, during which all construction must be halted.

3. Wood framing is completely soaked with moisture during the drying period and may warp.

Gypsum lath measuring $\frac{3}{8}$" × 16" × 4' (usually referred to as *rocklath*) or 27" × 8' metal lath is nailed to the studs and joists as a base for the plaster. Notice in Figure 33–35 that a gypsum lath base requires strips of metal lath to reinforce the areas most susceptible to cracking: wall and ceiling intersections, the upper corners of door and window openings, and other openings such as electric outlets. Wood grounds, equal in thickness to the lath and plaster, are installed around openings and near the floor. They serve as a leveling guide for the plaster, and act as a nailing base for the finished trim and the baseboard. Steel edges must be used on outside corners to protect the edges from chipping.

Three coats of plaster—a scratch coat (so called because it is scratched to provide a rough bond with the next coat), a brown coat (which is leveled), and a finish coat—are used over metal lath. Two-coat, *double-up* plaster (the

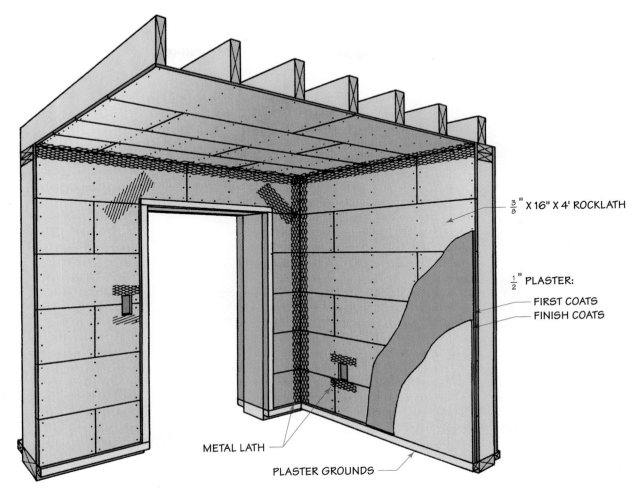

$\frac{3}{8}$" X 16" X 4' ROCKLATH

$\frac{1}{2}$" PLASTER:

FIRST COATS
FINISH COATS

METAL LATH

PLASTER GROUNDS

Figure 33–35   *Plastered finish.*

scratch coat and brown coat combined or *doubled up*) is used over rocklath. The finished coat may be a smooth, white coat that is painted or wallpapered, or it may be a textured coat (called *sand finish*), which usually has the color mixed into the plaster so that no finishing is necessary. A moisture-resistant plaster (called *Keene's cement*) is used in the kitchen and bathrooms. It is also possible to plaster with only a single $\frac{1}{4}$" coat of finish plaster when applied over a special $\frac{1}{2}$" × 4' × 8' gypsum lath.

Drywall

The most common type of drywall material is $\frac{3}{8}$" or $\frac{1}{2}$" gypsum board, as shown in Figure 33–36. When finished, this wall will look just like a plastered wall. Other kinds of drywall finishes, such as $\frac{1}{4}$" plywood panels (with a hardwood veneer) are used for special effects such as a single bedroom wall or fireplace wall. A den may be completely paneled, but take care that paneling is not overdone. For all drywall finishes, it is important that the studs or joists be carefully aligned.

Gypsum Board   Gypsum board consists of a cardboard sandwich with a gypsum filler. It is installed by nailing 4' × 8' sheets directly to the studs or joists, and slightly setting the nailheads. Joint cement then covers the nailheads; joint cement over perforated paper tape is used at the joints. When the joint cement has been sanded, a smooth wall results.

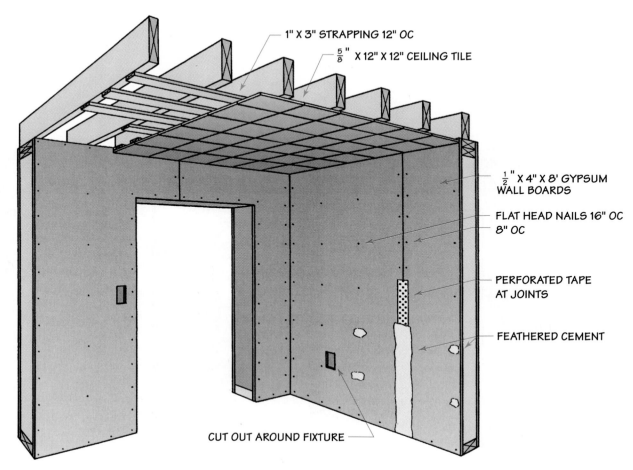

1" X 3" STRAPPING 12" OC

$\frac{5}{8}$" X 12" X 12" CEILING TILE

$\frac{1}{2}$" X 4" X 8' GYPSUM WALL BOARDS

FLAT HEAD NAILS 16" OC
8" OC

PERFORATED TAPE AT JOINTS

FEATHERED CEMENT

CUT OUT AROUND FIXTURE

Figure 33–36   Dry-wall finish.

OTHER CONSTRUCTION TYPES

Platform framing (Figure 33–37), in which framing studs only one-story high rest on a complete platform, has been discussed in detail. Other construction types are **balloon framing** and **plank-and-beam framing**

Balloon Framing

The balloon-framed house is characterized by studs resting directly on the sill and extending the full height of the stories, as shown in Figure 33–38. Second-floor joists rest on a ledger, which is spiked to the studs. The joists are also lapped and spiked to the studs. This type of framing has been largely replaced by platform framing, but balloon framing does have the advantage of lessening vertical shrinkage, and therefore is best for

two-story brick-veneer or stucco construction. However, additional fire stopping of 2" blocking must be provided to prevent air passage from one floor to another. Recently, balloon framing has been used in cold climates to minimize air infiltration, because balloon framing permits the vapor barrier to be installed without gaps at the tops and bottoms of walls.

Plank-and-beam Framing

A building method called *mill* construction has been used for years in factories and warehouses where the loads are heavy and the fire danger high. In mill construction, a few heavy posts and beams support a solid wood floor 3" to 6" thick—since a few large members will resist fire longer than many small members. Although mill construction has been largely replaced by

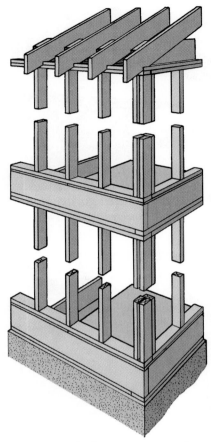

Figure 33–37   Platform construction.

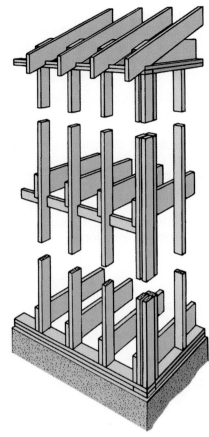

Figure 33–38   Balloon construction.

steel and concrete construction methods, an adaption has been widely used in residential construction. This is called plank-and-beam construction.

In plank-and-beam construction, 2"-thick tongued-and-grooved planks replace the 1" subflooring of conventional framing, 4" × 8" beams 4' to 7' oc replace conventional 2" joists 16" oc, and 4" × 4" posts under the beams replace the conventional 2" × 4" studs.

Plank-and-beam framing in residential construction developed from the trend toward picture windows and window walls, which made it necessary to frame a number of large openings in the exterior walls. Consequently, it is usually used in the construction of modern houses. Other advantages are:

1. A few large structural members replace many small members.

2. Planks and beams are left exposed, eliminating the need for additional interior finish such as plastering and cornices.

3. A saving is made on the total height of the building.

4. Fire hazard is reduced.

Some disadvantages of this type of construction are:

1. Special furring must be used to conceal pipes and electrical conduits installed on the ceiling.

2. Additional roof insulation must be used due to the elimination of dead air spaces between the roof and ceiling.

3. It is more difficult to control condensation. Exhaust fans are used to reduce moisture in the house to a minimum.

Figure 33–39 shows a typical plank-and-beam construction. The large open areas between posts may be used for window walls or may be enclosed with curtain walls. Combinations of plank-and-beam and conventional construction are possible, using the advantageous features of each type.

Theory of Plank-and-beam Construction Compare the deflection of single-span planks and continuous-span planks as shown in Figure 33–43 and notice that the continuous-span will deflect less. In fact, the continuous-span will have about twice the stiffness of the single-span. Of course it is important that the floor be constructed so that it acts as one homogeneous unit. This is done by using tongue-and-groove planks with staggered end joints.

Laminated Beams

Beams consisting of 2" thick laminations glued together under pressure are often used in place of solid lumber. Such glue-laminated beams are called *glulams* (Figures 33–40 and 33–41).

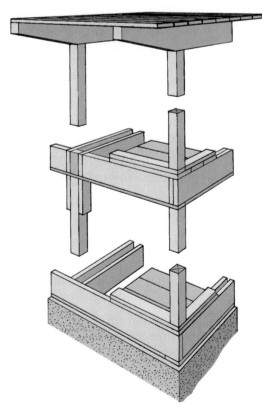

Figure 33–39 *Plank-and-beam construction.*

Figure 33–40 *Laminated garage door header.* (Courtesy APA—the Engineered Wood Association.)

Figure 33–41 *Rafters over glulam posts and beams.* (Courtesy APA—the Engineered Wood Association.)

Metal Framing

Metal framing is commonly used for constructing commercial and industrial buildings, and wood framing for residences. Wood members are widely available, easily fabricated, and economical, but they are combustible and may be weakened by decay or termites. Wood can be pressure-treated with fire-retardant chemicals or decay- and termite-resistant preservatives. Metal framing, however, is superior in many aspects of safety and durability.

Galvanized steel and aluminum components are manufactured for use as studs, sills, headers, joists, bridging, fascia, window framing, door framing, and doors. See Figure 33–42. Components are cut to length using a power saw with a metal cutting blade. Then they are assembled by snap-in clips, bolting, or welding. Stud depths are usually $2\frac{1}{2}$", 4", and 6". Joist

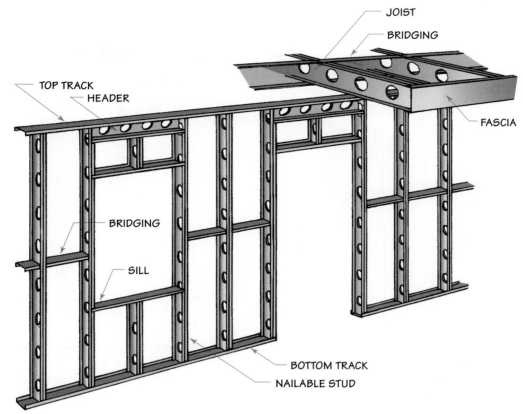

Figure 33–42   Metal framing components.

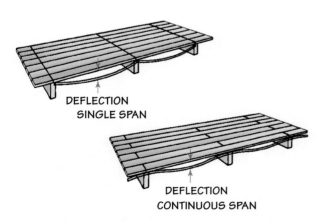

Figure 33–43   Plank-and-beam construction theory.

depths are 6", 8", 9", 10", and 12". Standard lengths of most sections are available up to 32' long, or they can be ordered cut to special lengths at the mill. Wall and ceiling surfaces can

be attached to nailable studs and joists by means of nailing grooves which hold spiral-shank nails tightly in place (Figure 33–44), or they can be attached to screw studs through knurled flanges using power driven self-drilling, self-tapping sheet metal screws (Figure 33–45).

In addition to the obvious advantage of incombustibility, metal framed structures have the advantage of easier piping and electrical installation through prepunched holes. However, grommets must be used to prevent pipe rattling and wire fraying.

FUTURE CONSTRUCTION METHODS AND MATERIALS

Today's methods of building construction are based mostly on lumber and nails put together in a specific way. It certainly does not take much foresight to realize that there will be revolutionary changes in building construction in the next few decades. It is quite possible that

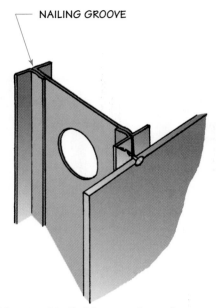

NAILING GROOVE

Figure 33–44 Nailable metal stud.

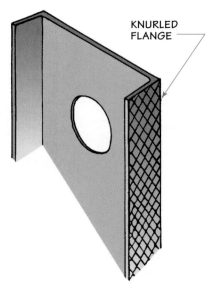

KNURLED FLANGE

Figure 33–45 Screw stud with knurled flanges.

some of these changes will be evident by the time you begin to practice architectural drafting. Already many companies are manufacturing *precut packages* (with lumber cut to the correct size), *panelized* buildings (with entire wall sections already factory assembled), and even entirely mass-produced buildings. Assembly-line methods are not the only answer, however, since most people want custom-built homes. Some companies have made complete breaks from traditional construction, and still more have launched full-scale research and development programs in the housing field.

Most mobile home manufacturers have departed from the traditional design to build *expandable* mobiles (having sections that can be telescoped or folded during transit), *double-wide* mobiles (two mobile units joined at the site), *sectional* mobiles (double-wide mobiles placed on a permanent foundation), and *modular* units (transportable units that can be connected side-by-side, end-to-end, or stacked several stories high in various combinations). The units are transported to the building site by truck and can be erected in a matter of days.

Although we can easily predict changes, it is impossible to predict their exact form and direction, since they will depend upon future engineering research and development. Be alert to these changes and accept them readily. For the present, though, learn all you can about current construction methods and materials.

Figure 33–46 Masonry corbeled for architectural effect. (Courtesy Andersen Windows, Inc.)

REVIEW QUESTIONS

1. What is the frost line?

2. What purpose does the footing serve and why must it be properly leveled after it is poured?

3. A residence built on firm ground has 10"-thick concrete block foundation walls. What would be the probable width and depth of its wall and column footings?

4. Under what conditions would you specify:
   a. Poured concrete foundation walls
   b. Concrete block foundation walls
   c. Concrete blocks filled solid with concrete
   d. Pilasters
   e. Reinforcing rods in footings
   f. Reinforcing rods in slabs
   g. Stepped footings

5. What is the difference between:
   a. Still and sole
   b. Ledger and girt
   c. Girder and joist
   d. Wall and partition
   e. Flooring and subflooring
   f. Rafter and truss
   g. Precut and prefabricated
   h. Header and rough sill

6. Give the normal oc spacing of:
   a. Anchor bolts
   b. Joists
   c. Studs
   d. Roof trusses

7. List the advantages of a built-up wood girder over a solid wood girder.

8. What is the reason for using:
   a. Cross-bridging
   b. Building paper
   c. The *key* between footing and the foundation wall

9. List the six methods used to frame joists to a beam.

10. What length lookouts should be specified for a 1'-8" second-story overhang?

11. When would the following types of cornice be used?
    a. Raised cornice
    b. Cornice with built-in gutter
    c. Cornice with wide overhang
    d. Flush cornice
    e. Plastic or metal cornice

12. Compare plaster and dry-wall finishes by listing the advantages of each.

13. List the advantages and disadvantages of:
    a. Plank-and-beam construction
    b. Metal framing

14. Distinguish between:

a. Expandable mobile homes

b. Double-wide mobiles

c. Sectional mobiles

d. Modular mobiles

ENRICHMENT QUESTIONS

1. Legally, why do you suppose it is a good idea to have a plot of land surveyed before purchasing or staking it?

2. There are many possible materials to use when siding a residence. List five types and discuss advantages and disadvantages of each. Be sure to consider cost.

3. A one-story flat-roof house with full basement has the dimensions of 30' × 40' with three columns spaced 10' oc. Estimate:

a. The total load to be carried by the foundation walls

b. The load on each column

4. For an average-sized one-story platform-framed house give the probable nominal size of:

a. Poured concrete foundation wall

b. Concrete block foundation wall

c. Anchor bolts

d. Sill

e. Header

f. Joists

g. Cross-bridging

h. Subflooring

i. Sole

j. Studs

k. Top plate

l. Sheathing

5. Computer-sketch or pencil-sketch three methods of framing joists level with a wood girder.

PRACTICAL EXERCISES

1. Draw or computer-produce a typical wall section of a one-story house with requirements as assigned by your instructor:

Foundation:

a. With basement

b. With crawl space

c. Without basement, floating slab

d. Without basement, perimeter foundation

Wall:

a. Wood frame

b. Brick veneer

c. Solid brick

d. Cavity brick

Roof:

a. Pitched 6 in 12

b. Pitched 4 in 12

c. Pitched 2 in 12

d. Flat

Cornice:

a. No overhang

b. 8" overhang

c. 4" overhang

d. Raised

e. Built-in gutter

f. Vinyl or aluminum

2. Construct a cutaway model showing the method of building:

a. Platform-framed house

b. Balloon-framed house

c. Plank-and-beam framed house

3. Construct a typical wall section of your original house design.

CHAPTER 34

Structural Calculations

OBJECTIVES

By completing this chapter you will be able to:

- 🏛 Tentatively size the following structural members for a residence or a small commercial building:

  - ■ Wall and column footings

  - ■ Wood and steel columns

  - ■ Wood and steel beams

  - ■ Joists and rafters

  - ■ Headers and lintels

- 🏛 Define terminology related to structural calculating.

KEY TERMS

Tributary area	Steel beams
Footings	Joists
Wood posts	Headers
Pipe columns	Lintels
Wood girders	

In professional offices, architectural design is the responsibility of registered architects, and structural design is the responsibility of registered architectural engineers. However, most architects and engineers appreciate the drafter who has the ability to make simple design and structural decisions independently. When the principal architect or engineer is busy or away from the office, many hours can be wasted if the drafter is not willing to make such decisions and continue working. Of course it is *very* important that the principal be informed of such tentative decisions at a more appropriate time for approval or revision.

STRUCTURAL MEMBERS

The tables in this chapter will enable architectural drafters to size the structural members in the average house or small building. However, for unusual framing methods or loading conditions, these tables will be inadequate (Figure 34–1). For more complete tables than are given in this book, see the *AISC* (American Institute of Steel Construction) *Handbook*, SFPA (Southern Forest Products Association) Span Tables, or *Architectural Graphic Standards*. Each of the

Figure 34–1 *Careful structural calculations are required for non-traditional, as well as traditional designs. (Courtesy of APA—the Engineered Wood Association.)*

following structural members will be considered in turn:

1. Footings:
   Wall footings
   Column footings
2. Columns:
   Wood posts
   Steel columns
3. Beams:
   Wood girders
   Steel beams
4. Joists:
   Floor joists
   Ceiling joists
   Rafters
5. Headers and lintels

Loads

The size and spacing required for any structural member depends upon a combination of elements—the distance spanned, the material used, and the load applied. The total applied load supported by the structural members of a house consists of a live load and a dead load. The live load is the weight supported by the house (furniture, people, wind and snow loads), and the dead load is the weight of the house itself. Although these loads may be fairly accurately calculated using the typical weights of materials given in *Architectural Graphic Standards*, a quicker method is to assume each floor has a live load of 40 lb./sq. ft. and a dead load of 20 lb./sq. ft. Attic live load may be assumed as 20 lb./sq. ft. if used for storage only, and a dead load of 10 lb./sq. ft. if not floored. When interior partitions exist, add an extra 20 lb./sq. ft. to the dead load of the whole floor upon which the partitions rest, remembering that these partitions also transmit any additional weight resting on them. Since roofs usually rest on exterior walls only, they should not normally add to the total girder or column load.

TABLE 34–1 *Residential Live and Dead Loads*

	Live	Dead
Each floor	40 lb./sq. ft.	20 lb./sq. ft.
Attic (storage only)	20 lb./sq. ft.	20 lb./sq. ft.
Attic (not floored)	0	10 lb./sq. ft.
Roof (built-up)	30 lb./sq. ft. (snow and wind load)	20 lb./sq. ft.
Roof (light-shingled)	30 lb./sq. ft. (snow and wind load)	10 lb./sq. ft.
Partitions	0	20 lb./sq. ft.

However, if they are included, add 30 lb./sq. ft. (snow and wind) live load and a dead load of 10 lb./sq. ft. (for light shingles) or 20 lb./sq. ft. (for built-up tar and gravel, slate, or tile).

For convenient reference, these loads are listed in Table 34–1.

Tributary Area

Each structural member supports a certain proportion of the total house weight or area, called the **tributary area**. The word *tributary* refers to the weight *contributed* to each member. Figure 34–2 illustrates the tributary area of a column and a beam of the Z residence. One beam

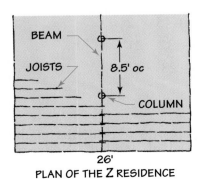

PLAN OF THE Z RESIDENCE

TRIBUTARY AREA OF COLUMN

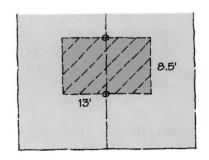

TRIBUTARY AREA OF BEAM

Figure 34–2 *Calculation of tributary areas.*

REAL WORLD APPLICATIONS

Order of Operations

As you look at this chapter, you may feel overwhelmed at the number of complex calculations you will be required to complete and comprehend. It may be a good idea to review a basic math concept: *order of operations*. Calculations must be performed in a specified order or the answers will be incorrect. For instance, you may want to know what your average score for this course is. You have taken three tests on which you obtained scores of 79, 85, and 85. Suppose you enter the following equation into your calculator: $79 + 2 \times 85 \div 3 = ?$. Your result is 2295. This is obviously not your average score. What was wrong with this equation? The way this equation was written instructed your calculator to first add 79 and 2 to obtain 81, and then to multiply 81 times 85 to obtain 6885, before dividing by 3 to obtain 2295.

How could you rewrite the equation to obtain a correct answer? It might look like this:

$$\frac{(2 \times 85) + 79}{3}$$

If you understood and followed the order of operations, you would first simplify within the parentheses by multiplying 2 times 85 to obtain 170, and then adding 79 to obtain 249, before dividing by 3 to obtain the correct average score of 83.

When performing calculations, do them in the following order:

1. First, *calculate any exponents or powers*. For example, you should simplify 3^4 to 81 or 12^2 to 144 before using these numbers in any calculations.

2. Next, *simplify within parentheses*. This means to obtain the result for calculations within parentheses first.

3. *Multiplication and division* calculations should be done as they appear from left to right. For instance, in the equation $(12 \div 3) + (6 \times 4) = ?$, the first calculation is $12 \div 3$. The next computation that should be completed is $6 \times 4 = 24$. These two results should then be added to obtain the final result of 28.

4. As a final step, *addition and subtraction* computations should be performed as they appear from left to right. Rewrite the following equations using parentheses and then calculate them according to the order of operations. Change the position of your parentheses and recalculate. You will notice that you will obtain different results based on the position of the parentheses.

   1. $55 + 4 \times 32 - 6^2 \div 4 =$
   2. $120 \div 4 + 65 \times 3 =$
   3. $121 + 5 - 64 + 45 \times 2^4 =$
   4. $3 \times 65 + 43 + 32 - 21 =$
   5. $65 + 75 \times 3 - 23 \div 6 =$

running the length of the house normally has a tributary area equal to half the total floor plan area. When columns are used, the tributary area is reduced in proportion to the reduction of the beam span. If joists run uncut across the beam, the beam supports five-eighths of the weight above instead of one-half. Notice in Table 34–1 that values are given in pounds per square foot. Therefore, the live and dead loads must be multiplied by the tributary square foot area to determine the total load supported by a member.

FOOTINGS

Using Tables 34–1, 34–2, and 34–3, it is possible to calculate the required sizes of **footings** for foundation walls, columns, and chimneys. The procedure is to calculate the total weight to be contributed to the footing using the Live and Dead Load table and the Weights of Materials table. When calculating tributary areas, remember it is the usual practice to support the roof by exterior walls, not interior partitions. The pressure

TABLE 34–2   *Weights of Materials*

Type of Material	Pounds per Cu. Ft.
Poured concrete	150
Concrete block, (including mortar)	85
Brick (including mortar)	120

TABLE 34–3   *Safe Ground Loadings*

Type of Ground	Pounds per Sq. Ft.
Ledge rock	30,000
Hardpan	20,000
Compact gravel	12,000
Loose gravel	8,000
Coarse sand	6,000
Fine sand	4,000
Stiff clay	8,000
Medium clay	4,000
Soft clay	2,000

of the footing on the ground may not be greater than the safe ground loadings given in Table 34–3.

Calculations for the footings of the Z residence will be used as an example.

FOUNDATION FOOTING CALCULATIONS
Find the load per running foot on the foundation footing. In Figure 34–3 notice that the live and dead loads have been multiplied by half the joist spans to obtain the total load.

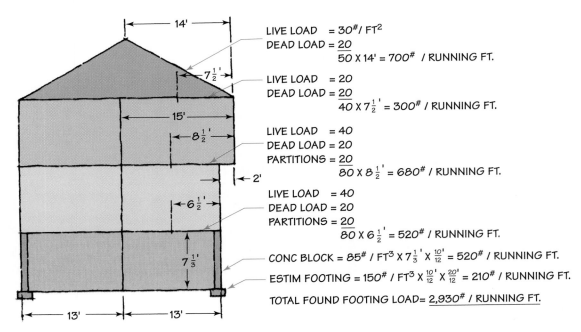

LIVE LOAD = 30#/ FT²
DEAD LOAD = 20
$\overline{50}$ X 14' = 700# / RUNNING FT.

LIVE LOAD = 20
DEAD LOAD = 20
$\overline{40}$ X 7½' = 300# / RUNNING FT.

LIVE LOAD = 40
DEAD LOAD = 20
PARTITIONS = 20
$\overline{80}$ X 8½' = 680# / RUNNING FT.

LIVE LOAD = 40
DEAD LOAD = 20
PARTITIONS = 20
$\overline{80}$ X 6½' = 520# / RUNNING FT.

CONC BLOCK = 85# / FT³ X 7⅓' X $\frac{10}{12}$ = 520# / RUNNING FT.

ESTIM FOOTING = 150# / FT³ X $\frac{10}{12}$ X $\frac{20}{12}$ = 210# / RUNNING FT.

TOTAL FOUND FOOTING LOAD= 2,930# / RUNNING FT.

USING A 20" FOOTING, THE LOAD PER SQ. FT = 2930 X $\frac{12}{20}$ = 1,760# / FT²

NOTICE THIS IS A SAFE LOAD FOR ANY TYPE OF GROUND LISTED IN TABLE 34-3.

Figure 34–3   Foundation footing calculations.

Figure 34–4   *Structural columns as a major design element.* (*Courtesy Scholz Design, Toledo, Ohio.*)

COLUMN FOOTING CALCULATIONS

Find the total load carried by a column footing. In Figure 34–5 notice that the average load per square foot has been multiplied by the tributary area to obtain the total load.

WOOD POSTS

Structural framing of a small house might consist of **wood posts** (usually 6" × 6" or 8" × 8") supporting the center girder (usually 8" × 8" or three 2" × 10"s spiked together), which in turn supports the floor joists. Table 34–4 is used to determine the post size required.

EXAMPLE

The load on each post of the Z residence was calculated in Figure 34–5 to be 22.1 kips (1 kip = 1,000 lb.). If 7'-long Douglas fir posts are to be used, a 6" × 6" size will be satisfactory, since they will safely support 29.0 kips.

STEEL PIPE COLUMNS

Wood post and girder framing is not practical for larger houses since the posts might have to be spaced so close together that use of the basement is restricted. Furthermore, deeper timbers for girders are too costly and either add to the house cubage or reduce headroom.

The solution, then, is to use steel columns and steel girders—as are specified in many building codes. Because steel is so much stronger than wood, steel columns can be spaced farther apart, giving more free basement area, and steel girders can be shallower than wood girders, saving on the cubic content or giving more basement headroom. W shapes, S shapes, and even channels are used for industrial building columns; round, hollow piping called **pipe columns** are used for houses. The hollow center of the pipe column may be filled with concrete to further increase its strength. Notice that a cap and base plate must be used (see Figure 33–14, page 406)).

The column diameter can be obtained by referring to a table published by the Lally Company, a popular manufacturer of pipe columns. A portion of this table is shown as Table 34–5.

EXAMPLE

In the Z residence, the previously calculated load of 22.1 kips on a 7'-long column will be amply supported by a $3\frac{1}{2}$" Lally column, since it will safely support 35.1 kips.

TABLE 34–4   *Safe Loads for Wood Posts (in kips)*

Lumber	Size	Unbraced Length			
		6'	7'	8'	9'
Spruce and pine*	6" × 6" ($5\frac{1}{2}$" × $5\frac{1}{2}$")	16.7	16.4	15.9	15.3
	8" × 8" ($7\frac{1}{2}$" × $7\frac{1}{2}$")	31.5	31.3	31.0	30.6
Douglas Fir**	6" × 6" ($5\frac{1}{2}$" × $5\frac{1}{2}$")	29.5	29.0	28.1	26.8
	8" × 8" ($7\frac{1}{2}$" × $7\frac{1}{2}$")	56.2	55.6	55.0	54.3

Adapted from *Light Frame House Construction*, U.S. Department of Health, Education and Welfare.

 * Red, White, and Sitka Spruce; the White Pines, No. 1 Common.

** Douglas Fir, Southern Pine, and North Carolina Pine, No. 1 Common.

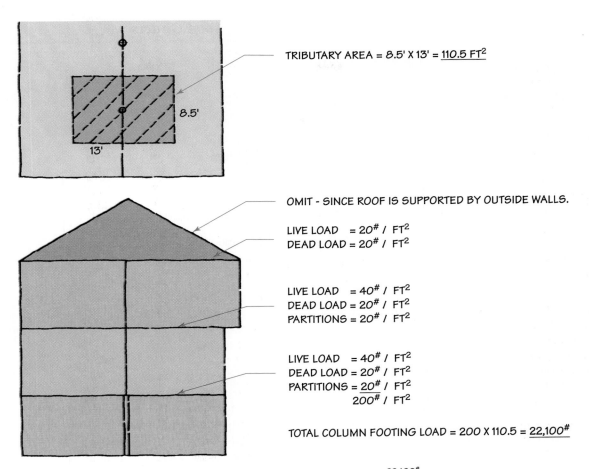

TRIBUTARY AREA = 8.5' X 13' = 110.5 FT²

8.5'

13'

OMIT - SINCE ROOF IS SUPPORTED BY OUTSIDE WALLS.

LIVE LOAD = 20# / FT²
DEAD LOAD = 20# / FT²

LIVE LOAD = 40# / FT²
DEAD LOAD = 20# / FT²
PARTITIONS = 20# / FT²

LIVE LOAD = 40# / FT²
DEAD LOAD = 20# / FT²
PARTITIONS = 20# / FT²
200# / FT²

TOTAL COLUMN FOOTING LOAD = 200 X 110.5 = 22,100#

IF GROUND IS MEDIUM-STIFF CLAY, COLUMN FOOTING AREA MUST BE $\frac{22,100^\#}{6,000^\#/ FT^2}$ = 3.7 FT²
∴ 20" SQ (2.8 FT²) IS TOO SMALL, BUT 24" SQ (4 FT²) IS O.K.

Figure 34–5 Column footing calculations.

TABLE 34–5 *Safe Loads for Heavyweight Columns (in kips)*

Column Diameter	Unbraced Length			
	6'	7'	8'	9'
3½"	37.9	35.1	32.3	29.4
4"	49.2	46.1	43.1	40.1
4½"	61.8	58.5	55.3	52.0

WOOD GIRDERS

Wood girders built up of 2" lumber spiked together or solid wood beams are often used in combination with steel columns. Table 34–6

may be used to determine the safe span of various sizes of wood girders.

EXAMPLE
For the Z residence, the structure width is approximately 26', and the building is two-story supporting a bearing partition. If we wish to space columns 8'-6" oc, we must use a wood girder greater than 6" × 12", because Table 34–6 shows that a 6" × 12" girder will safely span only 6'-2".

STEEL BEAMS

Steel beams of various sizes are especially useful when spanning distances greater than 10'. Steel beams are obtained cut to length (up to 60') and in various weights (an 8" W-shape beam weighing 20 lb./ft. will support more than an 8" W-shape beam weighing 17 lb./ft.). Only

TABLE 34–6   *Safe Spans* for Wood Girders*

Structure Type	Girder Size (solid or built-up)	22'–24'	26'–28'	30'–32'
			Structure Width	
One-story supporting	6" × 8"	8'-1"	7'-5"	6'-6"
Nonbearing partitions	6" × 10"	10'-4"	9'-6"	8'-4"
	6" × 12"	12'-7"	11'-7"	10'-2"
Two-story supporting	6" × 8"	4'-7"	—	—
Bearing partition	6" × 10"	6'-0	5'-1"	4'-6"
	6" × 12"	7'-2"	6'-2"	5'-5"

Adapted from *Manual of Acceptable Practices*, U.S. Department of Housing and Urban Development.
* Based upon allowable fiber stress of 1,500 psi (such as Douglas Fir or Southern Pine)

the more popular sizes stocked by suppliers are shown in Table 34–7.

EXAMPLE

For the Z residence, the desired span is 8'-6", and the load to be supported is found as shown in Figure 34–6 to be 22.1 kips. Reading between the 8' and 9' columns in Table 34-7, it will be seen that the W 8 × 17 beam will just support this weight. To be on the safe side, specify W 8 × 20, which will support approximately 26 kips.

JOISTS

The sizes of floor, ceiling, and roof **joists** and rafters are determined by Tables 34–8 through 34–10. Notice that you must know the

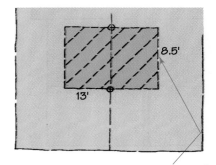

TRIBUTARY AREA = 8.5' X 13' = 110.5 FT²
UNIT LIVE & DEAD LOAD = 200# / FT²
TOTAL BEAM LOAD = 200 X 110.5 = 22,100# = <u>22.1 KIPS</u>

Figure 34–6   Steel beam calculations.

TABLE 34–7   *Safe Loads for Steel Beams (in kips)*

Size*	8'	9'	10'	11'	12'	13'	14'
				Span			
S 7 × 15.3	17.3	15.4	13.9	12.6			
W 8 × 17	24	21	18.8	17.1	15.7		
W 8 × 20	28	25	22.6	20.6	18.9	17.4	
W 10 × 21	36	32	29	26	24	22	21
W 10 × 25	44	39	35	32	29	27	25

Adapted from FHA standards. For more complete information, see *Minimum Property Standards for One and Two Living Units*, Federal Housing Administration.
* Note: S 7 × 15.3 refers to an S-shape I beam 7" high weighing 15.3 pounds per foot of length; W 8 × 17 refers to a W-shape beam 8" high weighing 17 pounds per foot of length.

span, spacing, and material. A spacing of 16" oc is more commonly used than 12" oc or 24" oc.

EXAMPLE

For the Z residence, No. 2 Southern Pine floor joists 16" oc are to span 13'. From Table 34–8, we see that 2" × 8" joists will span only 12'-4", and 2" × 10" joists will span 14'-8". Therefore, 2" × 10" should be used for first-floor joists, but 2" × 8" will be satisfactory for second-floor joists because of the slightly smaller second-floor load.

Ceiling joists over the master bedroom must span 15'. From Table 34–9, we see that 2" × 6" joists will span only 13'-6", and 2" × 8" joists will span 17'-5". Therefore use 2" × 8" ceiling joists 16" oc.

Rafters must span 14'. From Table 34–10, we see that 2" × 8" rafters will span 16'-2". Therefore, use 2" × 8" rafters 16" oc or 2" × 6" rafters of a higher grade than shown in the table.

LAMINATED DECKING

Laminated wood decking from $2\frac{1}{4}$"–$3\frac{3}{4}$" thick is often used for floors and roofs. Table 34–11 gives the maximum allowable uniformly

TABLE 34–8 *Floor Joist Sizes (40 psf live load; 20 psf dead load)*

Size	Spacing (oc)	Maximum Span		
		No. 1 Southern Pine	No. 2 Southern Pine	No. 3 Southern Pine
2" × 6"	16"	9'-11"	9'-6"	7'-5"
	12"	10'-11"	10'-9"	8'-6"
2" × 8"	16"	13'-1"	12'-4"	9'-5"
	12"	14'-5"	14'-2"	10'-10"
2" × 10"	16"	16'-4"	14'-8"	11'-1"
	12"	18'-5"	17'-0	12'-10"
2" × 12"	16"	19'-6"	17'-2"	13'-2"
	12"	22'-5"	19'-10"	15'-3"

TABLE 34–9 *Ceiling Joist Sizes (drywall ceiling, limited attic storage, 20 psf live load, 10 psf dead load)*

Size	Spacing (oc)	Maximum Span		
		No. 1 Southern Pine	No. 2 Southern Pine	No. 3 Southern Pine
2" × 4"	24"	8'-0	7'-8"	5'-9"
	16"	9'-1"	8'-11"	7'-1"
2" × 6"	24"	12'-6"	11'-0	8'-6"
	16"	14'-4"	13'-6"	10'-5"
2" × 8"	24"	15'-11"	14'-2"	10'-10"
	16"	18'-11"	17'-5"	13'-3"
2" × 10"	24"	18'-11"	17'-0	12'-10"
	16"	23'-2"	20'-9"	15'-8"

TABLE 34–10 *Rafter Sizes (drywall ceiling, light roofing, moderate snow load, 30 psf live load, 10 psf dead load)*

Size	Spacing (oc)	Maximum Span		
		No. 1 Southern Pine	No. 2 Southern Pine	No. 3 Southern Pine
2" × 6"	24"	10'-11"	10'-2"	7'-11"
	16"	12'-6"	12'-3"	9'-8"
2" × 8"	24"	14'-5"	13'-2"	10'-1"
	16"	16'-6"	16'-2"	12'-4"
2" × 10"	24"	17'-6"	15'-9"	11'-11"
	16"	21'-1"	19'-3"	14'-7"

Tables 34–8 through 34–10 adapted from *Southern Pine Maximum Spans for Joists and Rafters* of the Southern Forest Products Association, 1995.

TABLE 34–11 *Allowable Loads for Laminated Decking*

Finished Thickness*	Span	Allowable Uniformly Distributed Total Roof Load** (lb./sq. ft.)	
		Ponderosa Pine	Hem-fir
$2\frac{1}{4}$"	8'	101	142
	10'	52	72
	12'	30	42
3"	12'	71	100
	14'	45	63
	16'	30	42
$3\frac{3}{4}$"	16'	61	82
	18'	43	57
	20'	31	42

Adapted from *Koppers Company Design Manual.*
*Laminated decking sizes and load-carrying capacities may vary between manufacturers.
**Based upon controlled random layup continuous over at least three spans. Deflection limited to 1/240.

distributed load for floors or roofs up to 3-12 pitch.

EXAMPLE
A plank-and-beam building is designed on a 12' module. If the total live and dead roof load is determined to be 50 lb./sq. ft., a 3"-thick laminated Ponderosa pine roof deck is required, since a $2\frac{1}{4}$" deck will support only 30 lb./sq. ft.

ROOFS

It is good practice to use the exterior walls, and not interior partitions, for roof support. If the span between exterior walls is so great that ordinary rafters cannot be used, either wood trusses or steel beams are employed. Probably, a sloping roof would call for wood trusses, and a flat roof would call for steel open-web joists.

HEADERS

Large openings for windows and doors are spanned by wood or steel structural members. Wood members are called **headers**, and steel members are called **lintels**. Table 34–12 gives the maximum safe span for wood headers in light frame construction. For unusual loading conditions, special design is necessary.

EXAMPLE
Several 8'-wide window openings in the Z residence will require two 2" × 12" headers, since two 2" × 10" headers will span only 6'-6" safely.

TABLE 34–12   *Safe Spans for Headers*

Size	Span
Two 2" × 4" with plywood spacing	2'-6"
Two 2" × 6" " " "	3'-6"
Two 2" × 8" " " "	5'-0
Two 2" × 10" " " "	6'-6"
Two 2" × 12" " " "	8'-0

LINTELS

To span openings in masonry walls, wood headers are not satisfactory because wood shrinkage will cause cracks in the masonry above. Instead, steel angle lintels are used as shown in Figure 34–7. The $3\frac{1}{2}$" horizontal leg provides support for a $3\frac{3}{4}$"-wide brick, and the vertical leg provides the resistance to bending. Table 34–13 may be used to find the lintel size required in a simple masonry wall. If joists or other members frame into the wall above the lintel, their contributing weight must also be supported, and a larger size of lintel will be necessary.

EXAMPLE

In the Z residence, the angles needed to span the 3'-4" and 4'-0 fireplace openings are found to be 4" × $3\frac{1}{2}$" × $\frac{5}{16}$".

VENEER   10" CAVITY   8" SOLID   12" SOLID

Figure 34–7   Use of angle lintels in brick wall openings.

TABLE 34–13   *Safe Spans for Lintels in 4" Masonry ($3\frac{1}{2}$" leg horizontal)*

Size	Span
$3\frac{1}{2}$" × $3\frac{1}{2}$" × $\frac{1}{4}$"	3'
4" × $3\frac{1}{2}$" × $\frac{5}{16}$"	5'
5" × $3\frac{1}{2}$" × $\frac{5}{16}$"	6'

GENERAL EXAMPLE

Mr. and Mrs. X decide that they want a full basement under their 24' × 48' flat-roofed house. Design one girder running the length of the basement down its center, giving 12' floor joist spans. Find the size of the girder and the number and size of the columns.

Loads

First calculate the total live and dead load acting on the girder, remembering that in this plan the roof must be supported by interior partitions.

First-floor live load	40 lb./sq. ft.
First-floor dead load	20 lb./sq. ft.
First-floor partitions	20 lb./sq. ft.
Roof live load	30 lb./sq. ft.
Roof dead load	20 lb./sq. ft.
Unit live and dead load	130 lb./sq. ft.

The tributary area is 12' × 48' = 576 sq. ft. Therefore the total live and dead load acting on the beam is 130 lb./sq. ft. × 576 sq. ft. = 74,880 lb. = 75 kips.

Beam

Now determine the number of columns and size of the beam by trial and error, as shown in Figure 34–9 and the accompanying table.

The first three trials are no good, but the last three are possible. The final selection might be influenced by the future location of basement partitions or the difference in cost. (Since steel is sold by the pound, the lighter-weight beams should be compared with the cost of extra columns.)

Column

If the four-column solution is selected, the load on each column will be the tributary weight plus the weight of the girder:

$$\text{Column load} = 15 \text{ kips} + (10 \times 17 \text{ lb.})$$
$$= 15.2 \text{ kips}$$

Using Table 34–5, the $3\frac{1}{2}$"-diameter pipe column will be more than sufficient, and any length may be chosen ($6\frac{1}{2}$'-7' is common).

The footing, joist, and header sizes are calculated in the manner previously shown.

COMPUTER-AIDED STRUCTURAL PLANS

Specialized software programs are available for calculating and generating structural plans, elevations, sections, and pictorial illustrations. Softdesk's *Structural Plans & Elevations* (requires Softdesk® *Core* and *Building Base* to operate) is one example of such a program (Figures 34–10 and 34–11). Available construction

CAREER PROFILES

Architectural Engineer

Architectural engineers design and supervise construction of the engineering aspects of buildings, including structural, electrical, heating, and plumbing. An architectural engineer may accept employment as an architectural firm's engineer, supervise maintenance of a large structure, or establish private practice as a professional engineer.

Baccalaureate architectural engineering programs, like architecture, are five-year university programs. Typical courses include:

- 🏛 Architectural graphics
- 🏛 Computer applications
- 🏛 Statics
- 🏛 Dynamics
- 🏛 Strength of materials
- 🏛 Thermodynamics
- 🏛 Architectural building materials
- 🏛 Architectural design (three years)
- 🏛 Structural design

- 🏛 Electrical distribution and illumination
- 🏛 Plumbing and fire protection
- 🏛 Heating, ventilating, and air conditioning
- 🏛 Architectural acoustics
- 🏛 Environmental systems
- 🏛 Industrial relations
- 🏛 Professional conduct
- 🏛 Ethics

A two-year master's program or doctorate program in architectural engineering usually emphasizes advanced structural design, environmental systems, or construction management.

High-quality baccalaureate and master's programs in architectural engineering are approved by the Accreditation Board of Engineering and Technology (ABET). Currently, thirteen universities in the United States offer ABET-accredited majors. A list can be obtained from ABET, 111 Marketplace Suite 1050, Baltimore, Maryland, 21202. Phone: 410-347-7700. Additional information about ABET can be obtained through their Internet homepage at the following address: **http://www.abet.ba.md.us/**

To become registered as a professional architectural engineer, most states require graduation from an ABET-accredited program plus three years of varied experiences under the supervision of a registered architectural engineer and the successful completion of a comprehensive written examination. Lacking a degree, most states accept an additional five years of experience as a substitute.

materials include steel (rolled shapes and bar joists), concrete (poured and precast), timber, and glulams. Available dimensioning includes English and metric.

Figure 34–8 *A contemporary residence featuring laminated posts and beams. (Courtesy APA— the Engineered Wood Association.)*

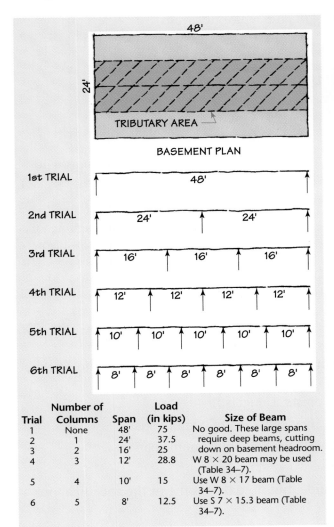

Figure 34–9 *Calculations for the X family.*

Trial	Number of Columns	Span	Load (in kips)	Size of Beam
1	None	48'	75	No good. These large spans require deep beams, cutting down on basement headroom.
2	1	24'	37.5	
3	2	16'	25	
4	3	12'	28.8	W 8 × 20 beam may be used (Table 34–7).
5	4	10'	15	Use W 8 × 17 beam (Table 34–7).
6	5	8'	12.5	Use S 7 × 15.3 beam (Table 34–7).

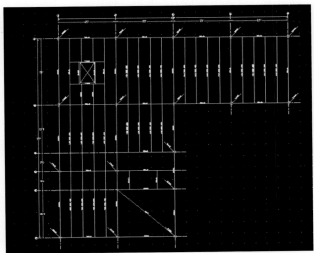

Figure 34–10 *Computer-produced structural steel plan for the commercial building shown in Figure 34–11. (Courtesy Softdesk, Inc.)*

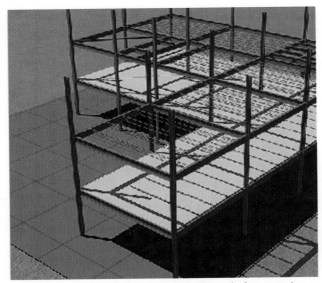

Figure 34–11 *Computer-produced three-point perspective of the structural steel framing for a commercial building. (Courtesy Softdesk, Inc.)*

REVIEW QUESTIONS

1. Name three factors that affect the size and spacing of structural members.
2. Distinguish between:
   a. Live load and dead load
   b. Wood beam, wood girder, and wood post
   c. Header and lintel
   d. Kip and ton
3. Give two methods used to determine dead loads.
4. What is meant by *tributary area*?
5. How much does a 20'-long, W 10 × 25.4 beam weigh?
6. What size header should be used for a 4' opening in a frame wall?
7. What size lintel should be used for a 4' opening in an 8" masonry wall?

ENRICHMENT QUESTIONS

1. Why are lintels used in place of headers to support masonry construction?
2. What size of (a) steel column and (b) Southern pine post would be specified to support a 40-kip tributary load if the unbraced length of column is 7'-6"?
3. What size steel beam would be specified to support a 25-kip tributary load over a 10' span?
4. What size No. 2 Southern pine floor joists would be specified to span 13'-6" in a residence if the joist spacing is to be 16" oc?

PRACTICAL EXERCISES

1. Calculate the minimum safe sizes of the structural members of the Z residence assuming the following specifications:
   a. Wall footing: medium clay ground
   b. Column footing: medium clay ground
   c. Steel column: 6'-8" unbraced length
   d. Wood girder: Douglas fir
   e. Joists: Douglas fir
   f. Headers: 7'-6" maximum span
   g. Lintels: 4'-4" maximum span

   (Use the examples given in this chapter as a guide, but note that some of the specifications have been changed.)
2. Calculate the sizes of the structural members of your original building design.

OBJECTIVES

By completing this chapter you will be able to:

🏛 Explain characteristics and problems associated with moisture control, insect infestation, decay, and natural disasters.

KEY TERMS

Water vapor	Carpenter ants
Vapor pressure	Powderpost beetles
Condensation	Deathwatch beetles
Vapor retarder	Decay
Ventilation	Corrosion
Ice dam	Galvanic corrosion
Termites	Radon

MOISTURE CONTROL

Two unseen elements play an important part in the life and livability of houses—temperature and moisture. Temperature is relatively easy to control through the use of adequate insulation and a properly designed and functioning heating system. Moisture also can be easily controlled through proper building design and construction (Figure 35–1). However, since the control of moisture is so little understood, moisture damage is undoubtedly the most prevalent of all problems connected with the home.

Water Vapor

Water vapor, moisture, humidity, and steam are different names given to water when it has evaporated to a gas state. Water vapor is invisible, but it is present to some degree in nearly all air. Two factors associated with water vapor should be clearly understood: vapor pressure and condensation.

Vapor pressure The water vapor in wet air always tries to flow toward drier air to mix with it. This vapor pressure causes the water vapor in a house to seek escape to the drier air outside, traveling through walls and roof (Figure 35–2). Water vapor is a gas, like air, and can move wherever air can move. It is not generally

Figure 35–1 *Ventilating skylights control temperature and humidity. (Courtesy Velux-America, Inc.)*

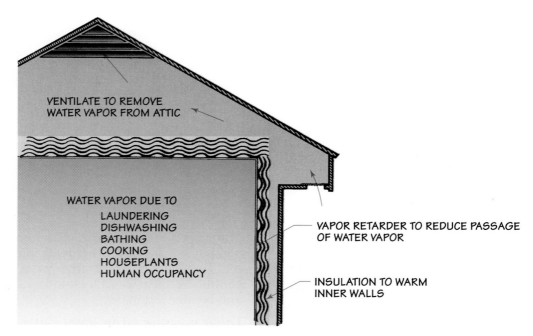

Figure 35–2 *Water vapor protection.*

understood that water vapor can travel through materials that air cannot readily penetrate, such as wood, brick, stone, concrete, and plaster. Under the vapor pressure of warm, moist air, the vapor constantly tries to escape through most building materials to the cooler outside where the pressure is lower.

The extent of this problem may be realized when you consider the numerous sources of water vapor in a home: laundering, dishwashing, bathing, cooking, house plants, and human occupancy. All this may amount to 20 gal. of water per week which enter the air and must escape from the home.

Condensation

When warm, moist air is cooled, as happens when it comes in contact with a cold surface, the water returns to a liquid state, or *condenses*. Figure 35–3 shows the **condensation** of water vapor into droplets of liquid water upon contact with the cold surface of a glass of iced tea. The same condensation will occur when water vapor, under pressure to escape from a warm house, comes in contact with cooler surfaces in attics, crawl spaces, and wall interiors. This

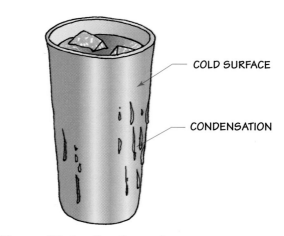

Figure 35–3 *Condensation.*

trapped condensation may cause mildewing and decay of structural members; damage to plaster, insulation, and roofing; blistering and peeling of paint; window sweating; musty odors; and many other problems.

Comfort

Although some people believe that high humidity is beneficial to health, this question is unsettled and probably academic. Florida is a health resort and has a high humidity; Arizona, too, is a health resort and has low humidity. Neither state shows superior advantages for all people. Moisture can cause more damage than its slight comfort benefits can offset. It is best to get rid of excess moisture in order to protect clothing, tools, and leather goods as well as the building itself.

Control

There are two principal ways of controlling water vapor and preventing moisture damage. One is to stop the passage of vapor through the structure so that it cannot reach a surface cold enough to cause condensation. This can be done by the use of *vapor retarders* on the inside (warm side) of walls. The second is to *ventilate* the outer walls and ceilings so that any water vapor reaching the outer surfaces of the structure can continue flowing to the colder outside air.

Vapor Retarders

A **vapor retarder** is any material that resists the passage of water vapor. Among the commonly used vapor retarders are:

1. Membrane vapor retarders which cover the entire wall and ceiling with openings only at the windows, doors, and electric boxes

2. Blanket insulation containing a retarder on one side

3. Aluminum-foil reflective insulation

4. Aluminum primer under wall paint

5. Asphalt coating on the back of interior wall surfaces

In every case, the vapor retarder must be kept warm by installation on the warm side of the wall. If it is installed on the cold side, condensation will form immediately. Storm windows and double-glazed windows operate on the same principle—condensation on the window is reduced in proportion to the temperature of the inside glass. Also, since no vapor retarder will completely prevent the passage of all water vapor, it is important that any penetrating vapor be permitted to continue flowing through the cold side of the wall to the outside air. This is the reason that building paper is made water-resistant but not vapor-resistant.

Ventilation   Although little can be done to reduce the amount of water vapor poured into a house, much of this vapor can be removed by **ventilation** in one or more of the following areas:

1. Openings at the eave and ridge of the roof. The ratios given in Figure 35–4 can be used to calculate the minimum net ventilation area required.

2. Crawl space ventilation together with a ground covering of waterproofed concrete or vapor barrier to reduce moisture entering from the ground (Figure 35–5).

3. Exhaust fans in kitchen, bathrooms, laundry, and basement

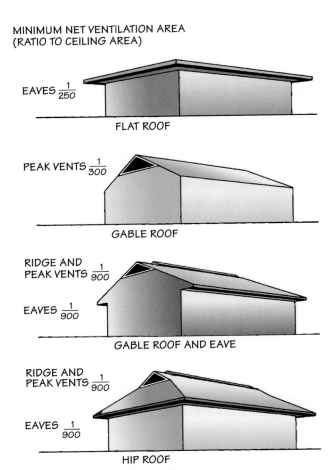

Figure 35–4   *Roof ventilation.*

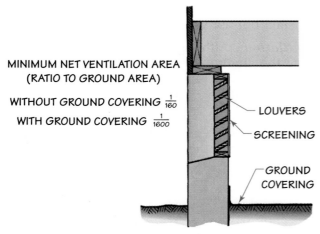

MINIMUM NET VENTILATION AREA
(RATIO TO GROUND AREA)

WITHOUT GROUND COVERING $\frac{1}{160}$

WITH GROUND COVERING $\frac{1}{1600}$

LOUVERS

SCREENING

GROUND COVERING

Figure 35–5 *Crawl space ventilation.*

4. Venting of gas appliances, since water vapor is a product of gas combustion

5. Outside cold-air intake on hot-air furnace, damper controlled

EXAMPLE

Find the size of ventilators for a 25' × 36' gable-roofed home with flush cornice.

Solution

1. Calculate the ceiling area: 25' × 36' = 900 sq. ft.

2. Using Figure 35–4, a gable house with a flush cornice, calculate the minimum net ventilation area:
$$\frac{1}{300} \times 900 = 3 \text{ sq. ft.}$$

3. Double this area due to louvers, reducing the actual opening area: 3 sq. ft. × 2 = 6 sq. ft.

4. Since two ventilators are used, each must be 3 sq. ft. minimum.

Preventing Ice Dams

An **ice dam** is a buildup of ice on the roof edge and gutter caused by freezing of roof runoff. This is undesirable because further runoff puddles on the roof and then leaks into the building. An ice dam is caused by inadequate roof insulation (causing snow on the roof to melt and run to the colder eave, where it refreezes) or by inadequate attic ventilation (allowing the attic to build up heat that again causes roof snow to melt). Inadequate attic ventilation is often caused by insulation installed at the attic's edge, interrupting air flow from under the roof eaves to the peak vents.

TERMITE CONTROL

Nature has provided insects whose primary function is to accelerate the reduction to dust of dead wood on the forest floor (or on a house). These insects, known as **termites**, are of two kinds: subterranean termites located in the southern part of the United States as shown in Figure 35–6, and dry-wood termites located only in southern California and southern Florida.

Dry-wood termites can live in wood without having contact with moisture or the ground. Fortunately though, they are fewer in number and less of a threat than the subterranean termites. Subterranean termites breed underground and then tunnel through the earth for the wood they need for food. They will even construct tunnels *above* ground, as shown in Figure 35–7, to reach wood within 18" of the

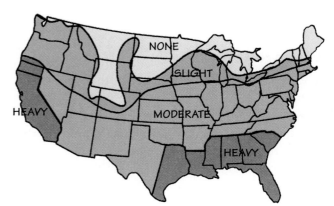

Figure 35–6 *Areas of termite danger.*

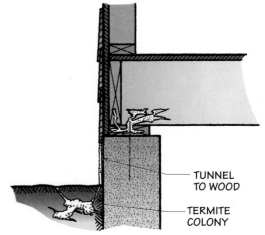

TUNNEL TO WOOD

TERMITE COLONY

Figure 35–7 *Termite damage.*

ground. In areas of termite hazard, the following precautions must be taken in building:

1. Keep wood members 18" above ground. Since this is not always possible, termite shields of 16-oz. copper (Figure 35–8) can be installed over the foundation walls and under the sill.

2. Solid concrete walls with reinforcing to prevent cracks are necessary, since termites are soft-bodied and able to squeeze through a paper-thin fissure. They may even tunnel through low-grade concrete.

3. If wood must be in contact with the ground (as in the case of wooden steps), use a concrete base under the wooden member, use the heartwood of redwood or cypress, and pressure-treat the lumber with chromated copper arsenate. **Warning:** Recent studies have called for a phase-out of arsenic-treated wood due to its toxic hazard. Decks that have been treated in this manner should be sealed and painted to reduce soil contamination.

4. Properly drain the area surrounding the house since subterranean termites need moisture to live.

5. Remove all dead wood from the building vicinity and treat the ground with antitermite chemicals.

6. Make sure that all areas such as crawl spaces are accessible for periodic inspections.

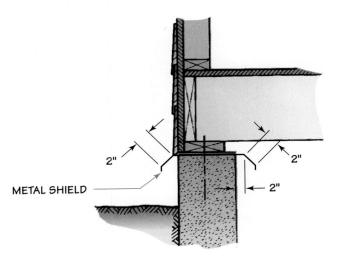

Figure 35–8 *Termite shield.*

7. Screen all openings (such as attic ventilation) with 20-mesh bronze screening.

8. Avoid rigid foam insulation in contact with the ground as it provides access which cannot be prevented with standard soil treatment.

CARPENTER ANT CONTROL

In addition to termites, **carpenter ants** also live in wood. Unlike termites, carpenter ants do not eat wood but merely chew enough room for nests. They may colonize and expand termite nests after the termites have been exterminated, or they may nest in existing cracks in beams or hollow-core exterior doors. Carpenter ants usually seek out moist wood that has begun to decay. Since they fly, they will chew nests at any height, giving an indication of a serious moisture problem in a building. Therefore, carpenter ants are best controlled by proper roofing and flashing to prevent water leaks. Other precautions are similar to those described above.

Additional wood-infesting insects are **powderpost beetles** and **deathwatch beetles**. Powderpost beetles feed in very dry wood, reducing it to powder. They are attracted to plywood and hardwood flooring. Deathwatch beetles prefer moist lumber, but their damage is the same. They are attracted to damp unventilated crawl spaces.

DECAY PREVENTION

Nature has also provided a low form of plant life, called fungus, which, like termites, serves to disintegrate wood. Fungus may be of a type that merely produces mold or stains (unsightly but not structurally dangerous), or of a type that produces actual **decay**.

Major conditions necessary for decay are:

1. Moisture in a moderate amount. Dry wood (less than 20 percent moisture content) and wood kept under water will not decay. Even so-called dry rot requires moisture.

2. Moderately high temperature (70°F.–85°F.). High temperature such as used in kiln drying will kill fungus, but low temperature merely retards it.

Since the conditions necessary for the survival of fungus are similar to those needed by termites, the preventive measures are fortunately the same.

1. Protect lumber from wetting by using wide cornice overhangs, proper construction and flashing, painting, and keeping all wooden members at least 8" above ground.

2. Use the heartwood of redwood, cedar, and cypress, pressure-treated with chromated copper arsenate. The heartwood of all lumber is more resistant to decay than the sapwood. Heartwood of redwood, cedar, and cypress is especially resistant to decay.

3. Make sure there is proper drainage of the ground surrounding the building. A vapor barrier, waterproofed concrete, or sand spread over crawl areas reduces the ground moisture evaporation.

4. Remove all dead wood from the area of the building since fungi spores float through the air from decayed wood.

5. Ventilate enclosed spaces.

CORROSION PREVENTION

Whereas wood is disintegrated by termites and decay, metal is disintegrated by rusting (**corrosion**). Rather than attempting to remove the conditions that cause corrosion, such as moisture, the best prevention is to use metals that are more corrosion-resistant.

Nonferrous metals such as bronze, copper, brass, and aluminum are corrosion resistant and therefore should be used for piping, flashing, guttering, and screening.

Ferrous metals such as steel, wrought iron, and cast iron are susceptible to corrosion but may be used for interior structural members without fear of their weakening. When they are exposed to moisture, however (as in exterior work), they must be constantly protected by painting or galvanizing (coating with zinc). Stainless steel can also be used.

Weathering steel is steel with a unique means of corrosion protection. Rather than painting or galvanizing, the steel's own corrosion serves as its protective coating. This coating forms in several years into a dense, hard oxide which effectively protects the steel from further corrosion. If the coating should be scratched, it is "self-healing." When first installed, weathering steel construction has a disappointingly rusty look. However, when exposed to the weather for a year or two, it darkens in color to a rich, dark brown patina. For uniform results, though, care must be taken during installation that the material is handled as a finished product. Rainwater running over weathering steel will stain other porous or light-colored building components. This can be prevented by proper design of overhangs as shown in Figure 35–9. The entire exterior of Pittsburgh's U.S. Steel Building (Figure 35–10) is clad in a curtain wall system of weathering steel.

Galvanic Corrosion

Galvanic corrosion occurs when different metals come in contact in the presence of an electrolyte, as in an electric battery. This process will cause one of the metals to corrode. Therefore certain combinations of metals should not be used in exterior construction, since water may act as a weak electrolyte. The lower-numbered metals in Table 35–1 will be corroded by contact with a higher-numbered metal; in addition, the corrosion will increase in proportion to the difference between their numbers.

EXAMPLE
1. If aluminum storm windows are installed over stainless steel casement windows, the aluminum will corrode.

Figure 35–9 *Design of overhang to prevent staining.* (Courtesy Bethlehem Steel Corporation.)

Figure 35–10 *A curtain wall of weathering steel. (Courtesy United States Steel Corporation.)*

TABLE 35–1 *Metals Subject to Galvanic Corrosion*

1. Aluminum
2. Zinc
3. Steel
4. Iron
5. Nickel
6. Tin
7. Lead
8. Stainless steel
9. Copper
10. Monel metal

2. If galvanized nails (iron and zinc) are used to fasten copper gutters, the nails will corrode.

LIGHTNING PROTECTION

In locations of frequent thunderstorms, a lightning conductor system should be installed to protect the occupants of a building and the building itself. Lightning protection consists of paths that permit lightning to enter or leave the earth without passing through a nonconducting part of the building such as wood, masonry, or concrete. These paths can be metal air terminals (often called *lightning rods*) with ground conductors or grounded structures of metal. There are many advantages to including lightning protection in the initial design of a building instead of adding it later. To ensure that the system will be effective, it should be designed by an architectural engineer. See the *National Fire Codes* for recommended practices.

HURRICANE PROTECTION

A hurricane is a violent, swirling storm that measures several hundred miles in diameter with winds from 74 to 150 miles per hour. Hurricane winds in the Northern Hemisphere blow counterclockwise around a calm eye about 20 miles in diameter. In addition to the destructive force of high winds, coastal buildings can experience flooding from a storm surge. Tornadoes may also be created by hurricane winds.

Sound building practices such as steel reinforcing of footings and foundations, strong tie-downs of sills, soles, joists, and rafters, cross-bracing of wall framing, heavy-duty siding and roof shingling, and proper fastening between all the building elements will permit most structures to withstand hurricanes. However, experience has shown that almost half the storm damage is caused by windows or doors destroyed by flying debris, which then permitted the rain to soak interiors and furnishings. Therefore, revised building codes in hurricane-prone areas (such as Florida) have specified high-performance windows, doors, and shutters. The nationwide Building Officials and Code Administrators International (BOCA) code also strengthened its window and door standards. Impact-resistant window glass is manufactured by laminating a PVB (polyvinyl butryal) film between two lites of $\frac{1}{8}$" to $\frac{1}{4}$" glass (similar to automobile windshields). The film holds the glass in place if broken by flying debris. Also, a special silicone sealant holds the broken glass in the frame. The Dade County, Florida, Building Code Compliance Department publishes a list of hundreds of hurricane-approved products (such as windows, doors, shutters, and shingles). The list is updated every two weeks.

ARCHINET   http://www.

New Product Information

When considering the structural design of a building, especially in hurricane and earthquake areas, it is important to investigate recent developments in the construction industry. Often, a new product can offer advantages that were not previously available. The Internet is a valuable resource for obtaining new product information.

For example, a recently introduced building system, termed Isorast, has the advantages of durability, ease of assembly, and high thermal insulation value. The hollow blocks are made of expanded polystyrene, and treated with a flame-retardant additive. These blocks are then erected LEGO fashion by means of their toothed tops and grooved undersides. Once erected, concrete is poured into the cavity of the construction formed by the blocks.

You can find more information on the Internet at the following address:
http://www.hurricane-homes.com/
On this page, do one of the following:

1. Click on the words "Product Catalog 1997."

2. Click on the words "We need your help," and then click on "Technical Information."

Both of these actions will (at this writing) take you to a page that will provide you with the information you need to answer questions such as:

1. How was Isorast developed?

2. What is Isorast's R-value?

3. What are Isorast's advantages and disadvantages when compared with traditional concrete block construction?

4. Write a summary of the durability of this product, including a discussion of aging, rot, fire resistance, mechanical fatigue, cracking, and withstanding strong winds, including hurricanes and tornadoes.

5. Now peruse other information on this homepage.

EARTHQUAKE PROTECTION

Earthquakes are caused by the breaking and sudden shifting of rock plates under the earth's surface. When the edges of the plates are smooth, the plates will slide slowly. But when the edges are rough, they lock together, allowing stress to build for years until they finally break apart and produce an earthquake. Often such earthquakes occur in cycles of 20 to 200 years, the longer cycles generating the largest quakes. Consequently some of the most serious earthquakes have occurred in areas thought to be dormant and safe.

Moderate earthquakes have occurred in nearly every state in the United States, and major (killer) earthquakes have occurred in Alaska, California, Hawaii, Idaho, Missouri, Montana, South Carolina, Utah, and Washington. Obviously then, both residential and commercial buildings must be designed to protect occupants from the hazards of earthquakes.

The most common earthquake damage to a single-family dwelling is:

1. The house shifting off its foundation.

2. The roof sliding off its walls.

3. Collapse of the house's chimney.

To resist damage, these building practices are recommended:

1. Bolt the sill to the foundation and nail wall sheathing to header and to sill.

2. Nail the wall sheathing to the top plates and to blocking between rafters.

3. Reinforce the chimney and anchor the chimney to the house framing.

Most building codes include seismic requirements. Many cities adopt model codes such as the Uniform Building Code published by the International Conference of Building Officials. The earthquake design rules in nearly all codes are similar. For nonengineered buildings such as single residences, the codes specify the construction details such as foundation reinforcing, quantity and size of anchor bolts, size and nailing of wood members, bracing of the frame, and chimney reinforcing. For commercial buildings, codes written before 1980 specify that buildings must be able to withstand a horizontal force equal to 8 percent of the building's total weight. (During an earthquake, the ground vibrates in both vertical and horizontal directions. Because the horizontal vibration is usually greater than the vertical, and because buildings are stiffer vertically, collapse during an earthquake is caused by the horizontal vibration.) More recent codes, however, specify that each building be specially designed to consider the amount of seismic hazard, ground conditions, the building's natural period of vibration, ductility of the building's members, and other factors.

RADON CONTROL

Radon is a harmful, radioactive gas produced in the ground by the decay of radium. It enters into buildings by leaks in basements and slabs such as cracks, joints, hollows in concrete blocks, floor drains, and plumbing penetrations. Consequently radon can be controlled by careful building design and construction such as:

1. Reducing or sealing all basement and slab joints and cracks with flexible, urethane caulking.

2. Sealing doors between basements and crawl spaces.

3. Sealing ductwork under slabs or through crawl spaces.

4. Installing removable stoppers in untrapped floor drains.

5. Installing 4" of gravel and a 6-mil continuous vapor barrier under all basements and slabs and over crawl spaces.

6. Installing perforated plastic (PVC) pipe under basements and slabs. Connect pipe to a roof vent.

7. Specifying concrete fill in the cores of the top course of concrete masonry basement walls.

8. Reversing air movement from inward to outward through a ventilation system that replaces and dilutes indoor air with outdoor air.

9. Reversing air movement from inward to outward through an air to air heat exchanger which replaces and dilutes indoor air with outdoor air. A heat exchanger is more efficient than a ventilation system because some heat is recovered from exhaust air and transferred to the incoming air.

10. Avoiding using backfill or building materials that contain excessive radium.

Radon control becomes particularly urgent when:

1. A sleeping area will be located in the lowest level of a building.

2. Children will occupy a building.

3. Occupants of a building have the habit of smoking.

Radon concentrations in buildings can be measured by charcoal detectors (for one-week testing) or track-etch detectors (for one- to twelve-month testing). Readings below 4 picocuries/liter of air are considered acceptable, but readings above 20 picocuries/liter of air require correction.

More detailed information can be obtained from the U.S. Environmental Protection Agency, Research Triangle Park, NC 27711.

REAL WORLD  APPLICATIONS

Radon

 Radon control is an important consideration in housing design and construction. Thus, as a future architect, it is beneficial for you to understand the origin and chemistry of radon.

The element radon is the heaviest known gas. It is colorless, odorless, and probably present where you are now, although you would never know it without a radon detector. Most radon is a decay product of the element radium. All soils are likely to hold some concentration of radium and uranium from which radon is a descendant. Radon is chemically inactive under normal conditions. Since it is heavy, it remains at ground level or lower floors of a house, unless carried elsewhere by air currents.

Though chemically inactive, radon atoms are radioactive, which means their nuclei release or radiate energy in the form of subatomic particles to form lighter elements. This process is ongoing and constant.

As radon decays and its subsequent "daughter" elements decay, charged alpha particles are emitted. The daughter atoms are solid and adhere to dust particles in the air. If breathed in, these particles could become trapped in the lungs and release the alpha particles into this sensitive tissue. If the body does not repair the damage caused by these particles, over time and constant exposure, cellular mutation may occur and become dangerous to your health.

As an architect, you may want to know how significant a problem radon is in your area. Also, if working on remodeling projects, your clients may want to have their radon level checked. For information on potentially harmful levels of radon and other facts on radon, you can call the National Safety Council's radon help line at 1-800-55-RADON. Call now and ask a specific question of interest to you.

Contributed by J. C. Wenger

REVIEW QUESTIONS

1. Give two methods to control:

   a. Temperature   b. Moisture

2. How does vapor pressure and condensation relate to the water vapor in a residence?

3. Why does water vapor in a building tend to escape to the outside?

4. List the sources of water vapor in the home.

5. What causes water vapor to condense?

6. List five commonly used vapor retarders.

7. Define ventilation.

8. List five commonly used methods of ventilation.

9. What steps can be taken to prevent a house from decay?

10. List the precautions to be taken:
    a. When building in high termite hazard areas
    b. For protection from decay
    c. To protect ferrous metal from rusting
    d. For lightning protection
    e. For earthquake protection
    f. For radon control
    g. For hurricane protection

ENRICHMENT QUESTIONS

1. Why is building paper made *water* resistant but not *vapor* resistant?

2. How has the process of making a house comfortable and controllable influenced the way people live?

3. Do you feel a house can be overprotected from events that rarely happen, such as lightning strikes, earthquakes, or floods? Explain your answer.

4. Why would it be poor practice to use lead flashing in direct contact with aluminum fittings?

5. A copper roof is desired above a porch. What kind of metal fasteners can and cannot be used?

PRACTICAL EXERCISES

1. Find the size of ventilators for:
   a. A 24' × 40' flat-roofed building
   b. A 30' × 40' gable-roofed home with eaves
   c. A 30' × 50' hip-roofed house

2. Calculate the ventilator sizes for each of the following and also indicate the type of vapor barrier to be used:
   a. The Z residence
   b. The building assigned by your instructor
   c. Your original house design

3. If you live in a part of the country that is susceptible to harsh weather, what type of precautions should be included on your original house design?

Electrical-Mechanical Systems

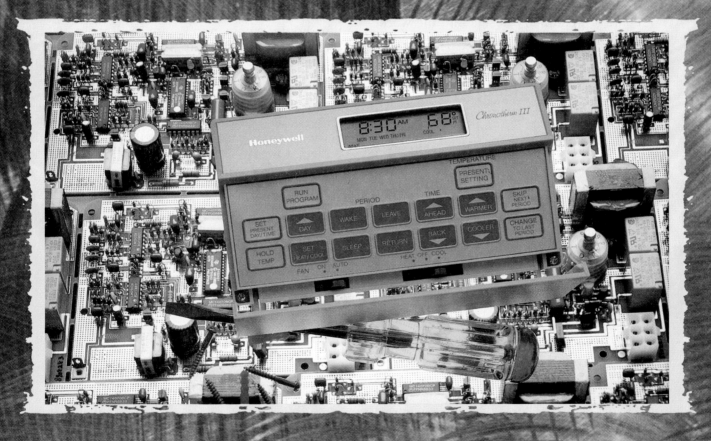

SECTION VI

36
Electrical Conventions

OBJECTIVES

By completing this chapter you will be able to:

🏛 Prepare a preliminary design and the working drawings for the lighting and other electrical requirements for a residence or small commercial building.

KEY TERMS

Fixture	Circuit
Switch	Circuit breakers
Outlet	Ground fault interrupters
Service	Smoke detectors

An electrical lighting system is an important part of a building, contributing both to its aesthetic effect and usefulness. The lighting system provides the major design feature during nighttime hours as shown in Figure 36–1 and may affect the mood in some rooms during daytime as well. Nearly all the mechanical servants in the home—necessities and luxuries—are operated by electricity.

To design an electrical system properly, the architectural drafter must be thoroughly familiar with the available fixtures, switches, and outlets; the symbols used for them; and how to combine these symbols properly.

Figure 36–1A *Exterior lighting as a major design element. (Courtesy California Redwood Association.)*

Figure 36–1B *Interior lighting can also be a major design element. (Courtesy Velux-America, Inc.)*

FIXTURES

The symbol used for an electric **fixture** is either a $\frac{3}{16}$"-diameter circle for an incandescent lamp or a $\frac{1}{8}$"-wide rectangle for a fluorescent lamp on a $\frac{1}{4}'$ = 1'-0 scale plan (Figure 36–2). These simple symbols may, however, represent a wide variety of fixture types and mountings, the exact design being further detailed to a larger scale and outlined in the specifications.

Fixture types may be either (1) direct lighting, as provided by the more commonly used fixtures, recessed lights, and spotlights, or (2) indirect lighting, as provided by valance and cove lighting. Some fixtures provide a combination of direct and indirect lighting.

Some common fixture types and mountings are given in Figure 36–3.

All of the fixture types shown in Figure 36–3 (except spotlighting) may be obtained with either

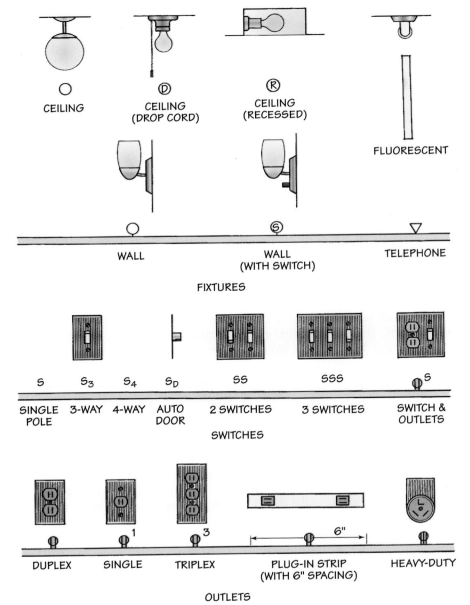

Figure 36–2 *Electrical symbols.*

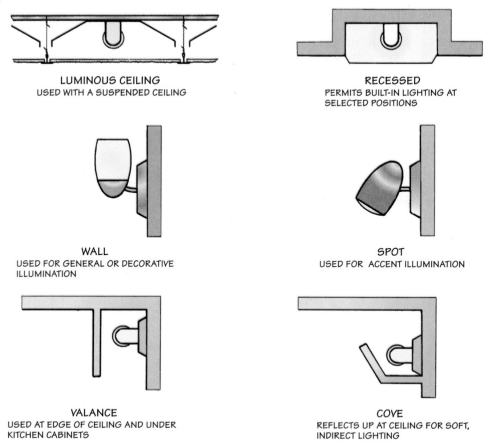

LUMINOUS CEILING
USED WITH A SUSPENDED CEILING

RECESSED
PERMITS BUILT-IN LIGHTING AT
SELECTED POSITIONS

WALL
USED FOR GENERAL OR DECORATIVE
ILLUMINATION

SPOT
USED FOR ACCENT ILLUMINATION

VALANCE
USED AT EDGE OF CEILING AND UNDER
KITCHEN CABINETS

COVE
REFLECTS UP AT CEILING FOR SOFT,
INDIRECT LIGHTING

Figure 36–3   Fixture types.

incandescent or fluorescent lamps. Fluorescent lamps are more efficient than incandescent lamps; that is, a fluorescent lamp emits many times more light than an incandescent lamp having the same wattage. Also, the average life of the fluorescent lamp is many times longer. Fluorescent lamps may be obtained in lengths between 6" and 24" at 3" intervals, and also in 3', 4', 5', and 8' lengths.

SWITCHES

Wall switches are often used to control the fixtures in rooms and halls. In living rooms and bedrooms, the switch may also control one or two convenience outlets. As in the case of fixtures, the symbol for a **switch** (the letter *S*, Figure 36–2) may represent a wide variety of switch types. The most inexpensive switch is a simple on-off *toggle switch*, although most persons prefer a *quiet switch* (very faint click) or a *mercury switch* (completely silent). *Pushbutton* types come in a wide range of button sizes ranging up to the *push plate* (in which the entire plate acts as the button). For special requirements, you may specify a *dimmer switch* (enabling a full range of light intensity to be dialed) or a *delayed-action switch* (which gives light for one minute after the switch has been turned off).

Switches are usually located 4' above the floor and a few inches from the doorknob side of each entrance door to a room. They are placed inside most rooms. Occasionally it may be convenient to place the switch outside the room entrance (as in a walk-in closet). Three-way switches are used when a room contains two entrances over 10' apart. Two sets of three-

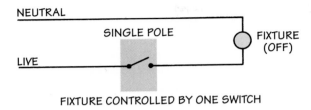

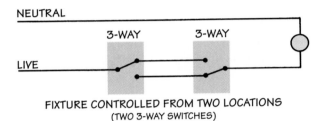

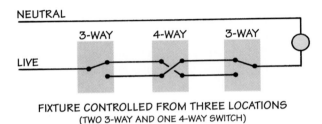

Figure 36–4 Theory of electric switches.

way switches are used at the head and foot of stairs: one set for the upstairs hall, and the other for the downstairs hall. When it is desirable to control a fixture from three or more locations, three-way switches are used at two locations, and four-way switches must be used for each additional location. (Figure 36–4 shows the wiring of these switches.) Automatic door switches for each clothes closet are useful. Closet ceiling lights, when installed above shelves, should be recessed fixtures, thus preventing accidental contact of combustible material against a hot light bulb.

OUTLETS

Room **outlets** should be duplex and specified no farther than 10' apart. They should be provided wherever they might be needed so that no extension cord longer than 6' need be used. Include short wall spaces, fireplace mantels, bathroom shaving, kitchen clock, and other special

requirements. Consider how the outlets will serve furniture groupings; also consider possible future changes in the use of the room. Hall outlets should be specified for every 15' of hall length. Outlets are located 18" above the floor except for higher positions in the kitchen and dining areas to accommodate counter appliances. Symbols for convenience outlets are given in Figure 36–2. The circles are $\frac{1}{8}$" in diameter on a $\frac{1}{4}$" = 1'-0 scale plan. The two-line symbol indicates a 120-V outlet, and the three-line symbol indicates a 240-V, heavy-duty outlet such as would be required for a range, oven, or dryer. Outdoor outlets for Christmas decorations or patio should be marked "WP" (weatherproof) on the plan.

In addition to individual outlets, *plug-in strips* are available containing outlets spaced 6", 18", 30", or 60" apart. These plug-in strips are often used behind kitchen counters when many outlets are required. They are also available as replacements for room baseboards as shown in Figure 36–5. Both individual outlets and plug-in strips may be wired to obtain constant electrical service on one of a set of dual outlets, and switch-controlled service on the other. A tabulation of the minimum electrical requirements (fixtures and outlets) for a residence is given in Table 36–1.

Figure 36–5 Plug-in strip.

TABLE 36–1 *Minimum Electrical Requirements*

Room	Minimum Fixtures	Minimum Outlets
Living area	Valance, cove, or accent fixtures	Outlets 10' apart plus smaller usable spaces One or several outlets controlled by switches Fireplace mantel outlet
Dining area	One ceiling fixture Possible valance or cove fixtures	Outlets 10' apart plus smaller usable spaces Table-height appliance outlets near buffet and table
Kitchen	Ceiling fixture for general illumination Fixture over sink, controlled by switch near sink Possible under-cabinet valance fixtures	Appliance outlets 4' apart Refrigerator-freezer outlet Dishwasher outlet Kitchen clock outlet Heavy-duty range outlet Heavy-duty oven outlet
Laundry	Ceiling fixture at each work center	Outlets for iron Heavy-duty outlets for washer and dryer
Bedroom	One ceiling fixture per 150 sq. ft. or major fraction, possibly controlled from bed location in addition to entrance door	Outlets 10' apart plus smaller usable spaces Possibly one or several outlets controlled by switch
Bathroom	One ceiling fixture if area greater than 60 sq. ft. Two wall fixtures on either side of mirror One vaporproof ceiling fixture in enclosed shower stall, sauna, or jacuzzi	One outlet for electric shaver
Hall	Ceiling fixtures 15' apart controlled from both directions	Outlets 15' apart
Stairways	One ceiling fixture on each floor or landing, controlled from both directions	None
Closets	One recessed ceiling fixture in each clothes closet controlled by wall switch or automatic door switch	None
Front and trades entrances	One or several weatherproof fixtures per entrance Bell or chime button	Weatherproof outlet for decorative lighting and electrical lawn care equipment
Recreation room	One flush ceiling fixture per 150 sq. ft. or major fraction Possible valance, cove, or wall fixtures	Outlets 10' apart plus smaller usable spaces
Porch, patio	One ceiling fixture per 150 sq. ft. or major fraction	Weatherproof outlets 15' apart
Utility room, basement	One ceiling fixture per 150 sq. ft. or major fraction Fixtures over workbenches	Workshop outlets as required Special outlets for heating and electric hot water heater
Garage	One ceiling fixture per two car	One weatherproof outlet per two cars
Attic	One ceiling fixture	One outlet

Figure 36–6   *Strip lighting over the kitchen sink. (Courtesy Pella Corporation.)*

TELEPHONES

Even though only one telephone is planned in a new home, it is well to run concealed wires to all possible future telephone locations. As a minimum, one telephone should be specified on each active floor level. Also consider the kitchen planning center, master bedroom, den or study, living room, teenager's bedroom, guest room, recreation room, patio, and workshop. Consider as well dual telephone lines to a home computer center to permit linking the home computer with the office computer simultaneously with telephone conversation. Some occasionally used rooms may be equipped for portable plug-in telephones. Most telephone companies will prewire residential and commercial structures while under construction upon request of the owner, unless concealed wiring can be provided after completion of construction. At added expense, telephone lines can be run underground from the street to the house.

USE OF CONVENTIONS

After you have decided upon the number of fixtures, switches, and outlets desired, together with their locations, the completion of the electrical plan is a simple matter. A freehand hidden line is shown connecting each fixture with its controlling switch(es). Remember this line does not represent an actual electric wire but merely indicates that a particular switch controls a particular fixture or outlet. Examples of some typical plans are shown in Figure 36–7. Study them carefully. Figure 36–8 (page 462) shows the electrical plan for the A residence.

COMPUTER-AIDED ELECTRICAL PLANS

If you use your computer graphics system to produce an electrical plan, check for available menus of electrical symbols. An ASG digitizer tablet menu of standard electrical symbols (Figure 36–9, page 463) includes lighting fixtures, outlets, switches, and exit and emergency equipment as well as data processing, communications, security, and fire alarm systems. These symbols can be moved and rotated into position to be included on plans or in pictorials. In addition to these standard symbols, you may wish to enter and store special symbols that you intend to use repetitively. Previously developed floor plans can be recalled from storage to serve as the basis (first layer) for electrical plans. When you are satisfied with a soft-copy electrical plan, use your printer or plotter to produce a hard copy, retaining the plan on diskette for possible future revision.

Figure 36–10 (page 463) shows an electric plan of ceiling fixtures and convenience outlets that was produced using Softdesk's *Electric* software.

SERVICE

Electricity is supplied to the home by means of an overhead three-wire **service** with a capacity for at least 150 amperes (A). The minimum wire size for this service is #2-gauge wire (see Figure 36–11, page 466). A large home (over 3,000 sq. ft.) or a home with electric heat would require a larger service of 200 A to 400 A. The required service is calculated from a NEC load formula which reflects the house size and electrical demand. At extra cost, unsightly overhead wire can be eliminated by underground conductors.

INTERPRETATION:
1 CEILING FIXTURE CONTROLLED BY
1 SWITCH.

RECESSED CLOSET LIGHTS
CONTROLLED BY AUTOMATIC DOOR
SWITCHES.

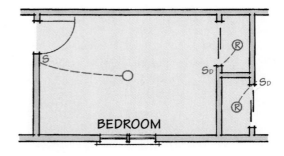

REQUIRED:
2 CEILING FIXTURES CONTROLLED BY
1 SWITCH

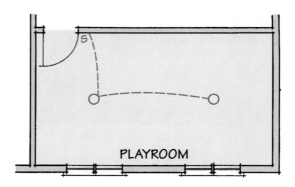

REQUIRED:
2 WALL FIXTURES CONTROLLED BY
1 SWITCH
          OR
2 WALL FIXTURES WITH INTEGRAL
SWITCH

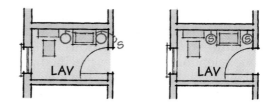

REQUIRED:
1 FLUORESCENT CEILING FIXTURE
CONTROLLED BY 2 SWITCHES

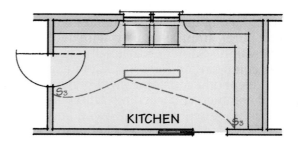

Figure 36–7   Typical electrical plans.

CIRCUITS

To distribute electricity throughout the house, branch **circuits** of various capacities are installed. Several outlets or fixtures may be placed on one branch circuit for protection by a common circuit breaker in the entrance panel. Normally house circuits are of three types:

1. *Light-duty* circuits (outlets, 2,400 watts [W] maximum: 20 A × 120 V = 2400 W).

REQUIRED:
1 VESTIBULE CEILING FIXTURE,
1 FAMILY ROOM CEILING FIXTURE,
2 OUTDOOR WALL FIXTURES,
CONTROLLED BY 3 SWITCHES IN 1 PLATE

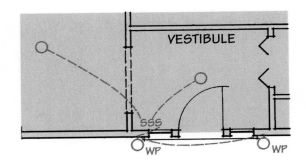

REQUIRED:
FIRST FLOOR HALL FIXTURE AND
SECOND FLOOR HALL FIXTURE
CONTROLLED BY SWITCHES ON
BOTH FLOORS

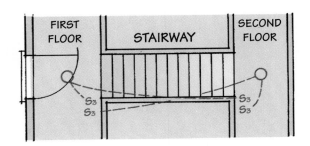

REQUIRED:
CONVENTIONAL NUMBER OF OUTLETS,
2 CONTROLLED BY SWITCH

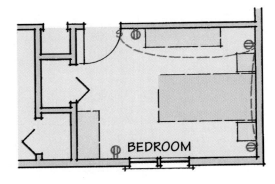

REQUIRED:
PLUG-IN STRIP WITH OUTLETS
SPACED 18" APART

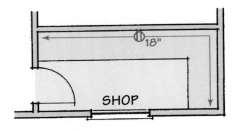

Figure 36–7 (continued)   Typical electrical plans.

Amperage: 20
Voltage: 120 (120–240 V, 115–230 V, or 110–220 V, depending upon the service supplied by the power company)
Wire size: #12-gauge wire
Description: Ordinary lights

2. *Appliance* circuits (one circuit for kitchen, one circuit for laundry, and so forth)
Amperage: 20
Voltage: 120 (120–240 V, 115–230 V, or 110–220 V, depending upon the service supplied by the power company)

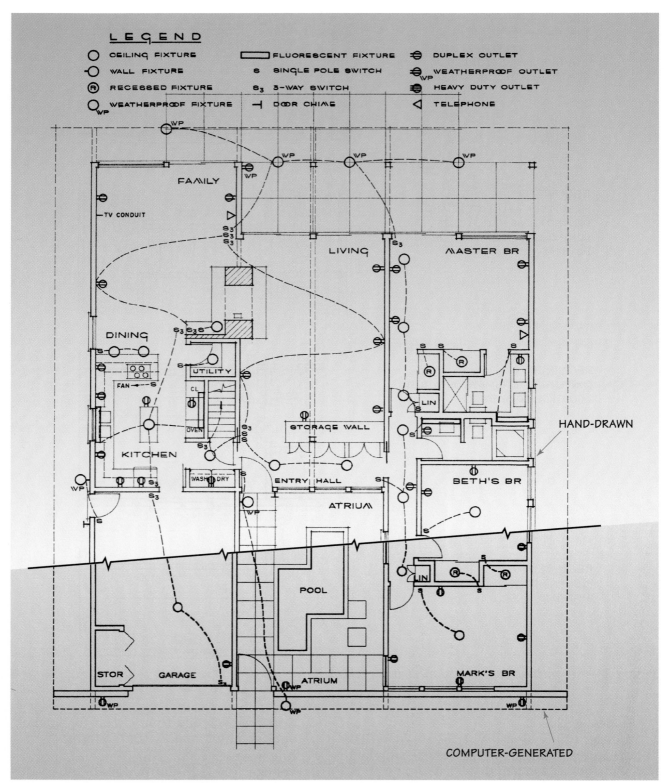

Figure 36–8 *Hand-drawn and computer-generated electrical plan for the A residence.*

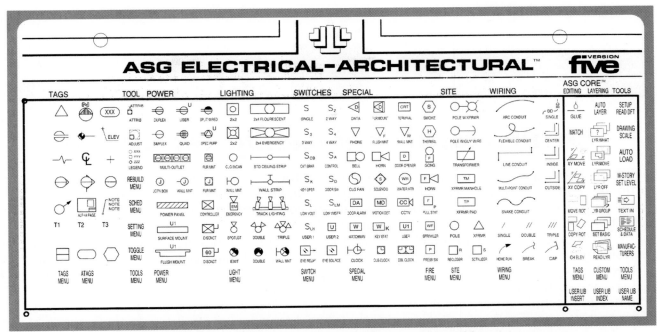

Figure 36–9 *Menu of ASG electrical symbols for AutoCAD digitizer tablet.* (Courtesy ASG.)

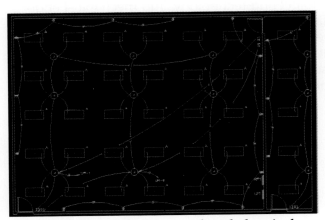

Figure 36–10 *Computer-produced electrical plan.* (Courtesy Softdesk, Inc.)

Wire size: #12-gauge wire
Description: Refrigerator, freezer, microwave oven, toaster, washer, TV, computer

3. *Heavy-duty* circuits (one individual circuit for each appliance)

Amperage: 30
Voltage: 240 (120–240 V, 115–230 V, or 110–220 V, depending upon the service supplied by the power company)
Wire size: #6-gauge wire

Description: Range top, oven, water heater, clothes dryer, air conditioner

Notice that the lower the wire gauge number, the heavier the wire. Copper wires offer resistance to the flow of electricity; the smaller the wire, the greater the resistance. Electric energy is necessary to overcome this resistance, and underdesigned wire will heat up, causing inefficient and expensive operation (a 10 percent voltage drop reduces the light from a lamp by 30 percent), and possibly a fire hazard.

Electrical wiring is usually of copper, but aluminum is occasionally used in industrial installations. Although aluminum wire is less costly than copper wire, it tends to loosen and oxidize at connections. This creates high resistance, which generates heat and is a fire hazard. For this reason, aluminum wiring is not used in residential work.

LOW-VOLTAGE SYSTEM

An electric system requiring a transformer to furnish low voltage (24 V) has been gaining popularity recently. The major advantage of this system (also called the *remote-control system*) is that a fixture may be controlled from a multitude of

CAREER PROFILES

Architectural Engineering Technician

When choosing a career, it is important to consider the job outlook for people in that occupation. According to the Bureau of Labor Statistics, one of the fastest-growing occupations in the next decade will be architectural engineering technicians. The magazine *Fortune* refers to technicians as "the rising stars of the labor force."

Architectural engineering technicians translate rough sketches and design concepts into working drawings and specifications. In addition to architectural drawings, they may produce electrical, mechanical, and structural plans. They also produce shop drawings, industrial catalogs, and technical illustrations. Skilled in both drafting and computer usage, architectural engineering technicians are employed in architectural offices as drafters, designers, estimators, inspectors, specifications writers, and technicians. Many are employed by manufacturers of building equipment to sell, install, and service architectural components. In addition to architectural technicians, larger architectural offices employ specialized technicians, primarily in the areas of civil engineering technology, electrical engineering technology, and mechanical engineering technology. Civil engineering technicians specialize in timber, steel, and concrete structures, electrical engineering technicians specialize in electrical distribution and illumination, and mechanical engineering technicians specialize in plumbing, HVAC, and refrigeration.

Architectural engineering technicians are usually graduates of two-year associate degree programs offered at community colleges, technical institutes, and universities. In the United

switch locations, the wiring being simpler than for the conventional system. Master switches may be installed that control all the house circuits. For example, all the house lights can be turned off by one master switch located in the master bedroom. Many other conveniences are possible, such as turning on the front entrance light from a kitchen location, turning off a radio or TV from the telephone locations, or controlling an attic fan from the lower house levels.

CIRCUIT BREAKERS

Circuit breakers, installed in the entrance panel, are protective devices that automatically interrupt (break) a circuit in the event of an electric overload or short circuit, thus prevent-

ing a fire hazard. Whenever a circuit breaker is tripped, the electrical problem must be located and corrected. Then the circuit breaker can be closed again by means of its switch.

GROUND FAULT INTERRUPTERS

Ground fault interrupters (GFIs), installed in the entrance panel or in individual outlets, are protective devices which automatically interrupt a circuit in the event of a line-to-ground fault, thus preventing an electrical shock. GFIs provide protection in addition to the protection provided by circuit breakers. A circuit breaker opens only when the current exceeds its rating; ratings of 15 and 20 amps are common. A GFI opens when a ground fault causes the frame of

States, there are approximately five hundred Accreditation Board for Engineering and Technology (ABET)-accredited associate programs at almost two hundred institutions. Some of the institutions that have a national reputation include DeVry, Franklin Institute, Milwaukee School of Engineering, Penn State University, Purdue University, Rochester Polytechnic, Southern Tech, SUNY, Texas A&M, and Wentworth Institute. A typical curriculum for an associate architectural engineering technology major would include these technical courses:

🏛 Architectural graphics

🏛 Computer-aided drafting

🏛 Architectural presentation

🏛 Building materials

🏛 Methods of construction

🏛 Site planning

🏛 Statics

🏛 Steel construction

🏛 Concrete construction

🏛 Building lighting and electrical layout

🏛 Heating ventilating and air conditioning

🏛 Architectural office practice

Other associate technology programs supply specialized graduates from majors such as civil engineering technology, electrical engineering technology, and mechanical engineering technology.

Certification as an engineering technician can be obtained from the National Institute for Certification in Engineering Technologies (NICET), 1420 King Street, Alexandria, Virginia 22314-2794. Phone: 703-684-2835. NICET offers four levels of certification: Technician Trainee, Associate Engineering Technician, Engineering Technician, and Senior Engineering Technician. Candidates for all levels of certification must successfully complete a written technical examination. In addition, two years of relevant engineering technician work experience are required for level 2, five years of experience for level 3, and ten years of experience plus significant involvement in a major project for level 4.

an appliance to become "hot." Such an electrical defect may result in only a small current flow, but an electrical shock of even a fraction of an ampere can be fatal.

The National Electrical Code requires that circuit breakers with GFIs be used for all 120 volt, single-phase, 15 and 20 amp bathroom outlets and outdoor outlets, and at any other potentially wet or damp locations.

SMOKE DETECTORS

Smoke detectors, while not infallible, offer the earliest warning of a potential fire. Photoelectric-type detectors are better at sensing slow smoldering fires than fast flaming fires; ionization-type detectors sense flaming fires better than smolder-

ing fires. The National Fire Protection Association recommends that smoke detectors be installed in the immediate vicinity outside of each separate sleeping area and on each additional story of a living unit, including the basement. When a detector is installed in a nonsleeping area such as a furnace room, the alarm should sound in the sleeping area.

CODES

In the design of any electric system, try to meet all the special requirements of your client. Also check that all wiring meets the requirements of the National Electric Code and existing state and municipal codes, as well as the requirements of the local utility company.

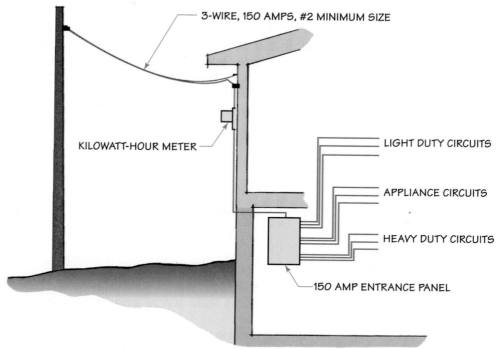

Figure 36-11 *Typical electrical supply to the home.*

ELECTROLUMINESCENCE

It is always interesting to speculate on future developments and improvements. In the area of electric lighting, possibly the electroluminescent lamp will become popular. This lamp can be roughly described as a thin sandwich of three sheets of material:

1. A rear sheet of metal

2. A middle sheet of plastic impregnated with zinc sulphide and purposely introduced *impurities*

3. A front sheet of transparent enamel

The front and rear sheets are electrically conductive. When alternating current is passed through them, a stream of electrons passes through the middle sheet, emitting light—thus the term *electroluminescent.*

The major advantages of this type of illumination are:

1. It provides glareless, uniform light distribution.

2. It is shockproof (eliminating periodic bulb replacement).

3. An infinite variety of sizes, shapes, and colors are possible, allowing entire walls or ceilings to be light-emitting surfaces. Experimentally produced lamps have been made in the general form of cloth and paper—suggesting drapes or wallpaper of light which may be changed in color by varying the electric frequency. It has also been suggested that an electroluminescent panel will provide the means of copying a television image—allowing flat TV screens to be hung on the walls of each room, like pictures.

A major disadvantage is that the frequencies and voltages necessary for operation are higher than are presently available in the home.

COMMERCIAL CONTROL

The electrical systems in large commercial buildings are monitored through a centralized

REAL WORLD APPLICATIONS

Calculating Wattage

Before a house can be inhabited, it must be inspected by an electrical inspector who certifies that the lights, outlets, circuit breakers, and other electrical devices are functioning correctly.

Electricity is measured in wattage (W) and is an indicator of the amount of current and voltage used by a device. For instance, if a microwave oven requires 6.25 amps (a) of current and is plugged into a typical outlet, how much wattage is used? Remember that most outlets supply 120 volts (v) of electricity. The formula and equation will look like this:

Power (watts) = Current (amps) ×
  Voltage (volts)
Power = 6.25a × 120v = 750W or .75
  kilowatt (kW)

Electricity is measured in the number of kilowatts per hour (kWh). If you use one kilowatt of electricity for one hour, it will cost about 9¢, although this amount varies widely across the United States. When discussing appliances with clients, many times they are concerned about how much it will cost to use those appliances.

If a refrigerator is on for ten hours, how much will it cost if electricity is 9¢ per kWh? To perform these calculations, you will need to determine how many amps the device uses. This information is usually located on the motor or in the user's manual. Assume the refrigerator uses 9.8 amps of electricity. Use the following formula to find the cost of the appliance.

Cost = v × a × 9¢ × hours of use/1000
  (v = volts, a = amps)
Cost = 120v × 9.8a × 9¢ × 10 hours/1000
  = 106 cents or about $1.06

control room. A computerized control panel is used to operate everything electrical, from light switching and air conditioning to fire alarm and heliport weather monitoring. Equipped to handle any emergency, a control center usually has a separate power source.

REVIEW QUESTIONS

1. State the two general types of electrical fixtures and give two specific examples of each.

2. Give six different classifications of electrical fixtures. Indicate where each type would be used.

3. Compare the advantages and disadvantages of incandescent and fluorescent lamps.

4. List seven types of switches, giving the uses of each.

5. When are the following pieces of electrical equipment used?
   a. Three-way switches
   b. Four-way switches
   c. Heavy-duty outlets and fixtures
   d. Weatherproof outlets and fixtures
   e. Plug-in strips
6. List nine rooms to be considered when planning telephone installations. Which are most practical?
7. List three types of house circuits, giving the amperage, voltage, and wire size of each.
8. Give two advantages of a low-voltage system.

ENRICHMENT QUESTIONS

1. List the fixtures and outlets in each major room of your home. Which of these fixtures and outlets are never used? What additional fixtures, outlets, or switches would be desirable?
2. How have electrical codes changed the methods by which houses are wired?
3. Will a #6-gauge wire or a #12-gauge wire have the greatest heat loss? Why?
4. After a thorough library search, give your estimate of the status of electricity in the home:
   a. Twenty-five years hence
   b. One hundred years hence
5. Why would low-voltage devices be more advantageous if the pool is 80' from the house?
6. How many three-way switches were used in the electrical plan in Figure 36–8?

PRACTICAL EXERCISES

1. Prepare a legend of the commonly used electrical symbols. List each type of electrical requirement and show its conventionally used symbol. Computer-produce or draw as appropriate.
2. Include in the legend of exercise 1 seldom-used electrical symbols. (Hint: Refer to a reference such as *Architectural Graphic Standards*.)
3. Draw or computer-produce the electrical plan for the A residence.
4. Draw or computer-produce the electrical plan for rooms of your own design with the following requirements:
   a. 14' × 23' living room with fireplace
   b. 14' × 14' dining room
   c. 15' × 27' playroom
   d. $3\frac{1}{2}'$ × 14' hall
   e. Front entrance vestibule with an outdoor fixture and outlet
   f. Stairways appearing on the first-floor plan (to the second floor and basement)
4. Draw or computer-produce the electrical plan for your original house design. To prevent cluttering the floor plan, the electrical plan is often a separate drawing—the wall, door, and window indications being *lightly* retraced with *darker* electrical symbols.

37 Plumbing

OBJECTIVES

By completing this chapter you will be able to:

🏛 Prepare a preliminary design and the working drawings of the plumbing requirements (water distribution and sewage disposal) for a residence or small commercial building.

KEY TERMS

Water distribution	Vent stacks
Sewage disposal	Waste stack
Fitting	Soil stack
Cold-water supply	House drain
Hot-water supply	House sewer
Fixture branches	Clean-out
Trap	House trap

An understanding of plumbing systems is important to the architectural designer and drafter. Since the plumbing and heating systems account for one-fifth of the total house cost, the designer is interested in obtaining an economical, as well as functional, design (Figure 37–1). The drafter often prepares mechanical plans—plumbing, heating, electrical, and structural—in addition to the architectural plans. Although plumbing plans may be omitted from the design of a small residence (leaving all decisions to the contractor), they are always included in the design of a larger building.

A plumbing system performs two major functions:

1. Water distribution
2. Sewage disposal

The **water distribution** system consists of the supply pipes that conduct water from the water main or other source to lavatories, bathtubs, showers, and toilets. A portion of this must be routed through a water heater to provide hot water. Most of the water piped into a building must also be drained out together with water-carried wastes. The **sewage disposal** system is composed of the waste pipes that conduct the water to the public sewer or disposal field.

WATER DISTRIBUTION

Piping

A wide variety of water supply pipes are available.

Copper tubing with soldered joints is often used in residential work. The nominal diameter indicates the approximate inside diameter of the tubing. The designations K, L, and M indicate wall thickness from heavy to light. Compare the following inside diameters (I.D.s) and outside diameters (O.D.s):

1" Cu. Tubing Type K: 0.995" I.D.   1.125" O.D.

1" Cu. Tubing Type L: 1.025" I.D.   1.125" O.D.

1" Cu. Tubing Type M: 1.055" I.D.   1.125" O.D.

The designation DWV (drainage, waste, and vent) indicates a still lighter tubing intended for sewage disposal only.

Figure 37–1   *This impressive island kitchen is the result of a well-designed plumbing plan.* (Courtesy Wood-Mode Cabinetry.)

Plastic pipe has become very popular in residential work, since revised plumbing codes permit it to be used for water supply as well as waste disposal. Plastic pipe and fittings are joined by solvent cement. Available nominal diameters (which approximate the inside diameters) are $\frac{3}{8}$", $\frac{1}{2}$", $\frac{3}{4}$", 1", $1\frac{1}{2}$", 2", 3", 4", and 6".

Brass pipe is more rigid than copper and plastic and is used with screwed fittings in large, expensive buildings.

Iron pipe is used for underground supply outside buildings, but is not used inside.

Steel pipe is inexpensive but not durable due to corrosion. Both iron and steel pipe must be galvanized or coated with hot pitch to reduce corrosion.

Fittings

Many different types of **fittings** are used to connect lengths of pipe and to provide control over the plumbing system. Pipe is joined by *couplings* to connect straight runs, or *elbows* to connect 45° or 90° bends (Figure 37–2). *Tees* are used for 45° and 90° branches. *Gate valves* are used to completely shut off the water supply for repair, *globe valves* to provide a range of water regulation from off to on (like a faucet). *Check valves* permit flow in one direction only

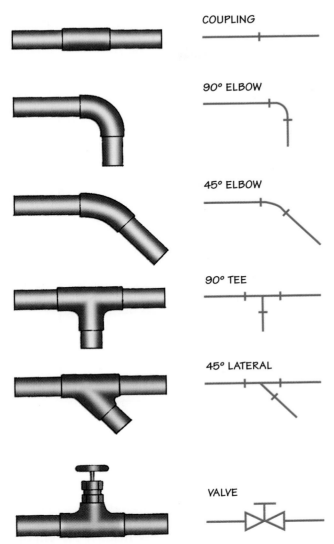

Figure 37–2 *Pipe fittings.*

COUPLING

90° ELBOW

45° ELBOW

90° TEE

45° LATERAL

VALVE

Figure 37–3 *Special plumbing design creates this 6-jet hydro-massage shower.* (*Courtesy Kohler Co.*)

and are used when there is a possibility of back-pressure. Valves are also called *cocks*, *bibbs*, and *faucets*.

Cold-water Supply

Let's trace the path of the **cold-water supply** from the street main to the house faucets as shown in Figure 37–4. Upon request, the city water department excavates the street to the public water main and installs a *tap* (pipe) to the property line. A gooseneck is included to allow for future settling of the pipe. Two cocks are installed on the tap—one close to the main,

called the *corporation cock*, and the other close to the property line, called the *curb cock*. The curb cock is attached to a long valve stem that reaches up to the ground so that water can be disconnected without another excavation.

The contractor connects $\frac{3}{4}$" or 1" copper tubing or PVC plastic pipe to the curb cock and runs it in a trench below the frost line to the building. It enters the building through a caulked pipe sleeve, and immediately the *service cock*, *water meter*, *check valve*, and *drain valve* are installed. The service cock is a gate valve which allows the owner to shut off water throughout the building. The water meter registers the quantity of water used. The check valve protects against a back flow of unpotable (undrinkable) water as a result of a break in the street main. It also protects the water meter from drainage as a result of a malfunctioning water heater. The drain valve is used when it is necessary to drain all water from the system. If a water softener is to be included, lines for hard-water hose bibbs are first connected.

A $\frac{3}{4}$" cold-water feeder line is then installed in the basement ceiling with $\frac{1}{2}$" risers running

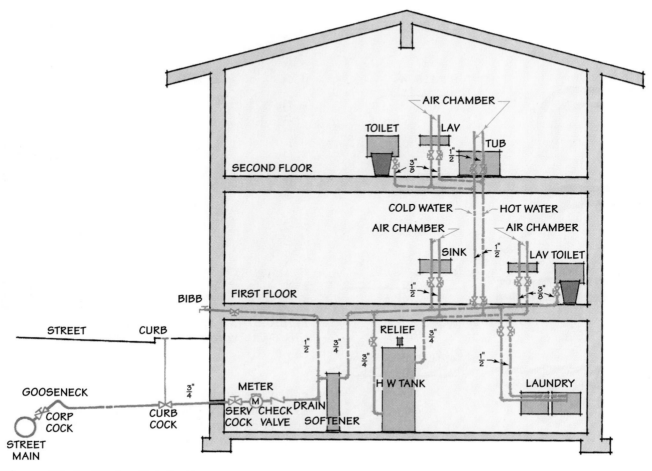

Figure 37–4 Water distribution.

directly to each fixture. Each riser is extended 2' higher than the fixture connection to provide an air chamber to reduce knocking (water hammer). Valves are installed at the bottom of each riser so that repairs can be made without shutting off the water for the entire house.

Hot-water Supply

The **hot-water supply** to a house is set up as follows: a tee is installed on the $\frac{3}{4}$" cold-water supply line so that some water is routed through the water heater—entering at 70°F and leaving at 130°F. A $\frac{3}{4}$" feeder and $\frac{1}{2}$" risers are again installed leading to each fixture (the lavatory, tub, and shower require hot and cold water, but a toilet requires only cold). A gate valve is installed at the entrance to the water heater so that it can be shut off for repair. Water may be

heated by the existing house heating system or independently by means of electric, gas, or solar heaters. A 66-gal. water heater is adequate for a family of three. A larger family should have an 80-gal. water heater. Electric water heaters are available as quick-recovery units and low-wattage units. To conserve energy, the low-wattage units are preferred.

Pipe Sizes

Proper pipe sizing depends upon a number of factors, such as the average water consumption, peak loads, available water pressure, and friction loss in long runs of pipe. For the average residence, however, this procedure may be simplified by the use of the minimum sizes recommended by the Federal Housing Administration as shown in Table 37–1. Notice that waste and

TABLE 37–1 *Minimum Pipe Sizes*

	Hot Water	Cold Water	Soil or Waste Branches	Vent
Supply lines	$\frac{3}{4}$"	$\frac{3}{4}$"		
Feeder lines				
Bathroom group plus one or more fixtures	$\frac{3}{4}$"	$\frac{3}{4}$"		
3 fixtures (other than bathroom group)	$\frac{3}{4}$"	$\frac{3}{4}$"		
Bathroom group	$\frac{1}{2}$"	$\frac{1}{2}$"		
2 fixtures	$\frac{1}{2}$"	$\frac{1}{2}$"		
Hose bibb plus one or more fixtures		$\frac{3}{4}$"		
Hose bibb		$\frac{1}{2}$"		
Fixture risers				
Toilet		$\frac{3}{8}$"	3"	2"
Bathtub	$\frac{1}{2}$"	$\frac{1}{2}$"	$1\frac{1}{2}$"	$1\frac{1}{4}$"
Shower	$\frac{1}{2}$"	$\frac{1}{2}$"	2"	$1\frac{1}{4}$"
Lavatory	$\frac{3}{8}$"	$\frac{3}{8}$"	$1\frac{1}{4}$"	$1\frac{1}{4}$"
Sink	$\frac{1}{2}$"	$\frac{1}{2}$"	$1\frac{1}{2}$"	$1\frac{1}{4}$"
Laundry tray	$\frac{1}{2}$"	$\frac{1}{2}$"	$1\frac{1}{2}$"	$1\frac{1}{4}$"
Sink and tray combination	$\frac{1}{2}$"	$\frac{1}{2}$"	$1\frac{1}{2}$"	$1\frac{1}{4}$"

vent stack sizes are also included to help in sizing the sewage disposal system.

SEWAGE DISPOSAL

The water distribution system just described is roughed in a new house before the interior walls are finished. The sewage disposal system which conducts waste water from the home is installed at the same time. The fixtures themselves are added after the interior walls are finished. Let us trace the path of the waste water from the fixtures to the public sewer (shown in Figure 37–5).

Fixture Branches

The **fixture branches** are nearly horizontal pipes which conduct the waste water from the fixtures to the vertical waste stacks (vertical pipes are called *stacks*). They are pitched $\frac{1}{8}$"–$\frac{1}{2}$" per foot away from the fixtures, and they should be as short as possible. Sizes are shown in Table 37–1.

Traps

To prevent sewer gases in the fixture branches from entering the living quarters, a U-shaped fitting called a **trap** is connected close to each fixture. This trap catches and holds waste water at each discharge, thus providing a water seal. Lavatory and bathtub traps like those shown in Figure 37–6 are installed in the fixture branch lines; toilets traps are cast as part of the fixture.

Vent Stacks

A sudden discharge of waste water causes a suction action which may empty the trap. To prevent this, vent pipes are connected beyond

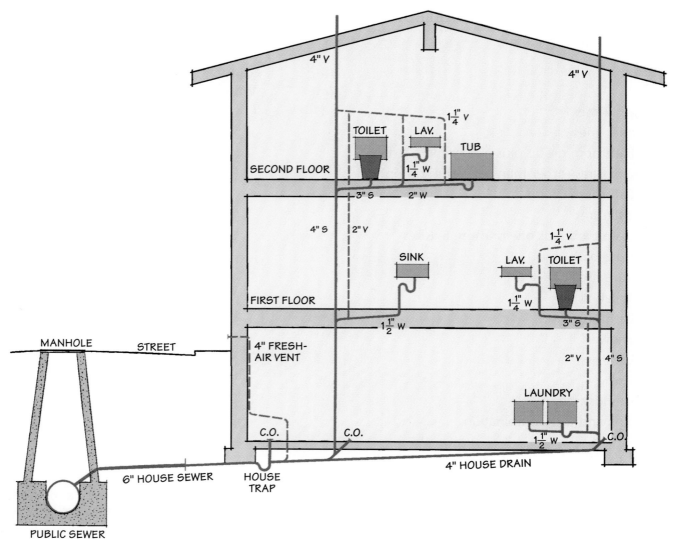

Figure 37–5   Sewage disposal.

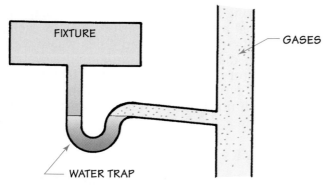

Figure 37–6   Fixture trap.

the trap and extended through the roof to open air. **Vent stacks** should be installed not less than 6" nor more than 5' from the trap. Vent stack sizes may be obtained from Table 37–1. The portion extending through the roof is increased to 4" in diameter to prevent stoppage by snow or frost. The 4" section should begin at least 1' below the roof, extend 1' above the roof, and be no closer than 12' to ventilators or higher windows.

Waste and Soil Stacks

It would be very costly to carry each fixture branch separately to the sewer; therefore the

REAL WORLD  APPLICATIONS

Lead Hazards

 Recently, the danger of lead contamination from solder in water pipes has been recognized by the plumbing industry. As someone involved in architectural design, it is important for you to consider such a hazard, especially during remodeling since some homes still have lead pipes. It is unfortunate we did not learn the harmful effects of lead from the Romans long ago. The link between lead hazards and the fall of Rome demonstrates the significance of historical facts on present-day practices.

Lead was abundant in the food and drink of the wealthy class of ancient Rome. Sources of the contamination included cookware, such as lead cauldrons and utensils, as well as water lines. Wine was prepared in lead-lined vats.

The wealthy ruling class was most affected because they had the means to indulge in the lead-laden food and beverages. It is believed that the Roman upper class not only suffered a reduced life span, but also experienced a decline in its ability to reproduce. This combination aided in the erosion of the rule of Rome over its empire.

Until recently, lead was also used extensively in paint, gasoline, and soldered food cans. It is apparent we may still have some lessons to learn from the Romans. What are the sources of lead that could affect us today? What considerations do you need to make in housing design and remodeling based on lead hazards?

More information on lead poisoning can be found in the following sources:

Hong, S. et al. (1994) "Greenland Ice Evidence of Hemispheric Lead Pollution Two Millennia Ago by Greek and Roman Civilizations" *Science*, 265.

National Safety Council's Lead Information Center, 1-800-LEAD-FYI.

Contributed by J. C. Wenger

---

fixture branches are connected at each floor level to a large vertical pipe. This pipe is called a **waste stack** if it receives discharge from any fixture except a toilet. It is called a **soil stack** if it receives discharge from a toilet, with or without other fixtures. Soil stacks are often 4" in diameter, waste stacks 3" in diameter. As with vent stacks, their upper ends should be 4" in diam-eter and extend 1' above the roof to open air. This retards the decomposition of organic matter since bacteria do not work in the presence of free oxygen. For maximum economy, fixtures should be grouped so that all fixture branches drain into only one or two stacks.

House Drain and House Sewer

The soil and waste stacks discharge into the **house drain**—a cast-iron, PVC plastic, or concrete pipe under the basement floor. It is a 4"-diameter pipe running under the footing and 5' past the foundation wall. The house drain is then connected to the **house sewer** which may be a 4" or 6" pipe. Both the house drain and house sewer are sloped $\frac{1}{4}$"/ft. to the public sewer.

Occasionally, the house drain and house sewer must exit through the basement wall *above* the basement floor in order to obtain the appropriate slope to the public sewer. When this occurs, the waste water from all basement fixtures must be pumped *up* to the house drain. A sump pump called a *sewage ejector* churns both the soil and waste water from a basement bathroom and pumps it to the house drain. A single basement laundry sink needs only a *laundry tray pump*. For a single basement toilet, install a power *up-flush* toilet.

Clean-outs

Clean-outs are elbows projecting through the basement floor to permit cleaning the house drain and sewer. They are installed in the house drain beyond the last stack, just inside the basement wall, and in between at points not over 50' apart. It is best to include clean-outs at the foot of each waste or soil stack and at each change of direction of the horizontal run. Threaded plugs are used to close the clean-outs.

House Trap

The trap installed next to the building in the house drain is called a **house trap**. It furnishes a water seal against the entrance of gas from the public sewer to the building piping. Clean-outs are located at the top of one or both sides of the trap.

Fresh-air Inlet

A 4"-diameter air vent installed next to the house trap admits fresh air to the house drain. The fresh-air inlet does not run through the roof, but rather to a place 6" above the ground. It is finished with a gooseneck bend or grille.

Septic Tank

In sparsely populated areas without public sewers, private disposal fields are used. Usually the sewage is directed to a 750-gal. underground septic tank where solid waste is decomposed by the bacteria contained in the sewage itself. The remaining liquid is distributed to the ground through porous pipes in the disposal field.

PIPING AND PLUMBING PLANS

The water distribution and sewage disposal systems may be shown in an elevation (as in Figures 37–4 and 37–5) or in an isometric drawing,

but are usually shown in plan (as in Figures 37–7 and 37–8). The water distribution system is shown in the *piping plan*, and the sewage disposal system is shown in the *plumbing plan*. Notice that since these plans are schematic drawings, it is not necessary to hold to scale. Some typical instructions appearing on such plans follow.

GENERAL NOTES

1. All underground piping should be type **K** copper (soft temper) with screwed-pressure-type joints.

2. All hot- and cold-water piping inside the building should be type **L** copper (hard-temper) with soldered joints.

3. All soil and waste piping above ground should be type **M** copper with soldered joints.

4. All soil and waste piping under ground should be heavy cast iron with lead and oakum bell and spigot joints.

5. Furnish and install stop valves in all hot- and cold-water lines before fixtures; if visible, valves to have same finish as fixture trim.

6. All fixtures shown on drawings to be furnished and completely connected with approved chrome-plated brass trim, traps, and suitable supports. Furnish and install chrome-plated escutcheon plates where pipes pierce finish walls.

For more detailed information than is given in this chapter, obtain the *ASHRAE Guide* from the American Society of Heating, Refrigerating, and Air Conditioning Engineers; the *National Plumbing Code* from the Government Printing Office in Washington, D.C.; or the *Uniform Plumbing Code* from the International Association of Plumbing and Mechanical Officials.

COMPUTER-AIDED PLUMBING PLANS

The same procedure used to obtain computer-aided electrical plans can be used to obtain computer-aided piping and plumbing plans. Check on the availability of diskette menus of plumbing symbols. An ASG digitizer tablet menu of standard piping symbols (Figure 37–9, page 480) includes pipes, valves, fittings, and equipment. The ASG menu of plumbing symbols includes sanitary fittings, fire protection equipment, and controls. These symbols can be moved and rotated into position to be included

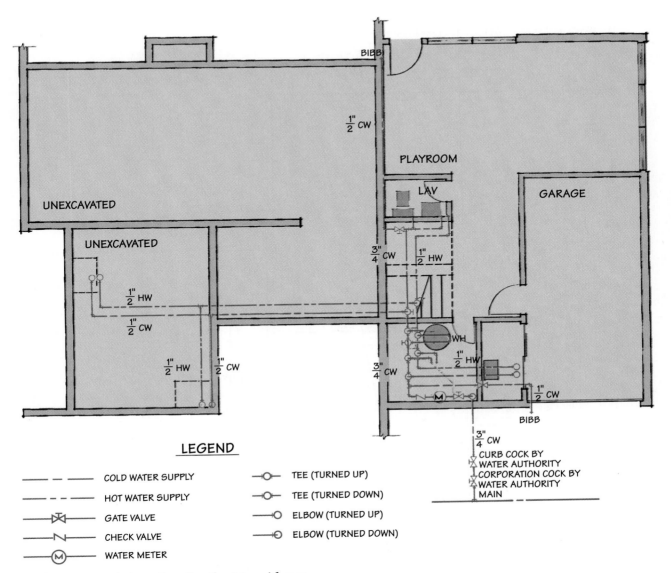

Figure 37–7   Piping plan for the M residence.

on plans or in pictorials. Previously developed floor plans can be recalled from storage to serve as the basis (first layer) for piping and plumbing plans. When satisfactory soft-copy piping and plumbing plans are shown on your screen, use your printer or plotter to produce hard copies, but also store copies on diskette for possible future revision.

Figures 37–10 and 37–11 (page 479) show a plumbing plan that was produced using Softdesk's *Plumbing* software, and Figure 37–12 (page 480) was produced using Softdesk's *Piping* software.

FUTURE TRENDS
Sovent Plumbing

A single-stack, self-venting sewage disposal system was developed recently and holds promise for multistory buildings. This system is called *sovent*, indicating a combination of *soil* stacks and *vent* stacks. The key to this system is an aerator fitting located at the connection of each fixture branch to the soil stack. This fitting limits the sewage flow velocity in the stack, thus reducing the suction which would siphon the traps. The sovent stack extends through the

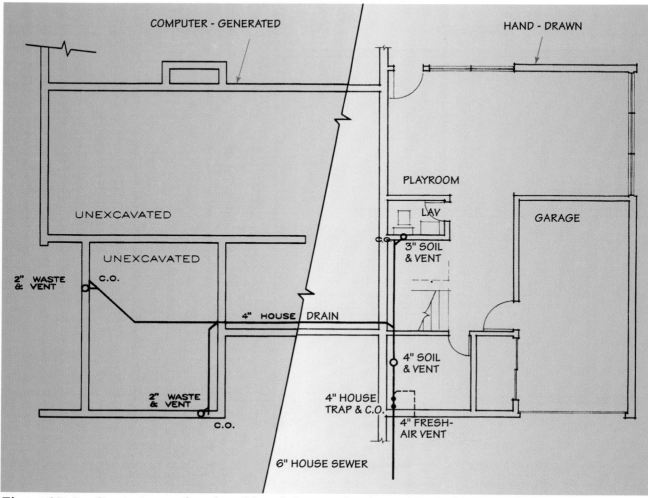

Figure 37–8 *Computer-produced and hand-drawn plumbing plan for the M residence.*

roof and acts both as a soil stack and as a vent stack.

Solar Water Heating

Solar hot water pre-heating has become quite common and is considered to be one of the most cost-effective means of using the sun's energy. This is because hot water is needed year-round, but space heating is needed only during cold weather. The principal elements of most solar hot water systems are a flat plate collector used to directly heat water, a hot water distribution system of piping, a hot water storage unit, appropriate automatic controls, and some method of protecting the water from freez-

ing. Several types of solar hot water collectors are described in the heating chapter. A solar hot water heating system, of course, does not require as much collector area as does a space heating system.

Solar heaters for swimming pools are also available in a wide variety of systems.

Solar Stills*

In some locations a supply of potable (drinkable) water has become a critical need. The solar still provides an effective solution. The first large solar still used to purify salt or

*Courtesy Horace McCracken, Solar Equipment Consultant.

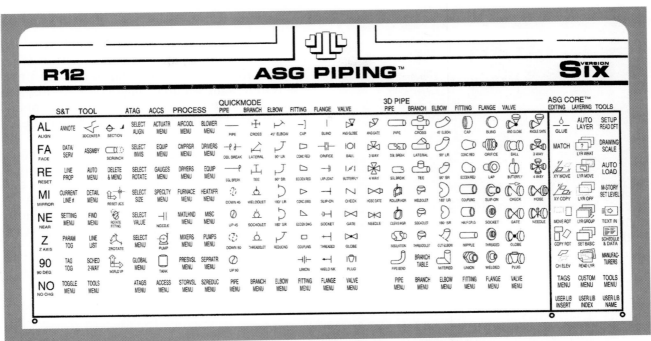

Figure 37–9   Menu of ASG piping symbols for AutoCAD digitizer tablet. (Courtesy ASG.)

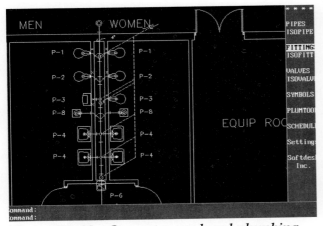

Figure 37–10   Computer-produced plumbing plan for a commercial restroom. (Courtesy Softdesk, Inc.)

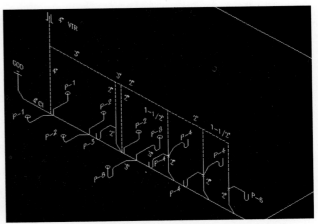

Figure 37–11   Computer-produced isometric drawing of the restroom plumbing shown in Figure 10. (Courtesy Softdesk, Inc.)

brackish water was built in Chile over a century ago (1872). In recent years more than twenty community-size solar stills, ranging from one hundred to several thousands gallons per day, operating on seawater or inland brackish water, have been built in ten different countries. This type of usage predictably will increase.

Small home stills to convert seawater to drinking water have been installed along the coasts of California and Mexico. These stills consist of shallow pans covered by glass panels. Installed on a flat roof, an automatic feed pump provides a constant supply of seawater. The daily output of such an installation is about $1\frac{1}{2}$ gal. of potable water.

Figure 37–12 *Computer-produced three-point "model" of the piping for an industrial plant.* (Courtesy Softdesk, Inc.)

Solar Farming

The development of solar stills has been expanded to provide the means to farm in areas where only saltwater is available. The process requires air-inflated plastic greenhouses in which the water transpired by vegetable plants is condensed on the plastic roof and then reused for root irrigation.

An integrated plant that supplies potable water and food as well as power was built in Abu Dhabi, which is both a member state and the capital city of the United Arab Emirates, by the University of Arizona. Among other benefits, the price of fresh vegetables was lowered enough to provide a new food source for the population of the state.

REVIEW QUESTIONS

1. What are the two functions a plumbing system performs?

2. Distinguish between:
   a. K, L, M, and DWV copper tubing
   b. Coupling, elbow, and tee
   c. Gate valve, globe valve, and check valve
   d. Corporation cock, curb cock, and service cock

3. Give the minimum pipe sizes for:
   a. Supply lines
   b. Feeder line to a bathroom
   c. Feeder line to a hose bibb
   d. Feeder line to a powder room
   e. Fixture risers to a lavatory
   f. Bathtub
   g. Shower

4. Give the principal reasons for using:
   a. Traps
   b. Vent stacks
   c. Clean-outs

5. Distinguish between:
   a. Soil stack and waste stack
   b. House drain and house sewer
   c. Piping plan and plumbing plan

6. What are some advantages and disadvantages of heating water using the sun's energy?

7. Why are notes commonly used on plumbing and piping plans?

REAL WORLD  APPLICATIONS

Percolation Test

 In most rural locations, municipal sewage service is not available for new home construction, so an on-lot sewage disposal system must be constructed. A percolation (perk) test is the first step in the process. Specific requirements for the perk test vary by location, but most municipalities follow this basic procedure.

On a given lot or property, there may be a few locations where a septic system would be suitable. Most municipalities require the system be built on less than a 10% grade and must be at least 50 feet from the property line. It is wise to consult the sewage officer for advice on a septic site before you decide the exact location for your house.

In order to perk a property, one must arrange for the use of a back hoe. The sewage officer will come to the property and will tell you where the back hoe operator is to dig a pit. The officer is looking for the depth of topsoil above the seasonal high water mark and the depth above the first impenetrable layer, such as shale or bedrock. These pits range between 3 and 5 feet in depth. If there is a layer of rock through which water cannot percolate, or the local water table is high, a permit for a standard septic system will not be issued, and an alternative sewage system will be needed. The particular variety of alternative systems will depend on the conditions. Other systems include sand mounds, discharge into flowing water, and above-ground spray discharge. The standard septic system is by far the most economical and easiest to maintain.

Once the sewage officer determines the type of sewage system suitable for the property, the actual percolation test must be performed. This test will determine how fast water will soak into the soil and hence how fast it will find its way back into the environment. Generally, six percolation holes are dug by hand around the site of the proposed septic system. These holes are 2 feet deep and 6 to 10 inches in diameter.

The night before the perk test, these holes are filled with water and left overnight to pre-saturate the ground. The next morning the holes are filled with 6 inches of water, and a measurement is taken every half hour to see how much water has soaked into the soil. After each measurement the holes are refilled to a depth of 6 inches. This process is repeated for six hours. With this data the sewage officer can determine the required size (square footage) of the septic system, given the size of the house to be built.

With the current emphasis on cleaning up the environment, one may ask how the continued discharge of sewage into the soil can be justified. Sewage systems rely on nature to manage the waste produced. First, sewage is piped into a septic tank after leaving the house. This large tank retards the flow of the sewage, allowing bacteria to act on the organic matter and also allowing any solid material to settle out. Thus, as sewage leaves the tank it has undergone significant degradation. The sewage then flows from the septic tank into the leach field. A leach field is a matrix of pipes containing small holes, which allow the sewage to be dispersed slowly over a large area. Once the sewage is dispersed, bacteria in the soil further degrade any remaining organic matter. Any toxins remaining, such as nitrates, are dispersed over a large area and are diluted with the ground water to a negligible concentration. With a properly designed septic system there is no contamination of the environment.

Contributed by M. A. Henck

ENRICHMENT QUESTIONS

1. Contact the local water and sewage utilities commission and find out the tap-in fees required for a new residence.

2. Find the local rates for water. Do you feel this is expensive? What can be done to conserve water in a typical residence?

3. Why is the location of the vent stack above the roof critical?

4. When designing a house, why is it advantageous to locate rooms that require water and drainage near each other?

PRACTICAL EXERCISES

1. Prepare a legend of the commonly used piping and plumbing symbols. List each requirement and show its conventionally used symbol. Computer-produce or draw as appropriate. (Hint: Refer to a reference such as *Architectural Graphic Standards*.)

2. Draw or computer-produce (a) the piping plan and (b) the plumbing plan for the A residence.

3. Draw or computer-produce a diagram showing:

   a. How water is distributed from street main to fixtures

   b. How waste water is carried from fixtures to the sewer

4. Draw or computer-produce (a) the piping plan and (b) the plumbing plan for your original building design. To prevent confusing the floor plan, the piping and plumbing plan is often a separate drawing—the wall, door, and window indications being *lightly* retraced with *darker* piping and plumbing symbols.

CHAPTER 38

Heating and Air Conditioning

OBJECTIVES

By completing this chapter you will be able to:

- ☖ Explain the differences between various types of heating systems.
- ☖ Explain the meaning of "air conditioning."
- ☖ Explain the differences between various types of air conditioning systems.
- ☖ Explain the different types of heat loss and heat gain and perform calculations for each.

KEY TERMS

Warm-air heating
Perimeter heating
Hot-water heating
Radiant panel heating
Electric heating
Heat pump
Solar heating

Passive solar
Active solar
Humidification
Dehumidification
Ventilation
Filtering
Air conditioning

The earliest known central heating systems were built by the Romans to heat their bathhouses. Tile hot-air ducts were used to heat the buildings, and lead pipes were used to conduct hot water to the baths. Although such systems have been used in palaces for several thousand years, they were not used for small-home heating until about a hundred years ago. Today we can choose from a multitude of systems which include cooling, humidity control, ventilation, and filtering, in addition to heating.

HEATING

The most expensive appliance in a building is the heating system—its initial cost amounts to about 10 percent of the total house cost and the fuel costs amount to several hundred dollars per year. A building's heating system is specified by an architectural engineer (specializing in heating) after consulting with the client. The architectural drafter must be able to prepare heating plans from the heating engineer's calculations and sketches. Therefore the drafter should be familiar with all heating systems, their design and layout, and how they are represented on the heating plan. We will consider heating systems of four general types:

1. Warm-air
2. Hot-water
3. Electric
4. Solar

Let us look at the operation of each, together with its advantages and disadvantages.

Warm-air Heating

In *forced* **warm-air heating**, warm air is circulated through sheet-metal supply ducts to the rooms, and cold air is pulled through return ducts to the furnace for reheating (usually to about 150°F). Duct work may consist of a number of small individual ducts leading to each room (*individual duct system*, Figure 38–1), or a

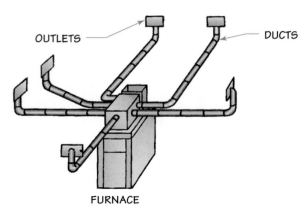

Figure 38–1   Individual duct system.

master duct which reduces in size as it branches off to feed the rooms (*trunk duct system,* Figure 38–2). Circular ducts are 4"-diameter tubes that slip together like stovepipe, with flexible elbows to form any angle up to 90°. Rectangular ducts are custom made to fit in spaces between joists and studs.

A thermostat is an instrument sensitive to changes in temperature. One type contains two metal strips having different temperature coefficients brazed together. One end of the strips is fixed, but the free end will move due to dissimilar expansion of the metals. This closes an electric contact to start a motor which controls the heating and cooling systems.

A fan, operating either continuously or intermittently, circulates the air. The thermostat controls the fan in the intermittent system. The fan may be similar to a common electric fan, but more probably it will be a centrifugal type called a *blower.*

For even temperature throughout the house, the warm air should be delivered to the places that lose heat fastest—the exterior walls. The warm air is supplied through grilles with manually operated louvers located in the exterior walls or floor, preferably under windows. These grilles are called *registers* or *diffusers.* In place of registers, warm-air *baseboard units* (Figure 38–3) may be used to distribute the heat along a wider portion of the exterior wall.

Since the heating element, air, is actually blown into the room, warm-air heating has the inherent disadvantage of distributing dust throughout the house. This is somewhat checked by *filters*—pads of spun metal or glass coated with oil to catch dust. These must be replaced or cleaned often to be effective. Another method is attracting dust by a high-voltage screen.

Perimeter Heating   This forced-warm-air system was developed for a special need—to heat the basementless house. As previously described, warm air is delivered to the exterior walls, or *perimeter,* of a building. **Perimeter heating,** however, has an additional function— to warm the floor itself, thus replacing the heat lost through the concrete slab floor or wood floor over a crawl space.

The recommended perimeter heating system for a slab floor is the *perimeter loop* (Figure 38–4). This consists of a 6"-diameter tile duct

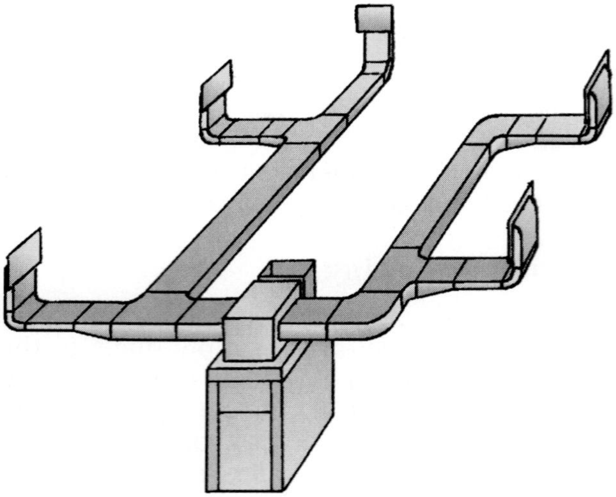

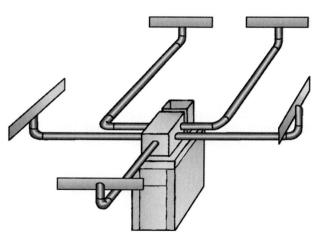

Figure 38–2   Trunk duct system.

Figure 38–3   Warm-air baseboard units.

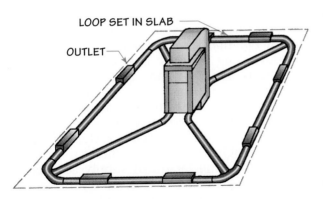

Figure 38–4   *Perimeter loop system.*

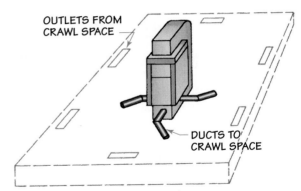

Figure 38–6   *Crawl-space plenum system.*

imbedded in the outer edge of the slab and supplied with warm air fed through radial ducts. The ducts warm the slab, and floor registers in the ducts supply heat to the rooms.

Another perimeter heating method is called the *perimeter radial* system (Figure 38–5). Radial ducts run directly to the floor registers with no outer loop. This system is often used in crawl spaces.

The *crawl-space plenum* system (Figure 38–6) utilizes the entire crawl space as a plenum (warm-air reservoir). Short ducts (6' minimum length) from the furnace heat the crawl-space plenum, which then supplies heat to the rooms through perimeter floor registers. All perimeter heating requires careful insulation of the foundation to prevent excessive heat loss.

Hot-water Heating

In *forced-circulation* **hot-water heating**, a hot-water boiler heats the circulating water to 200°F–215°F. (The entire system contains water

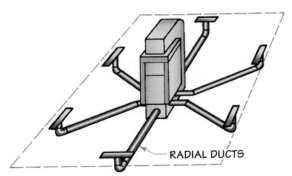

Figure 38–5   *Perimeter radial system.*

under slight pressure to prevent the formation of steam at 212°F.) An electric pump controlled by the thermostat circulates the water through narrow ($\frac{1}{2}$"–$\frac{3}{4}$") flexible tubing to radiators or convectors, giving up heat to the room. The temperature of the return water is about 20°F lower than the boiler delivery. The temperature of the delivery water can be automatically adjusted by a mixing valve that determines the amount of hot water required in the circulating system. Also, each radiator can be regulated by automatic or manual valves. Usually the boiler is fitted with additional heating coils to supply hot water, winter and summer, to the sinks, tubs, and laundry.

A hot-water system is also called *hydronic* heating. Three hot-water piping systems can be used:

1. Series loop

2. One-pipe system

3. Two-pipe system

Series Loop   The series loop system (Figure 38–7) is, in effect, a single baseboard radiator extending around the entire house and dropping under doorways and window walls. Hot water enters the baseboard near the boiler and travels through each baseboard section and back to the boiler for reheating. It is often used in small homes because it is inexpensive to install, but there can be no individual control of the heating units. Either the entire house is heated, or none of it. A compromise is the installation of *two* series loops in two zones, which may be independently controlled.

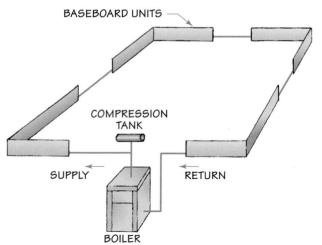

Figure 38–7   Series loop.

One-pipe System   The one-pipe system (Figure 38–8) is often specified for the average-size residence. Hot water is circulated through a main, special tee fittings diverting a portion of the water to each radiator. Radiators can be individually controlled by valves and located either above the main (*upfeed* system) or below it (*downfeed* system). Downfeed is less effective because it is difficult to coax the water into the branches.

Since water expands when heated, a compression tank is connected to the supply main. A cushion of air in the tank adjusts for the varying volume of water in the system as the water temperature changes.

Two-pipe System   The two-pipe system is used for large installations. The hot water is circulated by two main pipes, one for supply and one for return. The water is diverted from the supply main to a radiator, and then flows from the radiator to the return main. In this manner, all radiators receive hot water at maximum temperature. The *reversed return* system of Figure 38–9 is preferred to the *direct return* system of Figure 38–10. Piping is saved in the direct return system, but the total length of supply and return piping is the same in the reversed return system, assuring equal flow due to equal friction.

Piping in the one- and two-pipe systems needs no pitch except for drainage. All hot-

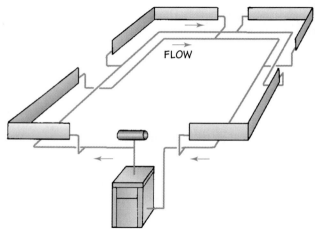

Figure 38–9   Two-pipe system, reversed return.

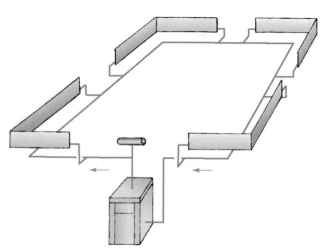

Figure 38–8   One-pipe system.

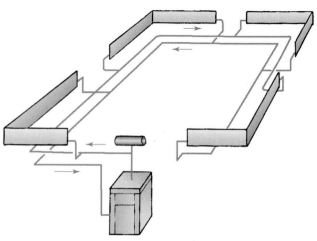

Figure 38–10   Two-pipe system, direct return.

water systems must be drained or kept operating to prevent their freezing when the house is not occupied.

Heating Units   The heating units used in hot-water heating are either radiators, convectors, or combinations of radiator and convector. A radiator has large, exposed surfaces to allow heat to *radiate* to the room. Radiant heat does not depend upon air movement; it passes through the air directly to any object. A convector, however, draws in cool air from the room at the bottom, warms it by contact with closely spaced fins, and forces it out into the room again. Heat is therefore circulated by air movement. The major disadvantage of convection is that the heat rises to the ceiling, leaving the floor cold.

Radiators and convectors (Figure 38–11) may be recessed into the wall to increase floor space. Baseboard heating may be radiant or convector. The distribution system is identical for conventional and baseboard heating.

Radiant Panel Heating   In a **radiant panel heating** system, the entire floor, the ceiling, or the walls serve as radiators. The heating element may be hot water or steam in tubing, hot air, or even electricity. In the hot-water system, prefabricated loops of tubing are imbedded into a concrete floor (Figure 38–12) or attached to the ceiling or walls before plastering (Figure 38–13). In drywall construction, the tubing is installed behind the wall or ceiling panels. In the warm-air system, ducts may be laid in the floor or ceiling, or the entire space above the ceiling may be heated by blowing warm air into it.

Electric Heat

Electric heating systems offer many advantages: low installation cost, no exposed heating elements, individual room control, cleanliness, and silent operation. The main disadvantage is

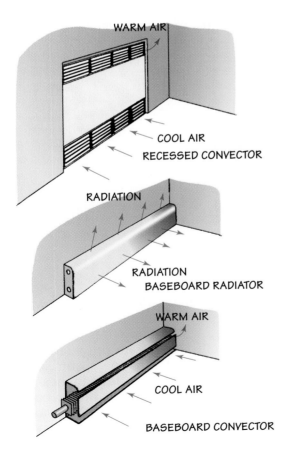

Figure 38–11   *Types of heating units.*

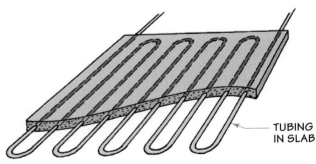

Figure 38–12   *Radiant floor panel.*

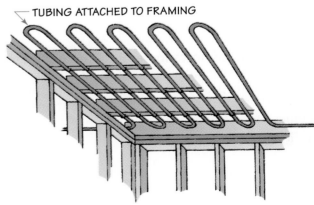

Figure 38–13   *Radiant ceiling panel.*

REAL WORLD APPLICATIONS

Graph Interpretation

A family of four have decided to heat their home with natural gas instead of electricity. Since the summers where they live are rather mild, there is no need for aid-conditioning and, therefore, they will use hot water baseboard heat rather than forced air.

The graph of their yearly bills is shown in the figure. The months are labeled on the horizontal x-axis with the letters representing the first letter of each month. The gas consumption, measured in million cubic feet (MCF), is shown on the vertical y-axis. The cost of natural gas in their part of the country is approximately $7.06/MCF.

For example, in March 1996 they spent $141.20 on gas. This is obtained by following these steps.

1. Locate the month.

2. Draw a horizontal line from the top of the bar to the point on the y-axis. March 1996 would be 20 MCF.

3. Multiply this amount (20) by the cost of the gas consumption ($7.06/MCF). 20 × $7.06 = $141.20.

Answer these questions regarding the graph.

1. What was the cost of the gas bill in these months?

   a. September 1995

   b. January 1996

   c. June 1995

2. Based on the graph, would you say 1995 or 1996 was a colder winter (January through March)? Why or why not?

3. Since the heater is not used in the summer months, why is gas being used?

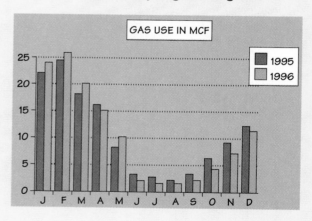

the high cost of operation in localities not offering low electric rates. Heavy insulation is required to keep operation costs to a minimum. The most popular electric heating systems for homes are:

1. Electric resistance cable

2. Electric panels

3. Electric baseboards

4. Heat pump

Electric Resistance Cable   Covered wire cables are heated by electricity and concealed in the ceiling or walls. The wires are manufactured to specific lengths to provide the rated wattage. They are stapled to gypsum lath in a gridlike pattern, only a few inches apart. Then they are covered by a $\frac{1}{2}$" brown coat and finish coat of plaster (Figure 38–14) or gypsum board (Figure 38–15). The temperature of each room can be individually controlled by thermostats.

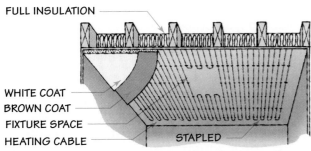

Figure 38–14   *Electric cable in plaster ceiling.*

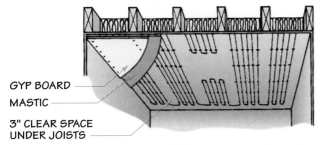

Figure 38–15   *Electric cable in gypsum-board ceiling.*

Electric Panels   Prefabricated ceiling panels (Figure 38–16) are only $\frac{1}{4}$" thick and constructed of a layer of rubber containing conductive material and backed with insulating material. They cover the entire ceiling and can be painted, plastered, or papered. Smaller glass wall panels are backed with an aluminum grid for the resistance element and are available in radiant, convection, and fan-forced types. These panels are set into the outside walls under the windows and are best suited for supplemental or occasional heating.

Electric Baseboard   Most electric baseboards are convection heaters (Figure 38–17). They

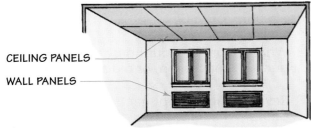

Figure 38–16   *Electric panels.*

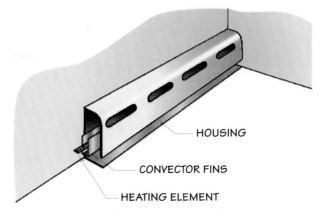

Figure 38–17   *Electric baseboard.*

consist of a heating element enclosed in a metal baseboard molding. Slots at the bottom and top of the baseboard permit the circulation of warmed air.

Heat Pump   A **heat pump** works on the same principle as a refrigerator, which takes heat out of the inside compartments and discharges it into the room. The heat pump takes heat from the outdoors and brings it into the house. It does this by further "refrigeration" of the outside air (or the ground). A refrigerant (which acts to absorb heat) circulates between two sets of coils—the evaporator and the condenser. During cold weather the outside air is blown over the evaporator coils. Although this air is cold, the refrigerant is much colder and thus absorbs heat as it changes from liquid to gas. The warmed refrigerant is then pumped to the condenser, where the room air is blown over the coils and is heated by the refrigerant as it changes back to a liquid. The still-colder outside air is blown back outdoors again. During warm weather the operation is reversed so that heat is removed from the room.

Heated or cooled air from a heat pump can be circulated through supply ducts (Figure 38–18) in the same manner as it is circulated in a warm-air system. Heat pump systems can be installed that also provide water heating.

In addition to air-source heat pump systems, *ground-source heat pumps* are becoming popular. A ground-source heat pump (also called a *geothermal heat pump*) does not remove heat from cold winter air and add heat to warm

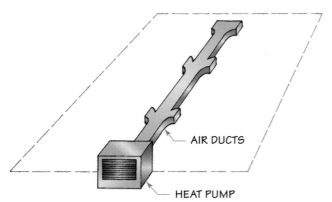

Figure 38–18   Air-source heat pump.

summer air. Rather, it moves heat into and out of the ground using an underground loop of plastic pipe buried in the earth that remains at a constant temperature (50°F) during the whole year. Long lengths of polyethylene pipe with heat-fused joints are laid in 6' deep horizontal trenches about 500' long. Or several vertical bore holes about 200' deep are drilled whenever the plot is small or when the ground contains much rock, as backfilling a trench with broken rock could puncture the piping. This long, continuous loop of plastic pipe contains the circulating coolant fluid, which collects heat from the ground in the winter and deposits heat into

the ground in the summer. It is common for the coolant to gain or lose about 8°F as it circulates. The coolant fluid is water with a nontoxic antifreeze, a dye, and a scent. The dye and scent are useful in locating leaks.

Although installation costs for ground-source heat pumps are high, annual heating and cooling costs are reduced. For more detailed information, contact the International Ground-Source Heat Pump Association at Oklahoma State University, Stillwater, OK 74078.

Solar Heat

A **solar heating** system uses the sun's rays as the heating source. Although this is a comparatively new concept, many homes are now being built with solar space heating and cooling or solar hot-water heating systems. Because of the diffuse nature of solar energy, most solar heating systems supply only about half of the space heating requirements of a building, the other half being supplied by auxiliary conventional heating systems. Solar heating systems are usually described by the method used to circulate heat through the building. When the heat flows naturally, the system is called *passive*; when the heat flow is forced by some mechanical means, the system is called *active*. Many solar buildings combine passive and active systems. For exam-

REAL WORLD   APPLICATIONS

Heat Energy

 Heating is an essential element for families who live in houses in many sections of the United States. A house can be heated in three different ways: convection, conduction, or radiation. Conduction is the transfer of heat through the contact of two materials. This occurs in hot-water baseboard heating when hot water is moved through a copper pipe and heat is transferred directly to the aluminum fins. The heated aluminum fins then use convec-

tion to transfer the heat into the air. Radiant heat is produced when a glass-walled room traps shortwave energy which warms surfaces in the room.

🏛 What features in your house utilize conduction, convection, or radiant heat?

🏛 Review the components of a heat pump. What type of heat energy does it produce? How is this heat produced?

ple, the natural heat flow of a passive system can be accelerated with the addition of a fan.

Passive solar heating systems are usually described by the path of the heat flow, while **active solar** heating systems are usually described by the type of collector used. Some of the most common types are:

🏛 Passive

    1. Direct gain

    2. Indirect gain

    3. Isolated gain

🏛 Active

    1. Flat-plate collectors
      A. Air-type collectors
      B. Liquid-type collectors
      C. Trickle collectors

    2. Focusing collectors

PASSIVE SYSTEMS

Direct Gain

The direct-gain passive system is the simplest and most common of solar heating systems. It consists of a large glass area on the southern wall to permit the sun's rays to *directly* heat the house. As shown in Figure 38–19, the roof overhang is designed to permit the winter sun swinging low on the horizon to warm the house, but exclude the summer sun, which is high in the sky. The exact angle of the sun by

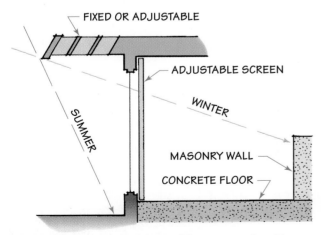

Figure 38–19   An adjustable system for direct solar heating.

hour, date, and latitude is given in standard reference books. Some form of adjustable screen is needed to prevent heat from escaping through the same glass areas during the evening and on cloudy days.

A south-facing window system should be designed to use available solar heat, but without overheating during midday. Some methods used are:

1. Interior designed to permit warm air to rise and heat additional rooms above.

2. Fans to circulate warm air to other parts of the building.

3. Windows facing east of true south to provide quick morning heat and less heat in the afternoon.

4. A heat storage system such as a masonry wall or concrete floor that absorbs the heat and returns it to the building for a few hours after sundown. Windows facing west of due south are best for this system because they extend the heat return later into the evening.

5. A separately zoned solar heated room such as a greenhouse or sun room, which can be closed off from the other rooms when desired.

Indirect Gain

The indirect-gain passive system consists of a dark-colored storage mass about 4" behind the southern glass area. The storage mass absorbs and stores heat, and the heat then flows from the mass to *indirectly* heat the house. This system is also called a *Trombe* system, after the French architect Felix Trombe who pioneered it. As shown in Figure 38–20, the heated air can be circulated by natural convection (called *thermosyphoning*), or it can be aided by a fan. Also, the indirect-gain system can be designed with adjustable vents at the top and bottom of the storage mass to control heat distribution. As with the direct-gain system, adjustable screens can be used to prevent the escape of heat at night.

Isolated Gain

The isolated-gain passive system consists of a collector and storage unit separated (thermally

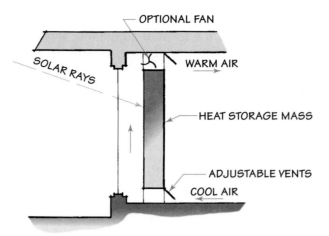

Figure 38–20   A thermosyphoning system for indirect solar heat.

Figure 38–21   The conversational area of a passive solar clubhouse. (Courtesy California Redwood Association.)

"isolated") from the area to be heated. The isolated-gain system is not as popular as the direct and indirect systems.

EXAMPLE

The S residence, shown in Figure 38–23 and in Appendix A, is a solar house with both direct and indirect passive space-heating systems, and an active, domestic hot-water heating system. The house is a three-level A-frame located on the site so that the front left corner faces south. This permits solar heat to be collected through both the front wall and the left wall. Features of this house are:

1. Solar heat is directly gained through large, glazed areas in the left (southwest) wall that warms the cathedral living level and loft sleeping level. Protection from summer heat is provided by the roof overhang and by tall, deciduous trees.

Figure 38–22   A dramatic design for a passive solar house. (Courtesy California Redwood Association.)

Figure 38–23   Cathedral living area of the S residence showing the southwest-facing window wall and one of two wood-burning stoves. (Courtesy Professors Richard E. and Mary J. Kummer, the Pennsylvania State University.)

2. Solar heat is indirectly gained through a greenhouse also on the left wall that takes cooler air from the basement level through operable windows and warms the living level through adjustable wall registers. The greenhouse has a brick floor and water storage containers to extend the time of heat release throughout the evening.

3. Additional early morning solar heat is gained through a sun trap on the front (southeast) wall that warms the basement and living levels, both through operable windows.

4. Two wood-burning stoves are located centrally, one on the living level and another in the basement level.

5. Auxiliary heat is provided as needed through conventional electric baseboard units. Approximately 50 percent of this home's space-heating requirements is provided by solar means, 40 percent through the two wood-burning stoves, and 10 percent by the auxiliary electric heat.

6. To reduce heat loss, the house is insulated by:
   (a) Earthen berms along the rear and right walls.
   (b) Full-thick (R-11) insulation in 4" walls sheathed with 1" (R-5) extruded styrofoam, giving a total insulating ability of R-16.

   (c) Ten-inch (R-30) insulation in 12" rafters with 2" vent space above.

7. Liquid-type flat-plate collectors are on the southeast roof to provide preheating for domestic hot water. Approximately 50 percent of the home's hot water heating requirements is provided by this active solar heating method.

ACTIVE SYSTEMS

Flat-plate Collectors

A flat-plate collector system consists of exterior solar collectors to heat air or liquids, which are then piped to a heat storage container. A second piping loop is used to pipe the heat from the storage container to the building. An auxiliary conventional heating system is included as shown in Figure 38–24.

For best operation, it is important that flat-plate collectors face the sun. For example, a flat-plate collector in Dallas, Texas (latitude 33°) should face south and be tilted at an angle of 57° (90° – 33° = 57°) with the horizontal.

Air-type Collectors

(Figures 38–25 and 38–26). An air-type collector consists of "flat" panels, each about 4" thick, 3' wide, and 7' long. A solar heat absorber is inside the panel. The absorber surface is made of a good heat conductor such as aluminum, copper, stainless steel, or galvanized steel. The panel is covered with a translucent sheet of tempered glass or plastic. This permits sunlight to enter, but prevents the radiation of longer wavelengths given off by the absorber from leaving. Thus, heat is trapped inside the collector. There is, of course, insulation at the bottom of the

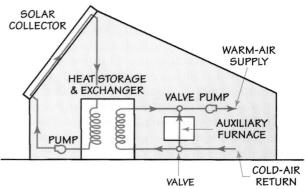

Figure 38–24   A flat-plate solar collector system.

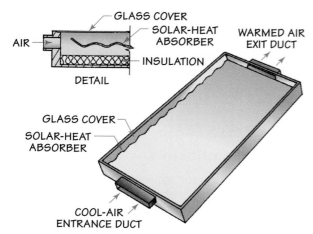

Figure 38–25 *An air-type solar collector.*

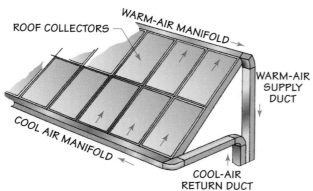

Figure 38–26 *An air-type assembly.*

panel. The panels can be mounted horizontally, vertically, or inclined. Cool air is drawn through the panel (under the absorber) by an automatic fan that operates only when the absorber temperature is sufficiently high. Usually the air enters at the bottom of an inclined or vertical panel and leaves at the top. The absorber surface is irregular in texture (such as corrugated) to cause the air to tumble. This increases the amount of heat transferred from the absorber to the air underneath. Also, the absorber metal usually has a special surface chemically or electrolytically applied to increase its absorption capacity. Viewed closely, it looks like the bluing on a gun barrel, but from a distance the collectors look like dark windows.

Air-type collector systems have lower installation costs than liquid-type collector systems.

They have no water freezing problems, and they are, in general, less complicated. However, they are not as efficient for hot-water heating, and they require hot-air ducts, which are many times larger than hot-water tubing. Also, rock storage containers (often used with air-type systems) require more space than hot-water storage (used with liquid-type systems).

Liquid-type Collectors

(Figures 38–27 and 38–28). A liquid-type collector is similar in operation to an air-type collector except that the heat transfer medium is water or antifreeze, which is pumped through copper tubing soldered to a copper absorber plate. The water system is drained during cold nights to prevent freezing. Some antifreezes are toxic and special precautions must be taken to prevent leakage into the hot-water supply. The water or antifreeze may be pumped either from the eave up to the ridge or from the ridge down to the eave.

Another form of liquid collector, the *evacuated tube*, is shown in Figure 38–29. This collector consists of three concentric metal or glass tubes. Cool liquid is supplied through the smallest feeder tube to the end of the assembly. The liquid picks up solar heat as it returns through the midsized absorber tube. There is a partial vacuum (an *evacuation*) between the absorber tube and the largest cover tube. The vacuum helps to reduce condensation and heat loss.

Trickle Collectors

(Figure 38–30). A trickle collector is a liquid-type collector that has cool liquid flowing from $\frac{1}{32}$" holes drilled in a header tube and trickling down open absorber channels rather than in closed absorber tubes. After the liquid is warmed, it is collected in an open gutter for distribution to storage. Compared to a closed system, the trickle collector is less costly to install and has fewer freezing problems, but has lower efficiency due to evaporation and condensation on the cover sheet.

Focusing Collectors

(Figure 38–31). Focusing collectors concentrate diffuse solar radiation onto a smaller absorber area and at a higher temperature. Figure 38–31

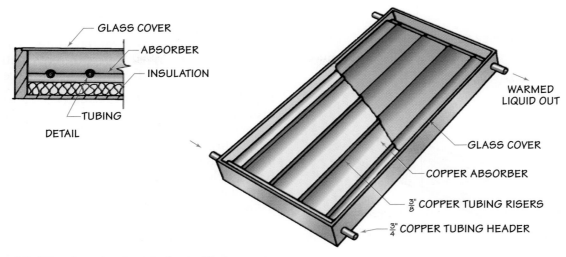

Figure 38–27   *A water-type solar collector.*

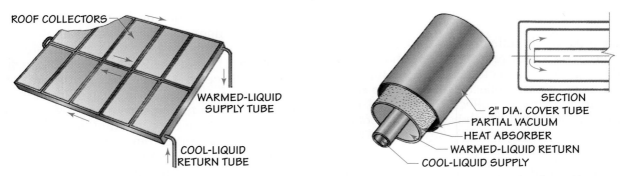

Figure 38–28   *A water-type assembly.*

Figure 38–29   *An evacuated-tube collector.*

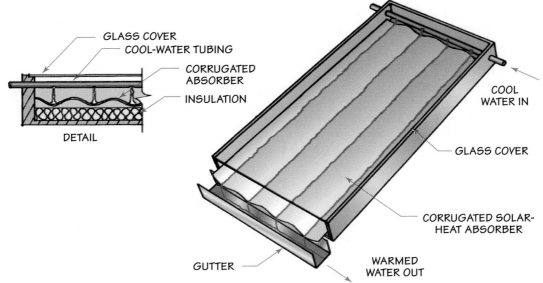

Figure 38–30   *A trickle-type solar collector.*

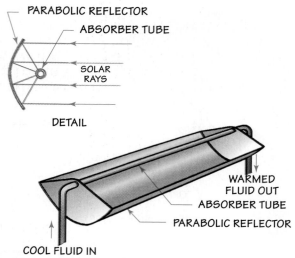

Figure 38–31 *A parabolic focusing solar collector.*

shows a parabolic reflector that focuses on the absorber pipe. Other reflector shapes have been used. Some systems include automatic tracking control to enable the collector to follow the sun's direction. A focusing collector connected to a heat pump can provide both heating and cooling.

Photovoltaic Cells

Photovoltaic cells convert solar radiation directly into electricity rather than into heat. Semiconductor chips are used which develop an electric field in the presence of sunlight. The chips are connected in a grid to produce electric current. This is a costly system, however, and consequently it is not often installed.

Heat Storage

In all solar heating systems, heat must be stored for future use in evenings and on cloudy days. Some of the most common types of heat storage are:

1. Rock storage
2. Liquid storage
3. Heat of fusion

Rock Storage Rock is usually used to store heat collected by air-type collectors. The heated air is drawn through an insulated container filled with pieces of rock or large-sized gravel.

As the air flows through the spaces between the rock, the rock is heated and some of that heat will remain for use several days later. The container, however, must be quite large—about 1 cu. ft. for each 2 sq. ft. of collector surface.

Liquid Storage Water is usually used to store heat collected by liquid-type collectors. This can be accomplished by either storing the heated liquid directly in an open-loop system or by circulating the heated liquid through a tank of water in a closed-loop system. The water tank will gain heat from the heated liquid, thus acting as a *heat exchanger*. A liquid storage system requires only about one-third the volume of a rock storage system.

Heat of Fusion The most promising system uses the *heat of fusion* to store heat. A great amount of heat is required to melt a solid into a liquid. When this melting, or *fusion*, occurs, the temperature does not change. For example, 1 Btu of heat is required to raise the temperature of 1 lb of water 1°F, but 144 Btu are needed to melt 1 lb of 32°F ice to 32°F water. In the heat-of-fusion system, heated air or heated liquid is circulated through tubing surrounded by a chemical compound having a melting point of about 90°F. As the chemical compound is melted from a solid to a liquid, it stores heat until the heat is released later by its changing back to a solid.

Heat Distribution

There are two common systems used to distribute stored heat in the building:

1. Forced warm-air
2. Forced-circulation hot-water

They differ from standard distribution systems only in that heat must be delivered in larger volume because it is at a comparatively low temperature. Conventional furnaces deliver warm air at about 150°F and hot water at about 190°F. In comparison, solar storage units deliver heat at about 100°F. Usually an auxiliary conventional heating system is provided which automatically cuts in as needed. If space cooling is desired in addition to space heating, in most cases a liquid system is better adapted than an air system.

Examples of Solar-heated Buildings The California residence shown in Figure 38–32 is heated

Figure 38–32   A sunroom formed by casements and skylights having internal, disappearing screens. (Courtesy Pella Corporation.)

Figure 38–33   An air-type solar collector. (Courtesy Peter Hollander.)

passively through automatically adjusted skylights. Heat is stored in interior masonry walls. Auxiliary heat is provided by electric resistance panels.

The Oregon residence in Figure 38–33 has an air-type solar collector on a roof sloping 60 degrees. The rock storage system contains 60 tons of 2" stone in a 6' × 12' × 18' insulated bin. Auxiliary heat is electric forced warm air.

The Oklahoma house in Figure 38–34 has a water-type solar collector on a roof sloping 50 degrees. The open-loop water storage system is a 2,000-gal. concrete basement tank. Auxiliary heat is electric hot water. Earth banked on several sides provides additional insulation.

The Colorado building of Figure 38–35 has a trickle collector on a 45-degree roof. The heat storage system is an 1,800-gal. steel water tank sized 4' × 5' × 12'. Auxiliary heat is provided by natural gas.

The accompanying table of advantages and disadvantages of different heating systems (Table 38–1) may help in the selection of a heating system for an individual case. You may want to add to it.

Figure 38–34   A water-type solar collector. (Courtesy Peter Hollander.)

Heat-loss Calculations

A heating system should be designed by a heating engineer specializing in the field. The drafter will prepare the heating plans, however, and should be familiar with the design process. Let us look, then, at the steps necessary to design a typical system.

Although every heating system must be individually designed, there are certain elements

Figure 38–35 *A trickle-type solar collector.*
(*Courtesy Peter Hollander.*)

common to all. For example, the total heat loss through the walls, ceiling, and floor of each room must be balanced by the heat delivered to each room. The first step, therefore, is to calculate the heat losses.

Heat Losses   Heat escapes from a room in two ways:

1. Transmission through walls, ceiling, and floor

2. Infiltration through cracks around windows and doors

Transmission Losses   Heat loss by transmission will increase in proportion to the area A of the surfaces of the room, the difference between the inside temperature and outside temperature $(t_i - t_o)$, and the coefficient of transmission U:

TABLE 38–1 *Comparison of Heating Systems*

	Advantages	Disadvantages
Warm-air	Quick heat	Ducts take up basement headroom
	No radiators or convectors to take up floor space	Ducts convey dust and sound
	Air conditioning and humidification possible	Flue action increases fire danger
	Cannot freeze	Separate hot-water heater required
	Low installation cost	
Hot-water	Low temperature heat possible for mild weather	Retains heat during periods when no longer required
		Slow to heat up
		Radiators require two lines
		Must be drained to avoid freezing when not in use
Radiant	No visible heating device	Slow response to heat needs
	Economical operation	Air conditioning must be separate unit
	Good temperature distribution	Repair costly
Electric	No visible heating device	Operation cost high in many locations
	Low installation cost	Heavy insulation required
	Individual room control	
	Clean, silent operation	
Solar	Low operation cost	Supplemental heating system necessary
		High installation cost
		System not fully developed

$$H = AU(t_i - t_o)$$

where:

H = Heat loss (Btu/hr)
A = Area (sq. ft.)
U = Coefficient of transmission (see Table 38–2)

t_i = Inside temperature (70°F is often assumed)
t_o = Outside temperature (assume 15°F above lowest recorded temperature; 0°F is used for New York, Boston, and Philadelphia)

TABLE 38–2   *Coefficients of Transmission (U) (Btu/hr./sq. ft./degree)*

Walls

Wood siding, plastered interior, no insulation	0.26
Wood siding, plastered interior, R-11 ($3\frac{1}{2}$") insulation	0.07
Wood siding, plastered interior, R-19 (6") insulation	0.05
Brick veneer, plastered interior, no insulation	0.26
Brick veneer, plastered interior, R-11 ($3\frac{1}{2}$") insulation	0.07
Brick veneer, plastered interior, R-19 (6") insulation	0.05
8" solid brick, no interior finish	0.50
8" solid brick, furred and plastered interior	0.31
12" solid brick, no interior finish	0.36
12" solid brick, furred and plastered interior	0.24
10" cavity brick, no interior finish	0.34
10" cavity brick, furred and plastered interior	0.24

Partitions

Wood frame, plastered, no insulation	0.34
4" solid brick, no finish	0.60
4" solid brick, plastered one side	0.51
4" solid brick, plastered both sides	0.44
6" solid brick, no finish	0.53
8" solid brick, no finish	0.48

Ceilings and Floors

Frame, plastered ceiling, no flooring, no insulation	0.61
Frame, plastered ceiling, no flooring, R-19 (6") insulation	0.05
Frame, plastered ceiling, no flooring, R-30 (9") insulation	0.03
Frame, no ceiling, wood flooring, no insulation	0.34
Frame, no ceiling, wood flooring, R-19 (6") insulation	0.05
Frame, no ceiling, wood flooring, R-30 (9") insulation	0.03
Frame, plastered ceiling, wood flooring, no insulation	0.28
Frame, plastered ceiling, wood flooring, R-19 (6") insulation	0.05
Frame, plastered ceiling, wood flooring, R-30 (9") insulation	0.03
3" bare concrete slab	0.68
3" concrete slab, parquet flooring	0.45
3" concrete slab, wood flooring on sleepers	0.25

Roofs

Asphalt-shingled pitched roof, no ceiling, no insulation	0.52
Asphalt-shingled pitched roof, plastered ceiling, no insulation	0.31
Asphalt-shingled pitched roof, plastered ceiling, R-19 (6") insulation	0.05
Asphalt-shingled pitched roof, plastered ceiling, R-30 (9") insulation	0.03
Built-up flat roof, no ceiling, no insulation	0.49
Built-up flat roof, no ceiling, 2" board insulation	0.12
Built-up flat roof, plastered ceiling, no insulation	0.31
Built-up flat roof, plastered ceiling, 2" board insulation	0.11

Windows and Doors

Single-glazed windows	1.13
Double-glazed windows	0.45
Triple-glazed windows	0.28
Glass blocks (8" × 8" × 4")	0.56
$1\frac{3}{4}$" solid wood doors	0.44
$1\frac{3}{4}$" solid wood door with storm door	0.27

The coefficient of transmission is a factor that indicates the amount of heat in British thermal units that will be transmitted through each square foot of surface in one hour for each degree of temperature difference. A poor insulator will have a high U value; a good insulator will have a low U value. Values of U have been computed by tests for a number of surfaces as shown in Table 38–2.

Air-infiltration Losses   In addition to the heat that is lost directly through the room surfaces, heat will escape through the cracks between windows and window frames and between doors and door frames according to the following formula:

$$H = 0.018LV(t_i - t_o)$$

where:

H = Heat loss (Btu/hr)
0.018 = 0.24×0.075 (0.24 is specific heat of air in Btu/lb.; 0.075 is density of air in lb./cu. ft.)
L = Length of all cracks (ft.)
V = Volume of air infiltration per foot of crack per hour. (See Table 38–3)
t_i = Inside temperature
t_o = Outside temperature

Values for the volume of air infiltration (V) for the average doors and windows are given in Table 38–3.

EXAMPLE
Find the hourly heat loss from the front bedroom of the M residence (Chapter 13) using the following data:
Door: Weatherstripped, $1\frac{3}{4}$" solid wood with storm door
Windows: Weatherstripped, double-glazed
Walls: Wood siding, plastered interior, $3\frac{1}{2}$" insulation
Floors: 6" insulation
Ceiling: 9" insulation

Solution

	A	U	$(t_i - t_o)$	$H = AU(t_i - t_o)$
Door	20 sq. ft.	0.27	(70 – 0)	378 Btu/hr.
Windows	24 sq. ft.	0.45	(70 – 0)	765 Btu/hr.
Walls	168 sq. ft.	0.09	(70 – 0)	1058 Btu/hr.
Floor	137 sq. ft.	0.05	(70 – 20)	343 Btu/hr.
Ceiling	137 sq. ft.	0.03	(70 – 0)	288 Btu/hr.
			Transmission losses =	2823 Btu/hr.

	L	V	$(t_i - t_o)$	$H = 0.018 LV(t_i - t_o)$
Door	19'	55	(70 – 0)	1317 Btu/hr.
Windows	34'	24	(70 – 0)	1028 Btu/hr.
			Infiltration losses =	2345 Btu/hr.

Total heat losses = 2823 + 2345 = 5168 Btu/hr.

TABLE 38–3   Volume of Infiltration (V) (cubic feet per foot of crack per hour)

Doors not weatherstripped	111
Doors weatherstripped	55
Windows not weatherstripped	39
Windows weatherstripped	34

System Design

After the total heat loss from each room has been calculated, a system must be designed that will replace the loss. For our example, we will assume that a series-loop baseboard, hydronic system is desired for the M residence, and that the baseboard heating units will be selected from the four types shown in Table 38–4. Two loops will be used on the main levels: one loop through the vestibule, kitchen-laundry, and dining-living areas; and a second loop through the bathroom, bedrooms, and study. The order of design is as follows:

1. Calculate the length of exterior wall available for baseboard heating units in each room and tabulate them in a form like that shown in Table 38–5.

2. Divide the heat loss by the available length of wall to find the minimum output required.

3. Select a baseboard type from Table 38–4 with a slightly greater output than required.

4. Divide the heat loss by the rated output of the baseboard selected to find the revised length of baseboard required.

TABLE 38–4   Radiant (R) and Radiant Convector (RC) Baseboard Outputs (average water temperature = 200°F)

Type	Height	Rated Output (Btu/hr./ft. of baseboard)
R	Low (7")	255
R	High (10")	365
RC	Low (8")	430
RC	High (10")	605

TABLE 38–5   *Baseboard Heating Design*

	Room	Heat Loss (Btu/hr.)	Exterior Wall Available (ft.)	Minimum Output (Btu/hr./ft.)	Baseboard Selected	Baseboard Length (ft.)	Correction Factor	Final Baseboard Length (ft.)
Loop 1	Vestibule	5,950	10	595	RC high 605	9.9	× 1	10
	Kitchen-laundry	6,240	13	480	RC high 605	10.3	× 1.075	11
	Dining-living	10,570	38	280	R high 365	29	× 1.15	33
Loop 2	Bathroom	2,150	4	540	RC high 605	3.6	× 1	4
	Bedroom	5,170	18	290	R high 365	14.2	× 1	14
	Master bedroom	8,000	14	570	RC high 605	13.2	× 1.075	14
	Study	3,670	14	260	R high 365	10	× 1.15	12
Loop 3	Lower level	17,500						

Total heat loss = 59,250 Btu/hr.

5. Divide the radiation approximately into thirds. Increase the middle third by 7.5 percent and the last third by 15 percent to compensate for the cooling of the water as it proceeds along its run.

6. Select a boiler with a capacity equal to the total heat loss (59,250 Btu/hr). Boilers for small buildings are usually rated on the basis of the net output rather than the gross output, so a 60,000-Btu/hr boiler will be large enough to include all heat losses and domestic hot-water demand.

7. Using manufacturers' catalogs, the sizes of the compression tank, pump, and main are determined. In this case, an 8-gal. compression tank and 1"-diameter main are selected. A separate 1" pump is installed in each of the three loops so that zone control of each loop is possible.

8. Lay out the system as shown in Figure 38–36. This layout can be completed using either hand-drawn or computer-assisted drafting.

AIR CONDITIONING

Air conditioning means different things to different people. The average person thinks of air conditioning as the cooling of a building on a hot day. The professional engineer, though, considers air conditioning to have a broader meaning: both heating and cooling in addition to humidity control, ventilation, filtering, and other processing. All of these elements are essential to human comfort and should be considered when planning a building.

Cooling

Heat will transfer from a warm surface to a cooler surface. Air cooling may be accomplished by withdrawing heat from the air by transferring it to the cooler surface of evaporator coils in a refrigerating unit. As described in the section on heat pumps, a *compressor* circulates a *refrigerant* between two sets of coils: *evaporator* coils and *condenser* coils. The refrigerant is a volatile liquid with such a low boiling point that it is a gas (*vapor*) under normal pressure and temperature. Common refrigerants are freon 11 (CCl_3F) and freon 12 (CCl_2F_2). As a vapor the freon is compressed in the compressor, and its temperature increases. At this high pressure and temperature, the vapor passes through the condenser coils to be cooled by the surrounding water and condensed into a liquid as shown in Figure 38–37.

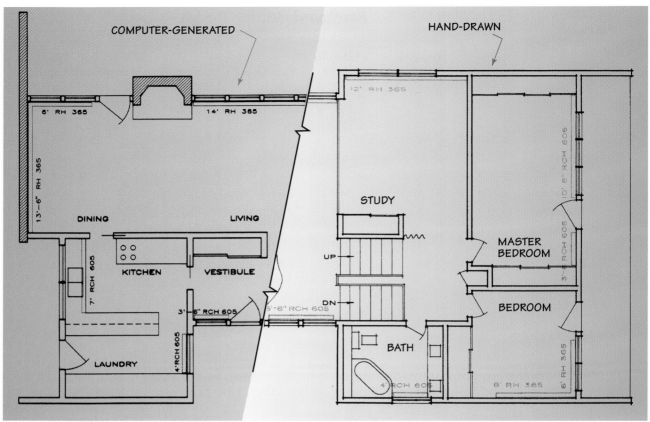

Figure 38–36 *Computer-produced and hand-drawn heating plan for the M residence.*

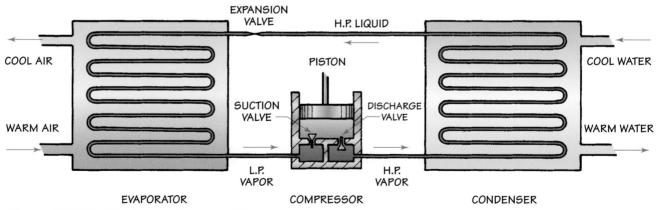

Figure 38–37 *Mechanical air conditioner.*

Still under pressure, the liquid refrigerant enters through the expansion valve into the evaporator, where the pressure is lowered by the suction stroke of the compressor. The boiling point of the liquid refrigerant drops, and the refrigerant changes into a gas (*vaporizes*). For this vaporization, a great deal of heat is withdrawn from the air or water surrounding the evaporator coils. The vaporized refrigerant is drawn back into the compressor through the suction valve to be compressed again, in a continuous cycle.

REAL WORLD APPLICATIONS

Heating a Greenhouse

 A greenhouse can be used to collect heat which can, in turn, be used to heat a home. Understanding the physical laws that govern this process will be beneficial as you consider a greenhouse design.

A greenhouse or sun room is heated through complex thermodynamics using sunlight as the energy source. In short, the light that passes through glass walls and ceilings is partially absorbed by the contents of the room. Light is energy and, as it is absorbed, it will again be radiated. The light is emitted as infrared heat waves, which cannot pass through glass and, thus are deflected back into the room. The room catches short waves of light energy and traps this energy by converting it to longer wavelength energy. Short-wavelength energy is ultraviolet and can go through glass while long-wavelength energy is infrared and cannot go through glass. It also helps that the glass walls and ceilings prevent heat loss through air circulation, so heat accumulates in the room.

The thermodynamic laws that govern this process of heating a sunroom are Wien's law and the Stefan-Boltzmann law. Go to your library and find information on these laws. What more have you learned? How can this information help you in designing a home using solar energy as a heat source?

Contributed by J. C. Wenger

In large air-conditioning systems, water is chilled in the evaporator and piped to the desired portion of the building. In smaller systems, such as unit air conditioners, the air to be cooled is allowed to enter the evaporator cabinet directly.

Humidity Control

As far as human comfort is concerned, temperature and humidity are inseparable. In dry air (low humidity), perspiration evaporates readily and cools the skin. Consequently, winter heating should be accompanied by **humidification**. In moist air (high humidity), perspiration will not evaporate, and the skin and clothing become wet and uncomfortable. Summer cooling, then, can be aided by **dehumidification**.

It is generally agreed that a comfortable winter temperature is 74°F with a relative humidity between 30 and 35 percent. (Relative humidity is the ratio of the quantity of water vapor actually present to the greatest amount possible at that temperature.) Lower humidity will dry furniture and house members, causing them to crack and warp. A higher humidity causes condensation on windows and possibly on walls. A summer temperature of 76°F at a relative humidity under 60 percent is desirable. Indoor temperatures in the summer are not lowered more than 15° below the outdoor temperatures to prevent an unpleasant chill upon entering or the feeling of intense heat upon leaving the building.

In addition to thermostats for controlling temperature, air conditioners are provided with *hygrostats* (also called *humidistats*) which are sensitive to and control the humidity of the air. Separate humidifiers and dehumidifiers in portable units are also available.

Ventilation

Temperature, humidity, and **ventilation** are all important to human comfort. A too-warm room having a gentle air motion may be more comfortable than a cooler room containing still, stale air. For air motion, a velocity of about 25 ft./min. is considered satisfactory. Much higher velocities cause uncomfortable drafts. Air-conditioning systems in large buildings continuously introduce some fresh outdoor air and exhaust stale air containing excess carbon dioxide, reduced oxygen, and unpleasant odors. Air from toilets, kitchen, and smoking and meeting rooms is not recirculated but exhausted directly. A complete air change every fifteen minutes is recommended for most activities. In uncrowded homes, natural infiltration provides a satisfactory amount of fresh outdoor air.

Filtering

Air contaminated with dust, smoke, and fumes can be purified by **filtering** it through filters and air washers of many designs. The most commonly used air filters are dry filters, viscous filters, and electric precipitators. Dry filters are pads of fibrous material such as spun glass or porous paper or cloth which must be cleaned or replaced to remain effective. Viscous filters are screens coated with viscous oil to trap dust. They may be cleaned by air or water and recharged by dipping in oil. Electric precipitators remove particles by passing the air through a high-voltage field. This charges the particles, which are then attracted to plates of opposite polarity.

Air Conditioning Systems

Air conditioning systems are designated *central* or *unit* systems. A central system may be designed as part of the heating system, using the same blower, filters, ducts or pipes, and registers. Or it may be separate from the heating system, having its own distribution method. In general, a single, combined, all-season system is more economical than two separate systems which must duplicate equipment.

For greater accuracy of control, it is often desirable to divide a building into zones for cooling as well as for heating. Frequently, the sections of a house vary in the amount of heating and cooling required due to different exposures to prevailing winds and sun, varying construction materials, and different uses. Thermostats are placed in each zone. The zones may be groups of rooms or individual rooms each with its own thermostat controlling the air conditioning.

Cooling systems may be combined with most heating systems. The warm-air system can supply cooled air as well. Chilled water can be circulated through the same pipes used in hot-water or steam heating systems. In this case the room convectors are equipped with blowers to circulate the warm room air over the chilled coils. The operation of the heat pump can be reversed to either supply or withdraw heat as required.

Self-contained room-sized *unit* systems are particularly effective in buildings with naturally defined zones. The units may be controlled automatically or manually as desired. Room units are built into the exterior walls in new construction and installed in window openings in existing dwellings.

Heat-gain Calculations

Summer heat enters a room just as winter heat escapes: by transmission through walls, ceiling, and floor and by air infiltration through cracks around windows and doors. But some additional factors, such as solar radiation, heat produced within the house, and latent heat, must also be considered.

Transmission Heat gained by transmission is calculated in the manner previously discussed for each area having a temperature difference between the inside of the room and the outside. For example, a wall adjacent to a garage without air conditioning should be included, but a below-grade wall should not.

Solar Radiation In addition to the normal transmission gain through glass, there is also a sun load through unshaded glass. Glass sun load is calculated only for the wall containing the largest area of unshaded glass, because the sun can shine directly on only one wall at a time. For an approximate calculation, assume that solar heat will enter through each square foot of glass on the east, south, and west walls at the rate of 100 Btu/hr. Double this amount for horizontal windows such as skylights. Windows facing north are omitted.

Infiltration Heat gained by air infiltration is calculated in the same manner as winter infiltration losses. The air conditioning system can be designed, however, to introduce sufficient outdoor air to maintain an indoor pressure capable of eliminating infiltration.

Occupants   The human body produces heat at an average rate of 300 Btu/hr. Therefore a heat gain for occupants is included by multiplying 300 times the assumed number of occupants.

Lighting   The equivalent heat of each watt of incandescent lighting is 3.4 Btu/hr. Therefore multiply 3.4 times the total wattage generally in use at one time. For fluorescent lights, use 4.0 in place of 3.4.

Appliances   The heat gain due to appliances can be estimated from Table 38–6. A figure of 1200 Btu/hr. is often used for the average residential kitchen.

Latent Heat   Latent heat gains must be included when water vapor has been added to the inside air. If the air conditioner is designed to condense an equal amount of moisture from the room, this factor is omitted.

Sizing the System

The total of all preceding heat gains is termed the *sensible heat gain.* The required size of the unit to be installed is found by multiplying the sensible heat gain by a performance factor of 1.3. A unit having this required heat removal rate in British thermal units per hour can then be selected from the manufacturers' catalogs. Cooling units may also be rated in tons of refrigeration. (A ton of refrigeration is the amount of refrigeration produced by melting a ton of ice in twenty-four hours.) A ton is equivalent to 12,000 Btu/hr. The size in tons, therefore, can be found by dividing the required size in British thermal units per hour by 12,000. A small house will usually require a 2- or 3-ton unit, a large house a 5-ton (60,000-Btu/hr.) unit.

A large commercial building might require several thousand tons of refrigeration. A 3,500-ton chiller and its computerized control center (Figure 38–38) are located on a mechanical floor (the sixty-third floor) of the U.S. Steel building in Pittsburgh. The computer center senses the solar energy being absorbed by the building and makes the required adjustments in the air conditioning.

Solar Air Conditioning

Research is now under way to perfect an air cooling system powered by some natural process rather than by a costly mechanical process. Recently an experimental solar air-conditioned home was built in Phoenix, Arizona. This building has a flat water-film roof in thermal contact with metal ceilings beneath and

TABLE 38–6   *Heat Emission of Appliances*

Electric oven	10,000 Btu/hr.
Electric range, no hood	4,000 Btu/hr./burner
Electric range, with hood	2,000 Btu/hr./burner
Electric warming compartment	1,000 Btu/hr.
Gas oven	10,000 Btu/hr.
Gas range, no hood	8,000 Btu/hr./burner
Gas range, with hood	4,000 Btu/hr./burner
Gas pilot	250 Btu/hr.
Electric motors	2,544 Btu/hr./horse-power

Figure 38–38   *Air conditioning control panel, U.S. Steel Building, Pittsburgh, Pennsylvania.*
(*Courtesy United States Steel Corporation.*)

covered by horizontal plastic panels. During winter daylight, the panels are retracted to allow the sun to heat the water and the house. The panels are retracted during summer nights also—but for a different reason. This allows the water film to evaporate, which cools the water and the house.

COMPUTER-AIDED HVAC PLANS

HVAC (heating, ventilating, and air conditioning) calculations and plans may be efficiently produced at your CAD workstation. An example of a typical HVAC computer program is Instant Software's *Solar Energy for the Home* which calculates heat loss and gain through south-facing windows. Previously-approved floor plans should be recalled from diskette storage to serve as the first layer on which to enter your HVAC design. Diskette menus of HVAC symbols can be helpful. An ASG digitizer tablet menu of standard HVAC symbols (Figure 38–39) includes round, rectangular, and flexible ductwork, registers, controls, fans, and accessory equipment. These symbols can be moved and rotated into position to be included on plans or in pictorials (Figure 38–40). In addition to these standard

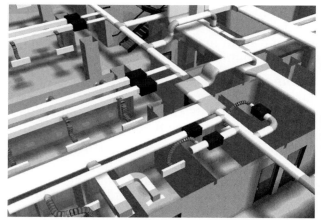

Figure 38–40 *Computer-produced three-point perspective of a commercial heat distribution design. (Courtesy Softdesk, Inc.)*

symbols, consider designing and storing any special symbols that you intend to use repetitively. When you have completed a satisfactory set of soft-copy HVAC plans, command your printer or plotter to produce hard copies. Also, store your plans on diskette for future reference.

Refer to Chapter 7 for operating details on providing computer-generated HVAC plans.

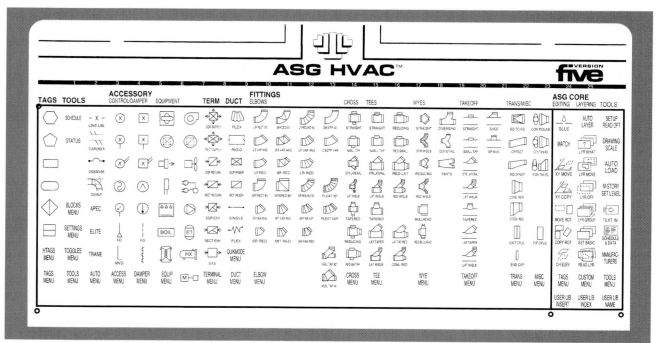

Figure 38–39 *Menu of ASG HVAC symbols for AutoCAD digitizer tablet. (Courtesy ASG.)*

REVIEW QUESTIONS

1. List four types of heating systems to consider when planning a new residence.
2. Distinguish between these forced warm-air systems:
   a. Individual-duct and trunk-duct
   b. Perimeter-loop and perimeter-radial
3. Distinguish between these forced hot-water systems:
   a. Series-loop, one-pipe, and two-pipe
   c. Direct-return and reversed-return
   b. Upfeed and downfeed
   d. Radiator and convector
4. What purpose does a thermostat serve?
5. Distinguish between these solar collectors:
   a. Air-type and liquid-type
   c. South-facing window and focusing-type
   b. Evacuated-tube and trickle-tube
6. Describe the method of operation of:
   a. The electric baseboard
   c. Solar heat (heat-of-fusion system)
   b. The heat pump
   d. The mechanical air conditioning system
7. Differentiate between passive and active solar heating systems.
8. Describe the process that a ground-source heat pump uses to heat and cool a house.
9. Give the equations that describe heat loss by transmission and by infiltration.
10. Distinguish between:
    a. Heat loss and heat gain
    d. Thermostat, hygrostat, and humidistat
    b. Central and unit air conditioning
    e. Infiltration and solar radiation
    c. Compressor, condenser, and evaporator

ENRICHMENT QUESTIONS

1. Why is a series-loop hot-water system not practical for larger houses?
2. List the advantages and disadvantages of each of these heating systems:
   a. Warm-air   b. Hot-water   c. Radiant   d. Electric   e. Solar
3. If a solar collector is sized 14' × 30', estimate the needed volume of:
   a. Rock storage
   b. Liquid storage
4. Describe two disadvantages that a flat-plate collector system would have if it were installed in a house in the forest.

PRACTICAL EXERCISES

1. Give the coefficient of transmission for:

   a. Brick-veneer wall, plastered interior, without insulation

   b. Brick-veneer wall, plastered interior, 2" insulation

   c. 10" cavity brick wall, without interior finish

   d. Wood-frame floor, without ceiling below, without insulation

   e. Wood-frame floor, without ceiling below, 4" insulation

2. Prepare a legend of the commonly used heating and air conditioning symbols. (Hint: Refer to a reference such as *Architectural Graphic Standards.*) Computer-produce or draw as appropriate.

3. For the A residence:

   a. Make the necessary calculations and draw or computer-produce the heating plan. Use the heating system of your choice.

   b. Include central air conditioning

4. Using the following data from the living room of the Z residence:

   🏛 Windows: Weatherstripped, double-glazed

   🏛 Unshaded area of east windows: 42 sq. ft.

   🏛 Unshaded area of south windows: 27 sq. ft.

   🏛 Unshaded area of west windows: 27 sq. ft.

   🏛 Walls: brick veneer, plastered interior, 2" insulation

   🏛 Floor: no insulation

   🏛 Outside temperature: 0°F

   🏛 First-floor temperature: 70°F

   🏛 Second-floor temperature: 70°F

   🏛 Basement temperature: 60°F

   🏛 Number of occupants: 5

   🏛 Wattage generally in use: 1,000

   a. Find the hourly heat loss   b. Find the hourly summer heat gain

5. For your original building design:

   a. Make the necessary calculations and draw or computer-produce the heating plan

   b. Include central air conditioning

CHAPTER 39

Energy Conservation

OBJECTIVES

By completing this chapter you will be able to:

- Identify and describe various forms of energy conservation techniques that can be incorporated into a residence or small commercial building.

KEY TERMS

Thermal insulation	Thermopane
R-values	Low-E glass
Vapor retarder	

Much of the energy used today is wasted. Obviously if the waste is reduced, the need for additional energy sources is also reduced. Some of the principal areas of residential energy conservation are in the reduction of waste in home heating and cooling, water heating, and electrical use.

HEAT CONSERVATION

Residential and commercial buildings require approximately one-third of the total energy we use. Most of this energy is needed to heat or cool buildings as shown in Figure 39–1, so the search for improvement has concentrated on improved insulation and other methods to save heat.

Thermal Insulation

Houses should be insulated to obtain the maximum efficiency from heating and cooling sys-
tems and to provide the comfort of a steady temperature. Properly installed **thermal insulation** can reduce heating and cooling costs by as much as 40 percent while at the same time making the house warm in winter and cool in summer.

Insulation is available in many forms. For new house construction, batts or blankets (rolls) are often specified. They are sized to fit snugly between joists or rafters and between studs. Most frame houses in cold climates are now being built with 2"× 6" studs spaced 24" oc (rather than 2"× 4" studs at 16" oc) to permit $5\frac{1}{2}$" of wall insulation, or with a staggered

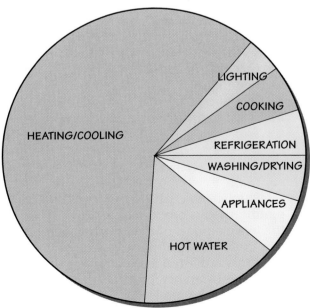

Figure 39–1 *Home energy consumption.*

double row of 2" × 4" studs to permit continuous insulation. Ceiling insulation is often 9" to 12" thick. However, it is important to understand that the insulating ability of different materials varies considerably. To provide a dependable system of comparison, insulating ability is specified by **R-values** (resistance values).

R-values

Prior to 1970 the insulating effectiveness of a building section was indicated by its thermal *conductivity* (U-factor). However, U-factors were confusing in that only a small difference in values (such as 0.03 and 0.04) indicated a large difference in insulating effectiveness. Conse-

quently Professor E. C. Shuman of the Pennsylvania State University proposed that insulating effectiveness be indicated by its thermal *resistance*, or R-value. An R-value is the reciprocal of a U-factor, so the above U-factors of 0.03 and 0.04 correspond to R-values of 33 and 25 which are obviously significantly different. Also, R-values are additive, while U-factors are not.

R-values are marked on all types of insulation and are the best indication of insulating effectiveness. Insulation with a high R-value has a higher insulating value than insulation with a low R-value. Two different types or brands of insulation with identical R-values will have the same insulating value, even if they have differing thicknesses. Insulation in batts and blankets

REAL WORLD *APPLICATIONS*

R-Values

In the past, the insulating effectiveness of a building section was indicated by its thermal conductivity (U-factor). More recently, the insulating effectiveness of a building section has been indicated by its thermal resistance, or R-value. An R-value is the reciprocal of a U-factor.

Calculating reciprocals is a math skill you learned in junior high school. To calculate a reciprocal of a number, divide it into 1. For instance, to find the reciprocal of .03, set up this formula: 1/.03 = 33.3. Conversely, finding the reciprocal of 33.3 can be done by dividing it into 1: 1/33 = .03. The numbers .03 and 33.3 are reciprocals of each other.

Practice calculating reciprocals by completing the accompanying table containing the five elements of an insulated wall. Calculate the R-values of the two elements whose U-factors are listed in the table. Then calculate the U-factors of the three elements whose R-values are listed in the table.

Material	R-value	U-factor
Brick	0.800	
$\frac{3}{4}$" Air space		1.090
$\frac{3}{4}$" plywood	0.930	
3" blanket insulation		0.091
$\frac{1}{2}$" drywall	0.450	
Total insulating factor		

Another advantage of R-values compared to U-factors is that they are additive. Prove this principle by totaling the R-value and the U-factor for the insulated wall. In the table, when the reciprocal of the total R-value is calculated you will notice that it is not equal to the total U-factor. The total R-value indicates the total insulating effectiveness, or thermal resistance of this wall, but the U-factor does not.

Calculate the R-value of the roof for residence A and your own residence design.

TABLE 39–1 *Common R-values for Mineral-fiber Batts*	
R-value	**Thickness**
R-11	$3\frac{1}{2}$"
R-19	6"
R-25	8"
R-30	10"
R-38	12"

TABLE 39–2 *Recommended Thermal Insulation in Cold Climates*	
Building Component	**Recommended Insulation**
Ceiling	R-30 or R-38
Walls	R-11 or R-19
Basement walls	R-11
Floor (over unheated space)	R-19
Slab on grade	R-7 (2" rigid insulation)

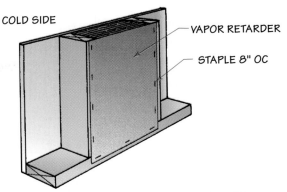

Figure 39–2   Nonfoil insulation should completely fill available space.

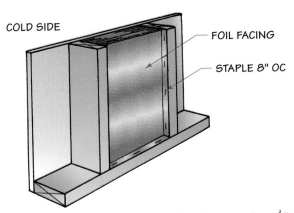

Figure 39–3   Foil-faced insulation requires $\frac{3}{4}$" air gap.

is commonly produced with R-values of 38, 30, 19, and 11 as shown in Table 39–1. R-values are additive. That is, two R-19 batts equal one R-38 batt. Table 39–2 indicates the amount of insulation recommended for houses in cold climates.

Installation

Obviously, insulation will not be very effective if carelessly installed. For example, batts and blankets should fit tightly without "fish-mouth" gaps between the insulation and the studs. Small spaces and cracks should be hand-packed with loose wool. As shown in Figure 39–2, install as much insulation as the available stud and joist depths allow. However, foil-faced insulation must be installed with a $\frac{3}{4}$" air space as shown in Figure 39–3 to benefit from the reflective value of the foil.

Most insulation batts and blankets are made with a **vapor retarder** on one side, which is in-stalled facing the inside (warm side) of the house to prevent condensation. (Often called *vapor barrier*, but the correct technical term is *vapor retarder* because the flow of vapor is merely retarded and not completely stopped.) Condensation occurs only when moisture reaches a cold surface through which it cannot readily pass, and the vapor retarder, when properly installed, does not become cold. If the insulation is improperly installed, with the vapor retarder toward the outside (cold side) of the house, moisture will condense upon contact with the cold vapor retarder, making the insulation ineffective. Also, any breaks in the vapor retarder will decrease its effectiveness. All such breaks should be repaired before the insulation is covered by the wall surface. Proper installation of insulation is important because condensed moisture cannot be dried in the summer. This leads to rotting of wood and rusting of metal.

For fire safety, do not insulate within 3" of recessed lighting fixtures or other heat-producing equipment.

Superinsulation

Buildings insulated more heavily than recommended in Table 39–2 are termed *superinsulated*. Residences have been built, for example, with R-30 wall insulation by staggering 2" × 4" studs within a 10" wall, thus preventing any heat loss by direct passage through the studs. See Figure 39–4.

Types of Insulation Two forms of insulation are commonly available: mineral wool (fiberglass and rock wool) and cellulose.

Mineral wool is manufactured in batts, blankets (rolls), and loose-fill. The loose-fill is installed by pouring or blowing under pressure. Mineral wool is preferred over other types of insulation because it is not combustible (since it is made from glass or rock) and does not slump under pressure.

Cellulose, on the other hand, is a shredded paper product and is chemically treated to make it fire-retardant. Unfortunately, the chemicals can be driven off by summer heat. Also, they leach out when the insulation becomes wet. Cellulose insulation slumps to about half its volume when wet.

Rigid insulation made of asphalt-impregnated fiberboard, and *foamed plastics*, such as Styrofoam® and urethane, are frequently used as insulation under built-up roofs. A common size for these materials is 24" × 48" in thicknesses of $\frac{1}{2}$" to 3" in $\frac{1}{2}$" increments. Foamed plastics are popular as perimeter insulation, installed between the foundation and the floor slab. Perimeter insulation is often 2" thick × 24" wide as shown in Figures 39–5 and 39–6. Insulation can also be sprayed on the interior of a building in thicknesses up to 2".

To insulate existing buildings, loose-fill insulation can be poured between studs and joists, or blowing wool can be blown in, filling the walls to the full depth of the studs and to the desired depth between joists.

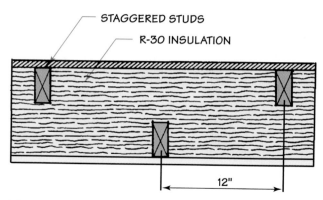

Figure 39–4 *Plan of a superinsulated wall.*

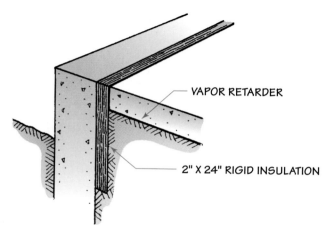

Figure 39–5 *Perimeter insulation for slab construction.*

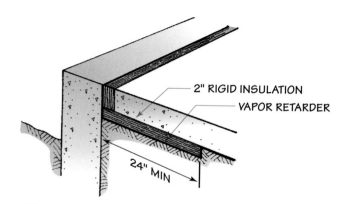

Figure 39–6 *Alternate method of installing perimeter insulation.*

Computer-aided Design for Insulation

Specialized computer programs are used by architectural offices to design buildings to desired levels of thermal insulation. For example, the transmission losses through walls, ceiling, and floor and infiltration losses around windows and doors can be determined by computer for each room and the entire building to arrive at recommended adjustments in insulation R-values. Some typical programs are *Building Performance Profile* by Compass Systems (calculates heat loss, heat gain, and energy consumption) and *Energy Audit Analysis* by Eloret Corporation (computes energy savings). Refer to the "Architectural Databases" sections of Chapter 6 for more detailed information on computer-aided design for energy considerations.

Additional Measures

In addition to insulation, many additional actions can help conserve heat. Some of these include:

1. Double glazing (such as **thermopanes** or storm windows over single panes) reduces heat loss through the windows by about 60 percent (Figure 39–7). Double glazing of **low-E glass** (low-emissivity glass manufactured

ARCHINET http://www.

Insulation Needs

Insulation needs of any new home are based on the size and location of the home. When designing a home for a consumer, it is important to check the Department of Energy's R-value recommendations for the home's location. The Department of Energy recommends a specific R-value for ZIP code areas based on method of heating. One convenient way to obtain this information is on the Internet.

The Owens Corning homepage is one place to start. The homepage address is: **http://www.owens-corning.com/ owens/around/insulation/howmuch/**

This address takes you to the "How much do I need?" page. Click "R-value recommendations for your area." On this new page click "Department of Energy recommendations." Once this site is located, type in your ZIP code at the prompt. Also, type in your method of heating. Copy or print this information. What R-values does the Depart-

ment of Energy recommend for a home in your area for the following: ceilings below ventilated attics, floors over unheated crawl spaces or basements, new construction exterior walls, and crawl space walls?

Now go back to the "How much do I need?" page on which your search began. Click "How much will you need to buy?" Follow the instructions provided to determine the amount of insulation you will need for your original house design. Finally, call three local dealers to compare prices for the insulation.

As someone interested in architecture you need to remain informed of the most recent technology. In the area of energy conservation, R-2000 homes are the latest in energy efficient housing in Canada. For information on these homes, access this Internet site sponsored by the University of Alberta: **http://www.ualberta.ca/ ~amulder/house**

Figure 39–7 A solar kitchen of insulated case-ments and venting skylights. (Courtesy Andersen Windows, Inc.)

with a microthin, transparent coating that re-flects inside heat in the winter and outside heat in the summer) reduces heat loss by about 66 percent. Triple glazing (storm win-dows over thermopanes) is even more effec-tive and reduces heat loss by about 75 per-cent. Double and triple glazing also reduce noise and condensation on the windows.

2. Appropriate glazing can also help cool the house in the summer. To reduce solar heat and glare, windows can be glazed with heat-absorbing glass. Such glass, combined with an ingenious design, reduced the solar-heat load by 75 percent in the Norfolk City Hall shown in Figures 39–8 and 39–9. An outer window wall is fastened 3' from the inner window wall with tubular steel trusses. This permits a cooling air flow between both lay-ers of glass as illustrated in Figure 39–10. The upper grid serves as a solar screen and is also used as a walkway for window washers.

3. Storm doors reduce transmitted heat loss through doors by about 40 percent. Storm doors also reduce the amount of heat lost when the door is opened.

4. Weatherstripping reduces heat lost by infil-tration around doors by about 50 percent. Windows also should be weatherstripped if they do not close tightly. Weatherstripping is available in foam rubber, felt strips, flexible vinyl, spring metals, and other materials.

Figure 39–8 Norfolk, Virginia, City Hall. (Courtesy American Iron and Steel Institute.)

Figure 39–9 Corner detail of Norfolk City Hall. (Courtesy American Iron and Steel Institute.)

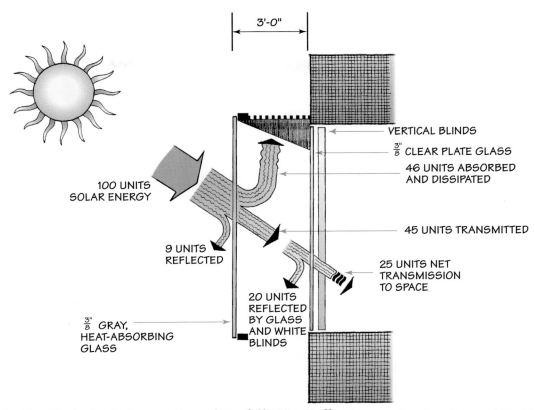

3'-0"

VERTICAL BLINDS

$\frac{3}{8}$" CLEAR PLATE GLASS

46 UNITS ABSORBED
AND DISSIPATED

100 UNITS
SOLAR ENERGY

45 UNITS TRANSMITTED

9 UNITS
REFLECTED

25 UNITS NET
TRANSMISSION
TO SPACE

20 UNITS
REFLECTED
BY GLASS
AND WHITE
BLINDS

$\frac{3}{8}$" GRAY,
HEAT-ABSORBING
GLASS

Figure 39–10   Typical window section of Norfolk City Hall. (*Courtesy American Iron and Steel Institute.*)

5. For maximum energy conservation, some houses are built partially or completely underground. The earth, with a more uniform temperature than air, protects against both the heat of summer and the cold of winter. At 10' below the surface, the temperature of the earth remains at about the average year-round temperature at that location. This is about 50°F in a temperate zone. *Underground* structures (also called *earth-sheltered* structures) are often built not entirely underground. Designs vary from side-hill exposures to atrium concepts which are open to the sky.

The principal advantages of underground structures are:

1. Energy savings of 50 percent or more due to the temperature moderation of the covering earth.

2. Natural air conditioning in hot summer months.

3. Insulation from outside noise.

4. Protection from fire, storm, and earthquake.

The principal disadvantages are:

1. Careful design is needed to reduce an isolated feeling.

2. An earth covering of over 8' is needed for constant temperature. This results in a roof load of 800 lb./ft^2.

3. Reinforced concrete construction is more costly than standard framing materials.

4. Underground repairs in a concrete wall are difficult.

5. Ground water conditions (such as a high water table) make many sites unsuitable for efficient underground construction.

6. Some additional energy is needed to heat, dehumidify, and light the building.

Earth-sheltered houses are often designed as passive solar houses. When this is done, the principal window wall should, of course, be facing south. Views from a house which is lower than normal are necessarily limited. Consequently, a site sloping down to the south offers the best opportunity to obtain both passive solar heat and a satisfactory view. When designing underground housing, it is important to remember that all building codes require an operable window or outside door in each bedroom for emergency use as fire exits.

REDUCING ENERGY WASTE

In addition to energy conservation by means of proper insulation and other design and construction features, there are many conservation habits that we should all develop. Most of these suggestions require no capital investment, just common sense.

Reducing Heat Waste

1. Keep windows and doors closed.

2. During cold weather, open drapes on south-facing walls to admit warm sunshine, but close drapes at night to keep out cold air. During warm weather, reverse the process. Close out the sun during the day and open drapes during cool nights.

3. Replace broken glass promptly.

4. Close the garage door of a garage attached to a house to reduce heat loss from that side of the house.

5. Keep the heating thermostat set as low as comfort permits. Each degree Fahrenheit lower reduces required heat energy by about 3 percent. Each degree Fahrenheit higher increases required heat energy by about 3 percent.

6. People generate heat. Therefore lower the thermostat about 5°F one hour before guests arrive.

7. Close radiators or heating ducts in unused rooms. Keep the door to these rooms closed.

8. Keep the fireplace damper closed when the fireplace is unused to prevent the natural draft of the chimney from pulling out warm air in the winter and cool, conditioned air in the summer.

9. Use kitchen and bathroom exhaust fans sparingly during the heating season—they

not only waste heat, but also decrease humidity. Low humidity is undesirable in the winter because then a higher temperature is needed for comfort. See the humidity control section in Chapter 38.

Reducing Air Conditioning Waste

1. Open windows for moderate cooling rather than operating equipment.

2. Close windows and doors when an air conditioner is operating.

3. Use a window air conditioner to cool the room occupied rather than the entire building. Close doors to other rooms.

4. Place window air conditioners in a north or shaded window. Direct sunlight increases the work load.

5. Do not purchase window air conditioners with energy efficiency ratios (EER) below 8. The EER is the cooling capacity in British thermal units per hour divided by the electrical power input in watts. Try to locate equipment with an EER of 9 or 10.

6. Draw drapes to close out hot sun.

7. Relieve attic heat by ventilation.

Reducing Hot-water Waste

1. Reduce standby heat loss by setting the water heater at the lowest acceptable temperature—130°F is adequate for most purposes.

2. Break the habit of running water needlessly, for instance while shaving or brushing teeth.

3. A shower requires only half the water of a bath.

4. Repair all faucet leaks promptly. A slow drip wastes about 100 gal. a month.

5. Insulate long lengths of hot-water pipe.

Reducing Electricity Waste

Water Heater Reduce electric water heater waste as just described. For a house that is not heated electrically, about half of the total electric energy is used to heat water.

Lighting

1. When purchasing incandescent lamps, consider their brightness (measured in lumens)

in addition to the electricity used (measured in watts). Remember that large-wattage lamps are more efficient (emit more lumens per watt) than small-wattage lamps. For example, one 75-W lamp requires the same wattage as three 25-W lamps, but it will emit about twice as much light. Only wattage is marked on the lamp, but look for the brightness in lumens, which is marked on the carton.

2. Also remember that long-life incandescent lamps emit only about 90 percent as much light as a standard bulb of the same wattage. Consequently they should be installed only when their durability is useful, such as in hard-to-reach places.

3. Choose fluorescent fixtures whenever possible, for fluorescent lamps are many times more efficient than incandescent lamps.

4. Keep lighting fixtures clean to obtain more light for the same wattage.

Refrigeration

1. To reduce the amount of cold air lost to the room, do not open a refrigerator or freezer door more often or longer than necessary. Plan ahead and remove all items for a meal at the same time.

2. Do not purchase a refrigerator or freezer larger than needed, for the energy used depends upon the size of the refrigerated space regardless of whether it is all or only partially used.

3. Choose a chest-type freezer over an upright, as less cold air is lost when the door is opened.

Cooking

1. Do not heat more water than needed at any one time.

2. A cooking utensil should completely cover the range-top burner to minimize the heat loss to the room.

3. Choose a microwave oven over a standard oven since only about half as much wattage is needed.

Laundry A partial load of laundry requires about the same wattage as a full load. Therefore, launder less frequently, but with full loads.

WATER CONSERVATION

The search for ways to better conserve home energy has expanded to conservation of water. One of the reasons that water shortages are predicted is that the same high-quality drinking and cooking water is also used for bathing, washing, and toilet flushing. It is possible to install a water supply system that recycles graywater (from tubs and sinks) for use in flushing. A filtration unit is a necessary part of the system. Another system recycles blackwater (flushing water) over and over again by directing it through a sewage treatment unit. Such a system, designed for use in commercial buildings, is available from Cromaglass Corporation, Box 3215, Williamsport, Pennsylvania 17701.

A third system reduces the amount of water required for each flushing from four gallons to only one gallon by using a high velocity, rather than a high volume, of flushing water. A major manufacturer is Briggs Plumbingware, Box 31622, Tampa, Florida 33631.

A fourth system uses no water for flushing at all. Rather, the toilet and kitchen wastes are collected in a form of compost pile. The wastes slide down an incline and are converted to humus. Although it takes about two years for the first humus to be produced, the process is then continuous and will produce about 5 gals./person/yr.

REVIEW QUESTIONS

1. Define conservation as it relates to energy efficiency.

2. What should be done first if a homeowner is considering ways to save energy?

3. List three different types of insulation.

4. Differentiate between conductivity and resistance as they relate to thermal insulation.

5. How would you reduce the following by half?

a. Transmission loss through single-glazed windows

b. Transmission loss through exterior doors

c. Infiltration loss around exterior doors

6. List three methods to conserve:

a. Heat

b. Hot water

c. Cold water

d. Electricity

e. Air conditioning

7. Distinguish between:

a. Double glazing and triple glazing

b. Graywater and blackwater

ENRICHMENT QUESTIONS

1. What action would you take to double the insulation value of an R-19 attic floor? Why would you do this?

2. Given the following data, calculate the probable earth temperature at 10' below the surface:

Average temperature during December to March: 35°F

Average temperature during April and May: 56°F

Average temperature during June to September: 68°F

Average temperature during October and November: 50°F

3. Define "breakeven point." Do you feel there is a breakeven point when insulating a residence?

PRACTICAL EXERCISES

1. Prepare the insulation specifications for the A residence.

2. Prepare the insulation specifications for your original house design.

3. Design a building using nontraditional methods to conserve energy. Computer-produce, draw, or sketch as appropriate.

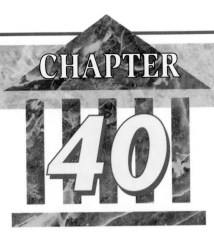

Energy Sources

OBJECTIVES

By completing this chapter you will be able to:

🏛 Describe the advantages and disadvantages of alternative wood, wind, and water forms of energy.

KEY TERMS

Fossil fuels	Wind energy
Wood energy	Water energy

The history of the progress of civilization parallels the history of its wise use of the energy available coupled with development of new energy sources. Early energy sources included wood fires for heating and cooking, animal power for farming and land transportation, and wind power for sea transportation. Later, **fossil fuels** (coal, oil, and gas) were used to power machines such as steam and gasoline engines. Reliance on fossil fuels has increased to the extent that they now provide 95 percent of our energy. But it has become evident that these fuels are being rapidly depleted and will have to be conserved and replaced. A worldwide search for new energy sources has concentrated on nuclear reactions, synthetic fuels, ocean energy, geothermal ("earth's heat") energy, solar energy, and others. Some of these sources (such as solar energy) are renewable. Others (such as geothermal energy) are not, but are seldom tapped. The search for new sources of energy for buildings has concentrated on solar energy (see Chapter 35), **wood energy**, **wind energy**, and **water energy**. In most instances, these sources are able to provide only a small portion of our energy needs. Consequently, the designer must carefully investigate all the consequences of using "free" energy. No energy is really "free" since most energy becomes quite costly during its conversion to useful service.

WOOD ENERGY

For centuries wood provided the major source of energy in the United States. Even as late as 1900 almost half of our heat and power was supplied by burning wood. Today, in spite of the increased processing of wood to lumber and paper, there is an abundance of many wood species. This is due partly to the decline of wood as a fuel and partly to many small Eastern farms being converted back to forest land. In the search for additional energy, wood provides a partial answer. Also, when wood is intelligently harvested, it can be a renewable energy resource. Approximately 5 cords of hardwood are needed to provide heat for an average home in a temperate climate. (A *cord* is a 4' × 4' × 8' pile of wood.) This could be provided by a woodlot of approximately 10 acres. On a larger scale, wood plantations have been used as a "living filter" for disposal of treated sewage effluent. This also increases wood growth.

Wood Furnaces

Wood-burning furnaces are available that provide hot water as well as hot air. They are fitted

REAL WORLD APPLICATIONS

Article Analysis

Your firm has received a large contract to design homes for a distinguished contractor in your area. One of the questions the contractor has is the energy source to be used in these homes. This contractor is interested in using an alternative energy source such as wind, solar, or wood. Your supervisor has assigned you the task of reading articles on these energy sources and writing an evaluative summary of your findings.

Read at least two articles about these energy sources and write a summary of each. Be sure to include an evaluation of not only the content of the article but also the writing style. Is the article meant to be persuasive? If so, how much of the article is fact versus opinion? Are the ideas presented in the article supported by evidence? For what audience is the article intended? Would you consider this a more technical article or is it a public interest article?

It may be helpful for you to access the following web sites for resources and information:

🏛 American Wind Energy Association (AWEA)—**http://www.igc.apc.org/awea**

🏛 National Renewable Energy Laboratory (NREL)—**http://www.nrel.gov**

---

with wood storage features that permit wood to feed down as needed during a period of about twelve hours. They often are installed with a supplemental oil burner that automatically ignites when the wood supply burns out. All sources of combustion require safety measures, but the flues of wood furnaces, especially, must be cleaned frequently to avoid flue fires. And emission limits for wood furnaces require them to have catalytic converters to reduce toxic smoke pollution.

Fireplaces

Much of the heat in conventional fireplaces is lost by convection up the flue, but there are several methods of increasing efficiency. The fireplace liner described in Chapter 29 increases heating capacity by drawing in room air, warming it, and discharging it back to the room or to other rooms. A leading manufacturer of heat-circulating fireplace liners is Heatilator Inc. Division, HON Industries, 1915 W. Saunders St., Mt. Pleasant, Iowa 52641.

Further efficiency can be achieved by designing the fireplace and chimney so that outside air is used to help support combustion rather than the already heated house air. When a fireplace is designed to operate on outside air (as shown in Figure 40–1), a tempered glass door is used to separate the outside air within the fireplace from the house air. The fireplace will then provide heat to the house by means of a fireplace liner such as just described.

The efficiency of existing traditional fireplaces can also be improved by partially enclosing the fireplace opening with glass doors as shown in Figure 40–2. The doors have air intake dampers below and air outlets above. Cool air is pulled in through the lower damper, heated by the fire, and then directed back to the room rather than up the flue. Also, sufficient air is

Another method of increasing the heating capacity of a traditional fireplace is by means of a hollow fireplace grate. As shown in Figure 40–3, cool room air is drawn into hollow pipe grates at the base, warmed by the fire, and discharged back to the room from the top.

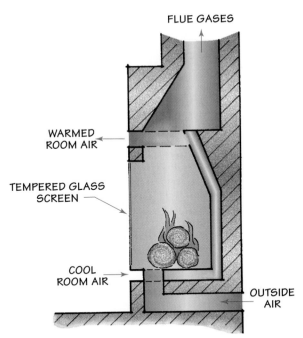

Figure 40–1   *A fireplace designed to use outside air.*

introduced into the fireplace to support combustion. A leading manufacturer is the American Stovalator Corporation, P.O. Box 1069, East Manchester Rd., Manchester Center, Vermont 05255.

Figure 40–3   *A hollow fireplace grate increases heating capacity.* (*Courtesy Thermograte.*)

Figure 40–2   *Tempered glass doors with air ducts increase fireplace efficiency.* (*Courtesy Thermograte.*)

Efficiency is further increased by grates with electrically driven blower units. These units are manufactured by Thermograte, 301 East Tennessee Street, Florence, Alabama 35631.

WIND ENERGY

For centuries wind was used to move ships, mill grain, and pump water. Today only about fifty thousand working windmills are operating in the United States. However, the windmill is experiencing a rebirth in the search for additional energy sources. Two principal manufacturers are Aermotor Division, Braden Industries, Box 1364, Conway, Arizona 72032; and Dempster Industries, P.O. Box 848, Beatrice, Nebraska 68310. Windmills are available with fans in sizes from 6' to 16' in diameter. Pumping capacities start at 100 gal./hr. and go up to 3,000 gal./hr.

There are many forms of wind machines that differ from the traditional water-pumping windmill of Figure 40–4. The *propeller-type* (Figure 40–5) has propeller blades similar to an aircraft's blades. There are upwind types (with tail vanes) that face the wind and downwind types (without tail vanes) that face away from the wind.

A *depression-type* wind machine (Figure 40–6) operates in yet another fashion. Its function is not to drive any mechanism by direct

Figure 40–4   Traditional water-pumping windmill.

gearing but rather to create a partial vacuum within the tubular tower so that air is drawn into the turbine situated in the base of the tower, thus driving an alternator. The propeller blades are hollow and have orifices at their tips through which the air contained in them is ex-

ARCHINET   http://www.

Cost Considerations in Heating

One substantial consideration of all persons building a new home is the type of fuel they will use in heating their home. As an architect, it is important that you be able to answer your clients' questions in this area. Therefore, you should be aware of various home heating options and their associated costs. One source of this information is the Internet. The following steps will enable you to find a table that compares costs of various fuels used in heating.

1. Type this address:

    http://www.hearth.com/hearth-net.html

    It will take you to the Hearthnet homepage.

Figure 40–5 *An upwind propeller-type wind machine.*

pelled by centrifugal force, thus reducing the pressure in the hub and tower interior.

The *Savonius rotor* (Figure 40–7) is balanced on a vertical shaft and has several curved vanes to catch the wind. A horizontal-shaft windmill must be able to rotate toward the wind, but a vertical-shaft rotor has the advantage that it always faces the wind. The *Darrieus rotor* is also balanced on a vertical shaft, but it has thin, curved blades which can rotate at high speed. A

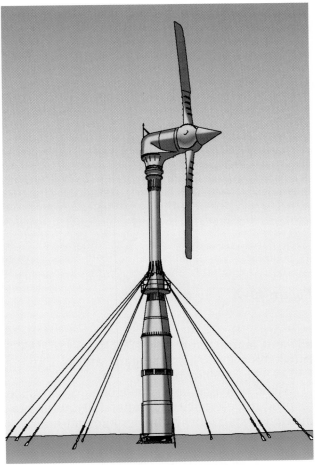

Figure 40–6 *A 100-kW depression-type generator. (Courtesy Delta Enfield Cables Limited.)*

2. On this page, click on "What's it all about?"

3. Click on "Specific Information on Hearth Fuels and Appliances," which is located at the top of this page.

4. Finally, click on "Compare the costs." This compares the costs of various home heating fuels using a table. Using the information from this table and information in this text, answer the following questions:

a. What does BTU mean?

b. How much does 1,000,000 BTUs of oil, propane, and wood cost?

c. What fuel is the least expensive? Most expensive?

d. What would cause these prices to change from year to year and season to season?

Now go back to the "Specific information" page and choose one of the fuels listed. Write a brief summary about this fuel's usage in home heating. The following questions may be useful. What heating appliances does this fuel use and how do they operate? What are the benefits and disadvantages of using this fuel?

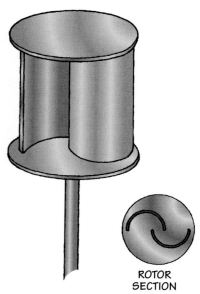

Figure 40–7 Savonius rotor.

65' unit manufactured by DAF Indal Ltd., 3570 Hawkstone Rd., Mississauga, Ontario, Canada L5C2V8 has a 50kw output.

Wind machine farms have been constructed to produce electricity commercially with production costs as low as 6¢ per kilowatt-hour.

If the wind is used to provide energy or pump water for a single building, it would be expected that the wind machine would be integrated into the overall design. Occasionally a wind energy collector and a solar energy collector will both be installed on a single building, for the wind often blows when the sun isn't shining. Of course all designs must be able to withstand high winds without damage. Also they must not disturb the occupants by excess noise or vibration.

WATER ENERGY

Hydroelectric Turbines

Although flowing river water produces the energy for many large-scale hydroelectric turbines, water is seldom used to provide energy for a single building since the number of suitable sites obviously is limited. Even those persons who are fortunate enough to have a home near a river with sufficient volume, velocity, and head (vertical drop) may not be permitted to interfere with the natural flow.

Hydraulic Rams

Many farms have hydraulic rams which use the energy of moving water to pump a portion of the water to storage at a higher level. As shown in Figure 40–8, when the rush of the supply water closes the waste valve, the momentum of the suddenly checked current opens the delivery valve and forces water into the air chamber, thus compressing the air. The compressed air shuts the delivery valve and forces water up to storage. When the waste valve drops open, the operation starts to repeat. A hydraulic ram having a 50-gal./min. intake with a head of 8' will pump about 9.6 gal./min. to a height of 25'. More delivery height can be obtained by increasing the intake volume, increasing the head, or decreasing the delivery volume. A principal manufacturer of hydraulic rams is Rife Hydraulic Engine Manufacturing Company, P.O. Box 790, Norristown, Pennsylvania 19401.

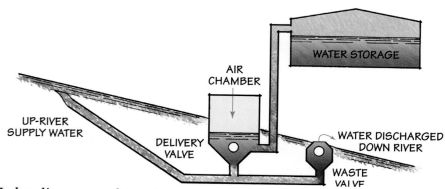

Figure 40–8 Hydraulic ram used to raise water.

REVIEW QUESTIONS

1. Describe briefly each of the following terms:
   a. Cord of wood
   b. Head of water
   c. Fossil fuels
   d. Living filter
   e. Hydraulic ram
   f. Turbine

2. What is the difference between:
   a. Solar energy and geothermal energy
   b. Propeller-type and depression-type wind machines
   c. Darrieus and Savonius rotors

3. List four methods of increasing fireplace efficiency.

4. List four types of nontraditional wind machines.

5. What is the advantage of:
   a. A vertical- over a horizontal-shaft wind machine
   b. A vertical- over a horizontal-shaft water turbine

ENRICHMENT QUESTIONS

1. Hardwood and softwoods are commonly used for burning. List five species of each and iden-tify which species burns the hottest and for the longest time.

2. Do you feel windmills are a practical source of energy for an individual residence?

3. Use a can to build an example of a Savonius rotor. Demonstrate how it functions for the class.

PRACTICAL EXERCISES

1. For your original house design, draw or computer-produce the details for:
   a. A supplemental wood-burning heating system
   b. A wind energy source
   c. A water energy source

2. Design a building using a nontraditional:
   a. Heating source
   b. Hot-water heating source
   c. Electrical power source

   Computer-produce, draw, or sketch as appropriate.

Presentation Drawing

SECTION VII

CHAPTER 41

Perspective

OBJECTIVES

By completing this chapter you will be able to:

🏛 Use three different methods to draw or computer-produce a one-point, two-point, or three-point perspective.

KEY TERMS

One-point perspective

Two-point perspective

Three-point perspective

In addition to *working drawings*, an architectural drafter must be able to prepare *presentation drawings*. A presentation drawing is used to help describe, or *present*, the proposed building to the client. For this purpose, a *perspective drawing* is nearly always used since it shows the appearance of the finished building exactly. Even persons trained in other types of drawing are able to visualize a design better in perspective. The architectural drafter, for example, will prepare thumbnail perspectives of each alternate scheme so that the most satisfactory design can be chosen.

A perspective drawing shows exactly how the building will appear to the eye or to a camera. The illustrator in Figure 41–1, drawing on a window with a wax pencil, will obtain the same perspective as a camera—as long as the relative positions of the observer, object, and picture plan are identical.

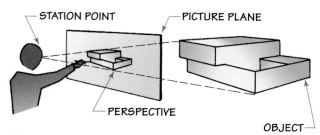

Figure 41–1   Perspective sketch.

THREE TYPES OF PERSPECTIVE

If the picture plane is placed parallel to a face of the object, the resulting perspective is called a **one-point perspective**. If it is placed parallel to one set of lines (usually vertical lines), the resulting perspective is called a **two-point perspective**. When the picture plane is oblique to all of the object's lines and faces, a **three-point perspective** results (see Figure 41–3).

The two-point perspective is more commonly used than either the one-point or three-point perspective. A one-point perspective of the exterior of a building is unsatisfactory for most purposes since it looks very much like a standard elevation drawing. Room interiors, however, may well be drawn in one-point perspective, as shown in Figure 41–2. A three-point perspective of a building exterior is not often used since it means that the observer must be looking *up* at the building (*worm's-eye view*) or *down* on the building (*bird's-eye view*, Figure 41–4). Obviously, neither of these is considered a normal line of sight.

There are many methods of obtaining a two-point perspective, but the most often used are

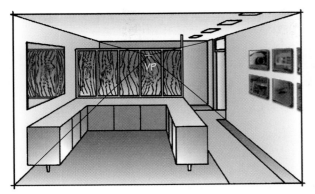

Figure 41–2   *One-point interior perspective.*

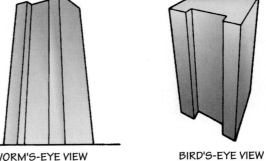

WORM'S-EYE VIEW   BIRD'S-EYE VIEW

Figure 41–4   *Three-point perspective.*

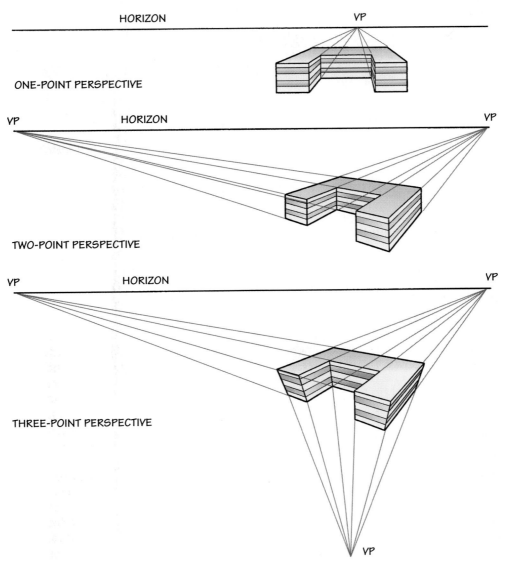

HORIZON   VP

ONE-POINT PERSPECTIVE

VP   HORIZON   VP

TWO-POINT PERSPECTIVE

VP   HORIZON   VP

THREE-POINT PERSPECTIVE

VP

Figure 41–3   *Types of perspective.*

the *common method* (Figures 41–5 through 41–13), the *direct projection method* (Figure 41–14), the *perspective grid method* (Figure 41–15), and *computer-assisted* (Figures 6–4 and 6–5). Let us look at each.

TWO-POINT PERSPECTIVE BY THE COMMON METHOD

Step 1: Locate the plan view behind the picture plane so that the front face is inclined 30° and the desired end face is inclined 60° to the picture plane. Actually, any set of complementary angles may be used, but the 30°–60° combination is most often used.

Step 2: Select any elevation view and position it on the ground line. Locate the horizon 6' (to scale) above the ground line. This means the eye of the observer is 6' above the ground. Other eye heights may be used if desired.

Step 3: Locate the station point (representing the location of the eye of the observer) directly in front of the plan view with a 45° maximum cone of vision. Occasionally a greater cone of vision angle is used to produce more dramatic results. The station point, however, should not be moved sideways because it will give a distorted perspective that does not represent the true proportions of the building.

Step 4: Find both vanishing points. In two-point perspective, a vanishing point is the perspective of the far end of an infinitely long horizontal line. The vanishing point for all horizontal lines extending 60° to the right (VPR) is found by drawing a parallel sight line 60° to the right until it intersects the picture plane. This is the plan view of the VPR. The perspective view of the VPR is found by projecting from the plan view down to the horizon. The VPL (vanishing point left) is located in a similar manner.

Step 5: Only lines located on the picture plane are shown true length in a perspective drawing. If a line is behind the picture plane, it will appear shorter; if it is in front, it will appear longer. Project the front corner that is located on the picture plane to the perspective view and lay off its true height by projecting from the elevation view.

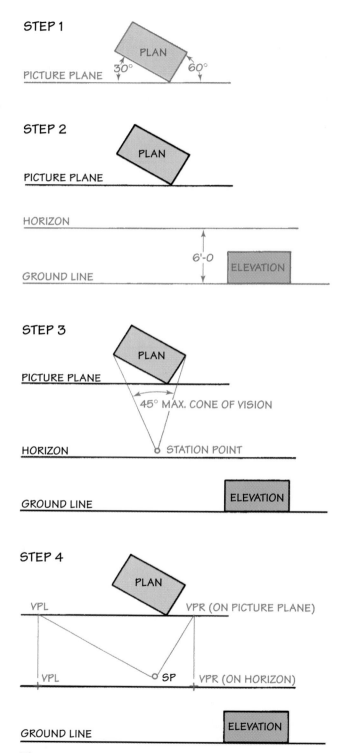

Figure 41–5 *Two-point perspective by the common method.*

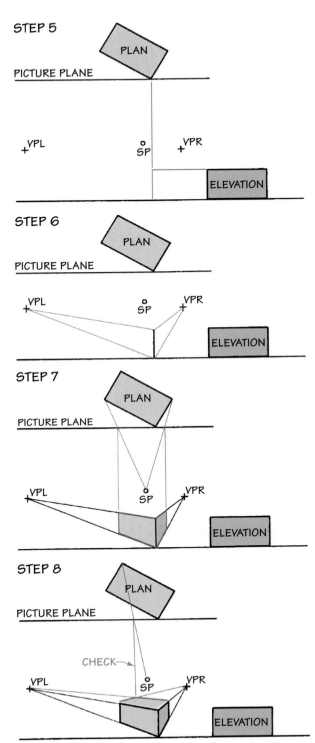

Step 6: Find the perspective of horizontal lines by projecting from the vertical true-length line to the vanishing points.

Step 7: Draw sight lines from the station point to the corners of the plan view. The intersections of these sight lines with the picture plane are projected to the perspective view to find the extreme corners.

Step 8: Complete the perspective by projecting from the extreme corners to the vanishing points. The far corner may be checked by projecting from the picture plane. Notice that invisible lines are omitted in a perspective drawing.

Measuring Line

In the preceding step-by-step illustration of the common method of drawing two-point perspective, the plan view was located so that the front corner touched the picture plane. This corner, then, was used to show the true height of the object in the perspective drawing. Occasionally *no* line in the plan view is located on the picture plane. When this occurs, as in Figure 41–6, a *measuring line* must be used. Simply imagine the plan view to be extended until a corner touches the picture plane. This corner is then projected to the perspective drawing for the true-height measurement.

Perspective of a Point

Figure 41–7 shows the method of finding the perspective of any point. This is a very useful

Figure 41–5 (continued)   Two-point perspective by the common method.

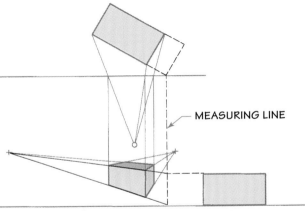

Figure 41–6   Use of the measuring line.

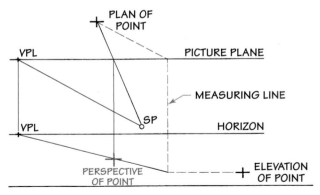

Figure 41–7   Perspective of a point.

exercise because the perspective of the most complicated object can be constructed merely by finding the perspectives of a sufficient number of points on the object. The perspective of an inclined line, for example, can be constructed by finding the perspective of both ends of the line. The perspective of a curved line can be constructed by finding the perspective of a number of points along the line.

Vanishing Points of Inclined Lines

Figure 41–8 shows that vanishing points of *inclined* lines (such as the inclined rafters of the shed roof) will lie directly above or below the vanishing point of horizontal lines that are in the same plane. Figure 41–8 also shows that vanishing points of all *horizontal* lines (such as the top and bottom rails of the open shed door) will always lie on the horizon.

Enlarging a Perspective Drawing

Often a perspective layout requires so much additional drawing space for the plan, elevation, and vanishing points that the resulting perspective drawing is smaller than desired. A number of "tricks" may be used to correct this.

1. Overlap the plan, elevation, and perspective drawings. Also, only an elevation *line* is needed—not an entire elevation view (Figure 41–9).

2. Place the plan view in *front* of the picture plane to obtain larger perspectives (Figure 41–10).

3. Allow the vanishing points to fall beyond the limits of the paper (Figure 41–11). The vanishing points may fall off the drawing table entirely. In this case, a curved cardboard

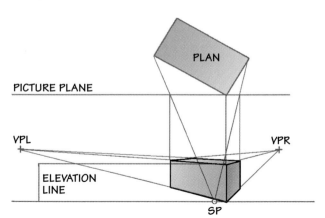

Figure 41–9   Overlapping views.

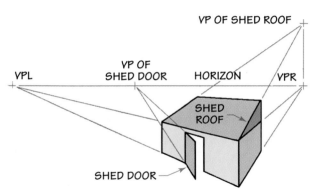

Figure 41–8   Finding vanishing points.

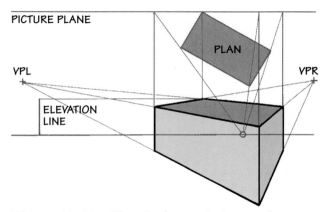

Figure 41–10   Plan in front of picture plane.

REAL WORLD *APPLICATIONS*

Reject Letter

As an independent architect, the jobs have been piling up. You receive a request from past clients to create a detailed perspective drawing of their parents' home. The drawing will be given to the couple as a fiftieth wedding anniversary present. Unfortunately, you realize it is not possible to take on another job at this time, even though it could be lucrative. Although you dislike having to turn down the job, it must be done. Reluctantly, you sit down to type a reject letter.

Before beginning your letter, consider proper business letter etiquette. The letter should be typed on company letterhead, aligned with the left margin, and have the date and customer's name and address on the top left. In the body of the letter, turn down the job in a respectful way so the opportunity for future business from this customer remains open. Be direct yet polite. You cannot afford for the client to misinterpret your letter or to take offense. Begin by stating that you are grateful they approached you. Then state concisely that due to time conflicts you cannot take their job proposal at this time. Before closing, again express appreciation to them for considering you for the job. In closing, assure them that you will be happy to consider their job proposal at a future time or that you hope they will consider you for future jobs. Be sure to use an appropriate salutation such as "Sincerely" or "Regretfully" above your signature and include a business card in the envelope.

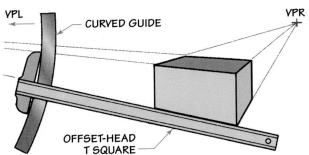

Figure 41–11   *Use of curved guide.*

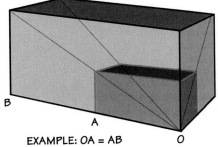

Figure 41–12   *Use of radiating lines.*

strip is prepared so that the T square will project nonparallel lines.

4. A perspective can be enlarged by the use of radiating lines. Each radiating line shown in Figure 41–12 has been doubled in size. There are many other enlarging devices (such as pantographs) that may be used.

Spacing of Lines

In perspective drawing, it is often desirable to show features that are evenly spaced without going through the trouble of projecting from the plan and elevation views. For example, a row of adjacent windows will appear smaller as they recede into the distance. If the first window is

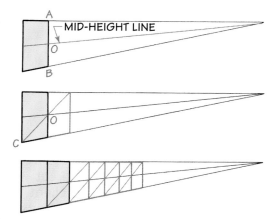

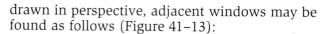

Figure 41–13   *Spacing of lines.*

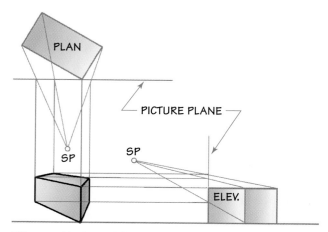

Figure 41–14   *Direct projection method.*

drawn in perspective, adjacent windows may be found as follows (Figure 41–13):

1. Draw a mid-height (*OA* = *OB*) line to the vanishing point.

2. Draw a diagonal (*CO*) and project it to locate the adjacent window.

3. Repeat as often as required to locate each window accurately.

DIRECT PROJECTION

The direct projection method of perspective drawing is similar to the common method with the exception of the addition of the end elevation view of the object, station point, and picture plane (Figure 41–14). Perspective widths are obtained as previously, by projecting from the plan view, and perspective heights are obtained by projecting from the end elevation view. This system has the advantage of not requiring a measuring line. In fact, even the vanishing points may be omitted if desired. In this case, however, very accurate projection is required.

PERSPECTIVE GRIDS

An easier method of constructing a perspective drawing is by the use of prepared grid sheets obtainable in a number of sizes (Figure 41–15). The perspective may be drawn directly on these grids or on tracing overlays.

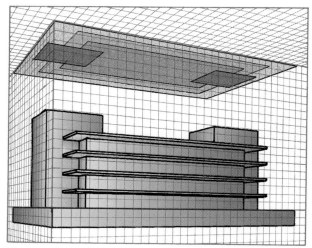

Figure 41–15   *Perspective grid method.*

COMPUTER-AIDED PERSPECTIVES

Your computer graphics system can be used to obtain a one-, two-, or three-point perspective of any building that has been entered in coordinate form. Using AutoCAD's "Viewpoint" and "DView" commands, your viewpoint can be moved around the building, up or down, and closer to or farther from the building, resulting in a variety of wire-frame perspectives. A wire-frame perspective shows *both* visible and invisible lines as solid lines. Using the "Hide" command, a solid-model perspective results, showing *only* visible lines. If you have an interactive computer system, use your joystick or the arrow keys on your keyboard to rotate the perspective and obtain desired views. Refer to Chapter 7 for operating details on providing various computer-generated perspectives.

Figure 41–16 illustrates a solid-model three-point perspective created using Softdesk's *Building Base* software. The perspective was then rendered and shaded in variable intensity.

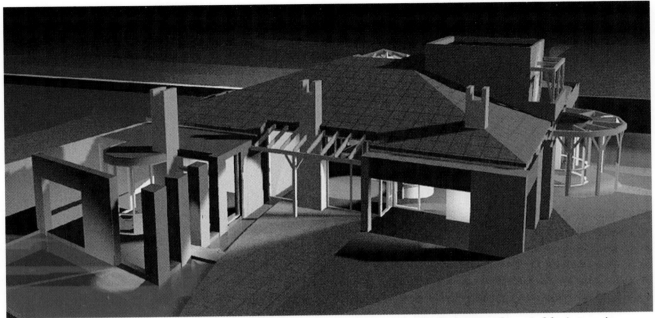

Figure 41–16 *Computer-produced three-point perspective with rendering and variable intensity shading. (Courtesy Softdesk, Inc.)*

REVIEW QUESTIONS

1. What is the purpose of a presentation drawing?

2. What is the principal difference between one-, two-, and three-point perspective? When is each used?

3. Define the following words as they relate to perspective drawings:

   a. Station point
   b. Perspective
   c. Picture plane
   d. Horizon
   e. Vanishing point

4. How is the perspective of an inclined line determined?

5. List four techniques used to enlarge a perspective drawing.

ENRICHMENT QUESTIONS

1. How will a perspective drawing change if the vanishing points were moved up or down?

2. Why are perspective drawings commonly used for presentation drawings in lieu of isometric or oblique drawings?

3. Sketch a two-point perspective layout of a simple object and label the picture plane, horizon, station point, right and left vanishing points, plan view, elevation view, and perspective view.

   a. Use the common method   b. Use the direct method

4. Sketch a two-point perspective layout illustrating the correct method to obtain the perspective of a horizontal line making an angle of 30° with the picture plane.

PRACTICAL EXERCISES

1. Draw or computer-produce two-point perspectives of the buildings shown in Figure 41–17. Use the method assigned by your instructor.

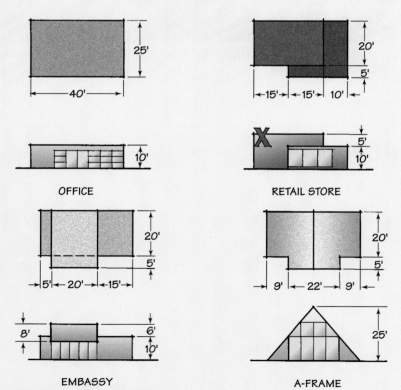

Figure 41–17   Buildings for Practical Exercise #1.

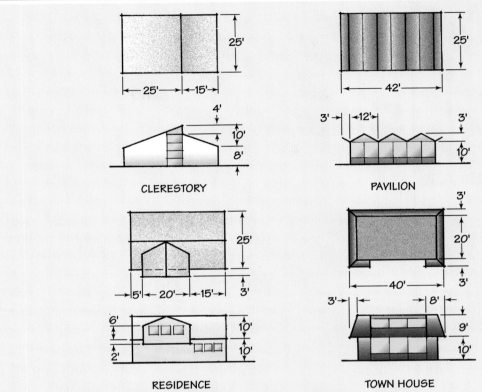

Figure 41–17 (continued)   Buildings for Practical Exercise #1.

2. Draw or computer-produce a two-point perspective of the A residence.

3. Draw or computer-produce a two-point perspective of your original house design.

CHAPTER 42

Shadows

OBJECTIVES

By completing this chapter you will be able to:

🏛 Differentiate between shade, shadow, and umbra.

🏛 Explain the principles and be able to include the shadows of a building in:

■ Multiview projection

■ Perspective using parallel light rays

■ Perspective using oblique light rays

KEY TERMS

Shade
Umbra
Shadow

A knowledge of shadows is important to both the architectural designer and the architectural drafter. The designer will consider the effect of shadows upon his or her proposals. The drafter will show these shadows on the presentation drawings to give them an extra three-dimensional quality.

Historically, shadows have greatly influenced architectural design. For example, Greek architecture evolved in a latitude where the bright sunlight exquisitely modeled the bas-relief carvings and fluted columns. Gothic cathedrals would lose their mystic beauty in the same climate. Today, much emphasis is given to the sun shielding and the patterns created by overhangs. In some instances, the shadows may be a major element in the final solution.

TERMINOLOGY

In a study of shadows, these terms should be understood (Figure 42–1):

Shade: A surface turned away from light

Umbra: The space from which light is excluded by the shaded surface

Shadow: A surface from which light is excluded by the shaded surface

Of the three, only the umbra is never shown on a drawing. Shaded surfaces are usually shown as a light gray, and shadows are shown as a darker gray or black. It is very easy to determine the surfaces in shade since they are merely the

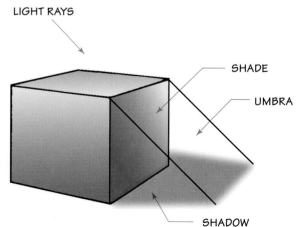

Figure 42–1 Terminology.

surfaces turned away from the direct rays of light. Shadows, though, are harder to find.

MULTIVIEW SHADOWS

Presentation drawings may include elevation views and a plot plan rendered to give a pictorial quality. Shadows are always included as shown in Figure 42–2 to increase the three-dimensional appearance.

In architectural renderings it is conventional to employ a distant source of light (the sun) as the basis for establishing the shade and shadow lines. The sun's rays are parallel and are usually assumed to have the direction of the diagonal of a cube extending from the upper left front to the lower right rear as shown in Figure 42–3. This is a convenient direction since the orthographic views of this diagonal are 45° lines.

To find shadows in multiview projection, two rules are used:

Rule 1 The shadow of a point upon a plane may be easily found in the view showing the plane as an edge.

Rule 2 The shadow of a line may be found by first finding the shadows of both ends of the line and then connecting these points.

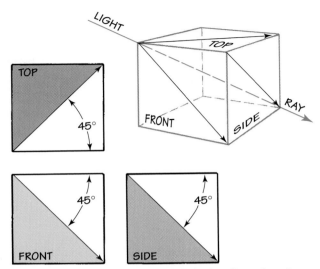

Figure 42–3 *Conventional light direction in multiview drawing.*

EXAMPLE

Find the shadow cast upon the ground by a flat-roofed structure (see Figure 42–4).

Step 1: The points *A*, *B*, and *C* will cast shadows upon the ground. Therefore draw lines through both views of these points parallel to the light direction (45°).

Step 2: Find where these 45° lines intersect the ground plane. This is found in the front view since the ground plane appears as an edge in the front view (rule 1). Project from the front view to the top view.

Step 3: Connect the shadows of points *A*, *B*, and *C* (rule 2).

A_sB_s represents the shadow of line *AB*.
B_sC_s represents the shadow of line *BC*.

Notice that the shadow of a line upon a parallel plane will appear as a parallel line.

AA_s and CC_s represent the shadows of vertical lines (rule 2).

Figures 42–5 through 42–10 show the multiview shadows cast by a variety of architectural shapes. Study each to find how the shadows were determined.

SHADOWS IN PERSPECTIVE— PARALLEL LIGHT RAYS

There are two methods of constructing the shadow of a building in a perspective drawing:

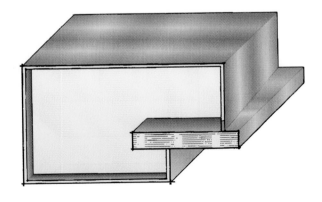

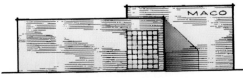

Figure 42–2 *Multiview shadows.*

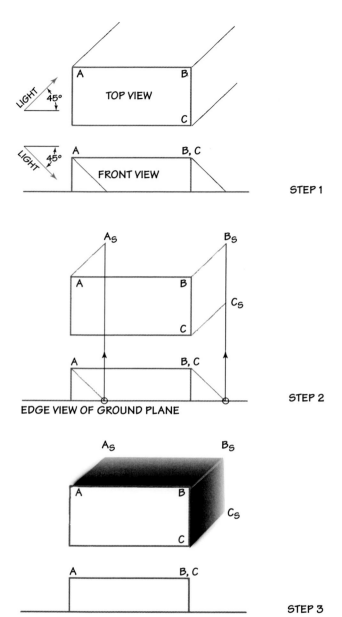

Figure 42-4 *Finding a shadow in a multiview projection (new steps in blue).*

1. The shadow may be first obtained in the multiview drawings and then projected to the perspective view in the same manner as any other line. This method is easy to understand but requires more work.

2. The shadow may be constructed directly in the perspective view. This method requires a bit of explanation but is often used since it is

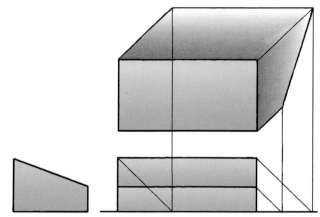

Figure 42-5   *Shed roof.*

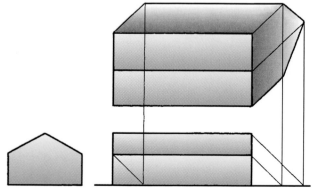

Figure 42-6   *Irregular plan.*

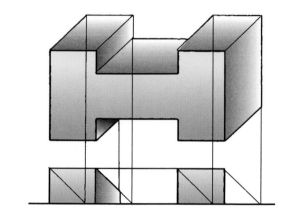

Figure 42-7   *Gable roof.*

shorter and more direct. To further simplify this method, the direction of light rays is assumed to be parallel to the picture plane and to make an angle of 45° (or some other con-

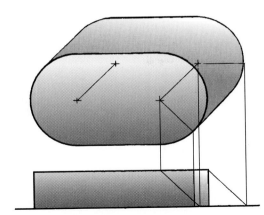

Figure 42–8   *Rounded plan.*

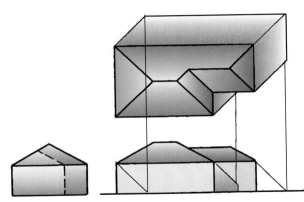

Figure 42–9   *Hip roof.*

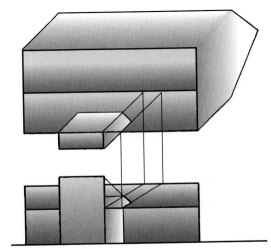

Figure 42–10   *Chimney.*

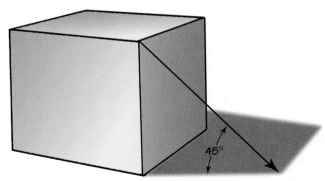

Figure 42–11   *Conventional light direction in perspective.*

venient angle) with the ground, as shown in Figure 42–11.

To find shadows in a perspective view, two rules are again used:

Rule 1 The shadow of a vertical line upon the ground (or any horizontal plane) is a horizontal line.

Rule 2 The shadow of a horizontal line upon the ground (or any horizontal plane) is parallel to the line.

EXAMPLE
Find the shadow cast upon the ground by a flat-roofed structure (Figure 42–12).

Step 1: The points *A*, *B*, and *C* will cast shadows upon the ground. Therefore draw 45° lines through these points parallel to the assumed light direction.

Step 2: Find where the 45° lines intersect the ground plane by drawing horizontal lines from points *D*, *E*, and *F* (rule 1).

> DA_S represents the shadow of line *DA*.
> EB_S represents the shadow of line *EB*.
> FC_S represents the shadow of line *FC*.

Step 3. Connect the shadows of points *A*, *B*, and *C*.

> A_SB_S represents the shadow of line *AB* and is parallel to line *AB* (rule 2).
> B_SC_S represents the shadow of line *BC* and is parallel to line *BC* (rule 2).

Notice that these lines will not be drawn *exactly* parallel to each other, but will be connected to their vanishing points.

Figures 42–13 through 42–18 show the perspective shadows cast by a variety of architectural

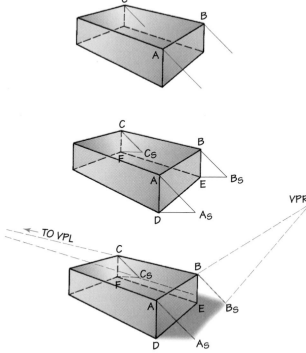

Figure 42-12 Shadow in perspective.

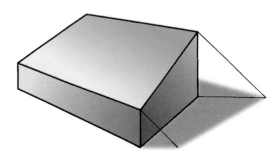

Figure 42-13 Shed roof.

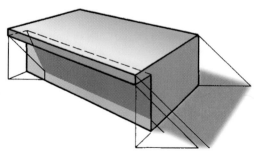

Figure 42-14 Flat-roof overhang.

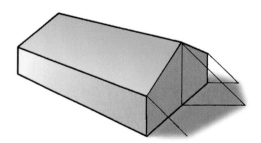

Figure 42-15 Gable roof.

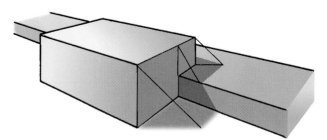

Figure 42-16 Offset surface.

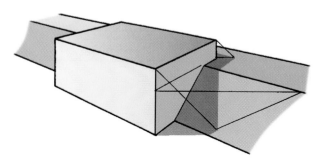

Figure 42-17 Sloping surface.

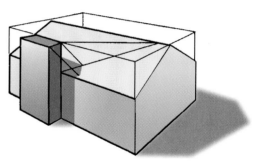

Figure 42-18 Chimney.

REAL WORLD  APPLICATIONS

Producing a Video

 You have been asked to teach new apprentices basic techniques of shadowing in architecture. It has been decided that producing a video will be most effective and cost-efficient. The video should explain the importance of shadowing in architecture, exhibit specific shadowing topics such as light direction, intensity and perspective shadows, and explain basic shadowing terminology.

Produce a video that accomplishes this task. You can work with a partner. When producing a video there are a number of basic guidelines to keep in mind.

1. Plan your video. Videotaping just to see how it will work is a waste of time and money. Prepare a script and follow it.

2. Avoid settings where there is little light or the main light source is emanating from the back of the main subject.

3. Be sure all presenters speak loudly and clearly. If you use wireless microphones they should be attached approximately 7 inches below the mouth, and, of course, contain fresh batteries.

4. Use slow camera movements. Panning the camera quickly produces jumpy images and makes the viewer uneasy.

5. Finally, practice, practice, and practice again. It is cheaper and less time-consuming to rehearse a scene without taping than it is to tape unwanted scenes.

---

shapes. As before, study each to learn how the shadows were determined.

SHADOWS IN PERSPECTIVE— OBLIQUE LIGHT RAYS

In perspective drawing, a shadow found by light rays parallel to the picture plane is not always entirely satisfactory. Although it is a difficult process, finding the shadow may be accomplished by using light rays oblique to the picture plane. The main difference is that the light rays must be drawn from their vanishing point rather than 45° with the ground. Figure 42–19 illustrates how the vanishing point of the light rays is found.

In actual practice, shadows are seldom determined by accurate construction since this is a time-consuming process. Rather, architectural drafters draw upon their knowledge of the general form of shadows to estimate the position

and shape of the shadows. After a little experience, a fairly accurate estimation is possible.

SHADOW INTENSITY

In addition to determining the location of a shadow, the architectural drafter must also make decisions concerning the intensity of the shadow, which may vary from black to a very light tone. This variation is due principally to the amount of light reflected upon the surface. This reflected light causes shadows to assume varying tones rather than one uniform tone. This variation may be useful for several purposes: to sharpen the shadow outline, to show changes in the direction of adjacent surfaces, to express changes in the depth of receding surfaces, and to show changes in surface texture. In general, the drafter must use a great deal of imagination and "artistic license" to obtain the desired result.

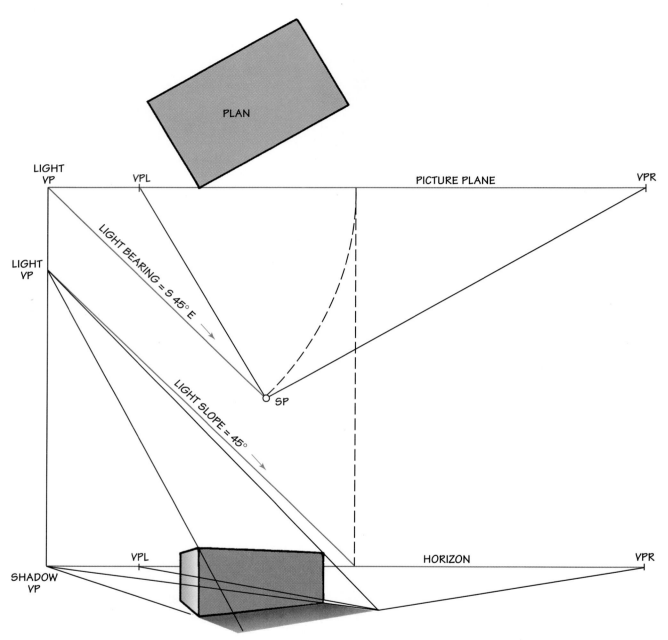

Figure 42–19 Shadow in perspective—oblique light rays.

COMPUTER-GENERATED SHADOWS

In addition to computer programs that produce rendered perspectives, advanced accessory programs are available that add shadow outlines and shadows rendered in varied intensities. Full-color capabilities and surface texture variations are also possible as illustrated in Figure 41–16. Refer to the "Computer-Assisted Pictorials" section of Chapter 6 and the "Three-Dimensional Drawing" section of Chapter 7 for more detailed information.

REVIEW QUESTIONS

1. Define each of the following:

   a. Shade   b. Umbra   c. Shadow

2. List two rules used to find shadows in a multiview projection.

3. What is meant by *conventional light direction*? Why is it used?

4. Give two methods that may be used to find the shadow of a building in a perspective drawing.

5. Give a method that may be used to find the shadow of any line.

6. Give four reasons to vary shadow intensity.

ENRICHMENT QUESTIONS

1. Why is it important to show the shadows of a building?

2. In what ways can showing shadows improve or diminish the quality of a drawing?

3. Identify different colors that can be used when shadowing. Redraw Figure 42–14 using various colors for the shadow. How does it change its appearance?

PRACTICAL EXERCISES

1. Add shadows to the two-point perspectives you drew for Practical Exercise #1 of Chapter 41. Use the method assigned by your instructor.

2. Add shadows to (a) the presentation drawing and (b) the perspective of the A residence.

3. Add shadows to (a) the presentation drawing and (b) the perspective of your original house design.

CHAPTER 43
Architectural Rendering

OBJECTIVES

By completing this chapter you will be able to:

🏛 Produce an architectural rendering of a residence or commercial building in the technique of:

- Pencil
- Pen-and-ink
- Scratch board
- Pressure-sensitive transfer
- Spray
- Combination of these techniques

🏛 Define various terms related to architectural rendering.

KEY TERMS

Rendering	Watercolor
Scratch board	Tempera
Pressure-sensitive transfer	

The ability to prepare a finished **rendering** is a desirable goal for every architectural drafter. Even a large architectural firm with a separate art department occasionally needs help to meet a deadline. Smaller offices must depend upon the regular staff for renderings or send the work out to companies specializing in presentation drawings. These companies produce excellent results but, of course, time and coordination must be considered.

Since a rendering shows a structure as it actually appears to the eye, it is used for presentation drawings that will be seen by clients who, in all likelihood, cannot read ordinary blueprints. As a matter of fact, the quality of the rendering may well influence the final decision of the client to continue with the proposed building. Most clients feel (and rightly so) that the finished product will look and function no better than the drawings. In addition to finished renderings, the architectural drafter should be able to make quick sketches of alternate solutions to a problem to show to the architect, who makes the final decision.

TYPES OF RENDERING

There are many kinds of renderings. Among the most common are:

Pencil	Tempera
Pen-and-ink	Spray
Marker	Photographic
Scratch board	Computer-produced
Pressure-sensitive transfer	Combinations of the preceding
Watercolor	

In this chapter we will study each of these techniques—putting the emphasis upon those that will be most useful to you. *Pencil rendering* is most often used since it requires no special equipment and is very versatile. Drawings from small, quick sketches all the way up to large, finished presentation displays can be rendered in pencil. *Ink*, however, is preferred for drawings to be reproduced in media such as newspapers, books, and brochures. *Pressure-sensitive transfers* offer a quick and effective method of applying a professional-looking shading. *Scratch board* is oc-

REAL WORLD APPLICATIONS

Business Cards

 As the newest member of an architectural firm, you have the opportunity to create your own personalized business cards. Business cards are used to represent you when you are not available or to give to prospective clients as an advertisement. They should include your name, address, phone number, electronic mail address, name of the firm, and a logo that represents you or your firm. Most business cards are 2" wide and $3\frac{1}{2}$" long and printed on a heavy colored card stock.

Keep these principles in mind when designing your individual business card.

🏛 Your name or your firm's name should be the most prominent item on the card.

🏛 All graphics should face toward the center of the card.

🏛 Use adjacent colors in an attractive design. Avoid bright opposite colors or fluorescent colors.

🏛 Keep it simple. Provide enough information for the holder to contact you by phone, fax, electronic mail, or letter. Too much text makes the card difficult to read.

casionally used, as is *watercolor*. *Tempera* applied by brush or spray is the normal medium used by the professional artist for a colored rendering.

PENCIL RENDERING

Pencil rendering is the most popular form of rendering because it is quick, errors are easily corrected, no special equipment is needed, and it is very versatile. Pencil renderings range from rough freehand sketches to accurate drawings using a straightedge for outlines and shading.

Pencils

The softer lead grades are used in pencil rendering. Although the final choice will depend upon the paper texture and personal preference, the following selection is ordinarily satisfactory:

2H for layout
F for medium-tone rendering
2B for dark tones

A sharp conical point is used for the layout work and detailing, but a flat chisel point is used to obtain the broad rendering strokes.

Paper

The paper selected should take pencil lines and erasures nicely. Bristol board, Strathmore paper, and tracing vellum meet these requirements. Fine- or coarse-textured paper may be used depending upon the result desired. During rendering, keep the paper clean by working with a paper shield under your hand. A fixative spray is used to protect the finished rendering from smudges.

Textures

The success of a pencil rendering depends, to a large extent, upon the ability of the drafter to indicate the proper texture of surfaces and materials. Some of the most common materials encountered are:

Glass and Mirrored Surfaces   Small areas are quite simply rendered by blackening with a 2B pencil. Larger window areas may be toned from a dark corner to an opposite light corner (Figure 43–1). Very large expanses of glass, however, require more thought. Normally some of the surroundings are shown reflected in the glass.

Figure 43–1   Glass renderings.

Highlights are made with an eraser and erasing shield. If desired, the glass rendering may be omitted in order to show interior features.

Brick   Brick surfaces are indicated by closely spaced lines made with a flat-pointed F pencil. The length of the lines does not matter since the horizontal brick joints are more important to show than the vertical joints (see Figure 43–2).

Stone   Stone surfaces are rendered by drawing a few individual stones, leaving large expanses of "white" space. In Figure 43–3, notice that the shadow line is darkest near its extremity.

Roof   Shingled roofs are shown by long, closely spaced lines—somewhat like brick. The lines may fade into light areas as in Figure 43–4. Built-up roofs, concrete, and plaster are stippled.

Foliage

A variety of foliage can be shown by varying the stroking and the pencil grade as shown in Figure 43–5. Usually, however, a single foliage type is selected and repeated in various sizes rather than mixing different types.

Figure 43–2   Brick rendering.

Procedure

The usual procedure for rendering in pencil follows. We will use the A residence for illustration (see Figure 43–6).

Figure 43–3   Stone rendering.

Figure 43–4   Shingled-roof rendering.

Figure 43–5   Pencil grades indicate depth and texture. (Courtesy Professor Milton S. Osborne, the Pennsylvania State University.)

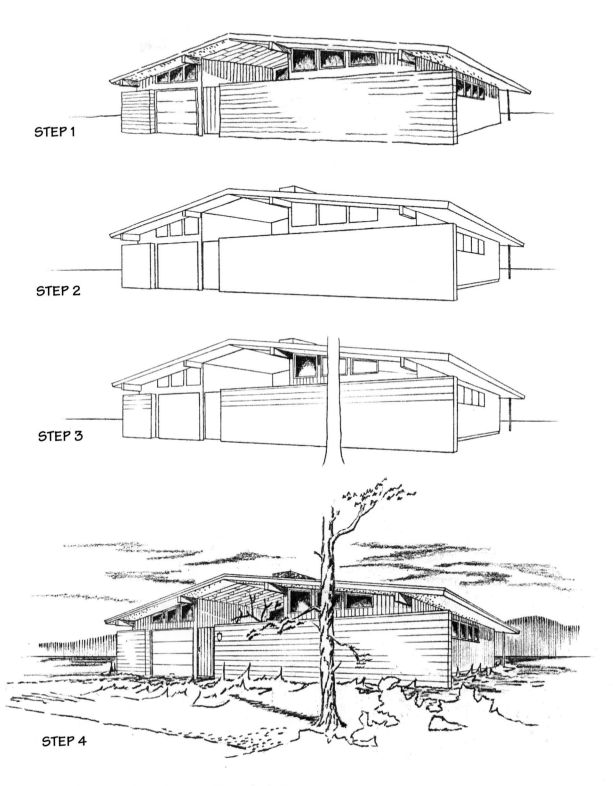

STEP 1

STEP 2

STEP 3

STEP 4

Figure 43–6 *The procedure for pencil rendering.*

Step 1: A rough pencil sketch is made to determine the most suitable perspective angle, light direction, and shadow locations. In general, there should be a balance between white, black, and gray areas.

Step 2: Transfer the mechanically drawn perspective to the finished paper. Outline with a sharp 2H line. Portions of these outlines may be erased later to provide highlights (bright areas).

Step 3: Render the central structure using an F or 2B pencil, depending upon the desired blackness. Windows are done first, followed by walls and roofs.

Step 4: Render the foreground details followed by the background elements. Use an eraser to provide highlights.

Study Figures 43–7 through 43–8 for examples of technique in pencil rendering.

Color

Pencil renderings may be colored with light watercolor washes. Select the colors carefully to

prevent disappointing results. Colored pencils are often used on colored mat boards. Monochromatic schemes (various tones and shades of

Figure 43–7   Pencil sketch of a summer camp. A successful sketch must blend perspective, shadows, and rendering. (Courtesy Professor Milton S. Osborne, the Pennsylvania State University.)

Figure 43–8   A pencil, airbrush, and watercolor rendering. (Courtesy Neff Kitchens.)

Figure 43–9   *An ink and watercolor rendering.* (Courtesy Scholz Design, Toledo, Ohio.)

one color) are particularly effective. White pencil or ink is used for highlighting.

INK RENDERING

Black ink rendering is the most suitable medium for drawings to be reproduced by any of the printing processes. Colored drawing inks are occasionally used for display drawings. Ink rendering differs from pencil rendering in that the shades of gray are more difficult to obtain. A wash of ink diluted with water may be used, but usually shading is done by varying the width of stroke or spacing between strokes as shown in Figures 43–10 and 43–11. The stippling technique (Figure 43–12) is also often used. Notice that the completely black shading in Figure 43–13 requires that the mortar joints be changed from black to white.

Paper

The paper used for ink rendering must take ink without fuzzing, allow ink erasures, and be smooth-surfaced. Some satisfactory materials are:

  Mat boards (available in white and colors)
  Bristol boards, plate surface
  Strathmore paper, plate surface

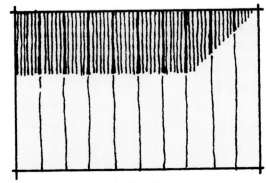

Figure 43–10   *Vertical siding.*

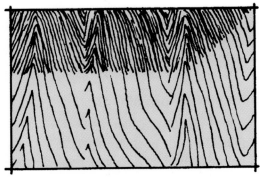

Figure 43–11   *Plywood.*

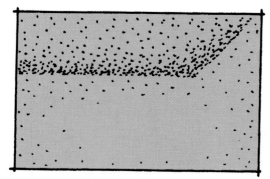

Figure 43–12   Concrete.

Figure 43–13   Stone.

Tracing paper (use only better-quality)
Tracing cloth

Pens

The technical fountain pen or ordinary pen nibs and holder may be used. A ruling pen (drafting pen), of course, is not used to draw freehand lines. Technical fountain pen points are sized:

00 Fine
 0 Medium fine
 1 Medium
 2 Medium heavy
 3 Heavy

The most popular pen nibs are:

Crow-quill: Hunt 102 or Gillott 659	Very fine
Hawk-quill: Hunt 107 or Gillott 837	Fine
Round-pointed: Hunt 99 or Gillott 170	Medium
Bowl-pointed: Hunt 512 or Speedball B-6	Heavy

Inks

Satisfactory drawing inks are produced by Higgins, Weber, and Pelikan. The following colored inks are available. Other shades are obtained by mixing.

Yellow	Violet	Brick red
Orange	Blue	Russet
Red orange	Turquoise	Brown
Red	Green	Indigo
Carmine red	Leaf green	White
Red violet	Neutral tint	Black

Ink erasing is done with a pencil eraser to prevent damage to the paper's surface. A sharp razor blade is used to pick off small ink portions.

Procedure

The usual procedure for an ink rendering is the following:

1. Make a charcoal study or rough ink sketch to determine the most suitable light direction and shadow locations. Try to obtain a balance between white, black, and gray areas.

2. Outline lightly in pencil on finished paper.

3. Ink in outlines with fine lines (unless the surfaces are to be defined by shading differences only).

4. Render each feature separately—first windows, then walls, and so on, working from foreground to background.

Students wishing to learn the technique of ink rendering would be well advised to copy a good rendering before attempting an original. Students should also start a collection of ink renderings by various artists so that they can study the various methods of handling the details. Ruled ink lines were used in Figure 43–9. Figures 43–14 and 43–17 show ink renderings drawn with a fine pen in a freehand style. In Figures 43–15 and 43–16, ink was used for outlines, but colored pencil was used for shading. In Figure 43–20, notice how the entourage is subordinated to direct your attention to the pavilion structure.

Figure 43–14 *Ink rendering.*

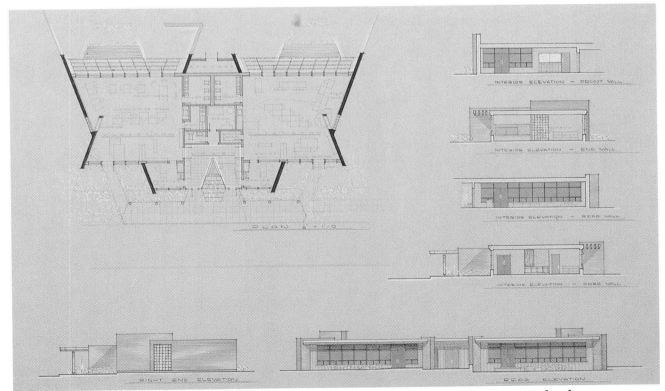

Figure 43–15 *Ink and colored pencil plan and elevations of a proposed nursery school.*

Figure 43–16   *Ink, colored pencil, and pressure-sensitive transfer perspective of the school in Figure 43–15.*

MARKER RENDERING

Markers are usually called Magic Markers (a trade name), but AD Markers, Design Art Markers, and other companies manufacture markers in over 100 colors, tints, and degrees of transparency. Point widths are available from very fine to broad brush. Markers can be used successfully on a medium of any color or thickness. They have the advantage of drying quickly and, unlike watercolors, they do not wrinkle tracing vellum.

SCRATCH BOARD

A simple method to create a rendering of white lines on a black background is provided by a **scratch board**, which is a board having a white, chalky surface. The surface is coated with black ink, and a rendering is obtained by drawing with a sharp-pointed stylus. The stylus will scratch through the black coating, leaving a white line. A knife-edged tool is used to whiten larger areas. If desired, large portions of the scratch board may be left white (not coated with black ink). In these areas, ink rendering is done in the usual manner.

The pictorial detail at the bottom of Figure 43–17 was done on scratch board.

PRESSURE-SENSITIVE TRANSFER

A **pressure-sensitive transfer** (appliqué) is a thin transparent plastic sheet with a printed pattern of black dots or lines. The sheet is coated on one side with a special adhesive to make it stick to the drawing. A wide variety of patterns is available; some are illustrated in Figure 43–18. Various shades of gray can be obtained by choosing the proper transfer or by applying several transfer sheets over each other for darker shades. In addition to patterns, a number of architectural symbols for trees, shrubbery, people, automobiles, and furniture may be obtained. Examples of such appliqué symbols are shown in Figure 43–19.

Procedure

The order of procedure is:

1. Outline in ink.

2. Select the desired transfer, remove its protective backing, and place it over the area to be rendered, adhesive side down. Rub

Figure 43–17 *The bottom portion is an example of scratch board rendering. (Courtesy Masonite Corporation.)*

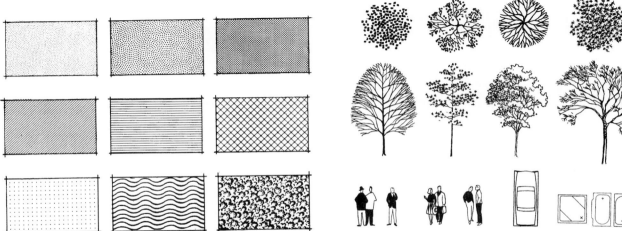

Figure 43–18 *Typical pressure-sensitive transfer patterns.*

Figure 43–19 *Typical pressure-sensitive transfer symbols.*

gently with your fingernail to increase adherence.

3. Lightly score around the outline with a razor blade without cutting the paper underneath.

4. Peel off the excess material and rub again with your fingernail to ensure permanent adherence.

Figure 43–20 shows a very simple—yet effective—presentation. One pressure-sensitive transfer sheet was the sole rendering medium.

WATERCOLOR AND TEMPERA RENDERING

Watercolor refers to a transparent water-based paint; **tempera** refers to an opaque water-based paint. Both are used (separately or together) to render large presentation drawings, watercolor giving a refined, artistic effect, tempera giving a more vivid and striking effect.

Watercolor may be applied with a brush using a very thin mixture of paint and water called a *wash*, or it may be applied directly from a *palette* used for mixing the colors (Figures 43–21 and 43–22). Professionals obtain watercolors in tubes and half-tubes rather than in dried cake form.

Tempera (known also as *poster paint*, *opaque watercolor*, and *showcard color*) is usually applied directly from the palette. Tempera is obtainable in jars and tubes. In addition to color illustration, tempera is very effective when used as a monotone medium. Black, white, and several tones of gray are often used. Figure 43–23 shows an example of such a rendering.

Figure 43–20 *Ink and pressure-sensitive transfer rendering.* (*Courtesy the Department of Architecture, the Pennsylvania State University.*)

Figure 43–21 *Professional watercolor and colored pencil rendering of a commercial building. (Courtesy John C. Haas Associates, Inc.)*

Figure 43–22 *Professional watercolor and ink rendering of the Convocation Center at the Pennsylvania State University. (Courtesy John C. Haas Associates, Inc.)*

Figure 43–23   *Professional tempera rendering of a proposed high school. (Courtesy Jack Risheberger.)*

SPRAY RENDERING

The smooth gradation of tones seen in professional renderings is usually obtained by spraying with an airbrush. This technique is excellent for the indication of smooth, glassy surfaces and background sky.

An airbrush is simply a nozzle that sprays a fine mixture of paint and air. The compressed air is obtained from a compressor or tank of carbonic gas. Since this equipment is expensive and not available to the average student, other methods are often used. Hand spray guns, pressurized spray cans, and even spattered paint (toothbrush rubbed on screening) have been used successfully. Tempera paint is used because it is fast-drying.

Procedure

The procedure for a typical spray rendering is as follows:

1. Block in the desired illustration using a sharp pencil line. The paper must be clean and smudge-free. A spray rendering is transparent and all smudges will show.

2. Select the area to be sprayed. Usually the structure is completed first, then sky, and finally the entourage. Mask out all other areas so that they will not be painted also. Transparent *frisket* paper, which is applied like an appliqué, is used.

3. Make a few test sprays upon scratch paper to determine the proper distance and motion of the spraying device. A light spray is preferred since a too-dark spray cannot be corrected. Carefully lift the frisket paper occasionally to compare relative shades and tones.

4. When the paint is dry, remove the frisket and move on to the next area.

5. Accent details with tempera using a fine brush.

See Figures 43–21 to 43–25 for examples of professional renderings that utilize a combination of media.

Figure 43-24   *Professional airbrush and tempera rendering. (Courtesy Jack Risheberger.)*

Figure 43-25   *A professional tempera rendering of an elementary school. (Courtesy Jack Risheberger.)*

ARCHINET http://www.

Career Mosaic— Job Search on the Internet

It is six months before graduation and you are ready to begin searching for employment. Now you need to do some homework on jobs available in your area of choice. Where do you begin? Suppose you wanted to live near Pittsburgh, Pennsylvania. Rather than subscribing to the local newspaper or paying a job search agency, you can access the Career Mosaic homepage at:

http://www.careermosaic.com/

This homepage offers many resources. You can post your resumé so a prospective employer can access it. The College Connection's Career Assistance Center offers sample resumés, cover letters, tips on job searching, links to professional associations, and career advice. A database of top employers in various fields, online job fairs, and information on international job opportunities are also available.

One of the most helpful resources on this page is a link to specific jobs available in various fields. The following steps will assist

you in performing a job search. First, click on J.O.B.S. Enter your career choice in the occupation field, and a city and state of your choice. How many positions are listed for your career in this city? Click on one of these titles to see more information about the job. Are there any listings that interest you?

Once you have made a list of tentative selections, do some research on each employer. Some information that will be useful in your decision-making process includes the size of the company, how long it has been in existence, the types of projects in which it specializes, and starting salary. Knowledge of the company will be beneficial as you write a cover letter for your resumé and prepare for an interview. Narrow your list based on this information and send resumés to those companies of your choice.

Another useful web site for choosing a career and finding employment is the Occupational Outlook Handbook homepage. The address for this is:

http://www.bls.gov/ocohome.htm

PHOTOGRAPHIC RENDERING

When a presentation model of a proposed structure is built, photographs of the model are also prepared for purposes of publication, ease of transportation, and file records. If no presentation model is available, a stage set cardboard model can be built using only those faces that will be seen by the camera. Often both panoramic photographs and close-ups are taken

and touched up using standard darkroom methods. See the chapter on Architectural History (Chapter 15) for several examples of photographic rendering.

COMPUTER-PRODUCED RENDERING

Depending upon the sophistication of your software and the storage capacity of your hardware, various computer-applied rendering (shading) techniques are available to you. Most CAD programs will let you render your pictorial as if lit by a light source directly behind the observer and some background lighting. The darkness intensity of each surface is in proportion to the angle of the surface to the observer. Your pictorial can be rotated slightly to prevent similar shading of surfaces at the same angle to the observer. Furthermore, each surface can be shaded with a graduated intensity that darkens a surface in proportion to its distance from the observer. Edge highlighting is also available.

Figure 43–26 shows an example of computer-produced architectural rendering.

Figure 43–26 *A computer-produced architectural rendering with entourage. (Courtesy Softdesk, Inc.)*

REVIEW QUESTIONS

1. Give the principal uses of each of the following types of renderings:
   a. Pencil
   b. Pen-and-ink
   c. Scratch board
   d. Pressure-sensitive transfer
   e. Watercolor
   f. Tempera
   g. Spray

2. What is meant by the texture of surfaces and materials?

3. In pencil rendering, how are each of the following materials treated:
   a. Windows
   b. Brick walls
   c. Stone walls
   d. Concrete walls
   e. Shingled roofs

4. What does appliqué mean?

5. In ink rendering, what is meant by the following terms:
   a. Wash
   b. Stipple
   c. Crow-quill

6. How does watercolor differ from tempera?

7. Give four methods of producing a spray rendering.

ENRICHMENT QUESTIONS

1. How would you explain to someone the difference between rendering and shadowing?

2. In what ways could rendering improve or diminish the appearance of a presentation drawing?

3. What effect can foliage have on an architectural rendering?

4. How are features of a residence subordinated so the viewer's attention is drawn to other areas of the drawing?

5. In what ways will computer-produced rendering change the process of rendering by hand?

PRACTICAL EXERCISES

1. Start a collection of renderings by various artists in the medium of your choice.

2. Prepare a *style sheet* showing how various materials and textures may be rendered. Use the medium of your choice.

3. Render (a) the presentation drawing and (b) the perspective of the A residence in a hand medium or computer medium.

4. Render (a) the presentation drawing and (b) the perspective of your original building design, using a hand medium or computer medium.

CHAPTER

44

Entourage

OBJECTIVES

By completing this chapter you will be able to:

- 🏛 Explain the importance of entourage in architectural drawing.
- 🏛 Trace from templates to include stylized entourage on the elevations, plans, and perspectives of a building.
- 🏛 Develop your own individual style of drawing the forms used in entourage.

KEY TERMS

Entourage

In architectural design, a structure is planned in relation to its environment. The characteristics of the land, trees, and surrounding buildings are all considered. The quantity and location of traffic—both vehicle and pedestrian—affect the design. To present the design in its proper context, then, an architectural rendering also includes an indication of the character and quantity of these surrounding elements. If this is done well, the client will be able to identify the surroundings and visualize the proposed building in them. If it is done poorly, the client may reject the entire plan. Subconsciously, she or he will feel that a poorly executed drawing means a poorly designed building.

In architectural drawing, the word used to describe the surroundings is **entourage**, pronounced *n-tur-ahge*. This is a French word with the accent on the last syllable, which is pronounced like the last three letters in the word *garage*.

Entourage is used to describe objects—trees, shrubbery, background mountains, human figures, and vehicles—that are included to increase the realistic appearance or decorative effect of the drawing. In addition to its use on perspective renderings, entourage may be used on presentation plans and elevations to increase their pictorial quality.

Contrary to what you might expect, entourage is drawn in a simplified, stylized manner rather than in a detailed, photographic style. There are three good reasons for this:

1. Only a trained artist can draw objects like trees, vehicles, and people to look exactly as they are.
2. There is a constant shortage of time in an architectural drafting room, so shortcut techniques must be used whenever possible.
3. Entourage drawn in great detail would detract from the central structure.

The stylized forms of trees, human figures, and vehicles shown in this chapter may be traced directly or increased in size to fit a particular requirement. You should, however, develop your own style by improving upon these drawings. In addition to entourage in perspective, the plan and elevation symbols are included, since a presentation drawing often contains the plan and elevations.

Nearly all of the plan-view trees in Figure 44–1 are based upon a lightly drawn circle. Remember to change the size of the circles for variety, but do not mix many different symbols on the same drawing.

Choose trees that will fit the location. Palm trees, for example, would be appropriate in

Figure 44–1   *Trees in plan view.*

SOFTWOOD TREES

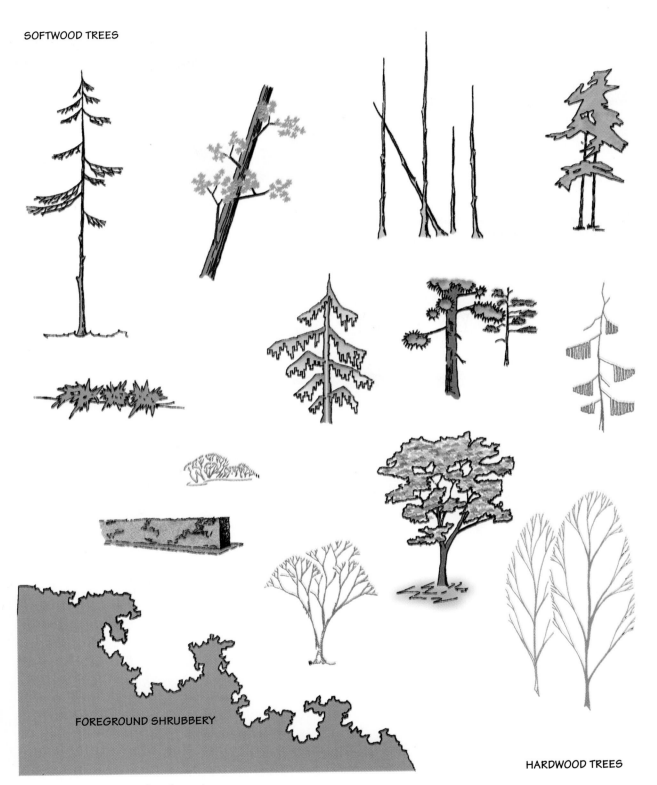

FOREGROUND SHRUBBERY

HARDWOOD TREES

Figure 44–2   *Trees in elevation.*

BACKGROUND

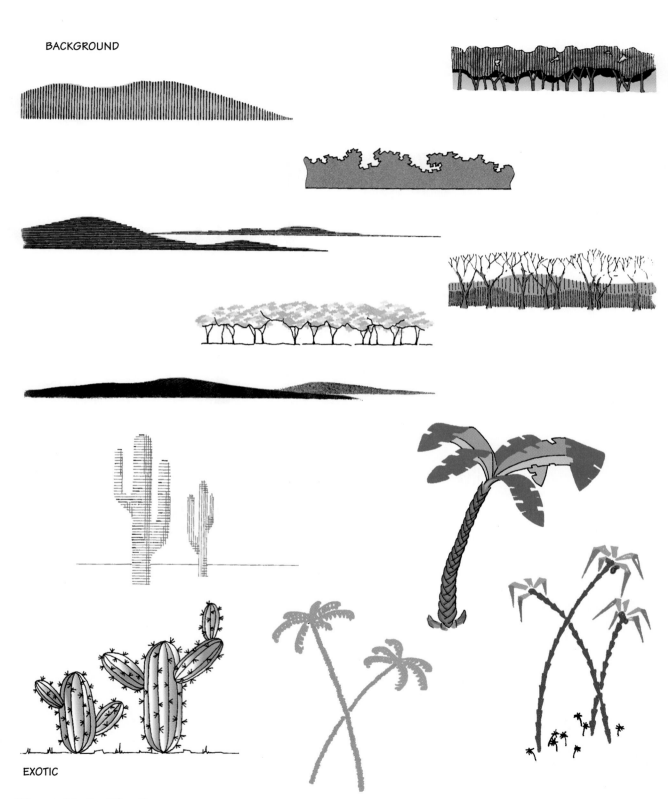

EXOTIC

Figure 44–3 Planting.

southern states, cacti in western states (Figure 44–3). The foreground in Figure 44–2 may be drawn across the bottom of a rendering, indicating shrubbery near the observer.

Choose a style of background mountains or trees from those shown in Figure 44–3 that is compatible with your tree style.

The perspective automobiles (Figure 44–4) are easily traced from advertisements found in magazines. The plan-view autos are drawn by adapting standard templates.

Human figures are most difficult to draw realistically, and therefore the outline forms shown in Figure 44–5 are used.

COMPUTER-PRODUCED ENTOURAGE

Menus of computer-produced entourage may be procured from software suppliers such as ASG and Softdesk. These menus include elevations, plans, and pictorial views of foreground and background planting, vehicles, and human figures. If you prefer, you can prepare your own customized menus by drafting and storing those items of entourage you will use repeatedly, or you can draft or sketch entourage as required at the moment. Figure 43–26 shows a computer-produced architectural rendering with entourage.

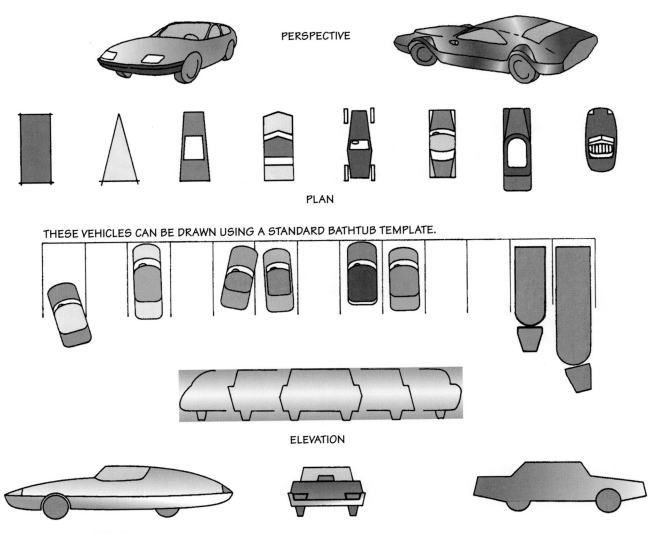

Figure 44–4   Vehicles.

Figure 44–5 *Human figures.*

CAREER PROFILES

Landscape Architect

Interested in the environment? Wondering how to tie this interest to architecture? Consider becoming a landscape architect. Landscape architects solve problems concerning relationships of the built and natural environments. You would work on projects ranging from the site design of a single building to large-scale developments having significant environmental impacts. Imagine yourself collaborating with other specialists on interdisciplinary teams to solve important environmental problems. Conserving and protecting natural resources would be your goal as you worked on projects involving forest and wilderness areas, coastal areas, or the restoration of misused lands.

Specifically, designing and supervising the development of land, including grading, planting, storm water management, and construction of structures, walls, pavements, steps, and seating would be your major tasks. You would help enrich our surroundings by designing and planning such projects as home landscaping, new communities, campuses, recreation areas, urban and regional parks, urban plazas, waterfront developments, playgrounds, gardens, zoos, and enclosed spaces. A notable project designed by a team of landscape architects at Penn State University is the Korean War Veterans Memorial on the Mall, Washington, D.C.

The term "landscape architect" was suggested by Frederick Law Olmsted when he designed New York City's Central Park in 1850. His followers founded the American Society of Landscape Architects (ASLA). As the official accrediting agency, ASLA has approved approximately fifty 5-year baccalaureate landscape architecture programs in the United States. A list of these programs can be obtained from ASLA, 4401 Connecticut Ave. N.W., Washington, DC 20008-2302. Phone:

Figure 44–5 (continued) Human figures.

202-686-2752. You can also obtain this information and much more on ASLA's homepage at:
http://www.asla.org/asla
This homepage includes access to the ASLA Job Link, news and events, education links, and discussion groups.

Typical courses in a landscape architecture program include:

- 🏛 History of landscape architecture
- 🏛 Western and Eastern art
- 🏛 Architectural graphics
- 🏛 Computer applications
- 🏛 Soils
- 🏛 Ornamental planting
- 🏛 Vegetation ecology
- 🏛 Landscape design
- 🏛 Site and regional planning
- 🏛 Professional practice
- 🏛 Ethics

A Master of Landscape Architecture program usually emphasizes advanced design projects or ecological issues.

Professional registration requirements for a landscape architect are similar to the requirements for an architect—graduation from an ASLA-accredited university, three years of practical experience under the supervision of a professional landscape architect, and the successful completion of a comprehensive written examination. Forty-five states require landscape architects to be licensed or registered. Licensing is based on the Landscape Architect Registration Examination (L.A.R.E.). Without a degree, a total of eight years of varied practical experience is usually accepted by the state as a substitute.

For more information about the occupation of landscape architect including job outlook, working conditions, more information on licensing and training requirements, and salary, go to the Occupational Outlook Handbook homepage at:
html://www.bls.gov/ocohome.htm
Click on Outlook for Specific Occupations. On this new page click Professional Specialty Occupations. Finally, click on Landscape Architects to retrieve the information you desire.

REAL WORLD APPLICATIONS

Perimeter Calculation

1/4"=1'-0

Clients have indicated that they want to outline a pond with miniature cherry trees. The pond is approximately 30 feet in diameter to the shores, but they want the new tree line not to extend closer to the shoreline than 5 feet. To do this you will need to calculate the perimeter of the pond. If the pond were rectangular, then the perimeter would be simple to calculate: add the length of all the sides. But since the shape is circular, use either of the formulas below.

Perimeter = $\pi \times D$ (D = Diameter)

or

Perimeter = $2 \times \pi \times r$ (r = radius)

If the cherry trees are to be spaced 9 feet apart and their branches spread about 3 feet from the trunk, and a 10' space should be left open to allow access to the pond, how many trees are needed?

REVIEW QUESTIONS

1. What is the meaning of the word *entourage*?

2. Why is it important to include entourage on a presentation drawing?

3. Give three reasons why entourage is drawn in a simplified rather than detailed manner.

4. List four types of entourage drawings that could be added to an elevation, plan, or perspective.

ENRICHMENT QUESTIONS

1. How does the entourage drawn on an elevation view differ from that drawn on a plan or perspective?

2. In what ways could the addition of entourage detract from or enhance the effectiveness of an elevation, plan, or perspective?

3. What rules could be established to determine if an elevation, plan, or perspective has too much entourage?

PRACTICAL EXERCISES

1. Draw or computer-produce a style sheet of your individual treatment of the following elements:

   a. Evergreen trees in plan

   b. Evergreen trees in elevation

   c. Deciduous trees in plan

   d. Deciduous trees in elevation

   e. Foreground shrubbery

   f. Background scenery

   g. Human figures

   h. Vehicles

2. Include entourage on (a) the presentation drawing and (b) the perspective of your original house design.

CHAPTER 45

Architectural Models

OBJECTIVES

By completing this chapter you will be able to:

- 🏛 Explain the difference between a study model and a presentation model.
- 🏛 Construct the following architectural scale models of a residence or a small commercial building:
  - ■ Study model
  - ■ Presentation model
  - ■ Cutaway model

KEY TERMS

Study model
Presentation model
Cutaway model

Figure 45–1 *Model of an Egyptian residence, the Cairo Museum. (Courtesy the Metropolitan Museum of Art.)*

Models have always been an important tool of the architect. Before the discovery of the blueprinting process, architects used models to describe their projects to builders. It was common practice to scale measurements directly from working models. Architects also used presentation models to attract new patrons, carrying models of their best works from town to town.

The model of an Egyptian residence shown in Figure 45–1 was taken from the XI Dynasty tomb of Meket-Re. It is one of the oldest known models dating to 2500 B.C. It was built of carved, painted wood. Walls were mitered together and mortised to the base. Tree branches, leaves, and fruit were made of doweled wood, and the atrium pool was lined with copper sheet to hold water.

Nearly everything we use in today's living was first constructed in model form. The ground, water, and air vehicles we ride in, the commercial, industrial, and educational structures we work in, and the residences we live in were all modeled before they were actually built. The construction of architectural models, therefore, is a definite part of the services performed by the architect and is a useful skill for the architectural drafter.

TYPES OF MODELS

Various kinds of architectural models are used for different purposes. In general, though, architectural models fall into two categories: study models and presentation models.

Study Model

The study model, as the name implies, is constructed by designers to help them *study* the function or appearance of a building. A study model is built during the planning stage and may be modified many times before a satisfactory solution is found. Obviously, then, a study model is not a carefully constructed finished model. In fact, very crude materials may be used. A *mass model*, for example, is used to study the general effect of the position (or *massing*) of architectural elements (see Figure 45–2). Modeling clay is an excellent medium for this, although balsa wood, Styrofoam, soap, and even raw potatoes are used. Other types of *thumbnail* models may be used to study plot contours, roof intersections, landscaping, interiors, and the like.

Presentation Model

A presentation model is a finished scale model showing *exactly* how a proposed building will appear and function. It may be built after the preliminary planning stage to help the architect explain the concept to the clients, or

it may be built after the finished plans are drawn to help the clients raise money for the actual construction. In larger offices, the presentation models are built by a special model-making department; smaller offices often contract with professional architectural model-building concerns.

In the following pages we will discuss some materials and methods used in building presentation models. First, however, two important rules in architectural model building—neatness and scale—should be emphasized.

Neatness

Take great care to be accurate and neat in all phases of the building. A sloppy modeling job is completely useless because it will not serve its main function—that of selling the design to the client.

Scale

Every feature of the structure, down to the smallest detail, must be built to the same scale. When this is carefully done, the model will have the professional look of a scale model; when it is not, it will look like a toy. The most commonly used scales are $\frac{1}{8}$" = 1' and $\frac{1}{4}$" = 1'.

MATERIALS

The most often used and versatile material for architectural modeling is heavy cardboard. Properly selected, cardboard can serve as finished walls and roofs, or as the base for the application of other finished materials. It is known as *mat board*, *mount board*, or *display cardboard*. Fourteen-ply cardboard ($\frac{1}{8}$"–$\frac{3}{16}$" thick) of the best quality should be used. These boards may be obtained permanently embossed with scaled surface detail of several types (see Table 45–1).

Another popular architectural modeling material is two- or three-ply *Strathmore drawing paper*, plate surface. It is often used for trims and overlays since it has a hard, tough surface, cuts cleanly, can be bent sharply without tearing, and will take paint without wrinkling.

These cardboards are cut with a new razor or razor knife, using a steel straightedge as a guide. The first cut should be fairly light, the cuts being repeated until the edge cuts through.

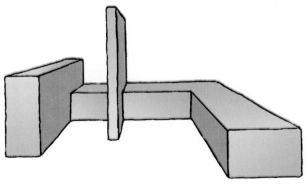

Figure 45–2 *Mass model.*

TABLE 45–1   *Available Mat Boards*

Surfaces	Sheet Size	Scale
Smooth surface	30" × 40"	
Pebbled surface	30" × 40"	
Embossed surfaces		
9" weatherboard or shiplap	3" × 22"	$\frac{1}{8}$"
12" weatherboard or shiplap	3" × 22"	$\frac{1}{8}$"
12" board and batten	3" × 22"	$\frac{1}{8}$"
brick siding	3" × 22"	$\frac{1}{8}$" and $\frac{1}{4}$"
concrete block siding	3" × 22"	$\frac{1}{8}$" and $\frac{1}{4}$"
stucco siding	3" × 22"	$\frac{1}{8}$" and $\frac{1}{4}$"
shake roofing	3" × 22"	$\frac{1}{8}$" and $\frac{1}{4}$"

The outside corners of walls must be carefully mitered. Good work can be done only with *sharp* instruments.

Some professional model makers prefer wood to cardboard. Basswood and ponderosa pine are used for their fine, even grain, and balsa wood is used for its extreme softness. Table 45–2 shows the available sizes of these woods in sheets, strips, and structural shapes. Thin wood may also be cut by razor, but heavier stock is best cut by a razor saw.

Household cement (such as Dupont Duco) may be used for cementing cardboard, paper, and wood. After setting one hour it will make a glue joint that is stronger than the materials joined. Rubber bands, drafting tape, straight pins, and small *lills* are used as clamps to hold the materials until set. To prevent model parts from adhering to a template, waxed paper may be used over the template.

Unless the model is to have a removable roof so that the interior can be studied, consider

TABLE 45–2   *Available Wood Sizes*

Sheet balsa (36" × 2", 3", 4", and 6")

Thicknesses: $\frac{1}{32}"$, $\frac{1}{16}"$, $\frac{3}{32}"$, $\frac{1}{8}"$, $\frac{3}{16}"$, $\frac{1}{4}"$

Strip balsa (36" long)

	$\frac{1}{16}"$	$\frac{3}{32}"$	$\frac{1}{8}"$	$\frac{3}{16}"$	$\frac{1}{4}"$
$\frac{1}{16}"$	✓		✓	✓	✓
$\frac{3}{32}"$		✓	✓		
$\frac{1}{8}"$	✓	✓	✓	✓	✓
$\frac{3}{16}"$	✓	✓	✓		
$\frac{1}{4}"$	✓	✓			✓

Sheet pine (22" × 1", 2", and 3")

Thicknesses: $\frac{1}{32}"$, $\frac{1}{16}"$, $\frac{3}{32}"$, $\frac{1}{8}"$, $\frac{3}{16}"$, $\frac{1}{4}"$

Strip basswood (24" long)

	$\frac{1}{32}"$	$\frac{1}{16}"$	$\frac{3}{32}"$	$\frac{1}{8}"$	$\frac{5}{32}"$	$\frac{3}{16}"$	$\frac{1}{4}"$
$\frac{1}{32}"$	✓	✓	✓	✓	✓	✓	✓
$\frac{1}{16}"$	✓	✓	✓	✓	✓	✓	✓
$\frac{3}{32}"$	✓	✓	✓	✓	✓	✓	✓
$\frac{1}{8}"$	✓	✓	✓	✓	✓	✓	✓
$\frac{5}{32}"$	✓	✓	✓	✓	✓	✓	✓
$\frac{3}{16}"$	✓	✓	✓	✓	✓	✓	✓
$\frac{1}{4}"$	✓	✓	✓	✓	✓	✓	✓

Milled basswood ($3\frac{1}{2}"$ × 24" sheets)

	$\frac{1}{32}"$	$\frac{1}{16}"$	$\frac{1}{8}"$
6" and 9" clapboard ($\frac{1}{8}"$ scale)		✓	
6" clapboard ($\frac{1}{4}"$ scale)			✓
12" board and batten ($\frac{1}{8}"$ scale)		✓	
9" board and batten ($\frac{1}{4}"$ scale)		✓	
3", 4", 6", 9", and 12" scribed planking ($\frac{1}{8}"$ scale)	✓	✓	
3", 4", 6", 9", and 12" scribed planking ($\frac{1}{4}"$ scale)	✓	✓	
shingled siding ($\frac{1}{4}"$ scale)			✓
brick siding ($\frac{1}{8}"$ and $\frac{1}{4}"$ scale)		✓	
concrete block siding ($\frac{1}{4}"$ scale)			✓
flagstone ($\frac{1}{8}"$ and $\frac{1}{4}"$ scale)		✓	✓

Structural shapes (24" long)

	$\frac{3}{64}"$	$\frac{1}{16}"$	$\frac{5}{64}"$	$\frac{3}{32}"$	$\frac{1}{8}"$	$\frac{3}{16}"$	$\frac{1}{4}"$
angles	✓	✓		✓	✓	✓	✓
tees	✓	✓		✓	✓	✓	✓
channels			✓	✓	✓	✓	✓
WF beams				✓	✓	✓	✓
I beams				✓	✓	✓	
quarter round		✓		✓	✓		

strengthening each inside corner with a $\frac{3}{16}$" square balsa strip.

BASE

Since a structure is designed in relation to its surroundings, the model is constructed showing some of its surrounding topography. The usual procedure is to select a base that is scaled to the plot size and shape. Plywood $\frac{1}{2}$" thick is quite satisfactory for this purpose.

Flat Plot

The plywood base may be finished directly with paint, flocking, or a loose material sprinkled over glue (such as dyed sawdust or sand). Sandpaper or black garnet paper may be glued face-up with rubber cement to give the effect of concrete or asphalt driveways.

Contoured Plot

To obtain an accurate reproduction of a contoured plot, the usual procedure is to build up successive layers of chipboard which have been cut to conform to the contour lines on the plot plan. Chipboard is an inexpensive cardboard available in thicknesses of $\frac{1}{16}$", $\frac{3}{32}$", $\frac{1}{8}$", and $\frac{5}{64}$". The thickness is selected to equal the scaled contour interval (vertical distance between contour lines). For example, the contour interval in Figure 45-3 is 2'. If the contour model is to be built to a $\frac{1}{8}$" = 1' scale, each layer of chipboard must be $\frac{1}{4}$" thick. Therefore, two $\frac{1}{8}$" chipboards should be used. Usually this kind of contour model is left in terraced steps for easy comparison with the plot plan rather than being smoothed to shape. It is finished in the same manner as the flat plot.

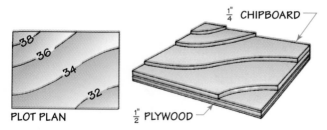

Figure 45–3   *Modeling a contoured plot.*

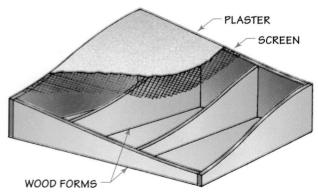

Figure 45–4   *Modeling a hillside plot.*

Hillside Plot

When the plot has a very steep slope, the contour plot method described requires too many layers of cardboard. The professional modeler then uses an alternate technique, building up the land model using bandsawed wood forms and covering them with wire screening as shown in Figure 45-4. The screening is plastered with a $\frac{1}{4}$"-thick mix of 50 percent (by volume) plaster, 50 percent mineral fibers, dry color, and water. Mineral fibers of the type used for insulation greatly increase the strength of the mix (like the aggregate in concrete). The dry colors are obtained from an art store and are sometimes called *earth colors*. A brown color is used so that the plaster will look like earth if it is chipped. This mix is troweled on roughly to indicate rock outcrops, and smoothed or stippled with a brush to indicate grass areas. Thin washes or oil stains are used for the final painting. The following colors are common:

> Grass: Color varies from pale yellow to olive green
> Brown soil: Burnt umber stain
> Reddish soil: Burnt and raw sienna stain
> Brown rock: Van Dyke brown stain
> Reddish rock: Van Dyke brown and burnt sienna stain
> Gray rock: Thin wash of black

WALLS

The selection of the method and materials used for wall construction depends upon the scale of the model. When models are built to a scale smaller than $\frac{1}{8}$" = 1', the walls are usually built of painted mat board. The recommended order of procedure is:

1. Accurately lay out all wall sizes in pencil on the mat board using a T square and triangles. Cut out with a *sharp* razor knife (Figure 45–5).

2. Paint the walls using poster paint (tempera) carefully mixed to the desired colors. When they are dry, rule on white or black lines (depending upon the color of the desired mortar mix) representing brick or stone joints and wood siding. These lines need not be completely ruled—only enough to give the impression of the surface (Figure 45–6).

3. Since stock windows are not available—or desirable—for such small scales, windows are painted on or outlined by scaled strip wood (Figure 45–7). The professional model builder uses a flat black or dark gray for glass areas, since this is the actual appearance of windows. Muntins and mullions are ruled on in white ink (such as Pelikan white

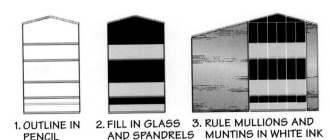

1. OUTLINE IN PENCIL   2. FILL IN GLASS AND SPANDRELS   3. RULE MULLIONS AND MUNTINS IN WHITE INK

Figure 45–7 A model window.

drawing ink) with a ruling pen or technical fountain pen.

4. Miter corners and assemble, being careful not to smudge the prepainted surfaces. All corners must be square (Figure 45–8).

Models built to a scale of $\frac{1}{8}$" or $\frac{1}{4}$" = 1' are somewhat simpler to construct since a great variety of stock material is available in these scales. These are common scales for residential models and probably the best ones for the beginner. Some plastic materials available in these scales are shown in Table 45–3, and some paper materials are shown in Table 45–4. The procedure for building these larger scaled models is as follows:

1. Select the desired material from Tables 45–1 through 45–4. Milled basswood, embossed mat board, and balsa wood are all satisfactory materials. If the model is to have a removable roof so that rooms inside can be viewed, the thickness of the material should scale to the actual wall thickness. Although it requires much more work, custom siding

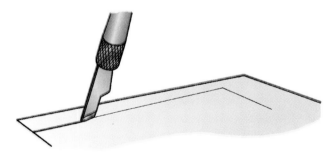

Figure 45–5 Use of the razor knife.

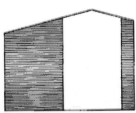

COMPLETELY RULED   PARTIALLY RULED

Figure 45–6 A model wall.

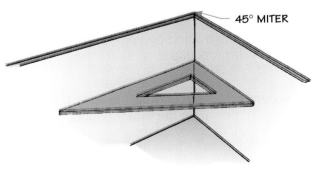

45° MITER

Figure 45–8 Miter and square corners.

TABLE 45–3   *Available Plastic Material*

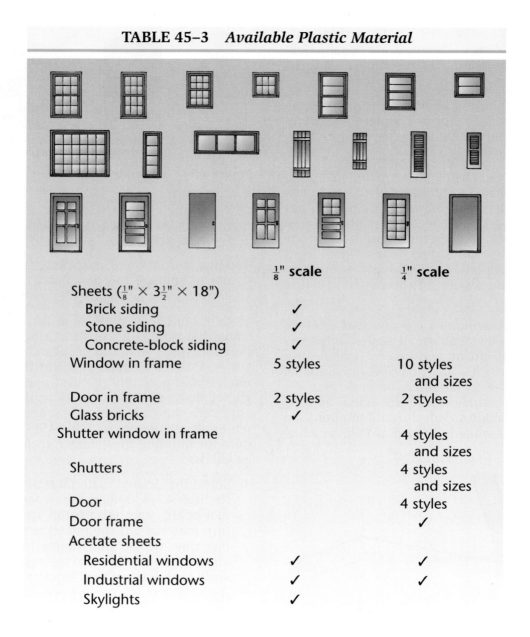

	$\frac{1}{8}$" scale	$\frac{1}{4}$" scale
Sheets ($\frac{1}{8}$" × $3\frac{1}{2}$" × 18")		
Brick siding	✓	
Stone siding	✓	
Concrete-block siding	✓	
Window in frame	5 styles	10 styles and sizes
Door in frame	2 styles	2 styles
Glass bricks	✓	
Shutter window in frame		4 styles and sizes
Shutters		4 styles and sizes
Door		4 styles
Door frame		✓
Acetate sheets		
Residential windows	✓	✓
Industrial windows	✓	✓
Skylights	✓	

can be constructed by cementing thin strips of Strathmore paper to sheet stock (Figure 45–9).

2. Accurately lay off all walls, windows, and doors in pencil on the siding (Figure 45–10). If stock windows and doors are to be used, some adjustment may have to be made to fit the available material to your plan. Cut out walls, windows, and doors using a *sharp* razor knife.

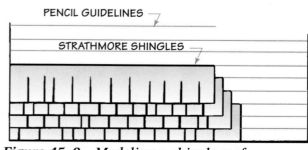

Figure 45–9   Modeling a shingle roof.

TABLE 45–4   *Available Building Paper*

	$\frac{1}{8}$" scale, 6" × 9" sheets	$\frac{1}{4}$" scale, 9" × 12" sheets
Red brick	✓	✓
Yellow brick	✓	✓
Brown fieldstone	✓	✓
Gray fieldstone	✓	✓
Gray flagstone	✓	✓
Brown ashlar stone	✓	✓
Grey ashlar stone	✓	✓
Yellow ashlar stone	✓	✓
Concrete block	✓	✓
Brown wood shingles	✓	✓
Grey wood shingles	✓	✓
Green wood shingles	✓	✓
Gray slate	✓	✓

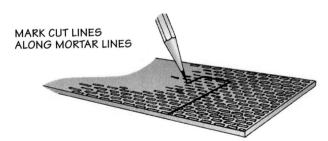

MARK CUT LINES
ALONG MORTAR LINES

Figure 45–10   Lay out on mortar lines.

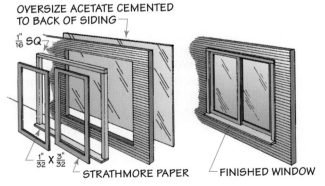

OVERSIZE ACETATE CEMENTED
TO BACK OF SIDING

$\frac{1}{16}$" SQ

$\frac{1}{32}$" X $\frac{3}{32}$"

STRATHMORE PAPER

FINISHED WINDOW

Figure 45–11   Custom-built window construction.

3. Windows and doors may also be custom-built (Figure 45–11). The construction of a sliding window is illustrated.

4. Carefully miter all corners and assemble the sides directly upon a floor plan cut from sheet stock. If this sheet stock is sufficiently thick, it can serve as the exposed portion of the foundation. Use pins or tape to hold the model until dry (Figure 45–12).

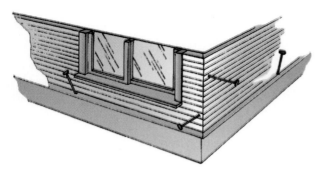

Figure 45–12   Pin corners until cement dries.

Figure 45–13   Built-up roofing simulation.

ROOFS

Flat built-up roofs of all scales can be made very simply of sandpaper glued face up to mat board and finished with the desired color latex paint. Marble chip roofs are simulated by sprinkling fine sand or salt on the wet white paint to add sparkle to the surface (see Figure 45–13). Preformed shingle roof material is available, but roofs may be custom-built by gluing strips of notched Strathmore paper to a mat board roof base.

Sheet metal roofs and flashing are modeled with strips of drafting tape cut to shape and painted copper or silver. A properly scaled fascia of strip basswood provides a finished look to any roof. Gutters and downspouts are built of properly scaled wire or strip basswood.

Chimneys should be cut from a balsa wood block with the horizontal brick joints scribed in (Figure 45–14). Bricks are painted on over a white undercoat to represent mortar. Of course, any of the stock brick sheet material may be used.

REAL WORLD *APPLICATIONS*

Opinion Surveys

You have created a model of your original house design to present to a client. This client is a builder who is interested in the ability to sell this design, so you decide to survey twenty consumers regarding their opinion of your model. Write a survey that will determine these consumers' opinions of the room design, floor plan, traffic flow, window selection, and any other features of interest to the builder.

The purpose of the survey is to show support for your model and its design, not to obtain feedback so that you can improve your design. Therefore, you will want to use mostly close-ended rather than open-ended questions. Close-ended questions provide a specific set of answers from which the respondents must choose; open-ended questions allow respondents to write their own answers.

As you write your survey, you need to consider the following guidelines:

1. Questions should be clear and concise.

2. Avoid negative questions such as "This floor plan is **not** sufficiently open for my taste."

3. The responses you provide for each question should be exhaustive. An ex-

Figure 45–14   Modeling a chimney.

ACCESSORIES

Accessories added to the structure (such as front door lamps), those added to the topography (such as shrubs and trees), and entourage (such as cars and people) may be custom built or purchased from stock. Some architectural modelers find it more convenient to obtain supplies from a model railroad supplier. For this purpose, it is necessary to know that O gauge is $\frac{1}{4}$" scale, HO gauge is approximately $\frac{1}{8}$" scale, and N gauge is approximately $\frac{1}{16}$" scale. See Table 45–5 for the exact conversion size.

Nearly any architectural modeling element may be purchased, but often custom-built accessories are less toylike and more profes-

TABLE 45–5   *Scale Conversions*

Model Railroad Gauge	Equivalent Architectural Scale
O gauge	1" = 1'-0
S gauge	$\frac{3}{16}$" = 1'-0
HO gauge*	$\frac{1}{8}$" = 1'-0
TT gauge	$\frac{1}{10}$" = 1'-0
N gauge*	$\frac{1}{16}$" = 1'-0
Z gauge*	$\frac{1}{20}$" = 1'-0

*Actually HO gauge is 3.5 mm = 1', which is approximately 1/7.3" =1'-0, N gauge is 1:160, which is approximately 1/13.3" = 1'-0, and Z gauge is 1:220, which is approximately 1/18.3" = 1'-0.

sional looking. Realistic trees may be modeled from dried yarrow or culex weeds dipped in shellac to prevent deterioration, shrubs and ground cover formed from lichen obtainable in many colors, hedges from sponge rubber or Styrofoam, people carved from erasers or soap, and stylized cars from balsa blocks. Care must be taken that all accessories are properly scaled.

haustive set of responses covers the complete range of answers that could be given to a question. For instance, if you are asking for a respondent's income, you would need to include all possible levels of income. Below are two examples of a set of responses to an income question. The first is exhaustive; the second is not. Why?

a. Below $10,000	a. $5,000–$9,999
b. $10,000–$20,000	b. $10,000–$19,999
c. $20,000–$30,000	c. $20,000–$29,999
d. $30,000–$40,000	d. $30,000–$39,999
e. $40,000–$50,000	e. $40,000–$49,999
f. Above $50,000	f. $50,000–$75,000

4. In addition to being exhaustive, the answers in the response set must be mutually exclusive. This means that no answer or category can include the same value as another. All answers or categories must be completely different so no respondent falls into two different ones. In the preceding example of income responses, the second list is mutually exclusive; the first is not. Why?

5. Avoid double-barreled questions. These questions are confusing to the respondent because they ask two questions. "Do you think the designs of the front and rear entrances are functional?" is a double-barreled question.

REAL WORLD   APPLICATIONS

Marketing Your Designs

The local home builders' show is approaching and you have rented an 8' × 10' booth. This is a great time to market your home designs. Now you must gather materials and organize your display. The booth display should attract attention, present your designs as aesthetically pleasing, and provide practical and accurate information. You will want to include renderings, floor plans, models, and photos of completed homes you have designed.

You will need to consider colors, placement of materials, lettering, and lighting when creating your display. Use colors that are attractive and complementary. Space the drawings and floor plans so they are easily viewed and do not appear crowded. Your firm's name should be prominent in the display. All lettering should be professional and pleasing. You may wish to take your own lamps to ensure appropriate lighting for viewing your designs.

Elevations, floor plans, models, and photos should be selected carefully. Display only one or two models that will catch potential consumers' eyes. Consider the following questions when choosing your material: What homes will sell best in your area? What price range will you include?

Your display booth should also have brochures or booklets and business cards,

set out in an organized fashion. Remember, you want people at the show to take these, so have them readily accessible. Incentives can be used to encourage potential consumers to take your material and view your display. These may include coupons, free consulting services, candy, pens, note pads, small gifts, or anything you feel will entice people to view your work. You may decide to put your information and material in plastic bags with a small gift and hand these to people as they approach your booth.

For fun and practice, create a booth display using the floor plan and model of your original house design. Include the business card you created as well as a sample brochure or booklet. Develop a firm name and present this within your display. Booths can be set up in your classroom so you can view your classmates' displays and gain ideas from one another. Which booth is most attractive? Why? Which booth makes most efficient use of space? How? Does material in any display look crowded? How could this be changed? Is the firm's name a prominent part of each booth's display? Are business cards and brochures organized and presented effectively? What incentives have your classmates included in their displays to attract potential consumers? How effective do you think the different incentives would be?

Some architectural models by students are shown in Figures 45–15 and 45–16.

CUTAWAY MODELS

A **cutaway model** of a structure is used to show actual construction methods and materi-

als. Such models are very popular classroom projects since they must be built up of studs, sheathing, and finish material in much the same manner as in current building practice. For example, a wall is constructed by pinning each stud and plate over a framing drawing. The framing is glued together rather than

Figure 45–15 *Solar-heated office building, student design by Todd Woodward. (Courtesy Professor Daniel E. Willis, the Pennsylvania State University.)*

Figure 45–17 *Cutaway residential model, student design by Mark Lukehart. (Courtesy Mark Lukehart.)*

Figure 45–16 *Low income health care facility, student study model by Jason Collins and Jacob Stevens. (Courtesy Jason S. Collins and Jacob G. Stevens.)*

nailed, with wax paper used to prevent it from sticking to the drawing. This is followed by wood sheathing, felt paper, and finish siding—each layer cut back so that the previous layer is left exposed for inspection. Insulation, heating ducts, and piping may also be shown in the walls. Other procedures are similar to those already described in this chapter. Recommended scales are $\frac{3}{8}$" = 1' and $\frac{3}{4}$" = 1' (Figure 45–17).

REVIEW QUESTIONS

1. What advantages and disadvantages do models have compared to presentation drawings?

2. How does a study model differ from a presentation model?

3. Why are neatness and scale important in building presentation models?

4. Match these modeling requirements with the materials to be used:

   a. Temporary fasteners   Lills

   b. Permanent fasteners   Acetate sheet

   c. Window and door trim   Mat board

   d. Window glass   Household cement

   e. Walls and roof   Strathmore drawing paper

5. What is a mass model?

6. Describe the modeling technique used to build the following bases:

   a. Flat plot   c. Hillside plot

   b. Contoured plot

7. Describe the technique used to build model walls to a scale of:

   a. $\frac{1}{16}$" = 1'

   b. $\frac{1}{4}$" = 1'

8. What materials are used to model:

   a. Built-up roofs   c. Flashing

   b. Shingle roofs   d. Guttering

9. Give the equivalent architectural scale for:

   a. O gauge   c. HO gauge

   b. S gauge   d. TT gauge

10. What materials are used to model these accessories:

    a. Trees   d. People

    b. Shrubs   e. Automobiles

    c. Hedges

ENRICHMENT QUESTIONS

1. Describe the process used to determine the size of the base of a model.

2. How will the construction of a model enhance the quality of a presentation to a client?

3. Prepare a list of materials and supplies that are needed to construct a model successfully.

4. Prepare fictitious situations in which a study model, a cutaway model, and a presentation model would and would not be appropriate.

PRACTICAL EXERCISES

1. Construct a study model of your original house design.

2. Construct a presentation model of:
   a. The A residence
   b. Your original house design

Commercial Design/Drafting

SECTION VIII

CHAPTER 46

Commercial Drafting

OBJECTIVES

By completing this chapter, you will be able to:

🏛 Explain the differences between the design, drafting, and construction of a residence and a commercial building.

🏛 Describe the construction of a commercial building by referring to a set of working drawings.

KEY TERMS

Sketch plans
Presentation drawings
Working drawings

There are important differences between the design, drafting, and construction of a residence and that of a commercial building:

1. *Design.* A single individual can design and draft the plans for a residence, but a number of specialists, including a registered architect and professional engineer, are needed to plan a commercial building. This involves all the consequent problems of coordination, cooperation, and human relations.

2. *Drafting.* More drawings are required for a commercial building due to its sheer size. But in addition, the plans must be more completely detailed to permit all interested builders to bid on a competitive basis.

3. *Construction.* A light wood-frame construction, as described in Chapter 30, is adequate for a residence, but a commercial building is usually framed in steel or reinforced concrete and finished with masonry or prefabricated composition panels.

A small commercial structure, the South Hills Office Building, recently designed and built in State College, Pennsylvania, will be used in this chapter to illustrate each of these aspects.

DESIGN

Three businesspeople—the owners of a law office, a real estate firm, and an advertising agency—joined together to solve a common problem—that of finding suitable space for their offices. They formed a partnership, purchased a half-acre site, and asked a local architect to design an accessible building to satisfy their needs. As is customary with most architects, the design procedure consisted of four major stages:

1. The program

2. Presentation drawings of several schemes

3. Preliminary drawing of the chosen scheme

4. Working drawings

At one of the early meetings, a *program* was prepared which indicated the owners' requirement of 4,000 sq. ft. of office space plus an additional 12,000 sq. ft. of rentable space for other companies. An attractive contemporary exterior appearance was considered to be an important requirement. A preliminary study of the zoning

REAL WORLD   APPLICATIONS

Leadership Skills

Having just completed your ninth year at an architectural firm, your boss asks you to represent the company on a community planning committee. The purpose of the committee is to develop some short and long range goals for improving the downtown business district. At the first meeting there are eight representatives of the business community: a banker, real estate agent, community official, and several retail business owners. You recognize that many of these individuals have little enthusiasm for the project and no one is assuming a leadership role. You report to your boss on these concerns. Her response is, "I guess you had better become the leader then!"

The next meeting will be in two weeks and will include selection of a chairperson.

Preceding this meeting, send a letter to each representative that contains a summary of the first meeting and includes an attached agenda. Finally, as you prepare for this meeting, remember to review the following key leadership skills:

- 🏛 Leading by example
- 🏛 Team building
- 🏛 Encouraging innovation
- 🏛 Interactive listening
- 🏛 Decision making
- 🏛 Delegating
- 🏛 Motivating people
- 🏛 Resolving conflicts

and building code requirements showed that building height would be limited to 55' (not including mechanical features occupying less than one-tenth of the roof area). It was also determined that one off-street parking stall would have to be provided for each office, for a total of twenty-four stalls.

The architect proceeded to develop **sketch plans** and **presentation drawings** of three alternate solutions: two-, three-, and five-story buildings. The solution preferred by the owners was a five-story basementless building with the first (plaza) level open to provide adequate parking. This solution included glass window walls and a central elevator shaft.

Preliminary drawings consisting of 4' modular plans and elevations were then prepared for this chosen solution. The framing was visualized as consisting of steel beams supported by ten steel columns. Each pair of columns would form a bent spanning 40', and adjacent bents would be 24' oc. Welded connections would be used where necessary to obtain a rigid frame. To provide uninterrupted glass window walls, the outer wall was cantilevered 4' beyond the columns, resulting in a rectangular floor plan of 48' × 104'. Stairwells were placed at both ends of the floor plan and were connected by a longitudinal corridor. An elevator was a "must" for accessibility. The elevator shaft, rest rooms, and maintenance room were placed in a central location. The remaining area was then available as clear floor space for maximum flexibility. A hydraulic piston elevator was chosen in preference to a hoist elevator to eliminate the unsightly elevator penthouse needed for the hoisting machinery. The piston elevator, however, does require that a piston shaft be drilled into the ground equal to the distance of the total lift. Piston elevators generally are not specified for lifts exceeding 60' in height. Heating and cooling were provided by electric air-conditioning space units installed in wall panels.

The preliminary drawings were approved by the owners with only one major change: the glass window walls were rejected due to the additional air-conditioning capacity required. The architect replaced the window walls with vertically aligned windows and air-conditioning units set in exterior brick walls. Vertical lines were emphasized by mullions framing each stack of windows.

Design for accessibility requirements resulted in 2'-8" door widths, 3'-6" minimum aisle widths, and special attention to restrooms, watercoolers, telephones, and tactile signs.

DRAFTING

Working drawings of the final solution were prepared as shown in Drawings No. 1–P2 (Figures 46–1 through 46–24). These drawings are nearly identical (some details such as stair and elevation sections are omitted) to the set of working drawings used to construct the South Hills Office Building and were redrawn in ink to assure good reproduction in this book. The north elevation Drawing No. 8, however, was computer-produced for you to compare with hand-produced drawings. Study these plans until you are confident that you understand how they are used to describe this project. The following remarks may help.

Drawing No. 1: Index (Figure 46–1) This is the cover sheet for the entire set of working drawings. In addition to the index of drawings, it includes a legend of all abbreviations and symbols used on the drawings. Some architectural offices also include a sketch of the building on the cover sheet.

Drawing No. 2: Plot Plan (Figure 46–2) This plot plan positions the building on the site, shows the existing and proposed land contour, landscaping, parking, walks, gas lines, water lines, and sanitary waste lines. The note "Swale to CB" indicates a downward slope to a catch basin. The designation "BC 100.5" fixes the bottom of the curb (road level) at an elevation of

100.5', and "TC 101.5" fixes the top of the curb (ground level) at an elevation of 101.5'.

Notice that two indications for north are given at the lower right corner. The large arrow enclosed in a circle is the direction of north, and "building north" shows the side of the building that is termed the *north elevation*. This is particularly important when a building is positioned such that two sides might both be considered north elevations.

Drawing No. 3: Foundation Plan (Figure 46–3) A 4' modular grid system is used with coordinate identification letters and numbers. This identification system helps to locate details on the plans, in the written specifications, and in the field. Notice that this system was adapted to use arrowheads to indicate both on-grid and off-grid dimensions. The callouts "0" and "6" refer to masonry courses, each 8" high. Thus the CMU (concrete masonry joint) wall marked "6" can be started 6 courses (48") above the walls marked "0." Refer to Drawing No. 11 for a better understanding of these masonry-course identification numbers.

A test boring was taken at each column location. Firm rock was only 6'-8' deep at the three "A" locations, but was 16'-18' deep at the seven "B" locations. Therefore two types of footings were designed. Notice on Drawing No. 4 that the contractor was given the option of using the reinforced footing "A" at all ten locations and chose that alternative in preference to driving piles as required for footing "B." To support walls, either 8"- or 12"-thick CMU foundations were used, depending upon the weight of the wall to be supported. This and all similar plans were originally drawn to a scale of $\frac{1}{4}$" = 1'-0.

Drawing No. 4: Footing Details (Figure 46–4) Four details are included on this sheet to show the reinforced concrete construction of footing "A," steel pile footing "B," the column waterproofing, and the reinforced concrete footing for the CMU walls. The note "HP 10 × 42" refers to a 10" × 10" bearing pile weighing 42 lb. per foot of length; "5-#11 bars" means five reinforcing steel bars, each $\frac{11}{8}$" in diameter. See Drawing No. 1 for the meaning of all abbreviations. These and similar details were originally drawn to a scale of $\frac{1}{2}$" = 1'-0.

CONSTRUCTION DRAWINGS

FOR THE

SOUTH HILLS OFFICE BUILDING

STATE COLLEGE, PENNSYLVANIA

INDEX OF DRAWINGS

1	INDEX
2	PLOT PLAN
3	FOUNDATION PLAN
4	FOOTING DETAILS
5	PLAZA FLOOR PLAN
6	FIRST FLOOR PLAN
7	SOUTH ELEVATION
8	NORTH ELEVATION
9	EAST & WEST ELEVATIONS
10	INTERIOR ELEVATIONS
11	LONGITUDINAL SECTION
12	TYPICAL SECTIONS
13	TYPICAL DETAILS
14	ROOM SCHEDULES
15	DOOR & WINDOW SCHEDULES
S1	FIRST FLOOR STRUCTURAL PLAN
S2	ROOF STRUCTURAL PLAN
S3	COLUMN SCHEDULE
S4	STRUCTURAL DETAILS
S5	CONCRETE SLAB PLAN
E1	ELECTRICAL PLAN
H1	HEATING-COOLING PLAN
P1	WATER SUPPLY PLAN
P2	SANITARY PLAN

LEGEND

—100—	EXIST CONTOUR
—100—	REVISED CONTOUR
	PROPERTY LINE
	BRICK
	CMU
	CRUSHED STONE
	EARTH
	STEEL
	CONCRETE SECTION
	IN PLAN
	BITUMINOUS
	RIGID INSULATION
	BATT
	ROUGH WOOD
	GYPSUM BOARD
	PLASTER
	CERAMIC TILE
	SWITCH LEG
	SWITCHED CIRCUIT
	BRANCH CIRCUIT
D-1	HOME RUN w/ CIRCUIT NO
	208V HOME RUN
	INCANDESCENT FIXTURE, CLG
	WALL
	FLUORESCENT "
	CONVENIENCE OUTLET
S	SWITCH
S3	THREE WAY SWITCH
S4	FOUR "
E	EMERGENCY CIRCUIT
	LIGHTING
	EXIT LIGHT, CKT# B1412
	" B1414
	HEATING-COOLING UNIT
	BASEBOARD HEATING "
	SHEET NO
	SECTION NO
	SHEET NO
	ELEVATION NO
	ROOM NO
	DOOR NO
	COLD WATER
	HOT WATER
	SANITARY WASTE
	VENT
A	WINDOW SYMBOL
	ROOM SYMBOL

ABBREVIATIONS

ACOUSTIC	AC		HOLLOW METAL	HM
ALUMINUM	ALUM		HOT	H
AMPERES	A		BEARING PILE	HP
ANGLE	L			
AT	@		INSULATION	INSUL
			IRON PIPE	IP
BEAM	B			
BEARING PILE	HP		JUNCTION BOX	JB
BITUMINOUS	BIT			
BOARD	BD		LAVATORY	L or LAV
BOTTOM OF CURB	BC		LIGHTING	LTG
BRITISH THERMAL UNIT	BTU			
BUILDING	BLDG		MANUFACTURER	MANUF
			MATERIAL	MAT'L
CABINET	CAB		MAXIMUM	MAX
CATCH BASIN	CB		METAL	MET
CEILING	CLG		MINIMUM	MIN
CEMENT	CEM			
CENTER LINE	℄		NORTH	N
CENTER TO CENTER	CC		NUMBER	NO or #
CERAMIC TILE	CER T			
CHANNEL	L or C		ON CENTER	OC
CLEAN OUT	CO			
CLEAR	CLR		PARTITION	PART
COLD	C		PHASE	Φ
COLUMN	COL		PLASTER	PLAST
CONCRETE	CONC		PLATE	PL
CONCRETE MASONRY UNIT	CMU		PORCELAIN	PORC
CONSTRUCTION	CONST		POUNDS PER SQUARE INCH	PSI
CUBIC FEET PER MINUTE	CFM		PUNCHED	PUN
DIAMETER	Φ		RAIN WATER CONDUIT	RWC
DITTO	DO or "		RECEPTACLE	REC
DOUBLE STUDS	DS		RIGHT OF WAY	R/W
DOWN	DN		RISER, RADIUS	R
EACH	EA		SHOCK ABSORBER	SA
EAST, EMERGENCY	E		SLOP SINK	SS
ELEVATION, ELEVATOR	ELEV		SOUTH	S
EQUAL	EQ		SPECIFICATIONS	SPECS
EXHAUST	EXH		STEEL	STL
EXISTING	EXIST		SYSTEM	SYS
EXPANSION JOINT	EXP JT			
EXTENDED	EXT		THRESHOLD	THRESH
			TOP OF CURB	TC
FINISH	FIN			
FLASHING	FLASH		UNPUNCHED	UNP
FLOOR	FL		URINAL	U
FOOTING	FTG			
FRESH AIR	FA		VINYL COMPOSITION TILE	VCT
FURRING	FUR		VOLTS	V
GYPSUM	GYP		WASTE, WATTS, WEST	W
GLASS	GL		WATER CLOSET	WC
			WATER HEATER	WTR HTR
HEATING	HTG		WIDE FLANGE	W
HEXAGONAL	HEX		WITH	w/

Figure 46–1   Drawing No. 1, Index of the South Hills Office Building.   (Courtesy Jack Risheberger.)

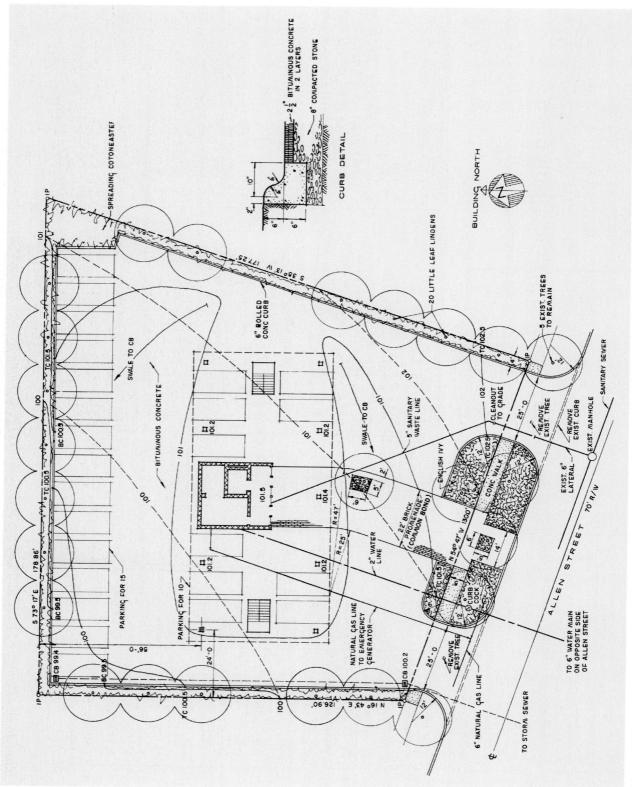

Figure 46-2   Drawing No. 2, Plot plan of the South Hills Office Building. (Courtesy Jack Risheberger.)

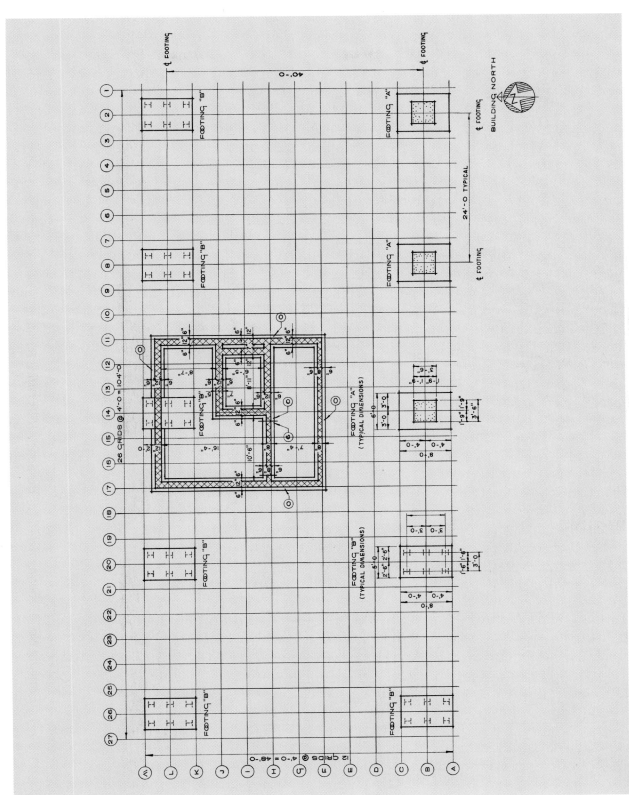

Figure 46-3 *Drawing No. 3, Foundation plan of the South Hills Office Building.* (*Courtesy Jack Risheberger.*)

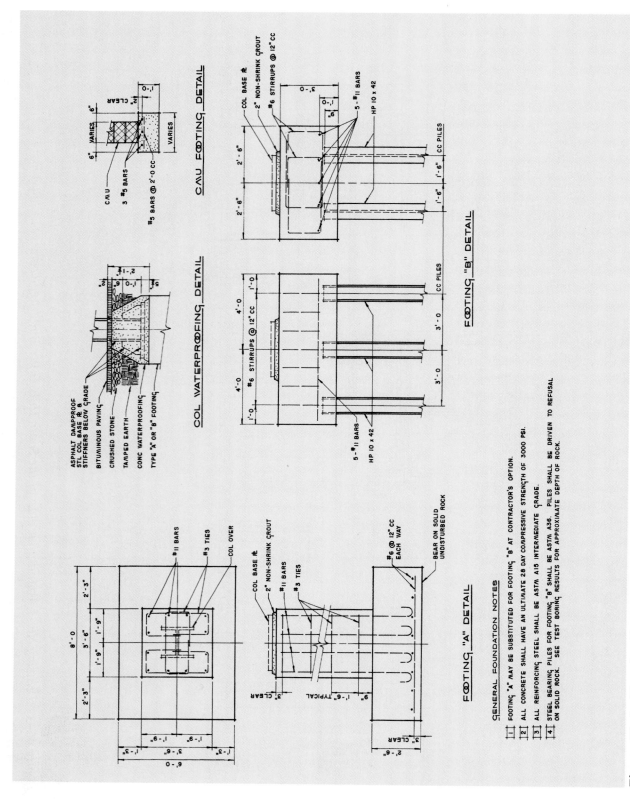

Figure 46–4 *Drawing No. 4, Footing details of the South Hills Office Building. (Courtesy Jack Risheberger.)*

Drawing No. 5: Plaza Floor Plan (Figure 46–5) Dashed lines are used to indicate overhead features such as the building line or overhead simulated beams. Each room, stairway, or corridor has an identification number (such as "P2"). Doors also have an identification number which is coded to the proper room (such as "P2/1" and "P2/2"). See the legend on Drawing No. 1 for the meaning of all such identification numbers.

Drawing No. 6: First-Floor Plan (Figure 46–6) Six air-conditioning units are located on the south wall, but only four units are required for the north wall because the south-wall cooling requirements are greater than the north-wall heating requirements.

Drawings No. 7–9: Elevations (Figures 46–7 through 46–9) Although 4' horizontal modules are used on the plans, 8" vertical modules are used on elevations to indicate courses of masonry 8" apart (a CMU course of 8" or three brick courses of $2'\text{-}\frac{2}{3}$"). For example, the balloon "39" means that the second floor is 39 courses, or 26' ($39 \times 8" = 26'$), above the top of the footing marked with a balloon "0." Control joints are formed by raking and caulking masonry joints. This directs any cracking along these joints rather than allowing it to occur at random. These elevations were originally drawn to a scale of $\frac{1}{4}" = 1'\text{-}0$.

Drawing No. 10: Interior Elevations (Figure 46–10) Interior elevations of all specially equipped rooms would be included in addition to these restroom elevations. The elevation identification "10/1" indicates Elevation No. 1 on Drawing No. 10.

Drawing No. 11: Longitudinal Section (Figure 46–11) This section is needed to explain the structural system and assure proper clearances and room heights. Only the more useful coordinate identification numbers are included. Note "AC CLG BD" is an abbreviation for acoustical ceiling boards.

Drawing No. 12: Typical Sections (Figure 46–12) Section 12/1 is a vertical section cut through a window (see Drawings No. 6 and 7). Section 12/2 is a vertical section through a simulated plaza roof beam (see Drawing No. 7). Section 12/3 is a horizontal section through a column (see Drawing No. 7). Multiple balloons such as "B" and "L" on Section 12/3 show that this section is typical of columns centered on both grid B and grid L.

Drawing No. 13: Typical Details (Figure 46–13) Plan detail 13/1 is a horizontal section cut through the window mullions (see Drawing No. 6). The two alternate details show the installation of panels and louvers. The callout "362 DS 16 PUN @ 16" oc" refers to $3\frac{5}{8}$" prefabricated metal studs as manufactured by the Keene Company: model no. 362, double stud, 16-gauge, punched, 16" on center.

Drawings No. 14 and 15: Schedules (Figures 46–14 and 46–15) Room finish information is contained in schedules such as shown on these two sheets. Complete schedules for all floors require many more pages. The written specifications contain even more detailed information.

Drawing No. S1: First-Floor Structural Plan (Figure 46–16) Each heavy line indicates the location of a steel member. The note "W 16 × 36 (–4)" refers to a 16"-wide flange beam weighing 36 lb./ft. with its upper flange 4" below the concrete slab surface. This and all similar plans were originally drawn to a scale of $\frac{1}{8}" = 1'\text{-}0$.

Drawing No. S2: Roof Structural Plan (Figure 46–17) Note "12 K5 EXT END (–$2\frac{1}{2}$)" refers to a 12"-deep, K5-series, open-web joist with an extended end and located $2\frac{1}{2}$" below the roof surface. Notation "DO" means ditto. The dashed lines show the location of cross-bridging. The note "ship lone" means that elevator beam 6 B 16 should be shipped without any shop connections because this beam is to be installed by the elevator technicians rather than the structural fabricators.

Structural steel designations were revised by the American Institute of Steel Construction in 1970. Although the "new" designations are used on the plans of the South Hills Office Building and throughout this book, the "old" designations are shown on Drawing No. S2. It would be

(continued page 609)

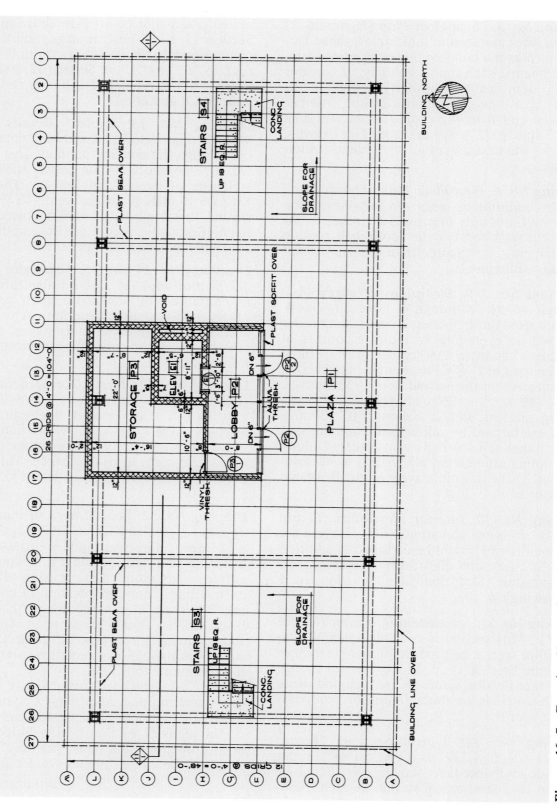

Figure 46–5 *Drawing No. 5, Plaza floor plan of the South Hills Office Building. (Courtesy Jack Risheberger.)*

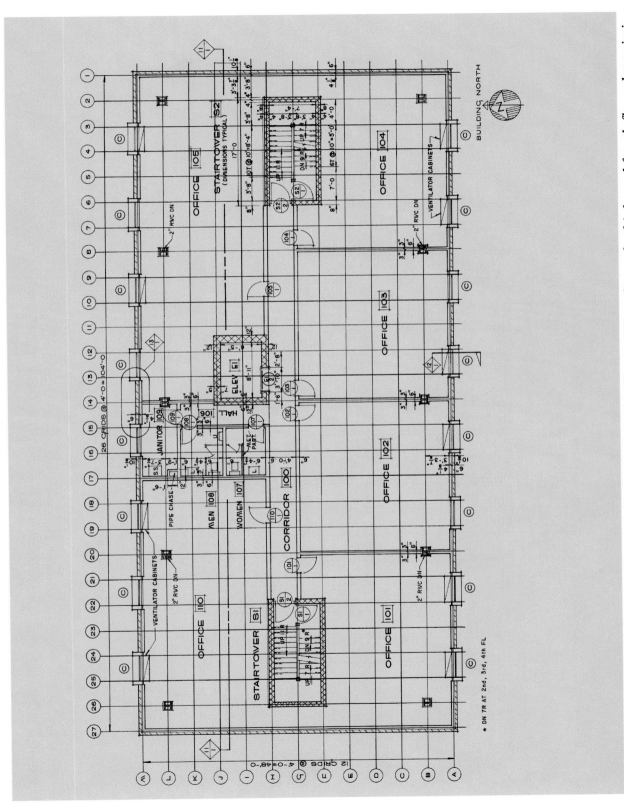

Figure 46–6 *Drawing No. 6, First-floor plan of the South Hills Office Building. (Second-, third-, and fourth-floor plans similar.) (Courtesy Jack Risheberger.)*

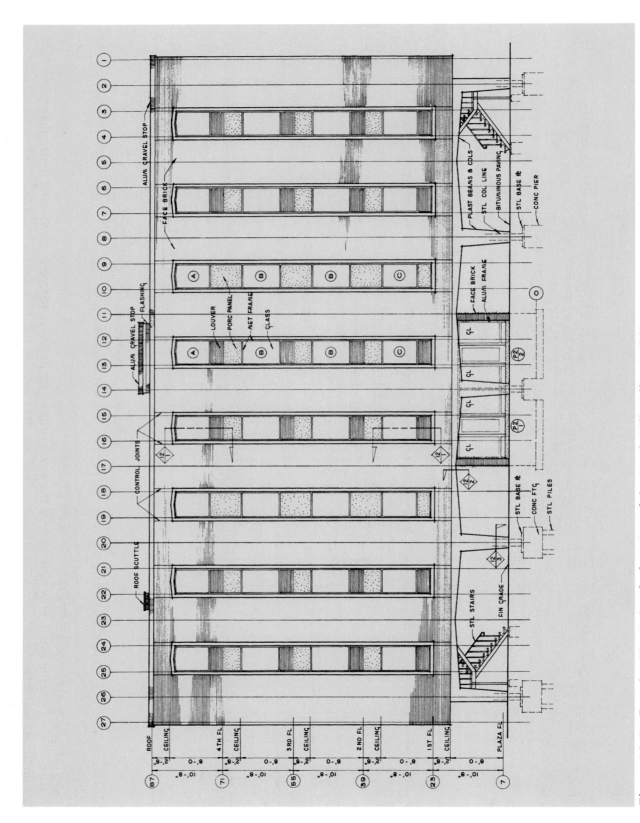

Figure 46–7   Drawing No. 7, South elevation of the South Hills Office Building. (Courtesy Jack Risheberger.)

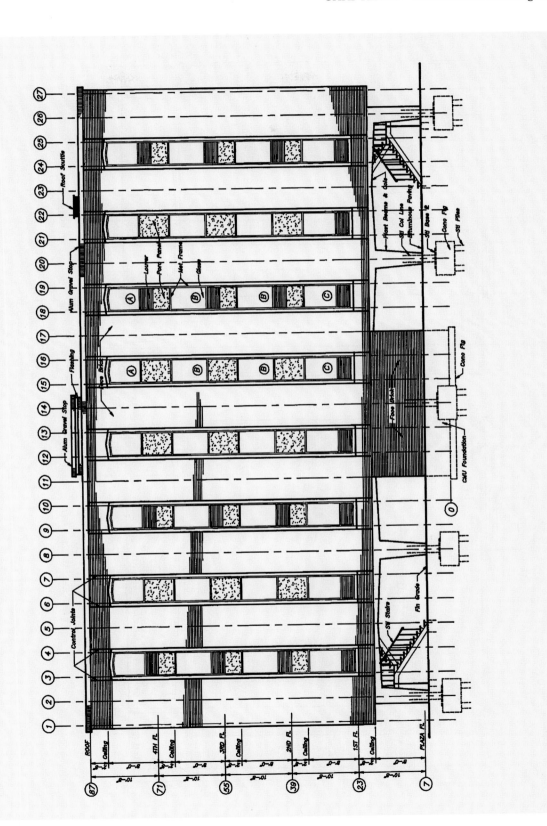

Figure 46–8 Drawing No. 8, North elevation (computer-generated) of the South Hills Office Building. (Courtesy Jack Risheberger.)

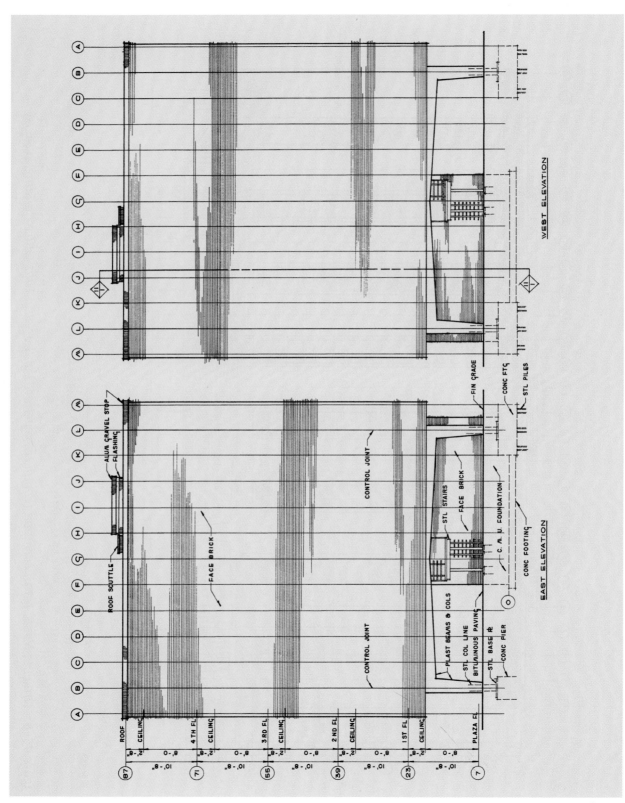

Figure 46–9   Drawing No. 9, East and West elevations of the South Hills Office Building.   (Courtesy Jack Risheberger.)

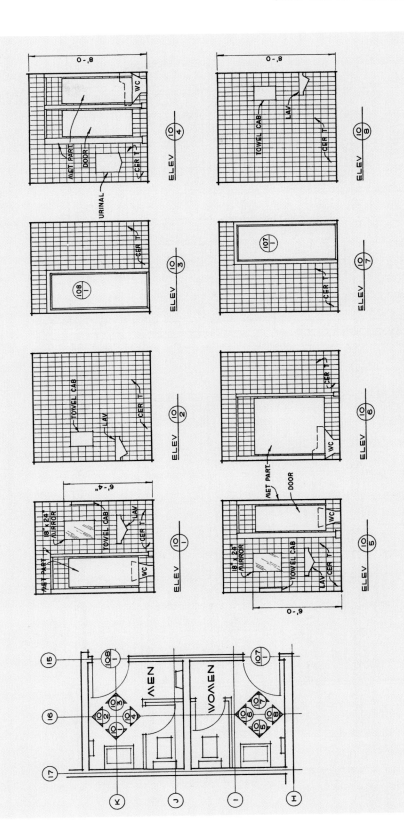

Figure 46-10 Drawing No. 10, Interior elevations of the South Hills Office Building. (Courtesy Jack Risheberger.)

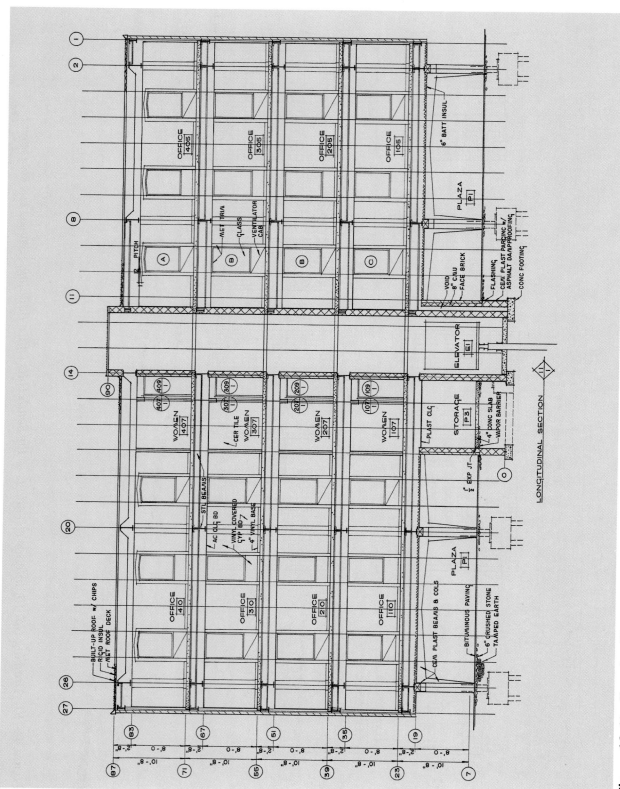

Figure 46–11 *Drawing No. 11, Longitudinal section of the South Hills Office Building. (Courtesy Jack Risheberger.)*

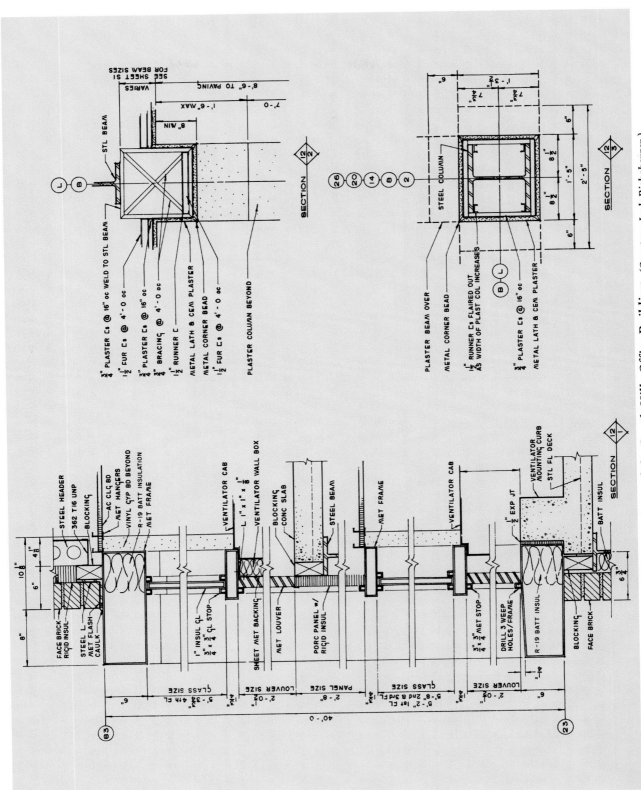

Figure 46–12   Drawing No. 12, Typical sections of the South Hills Office Building. (Courtesy Jack Risheberger.)

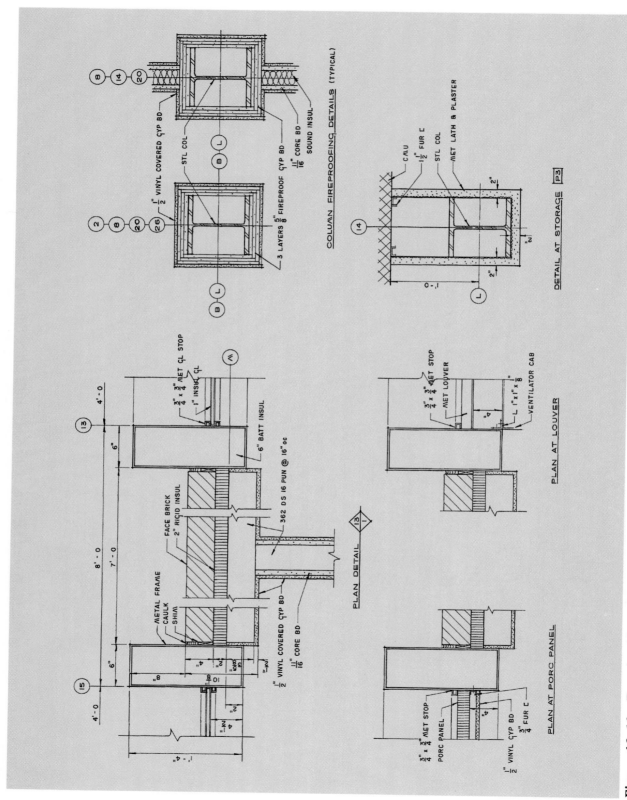

Figure 46–13 Drawing No. 13, Typical details of the South Hills Office Building. (Courtesy Jack Risheberger.)

LIGHTING FIXTURE SCHEDULE

NO	MANUFACTURER	CATALOG NO	FINISH	WATTS
A	LIGHTOLIER	7792		300
B	"	7794		100
C	"	7827 & 7821	WHITE	150
D	"	7827 & 7822		150 & 25
F	PRESCOLITE	WB-28-2	ALUM	150 & 100
G	LIGHTOLIER	81675	WHITE	4L-40
H	STONCO	QD8501	ALUM	500

MAIN DISTRIBUTION PANEL

400A-3P-SW-W/3-300A-FU- / PANEL A
400A-3P-SW-W/3-225A-FU- / PANEL B
400A-3P-SW-W/3-225A-FU- / PANEL C
400A-3P-SW-W/2-275A-FU- / PANEL D
100A-2P-SW-W/2-100A-FU- / PANEL E
60A-2P-SW-W/2-50A-FU- / PANEL EW SW
*200A-3P-SW-W/3-200A-FU- / ELEVATOR
*TIME DELAY FUSES (FUSETRON)
1200A-BUS 3Ø-4W-120/208V

ELECTRICAL PANEL SCHEDULES

PANEL D (TYPICAL)

31-20A-1P-CB-LTG, REC
9-30A-2P-CB-HTG, WTR HTR
3-20A-2P-CB-HTG
2-20A-2P-CB-SPARES
4-20A-1P-SPACE ONLY
400A-MLO   3Ø-4W-120/208V

PANEL E

9-20A-2P-CB-LTG,FA SYS EXH FAN
4-20A-1P-CB-LTG, REC
4-20A-1P-CB-SPARES
4-20A-1P-SPACE ONLY
100A-MLO   1Ø-3W-120/208V

PANEL EW

5-20A-1P-FU-LTG, ELEV JB
5-20A-1P-FU-SPARES
100A-MLO   1Ø-3W-120/208V

ELECTRIC HEATING-COOLING UNITS

NO	MANUFACTURER	HEATING		COOLING	
		BTU	WATTS	BTU	WATTS
EK-7S	REMINGTON	8400	2460	6500	1240
EK-10S	"	8400	2460	9000	1520
EK-10W	"	11330	3320	9000	1520
EK-12S	"	8470	2480	11700	1770
EK-12L	"	15300	4480	11700	1770
EK-15L	"	15370	4500	14100	2220

ELECTRIC BASEBOARD HEATING

NO	MANUFACTURER	CATALOG NO	BTU	WATTS
A	ELECTROMODE	8950-D	2560	750
B	"	8950-A	1707	500
C	"	8960-C	4439	1300

ROOM FINISH SCHEDULE

NO	NAME	FLOOR	BASE	WALL	TRIM	WINDOW STOOL	CEILING	HGT	REMARKS
P1	PLAZA	BITUM		BRICK 5	ALUM 9		PLAST 11	VARIES	CEM PLAST BEAMS & COLS
P2	LOBBY	VCT 1		" 5	" HM	9/10	" 13	8'-0	
P3	STORAGE	CONC 17		CMU 6	HM 10		" 11	8'-0	
100	CORRIDOR	VCT	VINYL 3	VINYL BRICK 7/5	HM	10	ACCLG BD 14	8'-0	NO BASE AT BRICK WALL
101	OFFICE	"	"	VINYL 7	ALUM	10	" 14	8'-0	
102	"	"	"	" 7	"	9	" 14	8'-0	
103	"	"	"	" 7	"	9	" 14	8'-0	
104	"	"	"	" 7	"	9	" 14	8'-0	
105	"	"	"	" 7	"	9	" 14	8'-0	
106	HALL	"	"	" 3	"	9	PLAST 12	8'-0	
107	WOMEN	CERT 2	CERT 8	CERT 8	"	10	" 12	8'-0	PROVIDE MIRROR, TOWEL CAB, NET PART
108	MEN	" 2	" 8	" 8	"	10	" 12	8'-0	
109	JANITOR	VCT 1	VINYL 3/6	VINYL CMU 7/6	ALUM	9	ACCLG BD 14	8'-0	
110	OFFICE	" 1	" 3	VINYL 7	HM	10	" 14	8'-0	

STAIRTOWER FINISH SCHEDULE

NO	RISER	TREAD	STRINGER	INTERMEDIATE FLOOR	FLOOR BASE	FLOOR LANDING	SOFFIT	CEILING	RAILING	WALL RAILING	WALLS
S1	STL 15	VCT 15	STL 15	VCT 1	VINYL 3	VINYL 3	PLAST 13	PLAST 13	VINYL 16	VINYL 16	CMU 16
S2	" 15	" 15	" 15	" 1	" 3	" 3	" 13	" 13	" 11	" 11	PLAST 11
S3	CONC 17	CONC 17		CONC 17	VCT 1	STL 15	" 11	" 11	STL 15	STL 15	" 11
S4	" 17	" 17		" 17							

INTERIOR MATERIAL SCHEDULE

NO	MATERIAL	SIZE	TYPE	FINISH
1	VINYL COMPOSITION TILE	9" x 9" x 1/8"	SEE SPECS	WAX
2	CERAMIC FLOOR TILE	1 1/16" x 1 1/16"	CERAMIC MOSAIC	FACTORY FINISH, UNGLAZED
3	VINYL COVE BASE	4" HIGH	COVE BASE	MATTE GLAZE
4	CERAMIC BASE TILE	4 1/4" x 6" x 5/16"	SEE SPECS, COMMON BOND	CONCAVE JOINT
5	BRICK	3 COURSES = 8"		3/8" CONCAVE JOINT
6	CONCRETE MASONRY UNIT	1 COURSE = 8"		PAINT
7	VINYL COVERED GYP BD	4'-0 x 8'-0 x 1/2 SHEETS		ALUM BATTENS
8	CERAMIC WALL TILE	4 1/4" x 6" x 5/16"	WALL TILE	FACTORY FINISH, MATTE GLAZE
9	ALUMINUM		SEE SPECS	PAINT
10	HOLLOW METAL			SPRAYED ON WHITE
11	PLASTER		CEMENT	WHITE COAT
12	"		KEENE CEMENT	PAINT
13	"		SAND FINISH GYPSUM	FACTORY FINISH
14	ACOUSTICAL CEILING BOARD	2'-0 x 2'-0 x 5/8"	SEE SPECS, EXPOSED "T" BARS	PAINT
15	STEEL		STEEL STAIR PARTS	FACTORY FINISH, PAINT BASE
16	STAIR RAILING	2" x 2" x 3/8" STEEL BASE PLATE	VINYL STAIR RAIL	SEAL w/ LIPIDOLITH
17	CONCRETE			

Figure 46-14   Drawing No. 14, Room schedules of the South Hills Office Building. (Courtesy Jack Risheberger.)

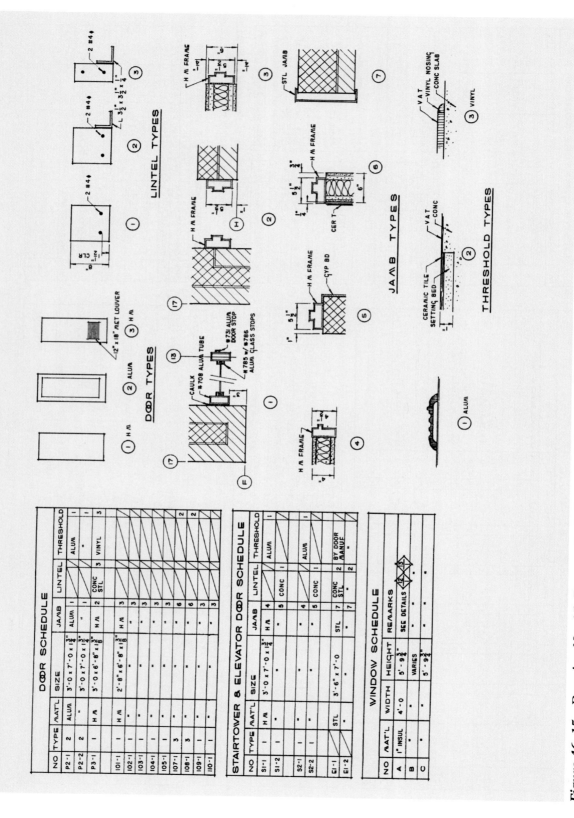

Figure 46–15 *Drawing No. 15, Door and window schedules of the South Hills Office Building.* (*Courtesy Jack Risheberger.*)

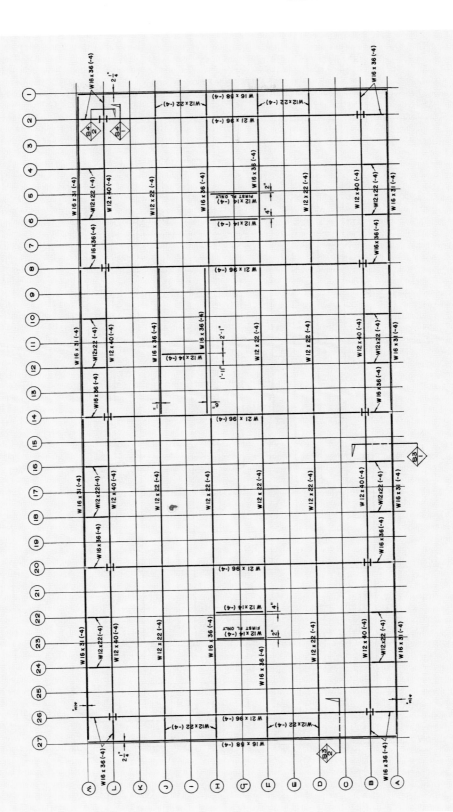

Figure 46–16 *Drawing No. S1, First-floor structural plan of the South Hills Office Building. (Second-, third-, and fourth-floor structural plans similar.) (Courtesy Jack Risheberger.)*

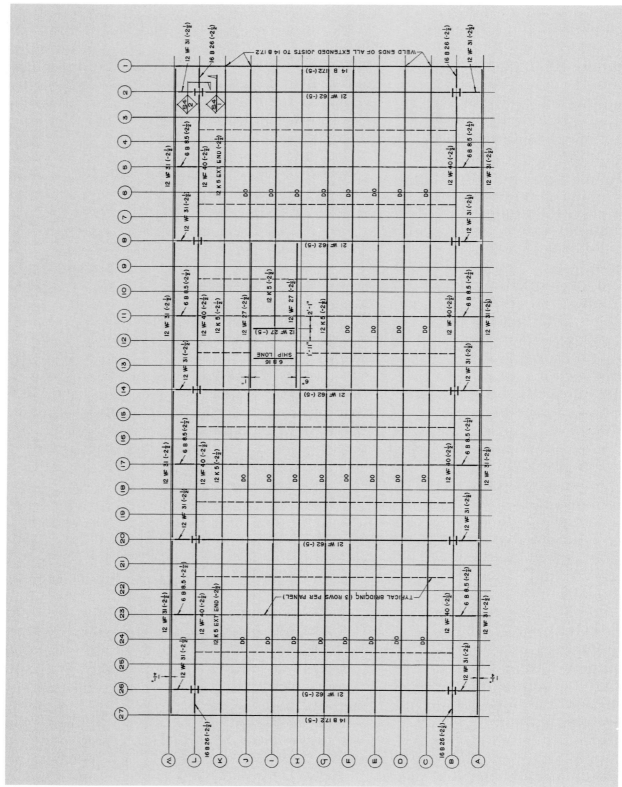

Figure 46–17 *Drawing No. S2, Roof structural plan of the South Hills Office Building (using "old" steel designations). (Courtesy Jack Risheberger.)*

well to become familiar with these earlier designations since they will still be seen on plans for many years. See Table 46–1.

Drawing No. S3: Column Schedule (Figure 46–18) Refer to Drawing No. S1 for an explanation of the double designations "S3/1" and "S3/2." The column schedule shows a typical bent. Notice that the column sections are spliced between floors where the bending moment is smaller.

Drawing No. S4: Structural Details (Figure 46–19) The location of Sections S4/1 and S4/2 is indicated on Drawings No. S1 and S2. The welding symbols used on this sheet include a closed triangle for a fillet weld, an open triangle for a vee weld, a closed circle for a field weld, and an open circle for an all-around weld.

Drawing No. S5: Concrete Slab Plan (Figure 46–20) Notation "#4@9 TOP" refers to $\frac{4}{8}$"-diameter steel reinforcing rods placed 9" apart and near the top surface of the concrete slab. "Granco" is the trade name for a decking manufacturer.

TABLE 46–1   *Structured Steel Designations*

Type of Shape	"Old" Designation	"New" Designation
W shape (formerly *wide flange*)	8 WF 31	W 8 × 31
W shape (formerly *light beam*)	8 B 20	W 8 × 20
S shape (formerly *American Standard I beam*)	8 I 18.4	S 8 × 18.4
American Standard channel	8 [11.5	C 8 × 11.5
Angle	∠4 × 4 × $\frac{1}{4}$	L 4 × 4 × $\frac{1}{4}$
HP shape (formerly *bearing pile*)	8 BP 36	HP 8 × 36

Note: Designations "8 WF 31" and "W 8 × 31" refer to a wide-flange beam 8" high weighing 31 lb. per foot of length.

Drawing No. E1: Electrical Plan (Figure 46–21) The dark rectangles represent the fluorescent ceiling fixtures, and dark circles represent incandescent fixtures. The letter within each fixture symbol identifies the type of fixture (see "Lighting Fixture Schedule" on Drawing No. 14). The alphameric designation at the end of each home run identifies the floor level and circuit number (see "Legend" on Drawing No. 1). An emergency lighting circuit is indicated by "E." See the legend for exit light information.

Drawing No. H1: Heating-cooling Plan (Figure 46–22) The dark rectangles with diagonal lines represent the Remington electric heating-cooling units, and the dark hexagons indicate Electromode electric baseboard heaters. See Drawing No. 14 for more detailed information. The two hash marks on each home run indicate 208-V circuits.

Drawings No. P1 and P2: Water Supply and Sanitary Plan (Figures 46–23 and 46–24) Drawing No. P1 shows the hot- and cold-water supply piping and the dry fire piping both in plan and pictorial projection. The plan also includes an air circulation system for the rest rooms. Drawing P2 shows the waste and soil disposal systems.

CONSTRUCTION

Careful design and detailing of the South Hills Office Building permitted the construction to be completed in four months without any major design changes or emergencies.

Although two column foundations were designed, the contractor was given the option of using the reinforced concrete foundation at all ten locations and chose that alternative.

The main steel members (Figure 46–25) were erected and held in place by temporary bolting until the weldments were made. Secondary members were fastened by high-strength bolts or unfinished bolts as specified. In Figure 46–26, notice that steel angles were welded to the exterior I beams to form a masonry shelf at each floor level. Also notice that intermediate floor beams are required to support the concrete floor at each level except the roof. At the roof, steel open-web joists are sufficient to support the roof deck

(continued page 618)

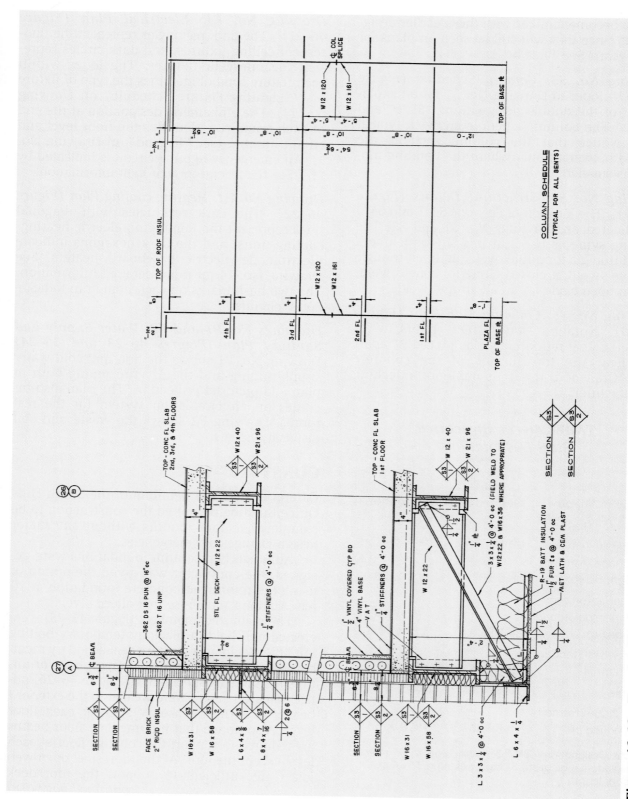

Figure 46–18   *Drawing No. S3, Structural section and column schedule of the South Hills Office Building.* (*Courtesy Jack Rishe-berger.*)

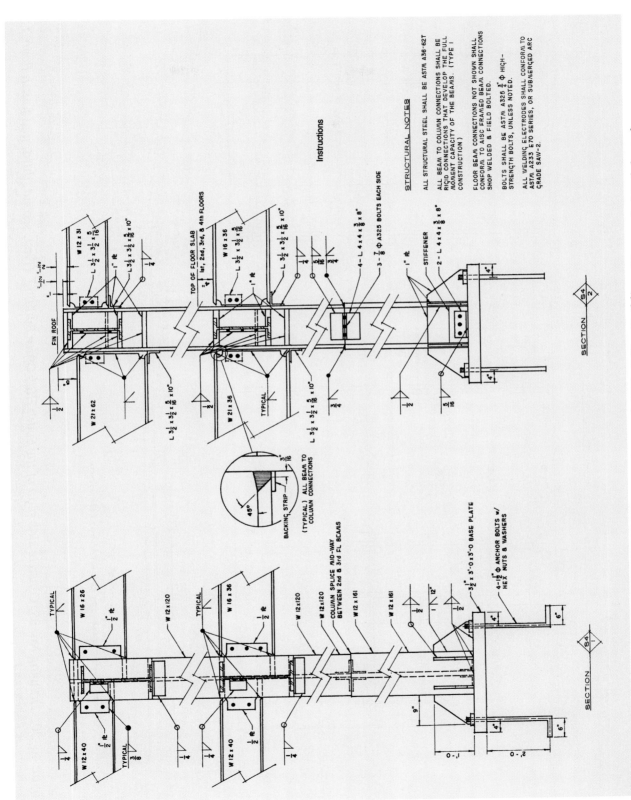

Figure 46–19   Drawing No. S4, Structural details of the South Hills Office Building.   (Courtesy Jack Risheberger.)

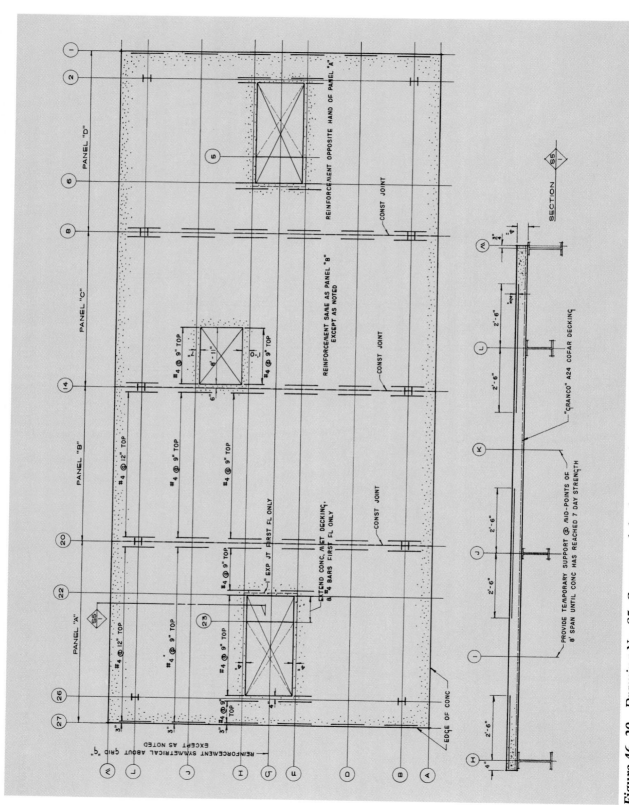

Figure 46–20 *Drawing No. S5, Concrete slab plan of the South Hills Office Building. (First, second, third, and fourth floors.) (Courtesy Jack Risheberger.)*

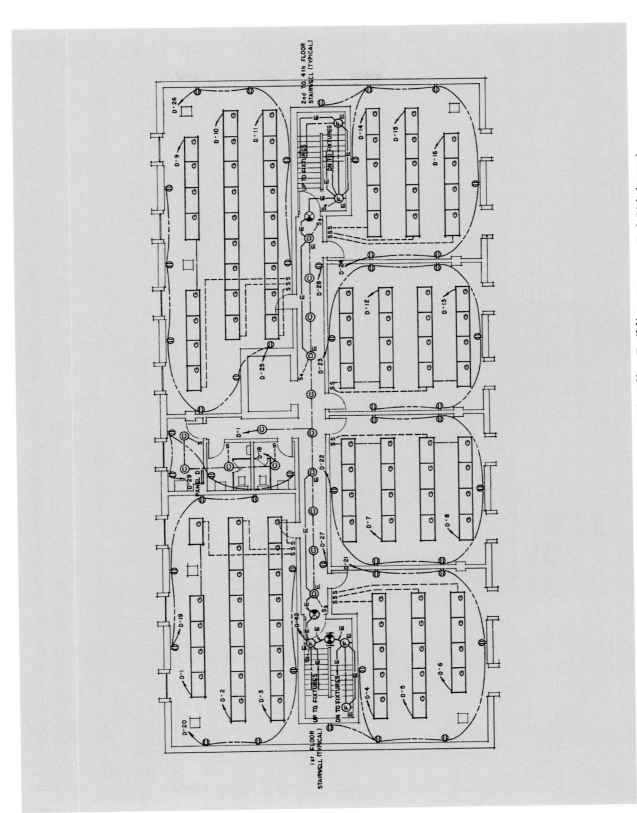

Figure 46–21   *Drawing No. E1, Electrical plan of the South Hills Office Building.* (*Courtesy Jack Risheberger.*)

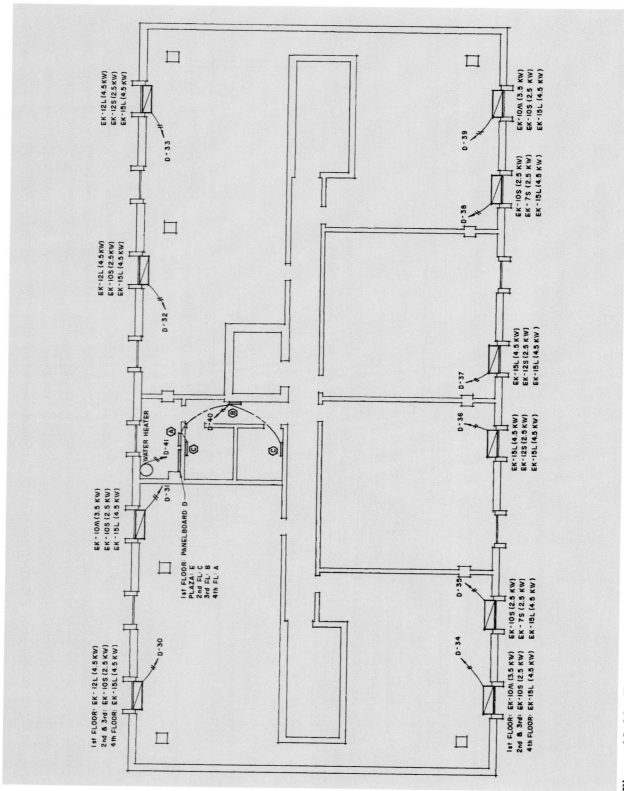

Figure 46–22   Drawing No. H1, Heating-cooling plan of the South Hills Office Building. (Courtesy Jack Risheberger.)

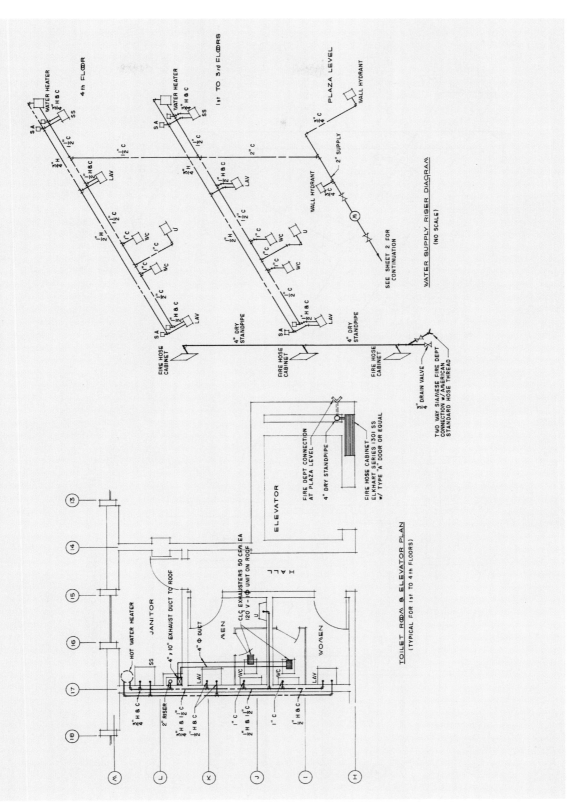

Figure 46–23   Drawing No. P1, Water supply plan of the South Hills Office Building.   (Courtesy Jack Risheberger.)

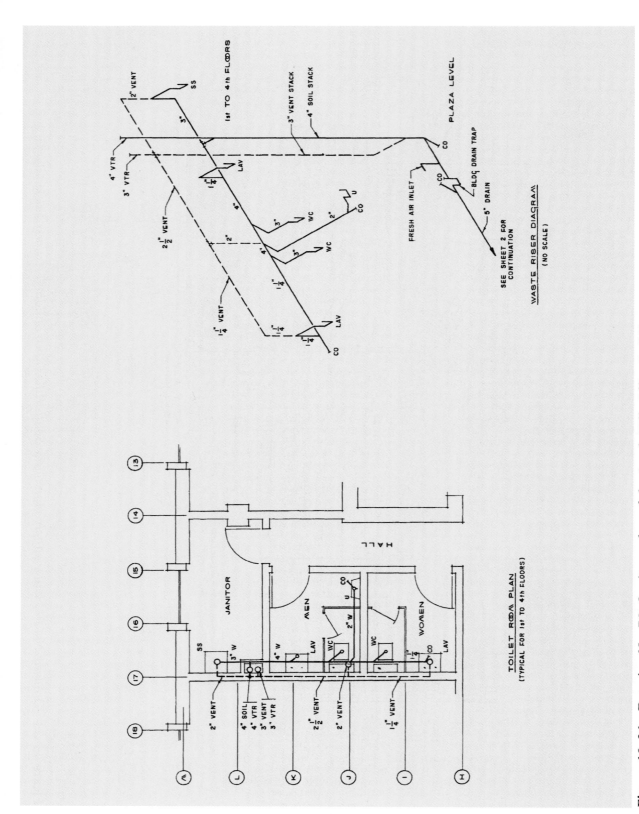

Figure 46–24   Drawing No. P2, Sanitary plan of the South Hills Office Building.   (Courtesy Jack Risheberger.)

Figure 46–25   *Steel framing of the South Hills Office Building.*

Figure 46–26   *Close-up of framing showing masonry shelves.*

(Figure 46–27). All structural steel was fireproofed as specified by the architect (Figure 46–28).

Six-inch batt insulation was installed with a special attention to the plaza roof. The plaza roof beams and tapered columns were simulated by light channels wired to shape (Figure 46–29) and covered with a metal lath base used under the final coating of cement plaster (Figure 46–30). Two-inch rigid insulation was used for the exterior walls.

A specialty metal company supplied the anodized aluminum components for window mullions which are so important to the exterior design. Figure 46–31 shows the entire four-floor section being field-fabricated before final erection. After erection, the electric heating-cooling

units and window frames were placed in the mullions (Figures 46–32 and 46–33).

Interior partitions were framed in lightweight metal as shown in Figure 46–34 using the system marketed by the Keene Company. The partition members were shop-welded into convenient wall sections and then field-welded in the final position. The vertical metal studs are supplied with a nailing groove to facilitate fastening the finished drywall. This groove is formed by two channels fastened together in such a way that a nail can be driven between them. The nail is not only held by friction but is also deformed when driven to provide greater holding power. The stairwells and elevator shaft were built of concrete masonry units. Wood strapping (Figure 46–35) was nailed

Figure 46–27   Corrugated decking over open-web roof joists.

Figure 46–29   Installing forms for simulated beams over plaza level.

Figure 46–28   Detail of fireproofing sprayed on steel beams.

Figure 46–30   Metal lath installed prior to plastering.

Figure 46–31 *Assembling aluminum window mullions before installation.*

Figure 46–33 *Outside wall detail after mullion installation.*

Figure 46–32 *Outside wall detail before mullion installation.*

Figure 46–34 *Steel interior partition framing.*

to the masonry units to provide a base for the drywall of $\frac{1}{2}$", vinyl-covered gypsum board. A suspended system of steel channels was used to support the finished ceiling panels (Figure 46–36), and vinyl composition tile flooring was laid directly over the concrete floor. The electrical and plumbing work was completed in appropriate steps during the various stages of construction. Figures 46–37 through 46–40 show the completed structure.

Study the photographs in this chapter until you are familiar with the main construction steps. Also try to visit construction sites near you at least once a week to become familiar with the latest construction techniques.

Figure 46–35 *Wood strapping nailed to masonry stairwell in preparation for drywall installation.*

Figure 46–36 *Hung metal channels will support ceiling panels.*

Figure 46–38 *The law office in the South Hills Office Building.*

Figure 46–37 *The real estate office in the South Hills Office Building.*

Figure 46–39 *Interior elevator area of the completed South Hills Office Building.*

REAL WORLD   APPLICATIONS

Creating Effective Visuals

Producing a set of plans for a commercial building is more time-consuming than for a residential project, yet the rewards for a "job well done" are extremely satisfying. One of the challenges for commercial drafting is that you are usually required to please more than just the client financing the project. Because of zoning ordinances and the effect public sentiment has on commercial development, you will be required to give persuasive presentations more than once for a single project. Developing effective visual aids for these presentations is essential!

After almost seven months of planning and revising a new five-story multipurpose commercial building, the client has finally agreed to the design. The next step is to convince the planning commission, the zoning board, the city council, and the general public that this is a viable project. Keep in mind the following tips for creating effective visuals as you prepare these presentations.

Figure 46–40   *Southwest elevation of the completed South Hills Office Building.* (*Courtesy Jack Risheberger.*)

1. Keep it simple. Visuals that contain a lot of text and graphics can be more distracting than informative. The purpose of a visual is to give the audience a general idea about what you are going to discuss.

2. Organize lists with bullets (•), checkmarks (✓), or numbers.

3. Limit the presentation to two type styles and fonts.

4. Words that need emphasis should be made **bold** or <u>underlined</u>.

5. Follow the rule of six: "Thou shalt not use more than six lines, and thou shalt not use more than six words per line."

6. Add a moderate amount of pizzazz by incorporating text that is in perspective, arched, circular, or outlined.

7. Visuals should lead the reader to the center of the diagram, so place them facing in on the left side.

8. Use color to persuade the audience. Red, orange, and yellow are associated with warmth, green means either "proceed" or "money," red means "caution," and blue, green, and violet refer to calm and cool situations.

Prepare a presentation that includes at least three different visuals. The presentation is intended to convince the audience to support your plans.

REVIEW QUESTIONS

1. Differentiate between the design, drafting, and construction of a residence and a commercial building.

2. Name four major steps in the design and drafting of a commercial structure.

3. List the drawings normally included in a set of commercial working drawings.

4. Identify the information that is typically located on a title page of a set of commercial construction drawings.

5. What is the principal reason for including details of alternate construction methods in the presentation drawings?

6. Describe briefly each term:
   a. Structural bent
   b. Swale
   c. Control joint
   d. Building north
   e. Roof scuttle

7. Give the meaning of these abbreviations:
   a. FTG, HTG, LTG
   b. HTR, WTR
   c. CMU, DO, GL, VT
   d. WC

8. Give the meaning of:
   a. Detail 12/3
   b. Door S2/2
   c. Elevation 10/6

9. Give *two* meanings for each of the following abbreviations:
   a. E
   b. ELEV
   c. W
   d. Ø
   e. #

ENRICHMENT QUESTIONS

1. On working drawings of the South Hills Office Building, the vertical modular grid system is different from the horizontal grid system. Why?

2. Are the same number of heating-cooling units used on opposite walls of the South Hills Office Building? Why?

3. Identify the primary considerations and produce a preliminary layout for the following commercial projects.

a. An 80' × 160' unheated storage warehouse for a building-supply distributor. Clear ceiling height should be 12'-0. Use masonry (10" CMU) or wood frame construction (2" × 6" studs 24" oc with corrugated aluminum siding). Provide a small heated office and lavatory, rail receiving dock, and truck shipping dock. Fire protection will include a dry sprinkler system.

b. A single-story retail candy store for a 24'-wide × 120'-deep commercial site. Use brick bearing-wall and steel-joist roof construction. Provide an attractive front elevation with display window, rear office of 200 sq. ft., and storage of 300 sq. ft. with delivery door. Show details of interior planning.

PRACTICAL EXERCISES

1. Draw or computer-produce the working drawings for the South Hills Office Building as assigned:

   a. Second-floor plan

   b. Roof plan

   c. Interior elevations of lobby

   d. Transverse section

   e. Stair section and details

   f. Elevator section and details

   g. Roof cornice detail section

   h. Lobby sill detail section

   i. Second-floor schedules

   j. Plaza electrical plan

   k. Telephone plan

   l. Fire alarm diagram

2. The elevation drawings of a community church are shown in Figures 46–41 and 46–42 page 624. Using 6" × 24" laminated roof beams 8' oc, draw or computer-produce the working drawings as assigned:

   a. Floor plans

   b. Elevations

   c. Transverse section through auditorium

   d. Typical wall sections

   e. Stair details

   f. Schedules

   g. Electrical plan

   h. Heating and air circulation plan

   i. Plumbing plan

3. Design and draw or computer-produce drawings for the project assigned:

   a. An innovative children's playground for a 50'-wide × 150'-deep urban site.

   b. A drive-in movie screen structure. The screen is to be 120' wide × 50' high and at an angle of 12 degrees from the vertical. The bottom of the screen is 12' above the ground level. Use timber or steel construction as assigned.

   c. A two-story community college academic building containing an auditorium seating two hundred students; five classrooms, each seating forty students; a drafting room for thirty students, eight two-person faculty offices, rest rooms, and maintenance. Use steel and masonry construction.

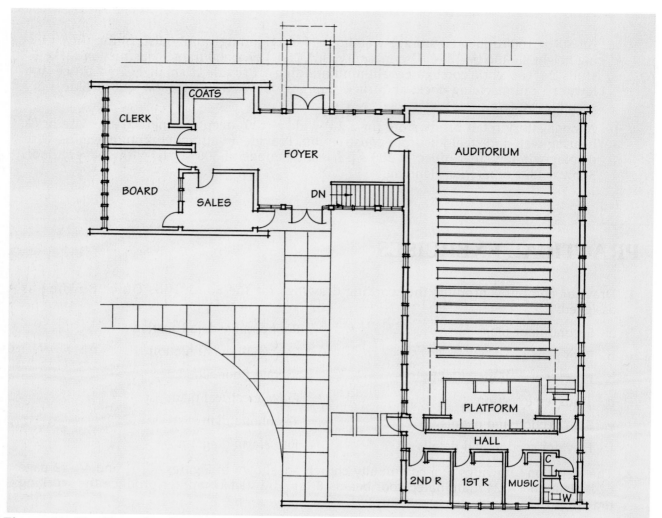

Figure 46–41 *Upper-level plan of a community church. (First Church of Christ. Scientist, State College, Pennsylvania, designed by Roy D. Murphy & Associates, architects, Urbana, Illinois.)*

Figure 46–42 *Front elevation of a community church. (First Church of Christ. Scientist, State College, Pennsylvania, designed by Roy D. Murphy & Associates, architects, Urbana, Illinois.)*

47 Design for Accessibility

OBJECTIVES

By completing this chapter you will be able to:

🏛 Describe the adaptations that need to be made to a building for an individual with:

■ Impaired mobility

■ Impaired sight

■ Impaired hearing

🏛 Identify the basic components of the Americans with Disabilities Act (ADA) of 1990.

KEY TERMS

Accessibility	ANSI standards
ADA	Curb cut

Figure 47–1 *A cantilevered accessible solution for a hillside site.*

For a public building to be used as intended, it must be accessible to *all* individuals, including those with walking, sight, and hearing impairments. Although the design and construction of a new building for **accessibility** is easily accomplished, it often is quite difficult to correct design errors and remove architectural barriers at a later time. In a few instances, accessibility features will help one group but hinder another. This, of course, is often true in architectural design, where no one solution is ideal for everyone. However, designers still have a moral, as well as a legal, obligation to include accessibility considerations in their designs.

The cost of accessibility varies greatly. Depending upon what is required, it can range from almost no additional cost to virtually prohibitive costs. Usually in a new building, accessibility can be designed into the job and the building can be constructed without major deviations from the original designs (Figure 47–1). Thus in new construction, provision for accessibility is accomplished at low cost. For most projects, additional costs incurred to provide accessible facilities, elevators, and ramps should not exceed from 0.5 to 5 percent of the total construction costs. In retrofitting existing facilities, the costs will depend on how easily accessibility can be attained.

STANDARDS

The basis for most federal, state, and local legislation on accessible design is standard ANSI A117.1 entitled *Specifications for Making Buildings and Facilities Accessible to and Usable by*

Physically Handicapped People. This standard establishes minimum design requirements for buildings and facilities so that they are usable by people with such physical disabilities as the inability to walk, difficulty in walking, reliance on walking aids, lack of coordination or stamina, reaching or manipulation difficulty, extremes of physical size, sight or hearing impairment, or difficulty in interpreting and reacting to sensory information. Accessibility and usability allow a disabled person to get to, enter, and use a building or facility. The Americans with Disabilities Act (**ADA**) of 1990 requires that all public and commercial buildings be accessible to impaired persons. Although these standards refer primarily to public buildings and facilities, many of the recommendations can be easily adapted to residential buildings and facilities as well.

The recommendations in this chapter are not entirely based on **ANSI standards**, but instead are the most common design considerations found in the many accessibility codes in effect throughout the Americas. These recommendations are not all-inclusive, and obviously a designer must consult the federal, state, and local accessibility regulations that apply to a particular project. If local zoning ordinances have accessibility requirements, those local requirements take precedence over ANSI standards. Failure to comply with the accessibility sections of the applicable codes may well result in the designer's facing legal action and the project being closed by court order until the violations are corrected.

COMPENSATING FOR IMPAIRED MOBILITY

Persons requiring wheelchairs do not constitute the majority of the disabled population, but since their requirements for mobility demand the most space (based on the size of the wheelchair, Figure 47–2, and its turning requirements, Figures 47–3 through 47–5), most spatial design criteria have been written with them in mind. Additional limitations such as maximum reach from a wheelchair further affect the criteria. It should be noted that electric wheelchairs may be slightly larger than the standard self-propelled model. With few exceptions, spatial considerations defined for chair-

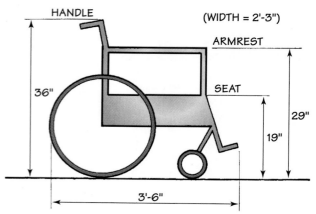

Figure 47–2   Dimensions of a typical wheelchair.

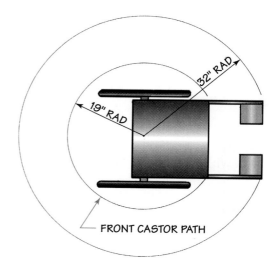

Figure 47–3   Turning radius of a typical wheelchair.

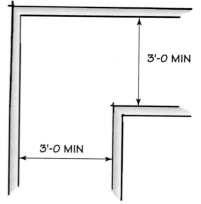

Figure 47–4   Minimum dimensions for a 90° corner.

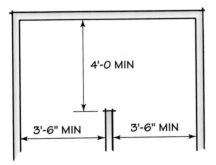

Figure 47–5 Minimum dimensions for a 180° turn about a partition.

bound individuals will benefit other disability groupings.

Parking Spaces

1. Parking space should be set aside and clearly marked for disabled use. The United States Architectural and Transportation Barriers Compliance Board has reviewed many of the various state and local building codes and has developed the recommendations indicated in Table 47–1. This table represents an adequate minimum for most types of facilities.

2. Parking should be located as closed as possible to an accessible entrance.

TABLE 47–1 *Recommended Number of Parking Spaces*

Total Parking in Lot	Required Minimum Number of Accessible Parking Spaces
1 to 25	1
26 to 50	2
51 to 75	3
76 to 100	4
101 to 150	5
151 to 200	6
201 to 300	7
301 to 400	8
401 to 500	9
501 to 1000	2% of total
Over 1000	20 plus 1 for each 100 over 1000

3. The approach from the parking space to any walkway should be ramped if necessary.

4. The width of each parking space should be at least 12'-0 to allow room for transfer from

REAL WORLD APPLICATIONS

The Americans with Disabilities Act

 New laws and regulations are introduced according to changes in a society's attitudes. The Americans with Disabilities Act of 1990 (ADA) resulted from a campaign by persons with disabilities and their advocates to inform the public of the difficulties they had in using public buildings. As the public became aware of the issue, appropriate legislation was introduced and approved.

As an architect, it is important for you to be aware of the public's and special interest groups' influence on building design. Study the ADA and provide an example of how it has affected architectural design in your community. What other regulations and public pressures have affected building design? Find at least one other example and explain its specific influence.

an automobile to a wheelchair. A minimum 13'-0 is more realistic to permit the use of vans equipped with side door lifts. This recommendation includes an 8'-0 wide access aisle adjacent to each van parking space instead of the typical 5'-0.

5. Lay out parking spaces so chairbound individuals need not travel behind or between other vehicles.

6. Vans fitted with raised roofs are increasingly being used by the disabled, so designers of parking garages must provide sufficient overhead clearance. Where accessible parking is provided in garages, a 9'-6" clear height should be provided for van headroom along the entire route between entry, parking spaces, and exit.

7. Accessible parking stalls should be on as level a surface as possible. In general the surface should not slope in excess of 1:50.

8. Surfaces of parking stalls and adjacent walkways should be stable, firm, and slip resistant. Graveled surfaces generally are not usable by the disabled.

Walks

1. Walkways should be a minimum 5'-0 wide with hard, nonskid surfaces. Some codes permit 3'-0 or 4'-0 wide walkways, but this is an insufficient width for two wheelchairs to pass each other. Furthermore, on the narrower walkways there is the danger of walking persons tripping over crutches, canes, or walkers used by semiambulatory persons.

2. Furniture and other obstacles such as benches, mailboxes, and refuse containers should not be within the minimum 5'-0 width of the walkway.

3. Surfaces should be level. Abrupt changes in level surface of over $\frac{1}{2}$" tend to jam the small front wheels of wheelchairs and to cause the unstable walker to trip.

4. Water, ice, and snow should be removed from outside walkways. In some climates this may require covered or heated walkways.

5. A minimum overhead clearance of 6'-8" should be maintained.

6. Below the overhead clearance, no object should overhang the side of the walkway by more than 4".

Ramps

1. The maximum slope of a ramp should be 8.3 percent (1:12).

2. The minimum width should be 4'-0 and is subject to the same considerations as walkway widths.

3. Ramps should have a continuous $1\frac{1}{2}$" diameter handrail 32" high. The handrail permits some chairbound persons to pull themselves up the ramp. Handrails are not necessary for ramps sloping less than 5 percent (1:20) with no drop-offs.

4. Handrails should extend 12" to 18" beyond the top and bottom of the ramp if they do not project into a main pathway.

5. If many children will be using the ramp, include an additional lower handrail at a height of 24".

6. Ramps with a drop-off should have a 2" curb.

7. Every 30' there should be a level area 3'-0 long to be used as a resting area (Figure 47–6).

8. Level landings should be provided at the top and bottom of the ramp. The landings should be at least 5'-0 long.

9. The ramp should have a nonskid surface and be kept free of water, ice, and snow.

Curb Cuts

1. **Curb cuts** should be placed where they will not be blocked by parked vehicles.

2. The width of the curb cut should be at least 3'-0 exclusive of flared sides (Figures 47–7 through 47–9), but 4'-0 is recommended.

3. A curb cut is a ramp, and the slope should not exceed 5 percent (1:20). However, when the curb cut is a short ramp, some states permit a slope up to 16.7 percent (1:6), although there is some danger of a wheelchair tipping over backwards at slopes greater than 8.3 percent (1:12).

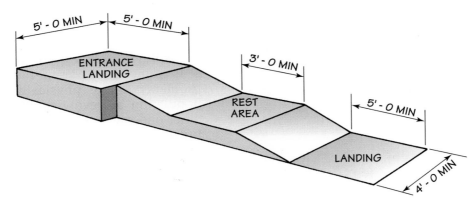

Figure 47–6   *Ramp dimensions.*

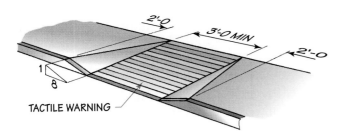

TACTILE WARNING SUCH AS
GRASS, BRICK, GROOVED
CONCRETE

Figure 47–7   *Radiused curb cut (preferred).*

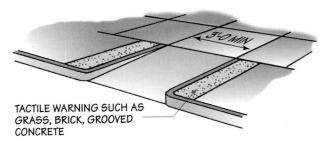

TACTILE WARNING

Figure 47–8   *Flared curb cut.*

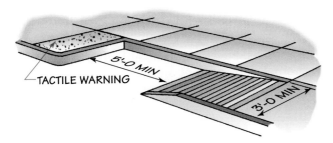

TACTILE WARNING

Figure 47–9   *Parallel curb cut used when depth is small.*

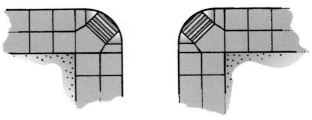

Figure 47–10   *Diagonal curb cuts.*

4. Care must be taken that the curb cut is not a hazard to pedestrians, particularly to persons with impaired sight. Thus the curb cut should be clearly marked with a planting strip, railing, or some form of tactile (capable of being *felt*) warning system.

5. A diagonal curb cut (Figure 47–10) is usable for both directions of travel and the hazardous, up-and-down effect of two adjacent ramps is avoided. Further, in parts of the country where snow removal is a concern, the diagonal ramps are more likely to be cleared, whereas ramps off to the side may be buried under snow.

   Persons with impaired sight frequently use curb cuts to orient themselves at intersections. Thus diagonal curb cuts may be confusing.

Entrances, Doors, and Doorways

1. Entrances should have a continuous level or ramped surface.

2. The outer surface of an entrance should have a slope no greater than 2 percent

(1:50) within 5'-0 of a door. Ideally there should be a level entranceway to permit the chairbound to conveniently manage the wheelchair and the door.

3. If an exterior threshold is absolutely necessary, it should have beveled edges and be a maximum $\frac{3}{4}$" high. Interior thresholds should be flush.

4. Doors should have a minimum 2'-8" clear passageway when the door is open (Figure 47–11). Double-leaf doors should have at least one door leaf meeting this requirement.

5. Doors placed in a series should open in the same direction or open away from the area between them. A minimum 4'-0 of clear space should be provided between such doors (Figure 47–11).

6. Revolving doors cannot be used at all by people in wheelchairs, and are difficult, if not impossible, for individuals with walkers, crutches, canes, or sight impairments. If a revolving door must be used to maintain an ef-

ficient air seal, a suitable accessible auxiliary entrance or exit must be provided nearby.

7. Turnstiles are often used to control pedestrian traffic. They present the same types of problems to the physically handicapped population as revolving doors. There are other one-way gates that perform the same function but that have wider openings and require less opening force. If turnstiles are necessary, some alternate type of entrance must be provided.

8. Door closures may be necessary for energy conservation, security, noise control, or fire control, and must have sufficient force to overcome friction, wind, and indoor and outdoor air pressure. If door closures are installed, they should require no more than 5 lbs. opening force on an exterior door. Automatic doors are recommended if greater opening forces will be required. Sliding automatic doors are preferred to swinging doors.

9. Manual doors should have single-action lever, toggle, or paddle-type handles. Some persons cannot grasp, pinch, or twist knobs.

10. Doors should have a smooth push panel extending at least 12" up from the bottom to permit pushing the door open with a wheelchair's foot pedal. This greatly minimizes damage to the door and helps by providing a smooth pushing surface.

11. Thick, bristly floor mats and plush carpeting should be avoided since they greatly increase the effort required to push a wheelchair.

12. Grates or grilles used in entrances to trap dirt and gravel are tripping hazards. Their openings may be rectangular if their smaller dimension (measured parallel to the traffic flow) is less than $\frac{3}{8}$" while the other dimension may be up to 4".

13. The disabled should be able to use the principal entrances of a building and not be restricted to using service entrances.

Stairs and Stairways

1. Tread run should not be less than 11", and tread rise should not exceed 7".

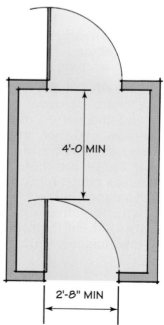

4'-0 MIN

2'-8" MIN

Figure 47–11   Minimum door dimensions for accessibility.

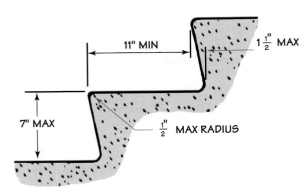

Figure 47–12   Stair dimensions for accessibility.

2. Tread nosing should not exceed $1\frac{1}{2}$", and the radius of the tread nose should not exceed $\frac{1}{2}$". (See Figure 47–12.)

3. Tread runs and rises should be consistent for the length of the stairway.

4. Continuous handrails should be provided, preferably on both sides of the stairway, and extend 12"–18" beyond the ends of the stairway.

5. Handrails should be 32" high and have $1\frac{1}{2}$" clearance between the railing and wall to prevent hands from getting caught.

6. Open stairways are very decorative. However, they are hazardous to those with leg braces who have difficulty clearing the stair nose. Persons with seriously impaired sight also have difficulty climbing an open staircase since they are unable to feel the front face of the next step with their shoe tips.

Elevators and Wheelchair Lifts

1. All elevators and elevator lobbies intended for use by the public and by employees should be accessible.

2. Elevators should be provided for all buildings of two or more levels.

3. The elevator should be large enough for at least one wheelchair to enter, turn 180 degrees, and exit.

4. The positioning of the control panel is important. It should be mounted on the front or side wall within diagonal reach from a wheelchair. Also, the uppermost control buttons should not be higher than 4'-6".

5. Controls should have tactile identification to be read by those with impaired sight, as well as tactile markings on both sides of the door jamb to identify the floor.

6. Emergency safety controls and devices should be located no higher than 3'-4".

7. The elevator should have automatic controls, with automatic open and close features as well as automatic reopening if the closing door strikes an obstacle.

8. The elevator should have automatic leveling within $\frac{1}{2}$" of the floor level.

9. Lobby call buttons should be centered no higher than 3'-6".

10. Visual and audible signals to indicate arriving elevator cars should be used, with one ring for an "up" car and two for a "down" car.

11. Ideally, visual and audible signals to identify each floor should be provided.

12. The wheelchair lift (Figure 47–13) is rather new in the building industry and is primarily used for less than two-floor operations when a ramp is impractical. These lifts have a capacity of 400 to 500 lbs. There are two types of wheelchair lifts, one that operates vertically like an elevator and the other that runs on an incline following the flight of stairs. However, some local codes do not permit their use.

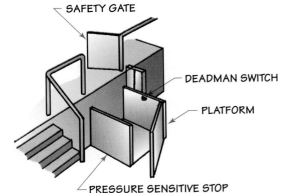

Figure 47–13   A vertical platform lift.

Building Controls

1. Building controls such as light switches, thermostats, air-conditioner controls, window and drapery hardware, call buttons, electrical outlets, and fire alarms, intended for use by visitors and employees, should be within reach of the chairbound.

2. These controls should be located no higher than 4'-0, but 3'-4" is recommended.

3. Electrical outlets should be located at least 18" high.

Restrooms

1. If public restrooms are provided, at least one restroom per sex should meet accessibility requirements.

2. When the toilet is enclosed in a stall, the stall should meet the following requirements:

   a. Minimum width 3'-0.

   b. Stall door 2'-8" wide that swings out or slides.

   c. Minimum depth 5'-0 if a wall-hung toilet is used; otherwise, minimum depth 6'-0.

   d. Handrails 2'-8" high on both sides, minimum 4'-0 long, $1\frac{1}{2}$" diameter, with $1\frac{1}{2}$" clearance from wall, capable of supporting 250 lbs. (See Figure 47–14.)

This minimum toilet stall arrangement requires transfer from the front or rear of the wheelchair to the toilet. Transfers of this type are difficult for many chairbound, and thus a layout such as shown in Figure 47–15 is recommended. This permits side transfers as well as the more awkward front/rear transfers.

3. The flushing control should be no higher than 3'-0 and easily reached.

4. The toilet tissue dispenser should be conveniently positioned.

5. The top of the lavatory should be no higher than 2'-8", and the lower edge should be no lower than 2'-3" (2'-5" is recommended). This will permit a wheelchair to wheel under the unit.

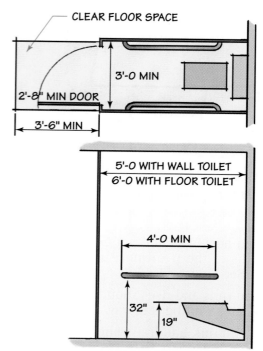

Figure 47–14   Stall dimensions for front/rear transfer.

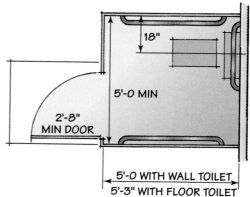

Figure 47–15   Stall dimensions for side transfer.

6. Single-lever faucet handles should be used.

7. All plumbing should be mounted close to the wall to provide for knee and toe clearance. Pipes carrying hot water should be insulated to protect those chairbound who have no sensation in their legs.

Figure 47–16   Accessible bathroom fixtures may also be attractive.

8. Accessories such as towel racks, towel dispensers, electric hand dryers, sanitary napkin dispensers, soap dispensers, refuse disposal units, vending machines, and shelves should have operating parts no higher than 3'-4".

9. The bottom of at least one mirror should be mounted no higher than 3'-2". Full-length mirrors are acceptable.

10. Baffle walls or partitions used to block public view of toilet facilities should not impede access.

Drinking Fountains and Watercoolers

1. The waterspout of a watercooler should be 2'-8" high. Sufficient clearance must be provided for the wheelchair to be wheeled under the cooler with appropriate knee and toe clearance, especially if the cooler is hung in an alcove.

2. The watercooler should have a front-position waterspout and water trajectory approximately parallel to the front of the cooler.

3. The watercooler should have controls that do not require manual dexterity to grasp, pinch, or twist. If a foot pedal is used, auxil-

iary single-action hand controls should be provided.

4. Cups should be in a convenient position.

Public Telephones

1. At least one telephone on each floor should be accessible to those with impaired mobility.

2. The height of the telephone's uppermost operating mechanism (usually the coin slots) should not exceed 4'-0.

3. Push-button telephones should be used whenever available.

4. The telephone receiver cord should be at least 2'-6" long.

5. The receiver should have a volume-control adjustment for use by individuals with impaired hearing.

6. The telephone book should be within reach.

7. The telephone should be mounted so that a wheelchair can make a parallel approach to the face of the phone.

8. The size of a telephone booth should be approximately 4'-0 × 5'-0 with at least a 2'-8" opening and a 3'-6" clear depth. A fixed seat would only be in the way of a wheelchair, so a hinged seat is preferred. The telephone should be mounted on a side wall of the booth enclosure or diagonally in a corner.

9. The telephone may be mounted in an alcove if sufficient knee and toe clearance is provided.

COMPENSATING FOR IMPAIRED SIGHT

Today most persons with seriously impaired sight use a long cane or a seeing-eye dog, enabling them to be more active than they were in the past. Typically, individuals who use a long cane will have sighted persons guide them through a building while they become familiar with room shapes and the positions of obstacles such as furniture and columns. Persons using seeing-eye dogs move about more easily, relying on dogs to direct them around obstacles.

The following are accessibility considerations for the sight-impaired:

1. Avoid low-hanging obstacles. A minimum overhead clearance of 6'-8" should be adhered to at all times.

2. Avoid curved plan features because they limit ability to direct the sight-impaired to other features.

3. Tactile senses are often used to compensate for impaired sight. Thus a change in floor surface, such as from carpet to tile, can be used to identify key areas such as restrooms or lobbies and danger areas such as stairs or ramps.

4. Persons with impaired sight can identify abrupt changes in surface such as a curb, but they are not always able to identify gradual changes such as a ramp. Thus some type of tactile surface change should occur 2'-4" before a potentially dangerous obstacle.

5. Handles and knobs on doors leading to danger areas such as boiler rooms, equipment rooms, fire escapes, stage doors, loading platforms, and the like should have knurled surfaces. This tactile identification serves as an immediate warning.

6. The sense of sound also is used to aid in moving about. An overly *live* acoustical environment can thus lead to confusion.

7. Ideally, all written directions and information should be accompanied by verbal instructions. Where this is impractical, signs should be constructed of raised capital letters and numerals at least $\frac{5}{8}$" high with color contrast between the letter and its background. This will permit the individual to use the sense of touch to read the sign. Furthermore, many persons have residual sight and can read signs when there is high contrast between the message and its background. To avoid confusion, keep signs brief.

8. All signs should be placed consistently between 4'-6" and 5'-6" high (Figure 47–17).

COMPENSATING FOR IMPAIRED HEARING

Most people with hearing impairment are able to hear some sounds, given a reasonable envi-

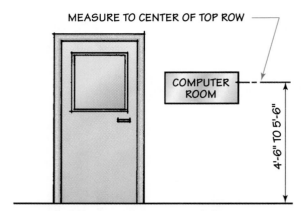

MEASURE TO CENTER OF TOP ROW

COMPUTER ROOM

4'-6" TO 5'-6"

Figure 47–17   Height of tactile signs.

ronment for sound transmission. Others are able to hear only certain frequencies of sound. Designers can accommodate both groups without extensive modifications. Some recommendations follow.

1. Fire alarms, telephones, doorbells, and other devices using sound signals should use sound frequencies that can attract the attention of individuals with partial hearing. For individuals with severe hearing impairment, visual signals such as blinking lights are the solution.

2. At least one telephone in a bank of public telephones should be equipped with an adjustable sound amplifier. This aid is available from all telephone companies. The Americans with Disabilities Act of 1990 requires telephone companies to provide relay services permitting persons with impaired hearing or speech to communicate by telephone. Also, private telephones for persons unable to hear the audible ring of a telephone should be equipped with a blinking light to indicate incoming calls.

3. Signs and directions should be composed with clarity of understanding so that the hearing-impaired individual need not ask questions.

4. Public address systems should not have excessively loud volume. The increased volume results in sound distortion for individuals with reduced hearing as well as for those with normal hearing.

COMPUTER-AIDED DESIGN FOR ACCESSIBILITY

Architectural offices use accessory computer programs to design residential and commercial buildings so they conform to building code and ADA requirements on accessibility. For example, the width of halls, stairs, and doorways as well as rest room clearances can be checked by computer for wheelchair accessibility. This alerts the designer to possible code violations, and the building is then redesigned to remove those violations. Refer to the "Architectural Databases" sections of Chapter 6 for more detailed information.

INTERNATIONAL SYMBOL OF ACCESS

The international symbol of access (Figure 47–18) is used to identify special facilities for the disabled. It should always be used in the design and proportions shown and in contrasting colors of black or dark blue and white. Recommended dimensions are shown in Table 47–2.

Figure 47–18 *International symbol of access.*

TABLE 47–2 *Dimensions for International Symbol of Access*

Size	Location	Viewing Distance
$2\frac{1}{2}$ in.	Interior	Up to 30 ft.
4 in.	Interior	Greater than 30 ft.
4 in.	Exterior	Up to 60 ft.
8 in.	Exterior	Greater than 60 ft.

REVIEW QUESTIONS

1. List three design considerations of aid to chair-bound persons:
   a. Parking an automobile
   b. Leaving an automobile
   c. Moving over a curb
   d. Moving down a level walk
   e. Moving up a ramp
   f. Moving through a doorway

2. List six design considerations of aid to persons with impaired sight.

3. List three design considerations of aid to persons with impaired hearing.

4. For accessibility, indicate the maximum recommended height for:
   a. Building controls
   b. Ramp handrails
   c. Elevator door controls
   d. Elevator emergency controls
   e. Top of lavatory

5. Indicate the minimum recommended height for overhead clearance.

6. For accessibility, indicate the minimum width of:
   a. Walkways
   b. Ramps
   c. Doorways

ENRICHMENT QUESTIONS

1. Do you feel the ADA has improved the accessibility of buildings or inhibited builders from designing new buildings because of added costs?

2. Write a paragraph that summarizes key points in the ADA.

3. Locate buildings within your community that have made alterations to their structure to accommodate individuals with physical limitations. Describe the changes that were made and how they have improved or diminished the buildings' appearance and functionality.

PRACTICAL EXERCISES

1. Check your original building design for compliance with local and statewide accessibility codes.

2. Draw or computer-produce the details for an original design to improve accessibility:
   a. Between floors for chair-bound persons
   b. Between rooms for persons with impaired sight
   c. For communications between rooms for persons with impaired hearing

CHAPTER 48

Design for Acoustics

OBJECTIVES

By completing this chapter you will be able to:

🏛 Describe the effects of sound transmission and reflection in the design of a building.

🏛 Identify methods of absorbing sound and the advantages and disadvantages of each method.

KEY TERMS

Sound transmission
Sound reflection
Sound absorption
Sound amplification
Sound masking

Proper environment benefits all human activities. It has long been recognized that proper lighting, decorating, and air conditioning will reduce stress, but sound conditioning has only been accepted as equally important since the 1960s, when minimum specifications for sound transmission were established by the Federal Housing Administration in its *Minimum Property Standards for Multifamily Housing*.

Sounds, as well as other factors affecting senses, can make us happy or annoyed, contented or distracted, efficient or careless. A building is a controlled environment and should be designed for sound control just as it is planned for durability and weather protection.

Obviously, theaters and auditoriums should be sound-conditioned, but attention should also be given to single- and multifamily buildings, offices, factories, and schools.

The acoustical design of an important building would be entrusted to a consulting firm specializing in architectural acoustics. Although architectural drafters would not be expected to be acoustical experts, they can prevent many noise problems by using common sense in their design and materials specifications. In planning a building for sound conditioning, designers have four factors under their control:

1. Layout
2. Sound transmission
3. Sound reflection
4. Sound absorption

Each of these elements will be studied in the same order in which they should be considered during design.

LAYOUT

Intelligent site selection and room layout can prevent many acoustical problems from developing. For example, a school should not be located adjacent to a superhighway or a hospital located near a jet air terminal. Also, quiet and noisy areas in the same building should be removed from each other or separated by buffer zones such as storage walls or corridors. For example, the placement of an office adjacent to a noisy manufacturing area (Figure 48–1) can be improved by using the storage room and lavatories

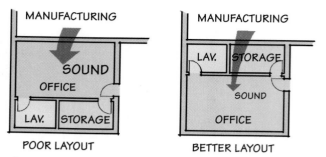

Figure 48–1 *Buffer zones reduce noise transmission.*

as a sound buffer. Figure 48–2 shows how office windows and doors can be separated to reduce sound transmission.

SOUND TRANSMISSION

Whenever it is not possible to separate quiet and noisy rooms, their common wall should be designed to reduce **sound transmission** to an acceptable level. It is not necessary to prevent *all* sound transmission, but rather to reduce it to a level below the normal sound level in the quieter room.

Sound travels through most building materials by causing vibrations in the material. The vibrations can be caused by *structure-borne* sound or *airborne* sound. Most sound transmitted through *floors* (such as footsteps and furniture scraping) is structure-borne. Structure-borne

sound can be reduced most economically by carpeting or resilient cork tiling. Most sound transmitted through *walls* (such as voice and typewriter clatter) is airborne. In general, heavy walls do not vibrate as readily as lighter walls and are therefore better barriers to airborne sound. Thus a solid brick wall would be more effective than a wood stud wall. Another method of reducing sound transmission is through the use of a *double wall*, that is, splitting a wall into two unconnected layers. In this manner, the sound vibrations are not directly transmitted from one room to another through the solid building material in a common wall. Figure 48–3 shows some methods of constructing double walls.

Table 48–1 can be used to compare the acoustical effectiveness of various types of walls. The ratings are given in *sound transmission classifications* (STCs), which can be considered as dimensionless numbers that rank the relative acoustical effectiveness of various wall types. For example, a plastered stud wall (STC = 35) is only slightly better than a dry-wall finished stud wall (STC = 32), but an 8" CMU wall (STC = 48) or 8" brick wall (STC = 49) are considerably better. Recommended minimum STC ratings for various conditions are given in Table 48–2 (page 640).

In addition to traveling through the walls and floors, sound will travel through any openings in the building materials and will also travel around them. An opening of only 1 sq. in.

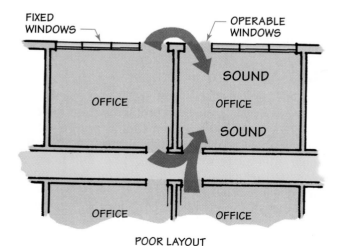

Figure 48–2 *Increased transmission paths reduce noise.*

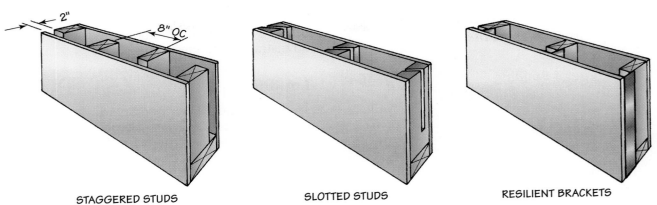

Figure 48–3 Construction of acoustical double walls.

in area will transmit as much sound as 100 sq. ft. of wall area. Some possible openings are ventilating ducts, oversized pipe openings, and

Type of Wall	STC
2" × 4" wood studs with dry-wall finish, both sides	32
2" × 4" staggered wood studs with dry-wall finish, both sides	41
2" × 4" wood studs with plastered finish, both sides	35
2" × 4" staggered wood studs with plastered finish, both sides	44
8" hollow CMU	48
8" hollow CMU with furred dry-wall finish, both sides	58
12" hollow CMU	53
12" doubled CMU (4" CMU + 4" air space + 4" CMU)	63
4" solid brick	41
8" solid brick	49
12" solid brick	54
6" reinforced concrete	46
8" reinforced concrete	51
12" reinforced concrete	56

TABLE 48–1 *Sound Transmission Classification (STC) of Walls*

door cracks. Back-to-back electric outlets or medicine cabinets will also transmit sound and therefore should be avoided.

Often a partition will extend only as high as a suspended ceiling, with the area above the ceiling serving as a plenum for wiring and piping (Figure 48–4, page 641). Sound will then travel right *over* the partition into the next room. Often such transmission by *flanking* is greater than transmission by wall vibration. This can be avoided by extending the partitions through the plenum to the floor above. Special closure panels are available for this purpose. Another solution is the use of acoustical insulation blankets laid over the top of the suspended ceiling and extending 3' beyond the partitions (Figure 48–5, page 641). Flanking paths *around* walls are also created by some contemporary construction methods that use continuous glass walls with interior partitions that do not completely connect to the window mullions. Even a crack between wall and floor offers a flanking path.

Special sound problems caused by vibrating machinery bolted to the structure can be solved by specifying resilient machinery mountings. Any piping or duct work connected to such machinery should be attached with flexible bellows or isolated from the structure with resilient gaskets. Figure 48–6 (page 641) illustrates these solutions.

SOUND REFLECTION

Special attention to acoustics is needed in the design of large assembly rooms where a principal function is listening to speech or music.

TABLE 48–2 *Recommended Sound Transmission Classifications*

Type of Building	STC
Private residence or apartment (same occupancy)	
Bathroom/living area	40
Living area/child's bedroom	40
Kitchen/living area	38
Kitchen/bedroom	38
Bedroom/bedroom	38
Other	32
Motel, hospital, or apartment (separate occupancy)	
Bathroom/living area	50
Bathroom/bedroom	50
Bathroom/bathroom	50
Living area/living area	50
Living area/bedroom	50
Bedroom/bedroom	50
Rooms/corridor	42
Other	45
Offices	
Washroom/private office	47
Washroom/general office	40
Private office/private office	45
Private office/general office	45
Other	38
Schools	
Music room/classroom	50
Shop/classroom	47
Mechanical equipment/classroom	45
Washroom/classroom	42
Classroom/classroom	40

Examples are auditoriums, theaters, churches, classrooms, gymnasiums, and courtrooms. When these rooms are properly designed, the audience is able to hear without difficulty: the sound will be loud enough, it will be evenly distributed throughout the room, and there will be no distracting echoes.

Loudness

The sound heard by members of an audience reaches them by at least two paths: (1) *directly* from the source and (2) by *reflection* from one or more surfaces as shown in Figure 48–7. Some surfaces, such as acoustical tile, absorb sound and reflect very little. But hard surfaces, such as wood, concrete, plaster, and glass, absorb very little and reflect nearly all sound, just as light reflects from a polished surface. The angle of **sound reflection** is determined by the same physical law governing light reflection: the angle of incidence equals the angle of reflection (Figure 48–8). To the listener, the reflected sound appears to come not from the actual source but rather from the sound *image*. When a room is being designed, the sound image is often plotted behind the reflecting surface to serve as a graphical shortcut to determine the path of the reflection.

The loudness of a direct sound depends upon the power output of the source and the distance from the source. The loudness of a reflected sound is always less than that of the direct sound because the path is longer and some of the sound is absorbed by the reflecting surface. But the total sound reaching the listener by direct and reflected paths may be substantially louder than the direct sound alone.

Distribution

An auditorium can be designed so that sound is evenly distributed throughout the audience, allowing those seated at the rear to hear as readily as those seated at the front. This is done by shaping the floor, ceiling, and walls to accomplish specific acoustical tasks.

A good rule of thumb in acoustics is that a poor sight line from stage to audience will also produce a poor hearing line. For example, direct sound to the audience seated at the rear of the level auditorium shown in Figure 48–9 (page 642) will diminish as it is absorbed by the people and upholstered seats in the front rows. This sound loss can be as much as 2 decibels per

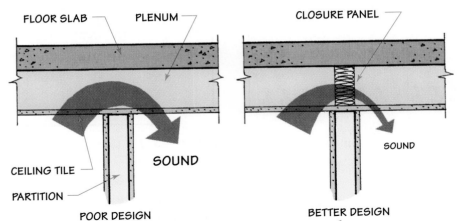

Figure 48–4 Reduction of sound through plenum by closure panel.

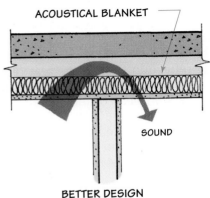

Figure 48–5 Reduction of sound through plenum by ceiling blanket.

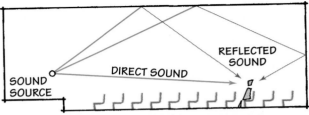

Figure 48–7 Direct and reflected sound.

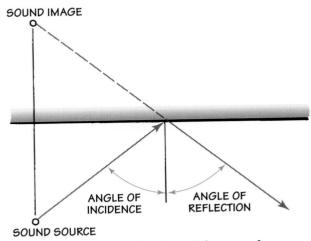

Figure 48–8 Sound reflected from a plane surface.

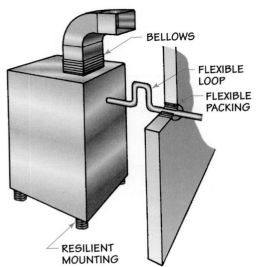

Figure 48–6 Reduction of sound from vibrating machinery.

row. (A decibel is a unit of measuring the relative loudness of sound and is equal to the smallest change in sound level detectable by the human ear.) Refer to Table 48–3 for an approximate indication of common sound levels in

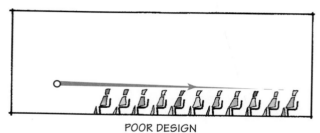

Figure 48–9   Sound distribution in a level auditorium.

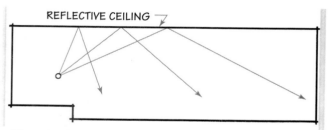

Figure 48–11   Sound reinforcement from a reflective ceiling.

TABLE 48–3   *Common Sound Levels (in decibels)*

120 decibels	Threshold of pain (ear damage)
100 decibels	Rock band
80 decibels	Factory
60 decibels	Office
40 decibels	Home
20 decibels	Whisper
0 decibels	Threshold of hearing

the ceiling and reinforces the direct sound. The ceiling is the most important reflective surface in an auditorium, and it is a good rule of thumb not to use sound-absorbing materials (such as acoustical tile) unless echoes must be prevented. Ceilings and other surfaces can be shaped to evenly distribute reflected sound. Figure 48–12 illustrates a ceiling designed to increase the reflection toward the rear of an auditorium. If a balcony is included, it is important that those persons seated under the balcony are able to "see" the ceiling (Figure 48–13) in order to receive sound reflected from it.

In addition to flat surfaces, convex and concave surfaces (Figure 48–14) may be desired.

decibels. A sound level change of 5 decibels is quite noticeable, and a change of 10 decibels appears to the listener to double (or halve) the sound. Figure 48–10 illustrates common solutions to this problem: the stage is raised or the auditorium floor is sloped (preferably both).

Another method of reinforcing the sound delivered to the rear seats is through the installation of hard, sound-reflecting ceilings. As shown in Figure 48–11, the sound reflects from

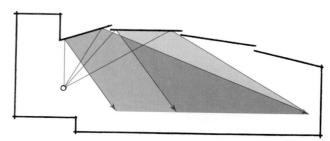

Figure 48–12   Ceiling designed to increase reflected sound delivered to the rear of the audience.

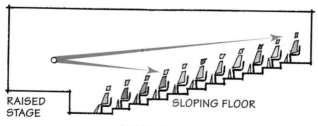

Figure 48–10   Sound distribution in a sloping auditorium.

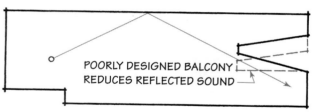

Figure 48–13   Balcony designed to permit reflected sound to the audience below.

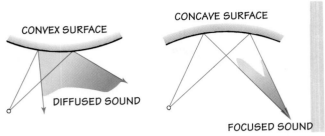

Figure 48–14 *Effect of convex and concave surfaces.*

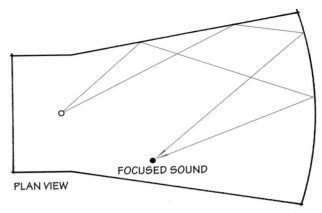

Figure 48–16 *Focusing effect of fan-shaped side wall combined with a concave rear wall.*

Deep convex surfaces of 6" or more (not fine ridges as in striated plywood) will diffuse sound, but concave surfaces will cause the reflected sound waves to focus on one spot—with a consequent loss of sound at other locations. Also, the reflected sound may be louder than the direct sound, which gives the impression that the sound is coming from the wrong direction. Curved rear walls often cause such a problem, but this can be corrected by segmenting the wall into separate panels positioned in different directions, or the wall can be treated with sound-absorbing materials.

Side walls also can reflect sound, resulting in undesirable effects. Figure 48–15 shows an auditorium with wide parallel side walls which cause the reflected sound to travel a great distance before reaching the listener and may result in echoes. Figure 48–16 illustrates the focusing effect of a fan-shaped side wall combined with a concave rear wall. Although there is no shape that is best for all conditions, Figure 48–17 shows the plan and elevation of an auditorium designed for even distribution of direct and reflected sound.

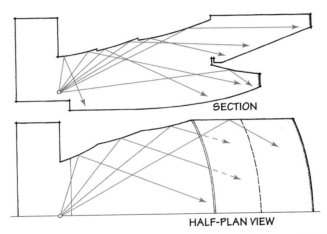

Figure 48–17 *Auditorium designed for sound distribution.*

Echoes

When the first reflection of a sound reaches the listener within $\frac{1}{20}$ second of the direct sound, it tends to blend with and reinforce the direct sound. But when a reflected sound reaches the listener after $\frac{1}{20}$ second, it is heard as a distinct repetition of the direct sound—called an *echo*. Reinforced sound is desirable, but echoes are undesirable. Sound travels through the air at a velocity of 1,125 ft./sec. Consequently, a reflected path that is about 50' longer than the direct path will produce echoes.

Example:
A person seated 15' from the front of a 45'-deep auditorium (Figure 48–18) will receive an echo from the rear

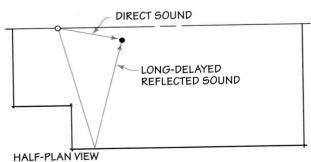

Figure 48–15 *Echo caused by wide side walls.*

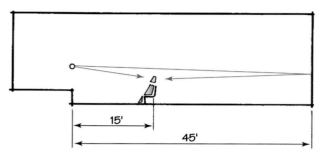

Figure 48–18   *Echo caused by reflected sound.*

wall since the reflected path is 75', the direct path is only 15', and the difference is 75' − 15' = 60'.

Surfaces that will produce echoes can be corrected by covering with sound-absorbing material. Also, the auditorium can be designed to reduce the length of the reflected sound paths by lowering the ceiling height near the stage and narrowing the auditorium width near the stage (Figure 48–17). Such a design is effective for medium-sized auditoriums (seating 500 to 2,000). For smaller auditoriums, the first reflections usually reach the audience quickly enough to prevent echoing. In larger auditoriums, it is nearly impossible to shape them to obtain even distribution of sound without echoing. Natural hearing in the front center section is nearly always more difficult than in the far-

REAL WORLD   APPLICATIONS

Cooperation on the Job

You have been approached by the local city council to remodel the town's 85-year-old theater. It is to be used once again for music and drama productions. This is an interesting yet challenging project because it will involve historic preservation and, most importantly, acoustical modifications. Not only will the city council oversee the project, but you will also be working with the future director of the theater, the contractor, the director of lighting, and the stage manager. You are only one member of this team, so make every effort to contribute peacefully to the team effort.

People do not instinctively know how to interact effectively on a team. Thus, there are some important small-group skills that you will need to put into practice. Setting group goals early in the project will help keep the group focused and united. Group goals on this project might include achiev-

ing high-quality sound, designing a flexible lighting system, and choosing an auditorium seating arrangement that meets the community's needs.

When deciding on goals and throughout the project, members of the team must be comfortable sharing ideas. This can be accomplished by giving equal time for everyone to speak, writing ideas anonymously, listening attentively, and getting to know and trust one another. In this way, respect is shown to all the persons on the team regardless of their background. Team members will also feel accepted and supported when the group considers each individual's ideas and opinions, encourages one another, and expresses appreciation for each others' contributions.

Effective communication is another key to successful cooperation. First, team members must communicate accurately and unambiguously. For instance, the stage man-

thest sections. Electronic sound amplification is the best solution.

SOUND ABSORPTION

In addition to room design, **sound absorption** can be used to control a room's acoustics. To do this, sound-absorbing materials are used in rooms to reduce noise or prevent echoes caused by reflected sound. They cannot reduce the level of direct sound, nor prevent sound transmission from another room (such as footsteps from the floor above). In practice, the addition of sound-absorbent material can reduce the reflected sound level in a room about 5 to 7 deci-

bels, but it can never reduce the total sound below the level of the direct sound itself.

The most efficient sound-absorbing materials are fiber blankets, porous ceiling tile, carpeting, drapes, and upholstered furniture. They work by converting sound into heat due to friction on the walls of the capillaries within the material. If the surface fissures of the sound-absorbent are blocked (by excessive painting, for example), the material will no longer be effective. Fiber products are usually applied by spray gun to a 3" depth or installed in 1"- or 2"-thick blankets. A *transparent* (acoustically speaking) facing may be installed over the fibers to achieve specific architectural effects. Some

ager and theater director must be as specific as possible in stating their acoustical expectations. This will enable the interior designer to select appropriate wall and floor coverings—these will greatly affect how the sound will be reflected or absorbed. If someone is unable to state opinions or ideas clearly, confusion may result. For this reason, it is important that the listener clarify what the speaker is saying by asking specific questions. Once the speaker has stated an idea, the listener should summarize what has been said to confirm understanding.

Almost inevitably, conflict will arise during the progress of the project, so team members must learn how to resolve conflicts. If you as the architect and the theater director differ in your opinions on the seating arrangement, you must compromise. Both of you must be willing to change your plans in some way to satisfy the other. If a compromise cannot be reached, both parties should present their ideas and justification to the city council and other team members, who can make the final decision by consensus. Once the design is chosen, it should be followed without argument.

Working with others is no easy task. No matter what the project, you will be work-

ing with a wide variety of architects, engineers, surveyors, technicians, drafters, and code inspectors, some having an even wider variety of personalities. Develop your skill in team cooperation as you continue to develop your drafting and computer skills.

One way to develop these skills is to form a group of three or four and use the example provided here (remodeling a local theater) as your project. Set goals, develop a work plan, divide tasks, and begin the project. The architect will design the project step by step, continually asking for input from other team members. Other team members can write down design ideas so they can readily share them with the architect and one another. Focus on acoustical considerations as you complete this project. Practice the cooperative skills discussed previously as you work. Evaluate your team's cooperation after completing the project. Did all members participate and share ideas equally? Were project goals set and used to direct progress? How effectively did the group communicate? How could communication be improved? What conflicts, if any, arose? Was a compromise reached?

possibilities are shown in Figure 48–19. Sound-absorbing materials are often specified in office buildings, schools, and auditoriums.

Office Buildings

Cover ceilings with absorbents. In large office areas, wall treatment is also recommended. In large conference rooms, provide a hard, sound-reflective panel in the center of the ceiling equal to 50 percent of the ceiling area. This panel aids conversation between opposite ends of the room. The remaining ceiling and walls should be treated.

Schools

In elementary school classrooms, cover ceilings with absorbents. In higher-level classrooms of more than 700 sq. ft. of area, provide a reflective ceiling panel equal to 50 percent of the ceiling area. A 2'-wide strip of absorbent above chalkboards is recommended on two adjacent walls.

Auditoriums

Place absorbents on surfaces that will cause echoes, but not on surfaces that will provide useful sound reinforcement. Usually ceilings over 25' high must be treated. In theaters, it may be desirable to adjust the sound-absorbent quality depending upon the performance (high reverberation for music but low for lecture). This can be done by designing flexible components such as retractable drapes as shown in Figure 48–20. For auditoriums, expert advice should always be obtained.

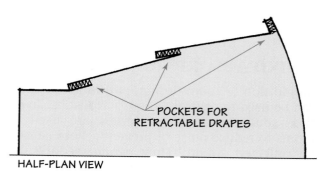

POCKETS FOR RETRACTABLE DRAPES

HALF-PLAN VIEW

Figure 48–20 Adjustable sound absorbency.

COMPUTER-AIDED DESIGN FOR ACOUSTICS

Architectural offices use specialized computer programs to design buildings to meet acceptable standards of sound transmission, reflection, and absorption. For example, the sound transmission classification of a proposed wall construction can be determined and compared with recommended values, and sound levels can be projected for each room shape and surface material. Refer to the "Architectural Databases" sections of Chapter 6 for more detailed information on computer-aided design for acoustical considerations.

AMPLIFICATION

In large rooms (over 100,000 cu. ft. volume), electronic **sound amplification** will generally be required. Amplification will also be useful in

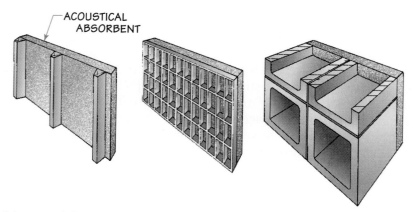

ACOUSTICAL ABSORBENT

Figure 48–19 Architectural facings over acoustical absorbents.

rooms like lecture halls down to a size of 20,000 cu. ft. Amplification is seldom required in smaller rooms.

Electronic amplification systems may be *central* or *distributed*. The central system is usually preferred because the loudspeaker (or group of loudspeakers) is located above the sound source, giving maximum realism of sound direction. The distributed system is used when a line of sight between the central loudspeaker and audience is not feasible. The distributed system consists of a number of loudspeakers distributed over the entire audience. This is often used in airport terminals and large convention halls.

The controls of amplified sound systems should be located at the rear of the audience rather than in a glassed-in booth. The sound system operator should be able to hear the sound just as it is heard by the audience.

SOUND MASKING

It is sometimes possible to **sound mask** undesirable noise by the addition of pleasant sound. This has been called *acoustic perfume*. Some examples are background music in commercial and industrial buildings, the sound of a water fountain in a busy lobby, and even the sound of air-conditioning units in motels located on noisy highways. It has been found that such a bland, continuous sound is more agreeable than intermittent noise.

REVIEW QUESTIONS

1. List the four major factors to be considered during the design of a building for sound conditioning.

2. List methods of reducing sound transmission:

    a. From room to room

    b. From vibrating machinery

3. List methods of increasing the sound delivered to the rear of an auditorium.

4. How does sound transmission influence building design?

5. Describe briefly each term:

    a. Sound transmission classification

    b. Flanking

    c. Sound image

    d. Acoustic perfume

    e. Decibel

ENRICHMENT QUESTIONS

1. Distinguish between the following by providing an example of each:

    a. Structure-borne and airborne sound

    b. Central and distributed amplification systems

2. Why does a reflected path 50' longer than the direct path produce an echo?

3. Explain where sound begins and how it changes as it travels through the air.

4. What positive and negative effects can buffer zones have on the design of a building?

5. Locate a building in your area and inspect it for acoustical flaws. How could the acoustics be improved?

6. Would you say your classrooms were designed with acoustics in mind? Why or why not?

PRACTICAL EXERCISES

1. Draw or computer-produce the sound conditioning details for:
   a. 12' × 28' conference room
   b. 30' × 40' high school classroom
   c. 600 seat high school auditorium

2. Draw or computer-produce the sound conditioning plan and details for:
   a. Your original building design
   b. The Z residence
   c. The South Hills Office Building in Chapter 46
   d. A four-unit condominium with each unit having 950 sq. ft.

CHAPTER
49

Fire Protection

OBJECTIVES

By completing this chapter you will be able to:

🏛 Define fire protection terminology.

🏛 Describe the impact building codes have on the design of new buildings.

🏛 Design and prepare a preliminary layout of a sprinkler extinguishing system.

KEY TERMS

Building classifications

Fire rating

Sprinkler systems

Standpipe

BUILDING CODES

The earliest record of an attempt to improve building safety is the Code of Hammurabi, a Babylonian king and lawmaker in 2100 B.C.

In the case of collapse of a defective building the architect is to be put to death if the owner is killed, and the architect's son if the owner's son is killed.

Laws governing building construction and land use were first introduced by the ancient Romans. During the reign of Julius Caesar, Rome grew rapidly, and tall, speculative apartments were built which often collapsed. Roman laws first limited heights to 70' and later reduced them to only 60'. In the fourteenth century the City of London adopted a law that prohibited the building of wooden chimneys. Building codes governing building construction methods and zoning ordinances governing land use were adopted in English, French, and Prussian cities by the nineteenth century and were accepted by all U.S. cities and most towns in the early twentieth century.

The basic concept of a building code or a zoning ordinance is that individual actions should be regulated in favor of the welfare of the general public. It has been shown many times over that such protection is necessary, and courts have supported the inherent power of the government to protect citizens from unsafe building practices. Building codes specify acceptable building materials and construction methods, allowable loads and stresses, mechanical and electrical requirements, and other specifications for health and safety. Architects have both a moral and a legal obligation to study and follow the building code requirements of the city in which they build. Building codes are *minimum* requirements and may be outdated. When this occurs, architects, as professionals, are expected to design at the current state of the art. Architectural drafters, also, are more effective when they understand some of these requirements.

Among the most important portions of any building code are the sections on fire protection. One need not experience the terror of fire to realize how necessary it is to design buildings that will not be a hazard to their occupants. In the United States over ten thousand persons are killed each year in fires. Many of these deaths occur in buildings that are in violation of fire protection codes.

BUILDING CLASSIFICATIONS

Building code requirements vary depending upon such factors as type of occupancy, building contents, type of construction, location, and fire-extinguishing systems. The codes permit "tradeoffs" between these **building classifications** with the goal of obtaining that degree of public safety as can be reasonably expected. For example, greater fire protection is required for a building that will be high-rise, densely occupied, constructed of flammable materials, or have hazardous contents. Building codes try to avoid requirements that involve unnecessary inconven-ience or interference with the normal use of a building. However, the codes do set minimum standards for public safety that must be followed even though a financial hardship may be imposed upon individuals or groups.

Classification by Occupancy

The Life Safety Code is only one portion of the ten-volume *National Fire Codes*. These codes are purely advisory but are widely used as a basis for establishing local or state building codes. Developed by the National Fire Protection Association, the Life Safety Code classifies buildings by eight types of occupancy:

1. Assembly (theatres, restaurants, churches, and museums)
2. Educational
3. Institutional (hospitals and prisons)
4. Residential (hotels, apartments, and dwellings)
5. Mercantile
6. Offices
7. Industrial
8. Storage

The code deals with the design of various types of buildings to reduce the danger from fire, panic, fumes, and smoke. It specifies the number, size, and arrangement of exits to allow prompt escape from buildings of each occupancy type. The code recognizes that safety is more than a matter of exits and therefore recommends a number of additional requirements. Following are some abstracts from the Residential and Assembly sections of this code to give you an idea of the variety of regulations that have been included. This is only a partial list of requirements, and local building codes should always be consulted for complete, updated requirements.

ONE- AND TWO-FAMILY DWELLINGS

The requirements for residences are far short of complete requirements for fire safety, but are those which can reasonably be enforced by law. Some of these requirements are:

1. In all residences, every occupied room (except storage rooms) must have at least two means of exit (such as a doorway or window). At least one exit must be a doorway. Below-grade sleeping areas must have direct access to the outside.

2. Exit doors must be at least 24" wide (30" preferred).

3. Occupied rooms must not be accessible only by folding stairs, trapdoor, or ladder.

4. All door-locking devices must be such that they can be easily disengaged from the inside by quick-release catches. All closet door latches must be such that they can be easily opened by children from inside the closet. All bathroom door locks must be such that they can be opened from the outside without the use of a special key.

5. The path of travel from any room to an exit must not be through a room controlled by another family, nor through a bathroom or other space subject to locking.

6. Passages from sleeping rooms to exits must be at least 3'-0 wide.

7. Stairs must be at least 3'-0 wide with risers no greater than 8" and treads not under 9".

8. Every sleeping room, unless it has a direct exterior exit or two interior exits, must have a window which can be easily opened from the inside without use of tools. This win-

dow must provide a clear opening of at least 5 sq. ft. with not less than 22" in the least dimension. The bottom of the opening must not be more than 4'-0 above the floor. Awning and hopper windows must be designed to permit full opening.

9. Storm windows, screens, and burglar guards must have quick-opening devices.

10. Combustion heaters and stoves must not be so located as to block escape in case of a malfunction.

11. Smoke detectors should be installed in the immediate vicinity outside of each separate sleeping area and on each additional story of a living unit including the basement.

See the *National Fire Codes* or local codes for more detailed requirements.

ASSEMBLY BUILDINGS

1. An assembly area must be at least 15 sq. ft. per person. Seating area must be at least 7 sq. ft. per person. Standing area (such as waiting rooms) must be at least 3 sq. ft. per person.

2. A satisfactory grade-level door must be provided for each one hundred persons. A door to a stair or fire escape must be provided for each seventy-five persons.

3. Assembly buildings with a capacity of over one thousand persons must have at least four exits widely separated from each other. Assembly buildings with a capacity of over six hundred persons must have at least three widely separated exits. Smaller assembly buildings must have at least two widely separated exits.

4. Exits must be arranged so that the total length of travel from any point to the nearest exit does not exceed 150' for unsprinklered areas, and 200' in areas protected by automatic sprinklers.

5. Exit doors must be at least 2'-4" wide. The floor on both sides of a door must be level and at the same elevation for a distance at least equal to the width of the door; except

exterior doors which may be one step ($7\frac{1}{2}$") higher inside.

6. Exit doors must swing in the direction of travel. Screen or storm doors must also swing in the direction of travel. Sliding, rolling, or folding doors must not be used.

7. Exit doors must be readily opened from the inside of the building. Latches must be simple and easily operable, even in darkness. Conventional hardware such as panic bars or doorknobs are satisfactory, but an unfamiliar method of operation (such as a blow to break glass) is prohibited. Exit doors of assembly buildings with a capacity over one hundred persons must have panic bars.

8. Locks must not require a key to operate.

9. Doors to stair enclosures and smoke stop doors must be provided with reliable self-closing mechanisms and never secured open unless provided with a reliable release device.

10. No mirrors shall be placed on exit doors. Doors must not harmonize in appearance with the rest of the wall.

11. Revolving doors must never be installed at the foot or top of stairs. They must not be considered as a portion of the required exits. No turnstiles which restrict exit are permitted.

12. Approved exit signs, lighting, and emergency lighting must be provided at all exits and approaches. Doors which lead to dead-end areas must be identified by signs indicating their character (such as "linen closet").

13. No open flames (such as candles) are permitted unless adequate precautions are made to assure that no other material is ignited.

14. Assembly buildings must be designed so the principal floor is not below grade unless protected by automatic sprinklers. Non-fire-resistive assembly buildings must have the principal floor not more than 28' above grade.

15. All interior stairways must be enclosed to prevent spread of fire.

16. All interior decorations must be of fire-resistive or nonflammable materials.

17. A row of seats between aisles must not exceed fourteen seats. A row of seats opening to an aisle at one end only must not exceed seven seats.

18. Seats must be at least 18" wide, spaced at least 33" between rows, with at least 12" leg room (measured between plumb lines).

19. Aisles must be at least 3'-0 wide. Steps must not be used in aisles unless the slope exceeds 1' rise in 8' run. Ramps must not exceed 1' rise in 8' run.

20. Balcony rails must be substantial and at least 26" high, at least 30" high at the foot of an aisle, and at least 36" high at the foot of a stepped aisle.

21. Rooms containing pressure boilers, refrigerating machinery, transformers, or other service equipment subject to possible explosion must not be located adjacent to or under the exits.

22. Special regulations govern air conditioning, ventilating, and heating equipment. For example, automatic devices must be provided to prevent circulation of smoke through ductwork.

23. Areas used for painting or repair must be effectively cut off from assembly areas or protected by automatic sprinklers.

24. Fire alarm systems must be visual and coded to alert employees rather than audible to alert the entire audience. Audible devices such as gongs or sirens may create panic in conditions where fire drills are not feasible. Employees must be drilled and present when the building is occupied by the public.

25. Automatic sprinklers are required for any stage rigged for movable scenery, as well as for understage areas, dressing rooms, and storerooms. An approved fire-resisting curtain with an emergency closing device must be provided. The stage roof must contain an approved, operable ventilator having a

free-opening area at least 5 percent of the stage floor area.

26. Motion picture projection apparatus must be enclosed by a fixed, fire-resistive booth.

See the *National Fire Codes* or local codes for more detailed requirements.

Classification by Contents

The Life Safety Code also classifies buildings according to their contents by three ratings: ordinary-hazard contents, extra-hazard contents, and light-hazard contents.

1. *Ordinary-hazard contents* represents the conditions found in most buildings having contents that are moderately combustible but that are not explosive and will not release poisonous fumes.

2. *Extra-hazard contents* are liable to burn rapidly, explode (such as gasoline), or release poisonous fumes. All extra-hazard-contents buildings must have sufficient exits to allow occupants to escape with a travel distance not over 75'. It is assumed that this distance can be traveled in 10 sec., which is the time normal individuals can hold their breath.

3. *Light-hazard contents* have low combustibility and, consequently, the primary danger will be from panic.

Classification by Construction Type

The *National Fire Codes* of the National Fire Protection Association classify buildings into five principal construction types as follows:

Type I: Fire-resistive Construction   Members are of noncombustible materials with fire ratings not less than:

Four hours for bearing walls
Four hours for columns and beams supporting more than one floor
Three hours for columns and beams supporting only one floor
Two hours for interior partitions

Type II: Heavy Timber Construction   Bearing walls are of noncombustible materials (usually masonry) with a minimum two-hour fire rating,

Fire Safety

 In March of 1993, a fire started in the attic of an apartment building owned by Mr. and Mrs. Ronald Shepherd. The cause of the fire was old and deteriorated electrical wiring. Fortunately, no one was injured, but the damage to the building was substantial. All seven apartments remained uninhabited for over eighteen months. The cost to the owners in construction and fire protection/prevention alone was over $130,000.

In order to receive the Department of Labor and Industry's approval, the following fire protection and prevention measures had to be completed.

🏛 Install new electrical wires and circuit breaker boxes.

🏛 Fog each room to remove smoke odor.

🏛 Install handrails down both rear stairways.

🏛 Install fire guard ceiling tile.

🏛 Install $\frac{5}{8}$" fire code drywall to ceiling joists, cut around pipe, and drop flex pipe through openings.

🏛 Relocate wall a minimum of 1' back from furnace toward front of building so access can be made on furnace repairs.

🏛 Install metal frames with wire glass in the five rear windows by the stairwell. Existing windows are not acceptable.

🏛 Each apartment will need an individual fire extinguisher, smoke alarm, and smoke detector.

🏛 All doors leading to the outside must have documents stating they are fire code units.

🏛 Front foyer needs lit exit sign above door, or exit sign with light directly above.

🏛 Install handrail on left front concrete porch. Existing handrail is too weak for acceptance.

Now take a look at your original house design. Is it designed so fire damage would be minimized? For additional information contact the United States Fire Administration. The Internet homepage address is: **http://www.usfa.fema.gov/fd5.htm** When you are there, find the approximate number of fires in the United States last year. How much did this cost property owners? What are the four main causes of fire in a residence? Does your original house design avoid any of these potentially fatal problems? What percentage of the fires that kill young children are started by children playing with fire? Finally, and most importantly, what measures can save your life in a residential fire? Design a fire escape route for your original house design.

National Fire Protection Association
Batterymarch Park
Quincy, MA 02269
Phone: (914) 273-6780
Fax: (914) 273-6785

and laminated or solid wood members are not less than:

   8" × 8" for columns

6" × 10" for beams
4" × 6" for trusses or arches supporting roof loads

4" for flooring
2" for roof deck

Type III: Noncombustible Construction   Structural members, walls, and partitions are of noncombustible construction such as unprotected steel.

When bearing walls are protected to a two-hour fire rating, and columns, floors, and roofs are protected to a one-hour fire rating, this is designated *Protected Noncombustible Construction.*

Type IV: Ordinary Construction   Exterior bearing walls are of noncombustible materials (usually masonry) with a minimum two-hour fire rating, and interior framing, roofs, and floors are combustible (usually wood).

When roofs, floors, and their supports have a one-hour fire rating, this is designated *Protected Ordinary Construction.*

Type V: Wood-frame Construction   All elements are of wood or other combustible material, but it does not qualify as heavy timber construction or ordinary construction.

When roofs, floors, and their supports have a one-hour fire rating, this is designated *Protected Wood-frame Construction.*

Table 49–1 simplifies these definitions.

TABLE 49–1   *Construction Classifications*

Type I.	Fire-resistive construction: noncombustible materials with four-hour bearing members
Type II.	Heavy timber construction: timber interior with two-hour masonry walls
Type III.	Noncombustible construction: unprotected steel
Type IV.	Ordinary construction: wood interior with two-hour masonry walls
Type V.	Wood-frame construction: wood interior and walls

Fire Ratings

The fire protection sections of building codes are based upon studies made by fire protection engineers who have tested various building methods to determine the fire resistance of each. The *standard fire test* (E119-58) of the American Society for Testing and Materials is the accepted standard for such tests. The degree of fire resistance of each building method is measured in terms of its ability to withstand fire from one to four hours. For example, a two-hour fire rating would indicate that a structural member could withstand the heat of fire (or the cooling of a fire hose) for two hours before serious weakening; or that a wall, floor, or roof would not allow passage of flame and hot gasses for two hours.

The fire resistance ratings for typical walls, columns, beams, floors, and roofs are given in Tables 49–2 through 49–6. Notice that the **fire ratings** of masonry and concrete walls can be improved simply by increasing their thickness (Tables 49–2 and 49–3). Steel members, however, must be screened by additional fire protection. Usually gypsum, perlite, vermiculite, or mineral fiber is used (Figures 49–1 through 49–11, pages 657 and 658).

After fire resistance ratings have been determined, the architect uses the *Fire Resistance Directory* published by Underwriters Laboratories (UL). This directory contains designs of all types of columns, beams, floors, walls, partitions, ceilings, and roofs. It also contains the names of products qualified to use UL's classification markings. Most code agencies require reference to UL classification on construction drawings.

Classification by Location

Buildings constructed in closely packed communities are a greater threat to the general public than buildings located in an open area. Therefore building codes establish *fire limits*, or *fire zones*. Within the limits of a fire zone, all buildings must be designed so that a fire will remain contained and not sweep on to adjacent building after building.

Originally, fire codes required masonry exterior walls to act as fire barriers, but present

TABLE 49-2   *Fire Resistance Ratings for Masonry Walls*

Type of Masonry Wall	Minimum Thickness for Ratings of:			
	4 hr.	*3 hr.*	*2 hr.*	*1 hr.*
Heavyweight concrete masonry units* (coarse aggregate, siliceous gravel)	6.7"	6.7"	4.5"	3"
Lightweight concrete masonry units* (coarse aggregate, unexpanded slag)	5.9"	5"	4"	2.7"
Lightweight concrete masonry units* (coarse aggregate, expanded slag)	4.7"	4"	3.2"	2.1"
Solid brick masonry**	8"	8"	8"	4" (nonbearing)
Clay tile masonry**	16"	12"	12"	8"
Solid stone masonry*	12"	12"	12"	8"

*Abstracted from National Building Code.
**Abstracted from Uniform Building Code.

TABLE 49-3   *Fire Resistance Ratings for Concrete Walls*

Type of Concrete Wall	Minimum Thickness for Ratings of:			
	4 hr.	*3 hr.*	*2 hr.*	*1 hr.*
Plain concrete	$7\frac{1}{2}"$	$6\frac{1}{2}"$	$5\frac{1}{2}"$	4" (nonbearing)
Reinforced concrete (unplastered)	$7\frac{1}{2}"$	$6\frac{1}{2}"$	$5\frac{1}{2}"$	4" (nonbearing)
Reinforced concrete ($\frac{3}{4}"$ portland cement or gypsum plaster, each side)	6"	5"	4"	3"

Abstracted from National Building Code.

TABLE 49-4   *Fire Resistance Ratings for Steel Columns*

Type of Column Protection	Minimum Thickness for Ratings of:			
	4 hr.	*3 hr.*	*2 hr.*	*1 hr.*
Vermiculite or perlite-gypsum plaster on self-furring metal lath (see Figure 49-1)	$1\frac{3}{4}"$	$1\frac{3}{8}"$		
Perlite-gypsum plaster on $\frac{3}{8}"$ perforated gypsum lath (see Figure 49-2)		$1\frac{3}{8}"$	1"	
Sprayed mineral fiber (see Figure 49-3)	$2\frac{1}{2}"$	2"	$1\frac{1}{2}"$	
Concrete encasement (see Figure 49-4)	3"	$2\frac{1}{2}"$	2"	$1\frac{1}{2}"$

Abstracted from *Fire-resistant Construction in Modern Steel-framed Buildings*, AISC.

TABLE 49–5 *Fire Resistance Ratings for Steel Beams, Girders, and Trusses*

Type of Beam Protection	Minimum Thickness for Ratings of:			
	4 hr.	3 hr.	2 hr.	1 hr.
Vermiculite or perlite-gypsum plaster on self-furring metal lath (see Figure 49–5)	$1\frac{1}{2}''$			
Sprayed mineral fiber (see Figure 49–6)	$1\frac{7}{8}''$	$1\frac{7}{16}''$	$1\frac{1}{8}''$	
Concrete encasement (see Figure 49–7)	$3''$	$2\frac{1}{2}''$	$2''$	$1\frac{1}{2}''$

Abstracted from *Fire-resistant Construction in Modern Steel-framed Buildings*, AISC.

TABLE 49–6 *Fire Resistance Ratings for Floor and Roof Systems*

Type of Floor and Roof Protection	Minimum Thickness for Ratings of:			
	4 hr.	3 hr.	2 hr.	1 hr.
Light-gauge steel, not fireproofed (see Figure 49–8) Sand-limestone concrete slab of thickness equal to:			$5\frac{1}{4}''$	$4\frac{1}{2}''$
Light-gauge steel, contact fireproofing (see Figure 49–9) $2\frac{1}{2}''$ sand-gravel slab with sprayed mineral fiber of thickness equal to:		$\frac{3}{4}''$	$\frac{1}{2}''$	
Light-gauge steel, membrane fireproofing (see Figure 49–10) 2" sand-gravel slab and 1" vermiculite-gypsum fireproofing on metal lath installed at a distance of:	$2''$	$15''$		
Precast cellular system, not fireproofed (see Figure 49–11) $1\frac{1}{2}''$ sand-gravel concrete topping over a limestone concrete precast unit of thickness equal to:		$6''$		

Abstracted from *Fire-resistant Construction in Modern Steel-framed Buildings*, AISC.

codes allow other construction methods having satisfactory wall fire ratings. Such walls must also be able to remain standing under fire conditions.

Extinguishing Systems

Building codes often require automatic water-sprinkler systems because they give excellent

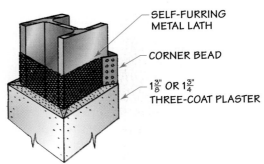

Figure 49–1   *Plaster-on-metal-lath fire protection of columns. (See Table 49–4.)*

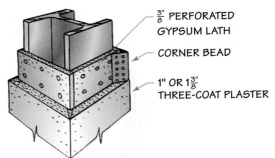

Figure 49–2   *Plaster-on-gypsum-lath fire protection of columns. (See Table 49–4.)*

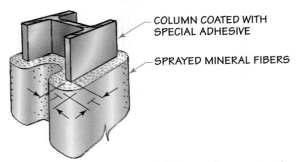

Figure 49–3   *Sprayed fibrous fire protection of columns. (See Table 49–4.)*

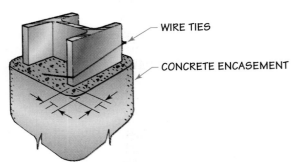

Figure 49–4   *Concrete fire protection of columns. (See Table 49–4.)*

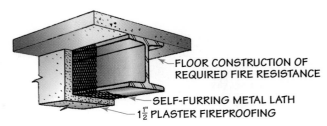

Figure 49–5   *Plaster fire protection of beams. (See Table 49–5.)*

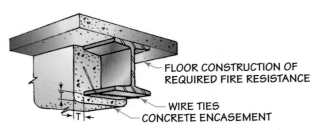

Figure 49–6   *Sprayed fibrous fire protection of beams. (See Table 49–5.)*

Figure 49–7   *Concrete fire protection of beams. (See Table 49–5.)*

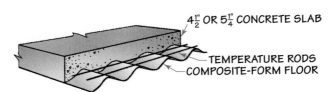

Figure 49–8   *Unprotected floors and roofs. (See Table 49–6.)*

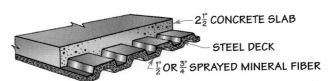

Figure 49–9   *Sprayed fibrous fire protection of floors and roofs. (See Table 49–6.)*

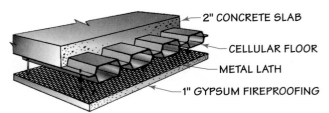

Figure 49–10 Membrane fire protection of floors and roofs. (See Table 49–6.)

Figure 49–11 Unprotected cellular floors and roofs. (See Table 49–6.)

fire protection in all types of buildings. Records show that when fires occurred in sprinkler-protected buildings, 80 percent of those fires were extinguished by the sprinklers, and another 18 percent held in check.

A **sprinkler system** consists of a network of piping placed under the ceiling and provided with a number of nozzles called *sprinklers* (Figure 49–12). When activated, the sprinklers spray water downward in a hemispherical pattern. Sprinkler system types are *fixed-temperature* and *rate-of-rise*.

Fixed-temperature Sprinkler Heads These heads are usually designed so temperatures of 135°F to 170°F will cause them to open auto-

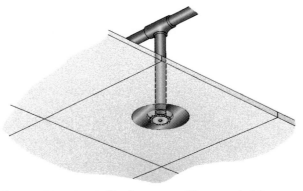

Figure 49–12 A flush-type ceiling sprinkler.

matically. Fixed-temperature sprinkler heads are color-coded to show their temperature ratings. Fixed-temperature sprinkler systems are *wet-pipe* when water is stored in the piping and *dry-pipe* when no water is in the piping.

The *wet-pipe system* is commonly used for most indoor conditions where temperatures will not fall below freezing. The water in the piping is kept under pressure behind each sprinkler. Sprinklers contain fusible links which are melted by heat and automatically open the sprinkler. Only sprinklers exposed to heat will open, thus preventing unnecessary water damage. A fire alarm sounds when the first sprinkler is opened. An antifreeze solution may be used in the piping for limited protection from freezing.

Buildings likely to have temperatures below freezing (such as unheated warehouses) can be protected by the *dry-pipe system.* The piping contains air under pressure rather than water. When heat from a fire opens one of the sprinklers, the air is released and water flows into the piping network and through any opened sprinklers. A fire alarm also sounds.

Rate-of-rise Sprinkler Systems Detectors open valves to the sprinkler piping rather than to the sprinkler heads. Rate-of-rise detectors open valves upon any abnormal increase of temperature. They are very sensitive and consequently give quick warning of a fire hazard. Rate-of-rise sprinkler systems are *deluge* and *preaction*.

The *deluge system* is used for extra-hazard conditions. All sprinkler heads are open, but the piping is dry. When a rate-of-rise detector opens the water supply valve, water rushes into the piping and out through all heads simultaneously, giving better protection for difficult conditions such as flammable liquid fires. An alarm also sounds.

The *preaction system* is used when it is important to reduce the possibility of accidental water damage. The principal difference between a preaction and a standard dry-pipe system is that in a preaction system, the water supply valve operates independently of the sprinkler heads; that is, a rate-of-rise detector first opens the valve and sounds an alarm. The fixed-temperature sprinkler heads do not open until their temperature ratings are reached. This gives time for small fires to be extinguished manually before the heads open.

CAREER PROFILES

Name: Danny R. Sacco   *Age:* 49
Gender: Male
Occupational Title: Director, Safety and Security
Employer: Indiana Hospital
           Indiana, PA 15701
Years of Experience: 28
Educational Background:
       Associate Degree in Design/Drafting
       Paramedic Certification, Indiana University of Pennsylvania (IUP)
       Bachelor's degree at IUP
Starting Salary:
       $35,000 to $45,000 plus benefits

JOB OVERVIEW:
My work day begins promptly at 8:00 a.m. Before beginning any tasks I review orders from the previous day and develop a work plan. Some of my daily tasks include attending meetings, supervising professional and technical staff, providing inservices, counseling employees, managing requests for special services, and responding to emergencies as they occur. On my job, no two days are the same. I leave work at 4:30 p.m., although many nights I attend work-related meetings. Primary responsibilities of my job fall into five areas: safety, security, disaster response, hazardous material and waste program, and emergency medical services clinical continuing education program. Years of on-the-job training in emergency medicine, fire rescue, fire safety, security, disaster management and planning, and safety sciences, including hazardous material and waste management have enabled me to meet these responsibilities.

One major project I completed for which the hospital received a state award was the development of a fire safety program for the operating room. I also revised the hospital's emergency response plan and am responsible for informing staff of these revisions through inservice training. In my position, state inspections are extremely important. I am pleased that in my area of responsibility we passed the Joint Commission inspection of the hospital with 100 percent compliance.

The major benefit of my job is that I like what I do and can excel at the things I like to do best. In addition, the job is never boring—there is always a new challenge to undertake and something different to do. One of the greatest challenges of my job is balancing cost with a high standard of safety. Another challenge as a director is maintaining a good staff and keeping them trained. A tip I would give to students considering a career path is: Choose a career doing the things you enjoy the most and you will never "work" again.

Sprinkler Layout The layout of a sprinkler system is performed by a professional engineer using established standards as a guide. Sprinkler layout depends upon the building classifications. For example, the *National Fire Codes* specify the following under a smooth ceiling construction:

1. *Light hazard:* The protection area per sprinkler must not exceed 200 sq. ft. The maximum

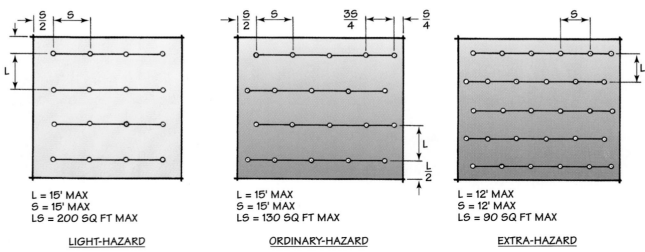

Figure 49–13 Sprinkler layouts for smooth ceiling.

distance between lines and between sprinklers on lines is 15 ft. Sprinklers need not be staggered. See Figure 49–13.

2. *Ordinary hazard:* The protection area per sprinkler must not exceed 130 sq. ft. The maximum distance between lines or between sprinklers on lines is 15 ft. Sprinklers on alternate lines must be staggered if the distance between sprinklers on lines exceeds 12 ft.

3. *Extra hazard:* The protection area per sprinkler must not exceed 90 sq. ft. The maximum distance between lines and between sprinklers on lines is 12 ft. The sprinklers on alternate lines must be staggered.

See the *National Fire Codes* or local codes for more detailed requirements.

Standpipes **Standpipes** are vertical water pipes with fire-hose outlets at each floor. They can be designed for small-hose ($1\frac{1}{2}$") to be used by the building occupants in the event of fire or large-hose ($2\frac{1}{2}$") to be used by fire departments—or both. Standpipes are usually wet-pipe rather than dry-pipe. At ground level, branches extend outside the building and are finished with *Siamese connections*. Should there be insufficient pressure in the public water system, the fire department can pump water into the standpipe through these connections to increase the pressure. Check valves relieve the pumps from back-pressure.

Standpipes are located so that any fire can be reached by a stream from not more than 75' of small-hose or 100' of large-hose.

See the *National Fire Codes* or local codes for more detailed requirements.

Some other extinguishing systems are foam, carbon dioxide, halons, and dry chemical. Foam is an aggregate of tiny gas-filled or air-filled bubbles used to smother fire by excluding air. Because foam contains water, it also has cooling properties. The principal use for foam is in fighting fires involving flammable liquids. Carbon dioxide, the halons, and dry chemical are nonconductive and therefore can be used on electrical fires as well as on flammable-liquid fires.

High-rise Buildings

Research is constantly being conducted to find better ways to prevent or control fires in buildings. Special attention is being given to the problem of fire safety in high-rise buildings, for their construction is increasing—some to heights of 1,000 ft. and more. These buildings may contain more than twenty-five thousand persons—equivalent to the population of a small city. Special precaution for fire protection must be taken in such buildings, for prompt evacuation is usually not possible. In addition, the building height may contribute to a stack effect, and many floors may be beyond the reach of fire department aerial equipment. Therefore, fire must be controlled and fought internally. A combination of three methods is usually used:

1. All building materials and furnishings selected to provide no potential fuel for a fire, including no potential for emitting smoke or toxic gasses

2. Compartmented structures capable of resisting and containing a fire within a relatively small portion of the building

3. Automatic fire-extinguishing systems capable of prompt and effective operation

An innovative fire protection system was used in the U.S. Steel Building of Pittsburgh, Pennsylvania (Figure 49–14). Conventional sprayed cementitious fire protection was used for interior columns, beams, and floors. However, exterior columns were protected by using hollow box-columns of weathering steel filled with water plus antifreeze and corrosion-inhibiting additives. In the event of fire, the water will absorb heat and keep the temperature of the columns below a critical point. Any steam generated escapes through vents. To prevent excessive hydrostatic pressure, the columns are divided into four separate sections, each sixteen stories high.

Figure 49–14 *U.S. Steel Building contains water-filled columns for fire protection.* (Courtesy United States Steel Corporation.)

COMPUTER-AIDED DESIGN FOR FIRE PROTECTION

Accessory computer programs are used by architectural offices to design residential and commercial buildings so they conform to building code requirements on fire safety. For example, the distance between each room and the nearest exit for safe evacuation can be compared by computer with fire protection standards. Also, the fire ratings of proposed construction material thicknesses can be checked for code violations, alerting the designer to correct oversights or errors in design. A typical program is Instant Software's *Hydraulic Calculations Sprinkler Systems*, which calculates water flow and pressure requirements for sprinkler systems. Refer to the "Architectural Databases" sections of Chapter 6 for more detailed information.

REVIEW QUESTIONS

1. What is the basic premise behind building codes?

2. List the eight building classifications by occupancy as established by the National Fire Protection Association.

3. List the three building classifications by contents as established by the National Fire Protection Association.

4. List the five building classifications by construction type as established by the National Fire Protection Association.

5. What is the difference between:

   a. Fire-resistive and noncombustible construction

   b. Heavy timber, ordinary, and wood frame construction

   c. Wet- and dry-pipe sprinkler systems

6. Give the meaning of *three-hour fire rating*.

7. Give the minimum thickness required to achieve a three-hour fire rating for:

   a. Lightweight CMU walls (with unexpanded slag)

   b. Solid brick masonry walls

   c. Reinforced concrete walls, unplastered

   d. Sprayed mineral fiber on steel columns

   e. Sprayed mineral fiber on steel beams

   f. Sprayed mineral fiber on $2\frac{1}{2}$" concrete slab poured on light-gauge steel decking

8. Differentiate between light-hazard, ordinary-hazard, and extra-hazard sprinkler systems.

9. What are the specifications for a standpipe?

ENRICHMENT QUESTIONS

1. What effects do you think the "building code" instituted by Hammurabi had on the types of structures that were built?

2. Should there be a limit to the amount of fire protection that should be imposed in existing buildings, regardless of cost? Why or why not?

3. Locate the fire codes for your community. Identify five codes that you feel significantly improve the welfare of the occupants.

4. Why are materials fire rated according to hours?

5. Why do you think many of the fire codes refer to sleeping areas as their location?

6. Differentiate between the different classifications of fires a fire extinguisher can put out.

PRACTICAL EXERCISES

1. Check your original building design for compliance with your local exits code or the National Building Exits Code.

2. Using local codes or the National Fire Protection Association codes, draw or computer-produce the fireproofing details for your original building design:

   a. Columns

   b. Beams or girders

   c. Trusses

   d. Floors

   e. Roof

3. Using the National Fire Protection Association codes, lay out the sprinkler system for your original building design. Computer-produce or draw as appropriate.

4. Using the National Fire Protection Association codes, design an NFPA sprinkler system for the first floor of the South Hills Office Building in Chapter 46.

CHAPTER 50

The Construction Documents

OBJECTIVES

By completing this chapter you will be able to:

🏛 Describe the documents and specifications needed for a large architectural project or a single residence.

KEY TERMS

Invitation to Bid

Instructions to Bidders

Bid Form

Owner-Architect Agreement

Performance Bond

Labor and Material Bond

General Conditions

Change Order

In addition to the working drawings, any large project also requires a number of written documents that are needed to advertise for and obtain bids, award a contract, and assure the satisfactory completion of the project. The index of a typical set of construction documents will give an idea of the many different documents required.

I INDEX

Some of the sections listed are comparatively short, simple documents; others (like the specifications) may consist of many hundreds of pages. The specifications are prepared by a *specs* writer who is specially trained to do this work. To give a better idea of the makeup of the various sections of the construction documents, let's look at each in more detail.

II INVITATION TO BID

In public work, bid invitations are mailed to all contractors who might be interested in the proposed project. Also, newspaper advertisements for bids are placed. (Occasionally in private work only selected contractors are invited to bid.) The advertisements are placed three times in three weeks, and they include a brief description of the work and location, together with the requirements (time and place) of bid delivery. A sample **Invitation to Bid** is shown in Figure 50–1. A sample Advertisement for Bids is shown in Figure 50–2.

III INSTRUCTIONS TO BIDDERS

This section gives more detailed information that a bidder needs to intelligently prepare

Jones and Brown, Architects
5555 Main Street
Smithville, Ohio
Phone: 999-555-4444

INVITATION TO BID
STATE UNIVERSITY SCIENCE BUILDING
Project 3813
October ___, _____

You are invited to bid on a General Contract, including mechanical and electrical work, for a two-story, thin-shell concrete, circular Science Building, approximately four hundred feet in diameter. All Bids must be on a lump sum basis; segregated Bids will not be accepted.

The State University Board of Governors will receive Bids until 3:00 p.m. Central Standard Time on Tuesday, November 8, _____, at 233 Uptown Street, Room 313, Smithville, Ohio. Bids received after this time will not be accepted. All interested parties are invited to attend; Bids will be opened publicly and read aloud.

Drawings and Specifications may be examined at the Architect's office and at:

The Plan Center
382 West Third Street
Smithville, Ohio

Associated Plan Bureau
1177 South Barnes
Smithville, Ohio

Copies of the above documents may be obtained at the office of the Architect in accord with the Instructions to Bidders upon depositing the sum of $100.00 for each set of documents.

Any bona-fide bidder, upon returning the documents in good condition immediately following the public opening of said bids, shall be returned his or her deposit in full. Any non-bidder returning the documents in good condition will be returned the sum of $75.00.

Bid Security in the amount of _____ percent of the Bid must accompany each Bid in accord with the Instructions to Bidders.

The Board of Governors reserves the right to waive irregularities and to reject Bids.

    By order of the Board of Governors

    State University
    Smithville, Ohio

    Hirmats J. Downe, Secretary

Figure 50-1   Sample Invitation to Bid. (*Courtesy the Construction Specifications Institute.*)

```
                Bids: November 8, __
                  STATE UNIVERSITY
                  SCIENCE BUILDING
                 SMITHVILLE, OHIO
                   Project 3813
                   October __, __
            Jones and Brown, Architects
                  5555 Main Street
                  Smithville, Ohio
                Phone 999-555-4444
```

The Board of Governors, State University, Smithville, Ohio will receive sealed bids on a General Contract, including mechanical and electrical work, for a two-story, thin-shell concrete, circular Science Building, approximately four hundred feet in diameter.

All Bids must be on a lump sum basis; segregated Bids will not be acceped.

The State University Board of Governors will receive Bids until 3:00 p.m. Central Standard Time on Tuesday, November 8, ____ at 233 Uptown Street, Room 313, Smithville, Ohio. Bids received after this time will not be accepted. All interested parties are invited to attend; Bids will be opened and publicly read aloud.

Drawings and Specifications may be examined at the Architect's office and at:

```
        The Plan Center
        382 West Third Street
        Smithville, Ohio
        Associated Plan Bureau
        117 South Barnes
        Smithville, Ohio
```

Copies of the above documents may be obtained at the office of the Architect in accord with the Instructions to Bidders upon depositing the sum of $100.00 for each set of documents.

Any bona-fide bidder, upon returning the documents in good condition immediately following the public opening of said bids, shall be returned his or her deposit in full. Any non-bidder returning the documents in good condition will be returned the sum of $75.00.

Contracts for work under this bid will obligate the Contractor and subcontractors not to discriminate in employment practices. Bidders must submit a compliance report in conformity with the President's Executive Order No. 11246.

This contract is Federally assisted. The Contractor must comply with the Davis-Bacon Act, the Anti-Kickback Act, and the Contract Work Hours Standards.

Bid Security in the amount of ____ percent of the Bid must accompany each Bid in accord with the Instructions to Bidders.

The Board of Governors reserves the right to waive irregularities and to reject Bids. By order of the Board of Governors

```
        STATE UNIVERSITY
        SMITHVILLE, OHIO
        October __, __
```

Figure 50–2   Sample Advertisement for Bids.
(*Courtesy the Construction Specifications Institute.*)

and submit a bid. The information includes the following:

1. Availability of construction documents
2. Examination of construction documents and site
3. Resolution of questions
4. Approval for substitution of materials
5. Basis of bids
6. Preparation of bids
7. Bid security information
8. Requirements for the Performance Bond and the Labor and Material Bond
9. Requirements for listing any subcontractors
10. Identification and submission of bid
11. Modification or withdrawal of bid
12. Disqualification of bidders
13. Governing laws and regulations
14. Opening of bids
15. Award of contract
16. Execution of contract

A sample **Instructions to Bidders** is shown in Figure 50–3 (pages 668–669).

IV   BID FORM

The **Bid Form** is a sample bidding letter from the bidder to the prospective owner. It contains blank spaces to be filled in by the bidder and a place for the bidder's signature (and for the seal of corporations) to indicate agreement with all provisions. The Bid Form includes the following:

1. Acknowledgment that all construction documents were received by bidder.

2. Agreement statements that bidder will hold bid open until a stated time and that bidder will abide by the Instructions to Bidders.

3. Price of project including price of any alternatives. Alternate bids should be included in addition to the base bid as a means of keeping the project cost within the budget and as a "keep-honest" feature; that is, the bid may

be higher if there are no allowable substitutes for materials for competitive bidding. Some alternate bids for a large project might be the following:

a. Asphalt tile as an alternative to rubber tile in corridors

b. Quarry tile as an alternative to terrazzo on interior floor slabs

c. Asphalt and slag roof as an alternative to a pitch-and-slag roof

d. Cold-mixed bituminous surfacing as an alternative to hot-mixed asphaltic concrete surfacing on driveways and service areas

4. Attachment statement that required information (such as a subcontractor listing or evidence of bidder's qualifications) is enclosed.

A sample Bid Form is shown in Figure 50–4 (page 670).

V AGREEMENT

The Agreement is one of several forms that are preprinted to simplify contract preparation. The contract forms supplied by the American Institute of Architects are commonly used, but many government agencies have developed standard contract forms for their own uses.

The **Owner-Architect Agreement** includes a statement of the architectural services ordinarily considered necessary and the owner's usual obligations. Many different forms of Agreement are used, but they differ mainly in the method by which the architect's compensation is determined. The fee can be as follows:

1. A percentage of the construction cost (usually from 5 to 15 percent)

2. A professional fee plus expenses

3. A multiple of personnel expenses

VI PERFORMANCE BOND

The **Performance Bond** is a guarantee to the client that the contractor will perform all the terms and conditions of the contract, and if defaulted will protect the client up to the bond penalty. The Performance Bond should be distinguished from the Labor and Material Bond, which protects the laborers and material men.

VII LABOR AND MATERIAL BOND

The **Labor and Material Bond** guarantees that the bills of the materials suppliers and subcontractors will be paid.

VIII ESTIMATE OF PAYMENT DUE

A first payment of 10 percent of the architect's fee is paid upon the execution of the Agreement. Additional payments of the fee are made monthly in proportion to the services performed. The total payments are increased to the following percentages at the completion of each phase:

1. Schematic design   15 percent

2. Design development   35 percent

3. Construction documents   75 percent

4. Receipt of bids   80 percent

5. Construction   100 percent

IX GENERAL CONDITIONS

This section is among the most important in the construction documents. The **General Conditions** contain additional contractual-legal requirements not covered by other contract forms. Whereas some architects use A.I.A. Document A-201 without change, others write the General Conditions to satisfy their own requirements. It is also possible to note only those modifications of the A.I.A. Document that apply to each job. Typical subsections are the following:

1. Definitions

2. Architect's supervision

3. Architect's decision

4. Notice

5. Separate contracts

6. Intent of plans and specifications

7. Errors and discrepancies

8. Drawings and specifications furnished to contractors

9. Approved drawings

10. Patents

Jones and Brown, Architects
5555 Main Street
Smithville, Ohio
Phone: 999-555-4444

INSTRUCTIONS TO BIDDERS
STATE UNIVERSITY SCIENCE BUILDING

To be considered, Bids must be made in accord with these instructions to Bidders.

DOCUMENTS. Bonafide prime bidders may obtain ____ sets of Drawings and Specifications from the Architect upon deposit of $ ____ per set. Those who submit prime bids may obtain refund of deposits by returning sets in good condition no more than ____ days after Bids have been opened. Those who do not submit prime bids will forfeit deposits unless sets are returned in good condition at least ____ days before Bids are opened. No partial sets will be issued; no sets will be issued to sub-bidders by the Architect. Prime bidders may obtain additional copies upon deposit of $ ____ per set.

EXAMINATION. Bidders shall carefully examine the documents and the construction site to obtain first-hand knowledge of existing conditions. Contractors will not be given extra payments for conditions which can be determined by examining the site and documents.

QUESTIONS. Submit all questions about the Drawings and Specifications to the Architect, in writing. Replies will be issued to all prime bidders of record as Addenda to the Drawings and Specifications and will become part of the Contract. The Architect and Owner will not be responsible for oral clarification. Questions received less than ____ hours before the bid opening cannot be answered.

SUBSTITUTIONS. To obtain approval to use unspecified products, bidders shall submit written requests at least ten days before the bid date and hour. Requests received after this time will not be considered. Requests shall clearly describe the product for which approval is asked, including all data necessary to demonstrate acceptability. If the product is acceptable, the Architect will approve it in an Addendum issued to all prime bidders on record.

BASIS OF BID. The bidder must include all unit cost items and all alternatives shown on the Bid Forms; failure to comply may be cause for rejection. No segregated Bids or assignments will be considered.

PREPARATION OF BIDS. Bids shall be made on unaltered Bid Forms furnished by the Architect. Fill in all blank spaces and submit two copies. Bids shall be signed with name typed below signature. Where bidder is a corporation, Bids must be signed with the legal name of the corporation followed by the name of the State of incorporation and legal signatures of an officer authorized to bind the corporation to a contract.

BID SECURITY. Bid Security shall be made payable to the Board of Governors, State University, in the amount of ____ percent of the Bid sum. Security shall be either certified check or bid bond issued by surety licensed to conduct business in the State of Ohio. The successful bidder's security will be retained until he or she has signed the Contract and furnished the required payment and performance bonds. The Owner reserves the right to retain the security of the next ____ bidders until the lowest bidder enters into contract or until ____ days after bid opening, whichever is the shorter. All other bid security will be returned as soon as practicable. If any bidder refuses to enter into a Contract, the Owner will retain his or her Bid Security as liquidated damages, but not as a penalty. The Bid Security is to be submitted ____ day(s) prior to the Submission of Bids.

Figure 50–3 *Sample Instructions to Bidders.* (*Courtesy the Construction Specifications Institute.*)

PERFORMANCE BOND AND LABOR AND MATERIAL PAYMENT BOND. Furnish and pay for bonds covering faithful performance of the Contract and payment of all obligations arising thereunder. Furnish bonds in such form as the Owner may prescribe and with a surety company acceptable to the Owner. The bidder shall deliver said bonds to the Owner not later than the date of execution of the Contract. Failure or neglecting to deliver said bonds, as specified, shall be considered as having abandoned the Contract and the Bid Security will be retained as liquidated damages.

SUBCONTRACTORS. Names of principal subcontractors must be listed and attached to the Bid. There shall be only one subcontractor named for each classification listed.

SUBMITTAL. Submit Bid and Subcontractor Listing in an opaque, sealed envelope. Identify the envelope with: (1) project name, (2) name of bidder. Submit Bids in accord with the Invitation to Bid.

MODIFICATION AND WITHDRAWAL. Bids may not be modified after submittal. Bidders may withdraw Bids at any time before bid opening, but may not resubmit them. No Bid may be withdrawn or modified after the bid opening except where the award of Contract has been delayed for _____ days.

DISQUALIFICATION. The Owner reserves the right to disqualify Bids, before or after opening, upon evidence of collusion with intent to defraud or other illegal practices upon the part of the bidder.

GOVERNING LAWS AND REGULATIONS
NON DISCRIMINATORY PRACTICES. Contracts for work under the bid will obligate the contractor and subcontractors not to discriminate in employment practices. Bidders must submit a compliance report in conformity with the President's Executive Order No. 11246.

U.S. GOVERNMENT REQUIREMENTS. This contract is Federally assisted. The Contractor must comply with the Davis-Bacon Act, the Anti-Kickback Act, and the Contract Work Hours Standards.

OHIO EXCISE TAX. Bidders should be aware of the Ohio Law (_____) as it relates to tax assessments on construction equipment.

OPENING. Bids will be opened as announced in the Invitation to Bid.

AWARD. The Contract will be awarded on the basis of low bid, including full consideration of unit prices and alternatives.

EXECUTION OF CONTRACT. The Owner reserves the right to accept any Bid, and to reject any and all Bids, or to negotiate Contract Terms with the various Bidders, when such is deemed by the Owner to be in his best interest.

Each Bidder shall be prepared, if so requested by the Owner, to present evidence of his or her experience, qualifications, and financial ability to carry out the terms of the Contract.

Notwithstanding any delay in the preparation and execution of the formal Contract Agreement, each Bidder shall be prepared, upon written notice of bid acceptance, to commence work within _____ days following receipt of official written order of the Owner to proceed, or on date stipulated in such order.

The accepted bidder shall assist and cooperate with the Owner in preparing the formal Contract Agreement, and within _____ days following its presentation shall execute same and return it to the Owner.

Figure 50–3 (continued)   Sample Instructions to Bidders.

TO: STATE UNIVERSITY SCIENCE BUILDING
The Board of Governors   Project 3813
State University
233 Uptown Street, Room 313
Smithville, Ohio

I have received the documents titled "Specifications for State University Science Building" and Drawings A-1 through A-27, S-1 through S-10, and M-1 through M-15. I have also received Addenda Nos. _____, and have included their provisions in my Bid. I have examined both the documents and the site and submit the following Bid:

In submitting this Bid, I agree:

   1. To hold my bid open until December 8, _____.

   2. To accept the provisions of the Instructions to Bidders regarding disposition of Bid Security.

   3. To enter into and execute a Contract, if awarded on the basis of this bid, and to furnish Guarantee Bonds in accord with Article 30 of the General Conditions of this Contract.

   4. To accomplish the work in accord with the Contract Documents.

   5. To complete the work by the time stipulated in the Supplementary Conditions.

I will construct this project for the lump-sum price of _____ dollars ($_____).

I will include the following alternatives as specified substitutes for the additional costs listed:

   1. Elevators Nos. 5 and 6   +$_____

   2. Steam pipe system   +$_____

If the following items, which are based on unit prices, vary more than 10 percent from the estimates furnished by the Architect, I will adjust the Contract Sum in accord with the following rates:

Concrete piling   +$_____   -$_____

Interior gypsum partitions,
including plaster and paint,
per square foot   +$_____   -$_____

I have attached the required Bid Security and Subcontractor Listing to this Bid.

    Date:_____   Signed:_____

Figure 50–4   Sample Bid Form. (*Courtesy the Construction Specifications Institute.*)

11. Permits, licenses, and certificates

12. Supervision and labor

13. Public safety and guards

14. Order of completion

15. Substitution of materials for those called for by specifications

16. Materials, equipment, and labor

17. Inspection

18. Defective work and materials

19. Failure to comply with orders of architect

20. Use of completed parts

21. Rights of various interests

22. Suspension of work due to unfavorable conditions

23. Suspension of work due to fault of contractor

24. Suspension of work due to unforeseen causes

25. Request for extension

26. Stoppage of work by architect

27. Default on part of contractor

28. Removal of equipment

29. Monthly estimates and payments

30. Acceptance and final payment

31. Deviations from contract requirements

32. Estoppel and waiver of legal rights

33. Approval of subcontractors and sources of material

34. Approval of material samples requiring laboratory tests

35. Arbitration

36. Bonds

37. Additional or substitute bonds

38. Public liability and property damage insurance

39. Workmen's Compensation Act

40. Fire insurance and damage due to other hazards

41. Explosives and blasting

42. Damages to property

43. Mutual responsibility of contractors

44. Contractor's liability

45. Familiarity with contract documents

46. Shop drawings

47. Guarantee of work

48. Cleanup

49. Competent workmen (state law)

50. Prevailing wage act (state law)

51. Residence of employees

52. Nondiscrimination in hiring employees (state law)

53. Preference to employment of war veterans (state law)

54. Hiring and conditions of employment (state law)

X   SUPPLEMENTARY CONDITIONS

The Supplementary Conditions contain special modifications to the basic articles of the General Conditions, together with any additional articles of a contractual legal nature that might be needed for a particular project.

XI   SPECIFICATIONS

The Specifications give detailed instructions on the required materials, finishes, and workmanship—all grouped by building trades. Each trade is included in the order of actual construction. Nearly all offices use the standardized specification system as recommended by the Construction Specifications Institute. This system is called the *CSI Format* and consists of sixteen *Divisions* (grouped by building trades) and a number of related *Broadscope Sections* (grouped by units of work). The *Divisions* are shown in Table 50–1, and the *Broadscope Sections* of the first ten Divisions are shown in Figure 50–5. The Broadscope Sections (printed in capital letters) are further refined into *Narrowscope Sections*. The *Narrowscope Sections* (printed in lowercase letters) of Division

TABLE 50–1 *The Divisions of the CSI Format*

DIVISION 1—GENERAL REQUIREMENTS
DIVISION 2—SITEWORK
DIVISION 3—CONCRETE
DIVISION 4—MASONRY
DIVISION 5—METALS
DIVISION 6—WOOD AND PLASTICS
DIVISION 7—THERMAL AND MOISTURE PROTECTION
DIVISION 8—DOORS AND WINDOWS
DIVISION 9—FINISHES
DIVISION 10—SPECIALTIES
DIVISION 11—EQUIPMENT
DIVISION 12—FURNISHINGS
DIVISION 13—SPECIAL CONSTRUCTION
DIVISION 14—CONVEYING SYSTEMS
DIVISION 15—MECHANICAL
DIVISION 16—ELECTRICAL

8 are shown in Figure 50–6. A specifications writer will select only those sections that apply to a particular job. Notice that a five-digit numbering system is used for the designation of all Broadscope and Narrowscope Sections. This helps offices that use automated printing and data retrieval systems.

The Construction Specifications Institute also recommends a uniform three-part approach for writing each section:

Part 1—General

Description

Quality Assurance

Submittals

Product Delivery, Storage, and Handling

Job Conditions

Alternatives

Guarantee

Part 2—Products

Materials

Mixes

Fabrication and Manufacture

Part 3—Execution

Inspection

Preparation

Installation/Application/Performance

Field Quality Control

Adjust and Clean

Schedules

Although architects still refer to "writing" specifications, actually the majority of sections are assembled from a computer-assisted library of carefully worded and approved paragraphs. See the specifications section in Chapter 6 (Computer-Aided Drafting) for additional detail.

Three basic sentence structures are commonly used in specifications to convey the architect's intent clearly and concisely: the indicative mood, the imperative mood, and streamlining.

The indicative mood, requiring the use of *shall* in nearly every sentence, is the traditional language of specs writing: "Two coats of paint shall be applied to each exposed surface."

The imperative mood is more concise. A verb begins the sentence and immediately defines the required action: "Apply two coats of paint to each exposed surface."

Streamlining is used to itemize products, materials, and reference standards:

"Materials shall meet the following requirements:

Portland cement: ASTM C 150, Type I.
Aggregate: ASTM C 33."

Some additional rules of thumb for specifications writing are the following:

1. Use short sentences and simple declarative statements.

2. Avoid complicated sentences whose meanings are so dependent on punctuation that inadvertent omission or insertion of punctuation changes the meaning or creates ambiguity.

SPECIFICATIONS

DIVISION 1—GENERAL REQUIREMENTS

01010	SUMMARY OF WORK
01020	ALLOWANCES
01025	MEASUREMENT AND PAYMENT
01030	ALTERNATES/ALTERNATIVES
01040	COORDINATION
01050	FIELD ENGINEERING
01060	REGULATORY REQUIREMENTS
01070	ABBREVIATIONS AND SYMBOLS
01080	IDENTIFICATION SYSTEMS
01090	REFERENCE STANDARDS
01100	SPECIAL PROJECT PROCEDURES
01200	PROJECT MEETINGS
01300	SUBMITTALS
01400	QUALITY CONTROL
01500	CONSTRUCTION FACILITIES AND TEMPORARY CONTROLS
01600	MATERIAL AND EQUIPMENT
01650	STARTING OF SYSTEMS/COMMISSIONING
01700	CONTRACT CLOSEOUT
01800	MAINTENANCE

DIVISION 2—SITEWORK

02010	SUBSURFACE INVESTIGATION
02050	DEMOLITION
02100	SITE PREPARATION
02140	DEWATERING
02150	SHORING AND UNDERPINNING
02160	EXCAVATION SUPPORT SYSTEMS
02170	COFFERDAMS
02200	EARTHWORK
02300	TUNNELING
02350	PILES AND CAISSONS
02450	RAILROAD WORK
02480	MARINE WORK
02500	PAVING AND SURFACING
02600	PIPED UTILITY MATERIALS
02660	WATER DISTRIBUTION
02680	FUEL DISTRIBUTION
02700	SEWERAGE AND DRAINAGE
02760	RESTORATION OF UNDERGROUND PIPELINES
02770	PONDS AND RESERVOIRS
02780	POWER AND COMMUNICATIONS
02800	SITE IMPROVEMENTS
02900	LANDSCAPING

DIVISION 3—CONCRETE

03100	CONCRETE FORMWORK
03200	CONCRETE REINFORCEMENT
03250	CONCRETE ACCESSORIES
03300	CAST-IN-PLACE CONCRETE
03370	CONCRETE CURING
03400	PRECAST CONCRETE
03500	CEMENTITIOUS DECKS
03600	GROUT
03700	CONCRETE RESTORATION AND CLEANING
03800	MASS CONCRETE

DIVISION 4—MASONRY

04100	MORTAR
04150	MASONRY ACCESSORIES
04200	UNIT MASONRY
04400	STONE
04500	MASONRY RESTORATION AND CLEANING
04550	REFRACTORIES
04600	CORROSION RESISTANT MASONRY

DIVISION 5—METALS

05010	METAL MATERIALS
05030	METAL FINISHES
05050	METAL FASTENING
05100	STRUCTURAL METAL FRAMING
05200	METAL JOISTS
05300	METAL DECKING
05400	COLD-FORMED METAL FRAMING
05500	METAL FABRICATIONS
05580	SHEET METAL FABRICATIONS
05700	ORNAMENTAL METAL
05800	EXPANSION CONTROL
05900	HYDRAULIC STRUCTURES

DIVISION 6—WOOD AND PLASTICS

06050	FASTENERS AND ADHESIVES
06100	ROUGH CARPENTRY
06130	HEAVY TIMBER CONSTRUCTION
06150	WOOD-METAL SYSTEMS
06170	PREFABRICATED STRUCTURAL WOOD
06200	FINISH CARPENTRY
06300	WOOD TREATMENT
06400	ARCHITECTURAL WOODWORK
06500	PREFABRICATED STRUCTURAL PLASTICS
06600	PLASTIC FABRICATIONS

DIVISION 7—THERMAL AND MOISTURE PROTECTION

07100	WATERPROOFING
07150	DAMPPROOFING
07190	VAPOR AND AIR RETARDERS
07200	INSULATION
07250	FIREPROOFING
07300	SHINGLES AND ROOFING TILES
07400	PREFORMED ROOFING AND CLADDING/SIDING
07500	MEMBRANE ROOFING
07570	TRAFFIC TOPPING
07600	FLASHING AND SHEET METAL
07700	ROOF SPECIALTIES AND ACCESSORIES
07800	SKYLIGHTS
07900	JOINT SEALERS

DIVISION 8—DOORS AND WINDOWS

08100	METAL DOORS AND FRAMES
08200	WOOD AND PLASTIC DOORS
08250	DOOR OPENING ASSEMBLIES
08300	SPECIAL DOORS
08400	ENTRANCES AND STOREFRONTS
08500	METAL WINDOWS
08600	WOOD AND PLASTIC WINDOWS
08650	SPECIAL WINDOWS
08700	HARDWARE
08800	GLAZING
08900	GLAZED CURTAIN WALLS

DIVISION 9—FINISHES

09100	METAL SUPPORT SYSTEMS
09200	LATH AND PLASTER
09230	AGGREGATE COATINGS
09250	GYPSUM BOARD
09300	TILE
09400	TERRAZZO
09500	ACOUSTICAL TREATMENT
09540	SPECIAL SURFACES
09550	WOOD FLOORING
09600	STONE FLOORING
09630	UNIT MASONRY FLOORING
09650	RESILIENT FLOORING
09680	CARPET
09700	SPECIAL FLOORING
09780	FLOOR TREATMENT
09800	SPECIAL COATINGS
09900	PAINTING
09950	WALL COVERINGS

DIVISION 10—SPECIALTIES

10100	CHALKBOARDS AND TACKBOARDS
10150	COMPARTMENTS AND CUBICLES
10200	LOUVERS AND VENTS
10240	GRILLES AND SCREENS
10250	SERVICE WALL SYSTEMS
10260	WALL AND CORNER GUARDS
10270	ACCESS FLOORING
10280	SPECIALTY MODULES
10290	PEST CONTROL
10300	FIREPLACES AND STOVES
10340	PREFABRICATED EXTERIOR SPECIALTIES
10350	FLAGPOLES
10400	IDENTIFYING DEVICES
10450	PEDESTRIAN CONTROL DEVICES
10500	LOCKERS
10520	FIRE PROTECTION SPECIALTIES
10530	PROTECTIVE COVERS
10550	POSTAL SPECIALTIES
10600	PARTITIONS
10650	OPERABLE PARTITIONS
10670	STORAGE SHELVING
10700	EXTERIOR SUN CONTROL DEVICES
10750	TELEPHONE SPECIALTIES
10800	TOILET AND BATH ACCESSORIES
10880	SCALES
10900	WARDROBE AND CLOSET SPECIALTIES

Figure 50–5   *Some Broadscope Sections of the CSI Format.* (Courtesy the Construction Specifications Institute.)

DIVISION 8—DOORS AND WINDOWS

Section Number	Title
08100	**METAL DOORS AND FRAMES**
-110	Steel Doors and Frames
	Standard Steel Doors and Frames
	Standard Steel Doors
	Standard Steel Frames
	Custom Steel Doors and Frames
	Custom Steel Doors
	Custom Steel Frames
-115	Packaged Steel Doors and Frames
-120	Aluminum Doors and Frames
-130	Stainless Steel Doors and Frames
-140	Bronze Doors and Frames
08200	**WOOD AND PLASTIC DOORS**
-210	Wood Doors
	Flush Wood Doors
	Plastic Faced Flush Wood Doors
	Metal Faced Wood Doors
	Panel Wood Doors
-220	Plastic Doors
08250	**DOOR OPENING ASSEMBLIES**
08300	**SPECIAL DOORS**
-305	Access Doors
-310	Sliding Doors
	Sliding Glass Doors
	Sliding Metal Fire Doors
	Sliding Grilles
-315	Blast-Resistant Doors
-316	Security Doors
-320	Metal-Clad Doors
-325	Cold Storage Doors
-330	Coiling Doors
	Coiling Counter Doors
	Overhead Coiling Doors
	Side Coiling Doors
-340	Coiling Grilles
	Overhead Coiling Grilles
	Side Coiling Grilles
-350	Folding Doors and Grilles
	Accordion Folding Doors
	Panel Folding Doors
	Accordion Folding Grilles
-355	Flexible Doors
-360	Sectional Overhead Doors
-365	Multi-leaf Vertical Lift Overhead Doors
	Vertical Lift Telescoping Doors
-370	Hangar Doors
-380	Sound Retardant Doors
-385	Safety Glass Doors
-390	Screen and Storm Doors
-395	Flood Barrier Doors
-396	Chain Closures
08400	**ENTRANCES AND STOREFRONTS**
-410	Aluminum Entrances and Storefronts
-420	Steel Entrances and Storefronts
-430	Stainless Steel Entrances and Storefronts
-440	Bronze Entrances and Storefronts
-450	All-Glass Entrances
-460	Automatic Entrance Doors
-470	Revolving Entrance Doors

Section Number	Title
08500	**METAL WINDOWS**
-510	Steel Windows
-520	Aluminum Windows
-530	Stainless Steel Windows
-540	Bronze Windows
-550	Metal Jalousie Windows
-560	Metal Storm Windows
08600	**WOOD AND PLASTIC WINDOWS**
-610	Wood Windows
	Metal Clad Wood Windows
	Plastic Clad Wood Windows
	Wood Storm Windows
-630	Plastic Windows
	Reinforced Plastic Windows
	Plastic Storm Windows
08650	**SPECIAL WINDOWS**
-655	Roof Windows
-660	Security Windows
-665	Pass Windows
08700	**HARDWARE**
-710	Finish Hardware
-720	Operators
	Automatic Door Operators
	Window Operators
-730	Weatherstripping and Seals
	Thresholds
-740	Electrical Locking Systems
-750	Door and Window Accessories
	Flood Barriers
08800	**GLAZING**
-810	Glass
	Mirror Glass
-840	Plastic Glazing
-850	Glazing Accessories
08900	**GLAZED CURTAIN WALLS**
-910	Glazed Steel Curtain Walls
-920	Glazed Aluminum Curtain Walls
-930	Glazed Stainless Steel Curtain Walls
-940	Glazed Bronze Curtain Walls
-950	Translucent Wall and Skylight Systems
-960	Sloped Glazing Systems
-970	Structural Glass Curtain Walls

Figure 50–6 *Some Narrowscope Sections of the CSI Format.* (*Courtesy the Construction Specifications Institute.*)

Name: Howard Abrams　　*Age:* 63
Gender: Male
Occupational Title: Zoning/Code Enforcement Officer
Employer: Indiana Borough
　　　　　Indiana, PA 15701
Years of Experience: $11\frac{1}{2}$ years
Educational Background:
　　　Bachelor of Arts, the Pennsylvania State University
Approximate Starting Salary: $25,000 plus benefits
JOB OVERVIEW:
　　I cannot describe a typical day on my job because no two days are the same. My responsibilities include issuing building permits, checking construction to make sure it meets code, inspecting unrelated individuals' dwelling units, checking for garbage and litter, and much more. I must also attend borough council, planning commission, and zoning hearing board meetings. Being a member of the Indiana Borough Planning Commission for twenty years provided me with experience needed to perform my present job.
　　The major benefit in my job is being involved in the development of the community. Helping to ensure that this development is completed in an orderly fashion is one of my greatest challenges. Another challenge is to protect property values.

3. Choose words and terms that are plain and well understood to convey the information. Avoid pompous or highly embellished language. For example, use "shall" rather than "it is incumbent upon" or "it is the duty." Use "the contractor may" rather than "if the contractor so elects, he may" or "the contractor is hereby authorized to." Use "means" rather than "shall be interpreted to mean." Use "by" rather than "by means of." Use "to" rather than "in order to." Never use "herein," "hereinbefore," "hereinafter," or "wherein." Avoid using "and/or," "etc.," and "as per."

4. Use "shall" for the work of the contractor. Use "will" for acts of the owner or architect. Do not use "must."

5. Use numerals (figures) instead of words for numbers over twelve. For example: one, six, twelve, 13, 18, 100. But use numerals for all sums of money, e.g., $1.00. Give numbers preceding a numeral as words, e.g., fifteen 8-hour days.

To get a better idea of the specifications, let's look at one of the sections in more detail. We have chosen Division 8 (Doors and Windows), Broadscope Section 08800 (Glazing) for this study.

Section 08800 Glazing

08801　STIPULATION
Applicable requirements of the "General Conditions" apply to this entire Specification, and shall have the same force and effect as if printed here in full.

08802　SCOPE OF WORK
The work covered by this Section consists of furnishing all labor, materials, equipment, and services necessary to

complete all glass and glazing required for the project, in strict accordance with this Section of the Specifications and the Drawings; including, but not limited to, the following:

a. Glazing of exterior doors, sidelights, transoms, and fixed metal window frames;
b. Glazing of interior doors, sidelights, and frames;
c. Mirrors

08803 WORK EXCLUDED

The following items are included in other sections of the General Contract Specifications:
a. All bank equipment shall be factory glazed.

08810 GLASS

All glass shall comply with Federal Specification DD-G-45a for glass, flat, for glazing purposes.

08813 TEMPERED GLASS (Exterior doors and side-lights at doors)

Tempered glass for the above locations shall be "Solarbronze Twindow" with $\frac{1}{4}$" polished plate Solarbronze exterior sheet and $\frac{1}{4}$" clear tempered plate interior sheet. Glass in doors shall be $\frac{13}{16}$" thick and $\frac{1}{4}$" air space. Other glass shall be $1\frac{1}{16}$" thick with $\frac{1}{2}$" air space. Set in metal glazing beads.

08823 INSULATING GLASS (Fixed exterior windows and transoms in aluminum frames)

Insulating glass shall be $1\frac{1}{16}$" thick "Solarbronze Twindow" set in metal glazing beads. Glass shall have a $\frac{1}{4}$" polished plate Solarbronze exterior sheet, $\frac{1}{2}$" air space, and $\frac{1}{4}$" clear polished plate interior sheet.

08830 MIRRORS

Over lavatories in toilet rooms, provide and install mirrors. Each mirror shall be of size indicated on the Drawings, equal to No. 53020, as manufactured by the Charles Parker Company, 50 Hanover Street, Meriden, Connecticut, complete with $\frac{1}{4}$" polished plate glass, moisture proof backing, removable back, narrow channel type plated brass or stainless steel frame, with concealed vandalproof mirror hangers. Mirrors shall be centered over lavatories, and set at height shown on Drawings or as directed by the Architect.

08840 GLAZING COMPOUND

Glazing compound for bedding glazing, Federal Specification TT-P-791a, Type I, elastic glazing compound. Glazing compound shall be specially prepared for the purpose, tinted to match frames, and shall remain plastic under a strong surface film similar to the product manufactured by "Tremco," "Pecora," or "Kuhls." *No putty will be accepted* (glass in doors and windows shall be set in glazing compound secured by glazing beads).

08841 SAMPLES

Samples of each type of glass and glazing compound shall be submitted for approval of the Architect.

08842 SETTING

All glass shall be properly bedded in glazing compound previously specified. Glazing compound shall not be applied in temperatures below 40°F, or during damp or rainy weather. Surfaces shall be dry and free of dust, dirt, or rust.

Glazing compound shall be used as it comes from the container without adulteration and only after thorough mixing. If thinning is required, use only such type of thinner as recommended by the manufacturer.

08843 REPLACEMENT AND CLEANING

Upon completion of the glazing, all glass shall be thoroughly cleaned, any paint spots and labels and other defacements removed, and all cracked, broken, and imperfect glass, or glass which cannot be properly cleaned, shall be replaced by perfect glass.

At the time of acceptance of the building, all glass shall be clean, whole, and in perfect condition, including glazing compound. Glazing compound applied after completion of painting shall be painted not less than two (2) coats.

08844 LABELS

Each light shall bear the manufacturer's label indicating the name of the manufacturer and the strength and quality of the glass. Labels shall remain in place until after final acceptance of the building, at which time the labels shall be removed and glass shall be given its final cleaning.

Computer-produced Specifications

Specifications are prepared, stored, and retrieved using word processing or computer equipment. Most architectural firms purchase diskettes of basic specifications such as *Spectext* by CSI and *Masterspec* by AIA. These commercial specifications are then revised to accommodate the specialties, experiences, and quality control of each firm. The specifications for each new building project are assembled from these master files with customized additions and deletions. After careful editing, dating, and an automatic spelling check, the completed specifications are printed in quantity.

XII WORKING DRAWINGS

The Working Drawings together with the Specifications are the most important parts of the documents constituting the contract. Information on the design, location, and dimensions of the elements of a building is found on Working Drawings, and information on the quality of materials and workmanship is found in the Specifications. A good Working Drawing gives the contractor the exact information he or she needs, is clear and simple, arranged in an orderly manner, and accurately drawn so that scaled measurements will agree with dimensions. (See Chapter 23 for more information.)

ADDENDA AND CHANGE ORDERS

Addenda and Change Orders are used to correct or change the original construction documents. The main difference between an Addendum and a Change Order is the timing. An Addendum revises the original construction documents *before* the contract is awarded, and a **Change Order** is a revision *after* award of the contract. A sample Addendum is shown in Figure 50–7, and a sample Change Order is shown in Figure 50–8.

CONSTRUCTION ADMINISTRATION

Construction administration is one of the major responsibilities of the architect. This includes periodic on-site inspections during construction, job conferences, review of Shop Drawings, Change Orders during construction, and, finally, the review and approval of Requisitions of Payment to the contractor.

REAL WORLD APPLICATIONS

Robert's Rules

Your first company meeting was so confusing you could barely understand anything that happened. You thought that they were going to discuss new changes to the company's established guidelines for the development of construction documents. Instead, people made motions, amended them, called for a question, said "second" when no one said "first," and some were even "out of order." What you heard were the proceedings of a meeting using Robert's Rules of Order. These practices were recommended by Major Henry Robert in 1876 to provide an orderly way to run a parliamentary meeting. Robert's Rules have gained world-wide acceptance for the conduct of all types of meetings, but unfortunately Robert's Rules may also be misused to defeat the will of the assembly. Major Robert foresaw this possibility and stated in the preface:

"While it is important to every person in a free country to know something of parliamentary law, this knowledge should be used only to help, not to hinder business. One who is constantly raising points of order and insisting upon a strict observance of every rule in a peaceable assembly in which most of the members are unfamiliar with these rules and customs, makes himself a nuisance, hinders business, and prejudices people against parlimentary law. Such a person either does not understand its real purpose or else wilfully misuses his knowledge."

Obtain a copy of *Robert's Rules of Order*. What is a quorum? What procedure is followed to make a motion, add an amendment to it, and then approve it? How is a meeting officially ended?

JONES AND SMITH, Architects: John Doe Bldg.
                    Washington, D. C.
First National Bank of Brownsville: Project No. 11863

ADDENDUM NO. 2: August 15, _____

To: All prime contract bidders of record.

This Addendum forms a part of the Contract Documents and modifies the original
specifications and drawings, dated July 1, _____, and Addendum 1, dated August
1, ____, as noted below. Acknowledge receipt of this Addendum in the space provided
on the Bid Form. Failure to do so may subject bidder to disqualification.

This Addendum consists of _____. (Indicate the number of pages and any
attachments or drawings forming a part of the addendum.)

ADDENDUM NO. 1

1. Drawings, page AD 1-1. In line 3, number of the referenced Drawing is changed
from "G-1" to "G-7."

INSTRUCTIONS TO BIDDERS
2. Proposals. The first sentence is changed to read: "Proposed substitutions must
be submitted in writing at least 15 days before the date for opening of bids."

GENERAL CONDITIONS
3. Article 13, Access to Work. The following sentence is added: "Upon completion
of work, the Contractor shall deliver to the Architect all required Certificates
of Inspection."

SUPPLEMENTARY CONDITIONS
4. Article 19, Correction of Work Before Substantial Completion. This Article is
deleted and the following is inserted in its place: "If proceeds of sale do not
cover expenses that the Contractor should have borne, the Contractor shall pay the
difference to the Owner."

SPECIFICATIONS
5. Division 7
Waterproofing: Page 4, following Paragraph 7C-02 Materials, add the following:
"(d) Option. Factory mixed waterproofing containing metallic waterproofing, sand
and cement, all meeting the above requirements, may be used in lieu of job-mixed
waterproofing."

6. Division 15
Refrigeration: Page 10, Paragraph 4—Chillers item "a" Line 4: Change total square
feet of surface from 298 to 316.

Liquid Heat Transfer: Page 17, Paragraph 10—Convectors item "b" Line 3: Delete "as
selected—or owner."
Page 23. Paragraph 13—Wall Fin: Omit entirely.

DRAWINGS
7. S-9, Beam Schedule. For B-15 the following is added: "Size, 12 × 26; Straight,
3 - #6; Bent, 2 - #8, Top Over Columns: 3 - #7."

8. M-1: At room 602 change 12 × 6 exhaust duct to 12 × 18; at room 602 add a roof
ventilator. See print H-1R attached and page 16, paragraph 13 Roof Ventilators
addenda above.

Figure 50–7   Sample Addendum. (Courtesy the Construction Specifications Institute.)

```
JONES AND SMITH, Architects/Engineers John Doe Bldg.,
                Washington, D.C.
CHANGE ORDER NO. 5: September 9, _____
JOB NO. 11863: First National Bank of Brownsville
OWNER: ABC Corp., Brownsville, Virginia
CONTRACTOR: Bildum Construction Co., Washington, D.C.
CONTRACT DATE: July 4, _____


TO THE CONTRACTOR: You are hereby authorized, subject to Contract
provisions, to make the following changes:
Bulletin No. 1                                    ADD $ 73.24
Bulletin No. 2                                    ADD   138.07
Bulletin No. 3 No Charge,/No Credit
Bulletin No. 4                          DEDUCT $ 75.32
Bulletin No. 5                          DEDUCT   36.99
TOTAL                                   DEDUCT $112.31   ADD $211.31
NET ADD $99.00


ORIGINAL CONTRACT AMOUNT:         $1,234,567.89
PRIOR CHANGE ORDERS (+, -):          +2,000.00
THIS CHANGE ORDER (+, -):               +99.00
REVISED CONTRACT AMOUNT:          $1,236,666.89


TIME EXTENSION/REDUCTION: None


OTHER CONTRACTS AFFECTED: None


SUBMITTED BY: _____   DATE: _____
                (arch/engr's signature)


APPROVED BY: _____   DATE: _____
               (owner's signature)


ACCEPTED BY: _____   DATE: _____
              (contractor's signature)


DISTRIBUTION: Owner, Contractor, Architect/Engineer,
              Field Representative, Other _____.
```

Figure 50–8   *Sample Change Order.* (*Courtesy the Construction Specifications Institute.*)

REVIEW QUESTIONS

1. Why is it important to prepare specifications for every architectural project?
2. What does it mean to have items on bid?
3. List the twelve documents included in a typical set of specifications.
4. List the sixteen divisions of a typical set of specifications.
5. Why must specifications be defined clearly?
6. Explain how specifications are categorized and numbered.
7. Give the reasons for including the following documents in a set of construction documents:
   a. Specifications
   b. Working drawings
   c. Agreement
   d. Bid form
   e. Performance bond
   f. Labor and material bond

ENRICHMENT QUESTIONS

1. What process would you use to obtain bidders for a single residence?
2. Obtain a set of specifications from a local architect for classroom study.
3. Contact a local attorney and interview them regarding the type of information that legally needs to accompany a set of construction documents.

PRACTICAL EXERCISES

1. Prepare an outline of the specifications for the building assigned by your instructor.
   a. Include the broadscope section headings only.
   b. Include the broadscope and narrowscope section headings.
   c. Write a detailed specification for the section assigned by your instructor.
   d. Computer-produce specifications as assigned by your instructor.

Appendix A
Plans of a Two-Story Residence and a Solar A-Frame

A single-story contemporary home designed for Mr. and Mrs. A has been used as an example throughout the preceding chapters. (Courtesy of Mr. and Mrs. Donald W. Hamer, State College, Pennsylvania.) In addition, the plans of a split-level home—also of contemporary design—for Mr. and Mrs. M were used to illustrate dimensioning in metric units in Chapter 14. (Courtesy of Professor Emeritus M. Eisenberg, the Pennsylvania State University.)

To provide a wider base of comparison, plans of two quite different homes follow. The Z residence is a two-story traditional house in an English garrison styling. The S residence is an A-framed solar house with direct passive, indirect passive, and active solar features. (Courtesy of Professors Richard E. and Mary J. Kummer, the Pennsylvania State University.)

Plans for a five-story commercial building of bolted and welded steel construction were used to illustrate commercial design-drafting in Chapter 46.

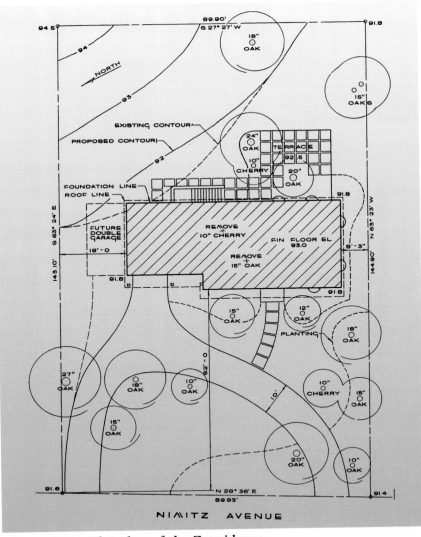

Figure A–1 *Plot plan of the Z residence.*

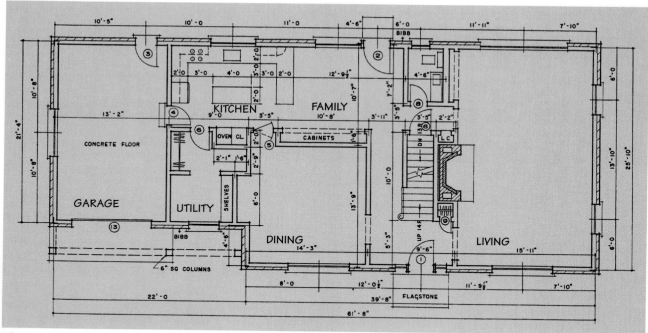

Figure A–2 *First-floor plan of the Z residence.*

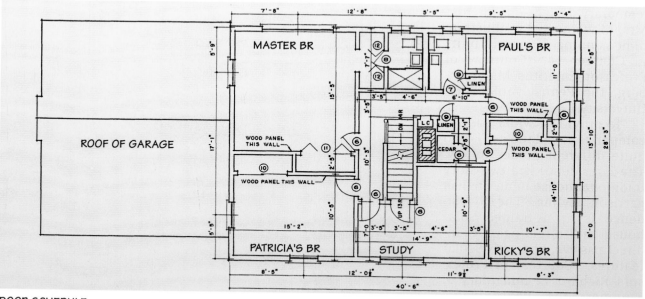

DOOR SCHEDULE

MK	NO	SIZE	DESCRIPTION
1	1	3'-0 x 6'-8" x 1 3/4"	14 PANEL WP, 4 LTS
2	1	2'-8" x 6'-8" x 1 3/4"	FLUSH WP, 1 LT
3	2	2'-8" x 6'-8" x 1 3/4"	2 PANEL WP, 3 LTS
4	1	2'-8" x 6'-8" x 1 3/4"	FLUSH BIRCH
5	1	2'-6" x 6'-8" x 1 3/8"	" " DOUBLE SWINGING
6	10	2'-6" x 6'-8" x 1 3/8"	" "
7	1	2'-4" x 6'-8" x 1 3/8"	" "
8	2	2'-2" x 6'-8" x 1 3/8"	" "
9	3	1'-8" x 6'-8" x 1 3/8"	" "
10	2	5'-4" x 6'-8" x 1 3/8"	" " DOUBLE SLIDING
11	1	6'-0 x 6'-8" x 1 3/8"	LOUVERED DOUBLE FOLDING, WP
12	2	3'-0 x 6'-8" x 1 3/8"	" SINGLE "
13	1	9'-0 x 6'-6" x 1 3/8"	18 PANEL WP, 6 LTS, OVERHD CAR

Figure A–3 *Second-floor plan of the Z residence.*

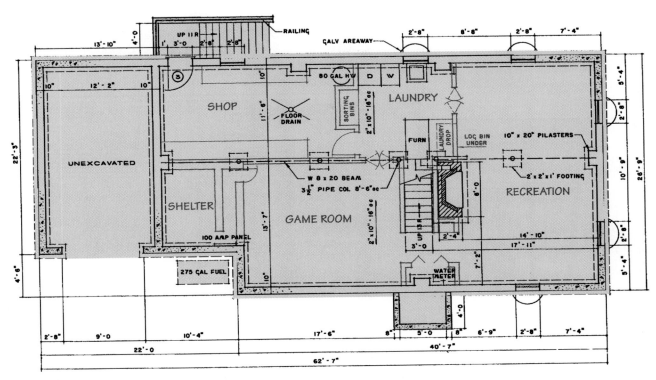

Figure A–4   *Basement plan of the Z residence.*

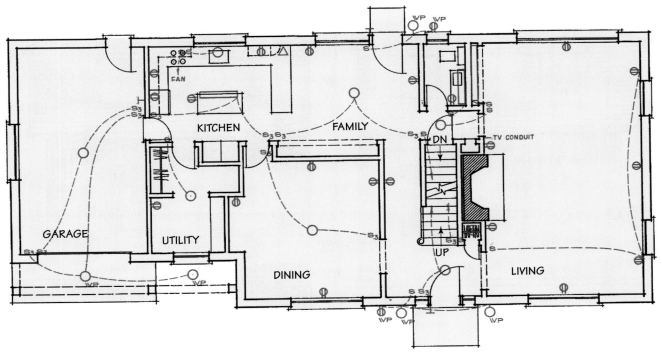

Figure A–5   *Electrical plan of the Z residence.*

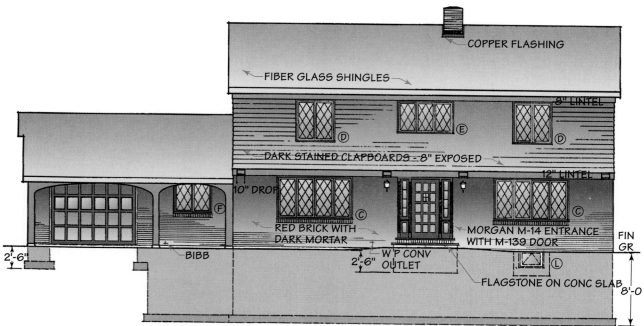

Figure A–6 *Front elevation of the Z residence.*

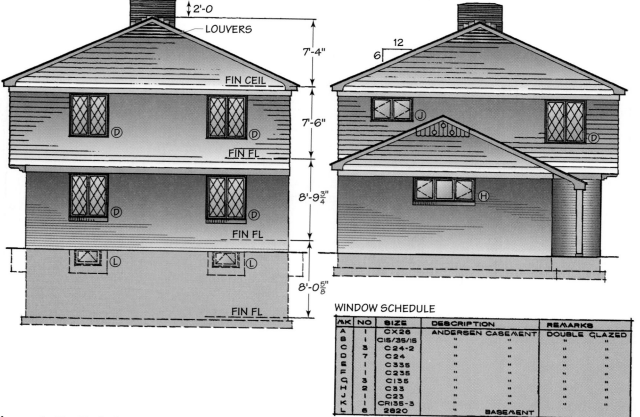

Figure A–7 *End elevations of the Z residence.*

WINDOW SCHEDULE

MK	NO	SIZE	DESCRIPTION	REMARKS
A	1	CX28	ANDERSEN CASEMENT	DOUBLE GLAZED
B	1	C15/35/15	"	"
C	3	C24-2	"	"
D	7	C24	"	"
E	1	C335	"	"
F	1	C235	"	"
G	3	C135	"	"
H	2	C33	"	"
J	1	C23	"	"
K	1	CR135-3	"	"
L	6	2820	"	BASEMENT

Figure A–8   *Rear elevation of the Z residence.*

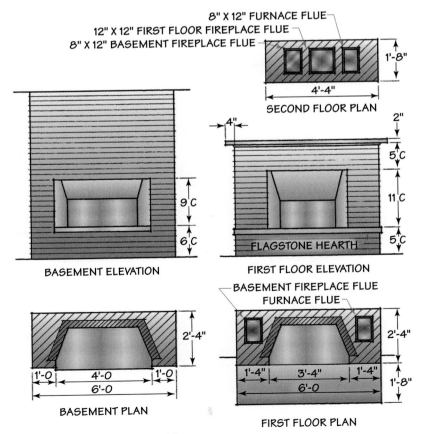

Figure A–9   *Fireplace details of the Z residence.*

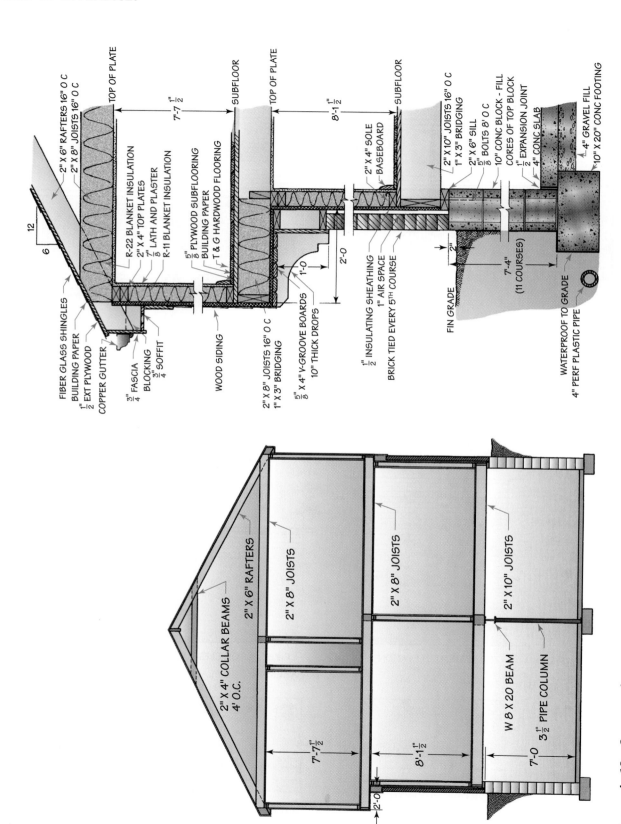

Figure A–11 Typical wall section of the Z residence.

Figure A–10 Structural section of the Z residence.

Figure A–12   *The Z residence as built in Pennsylvania.*

Figure A–13   *Front elevation of the Z residence.*

Figure A–14   *Entrance elevation of the Z residence.*

Figure A–15   *Garage entrances of the Z residence.*

Figure A–16   *Rear elevation of the Z residence.*

Figure A–17   *Patio of the Z residence.*

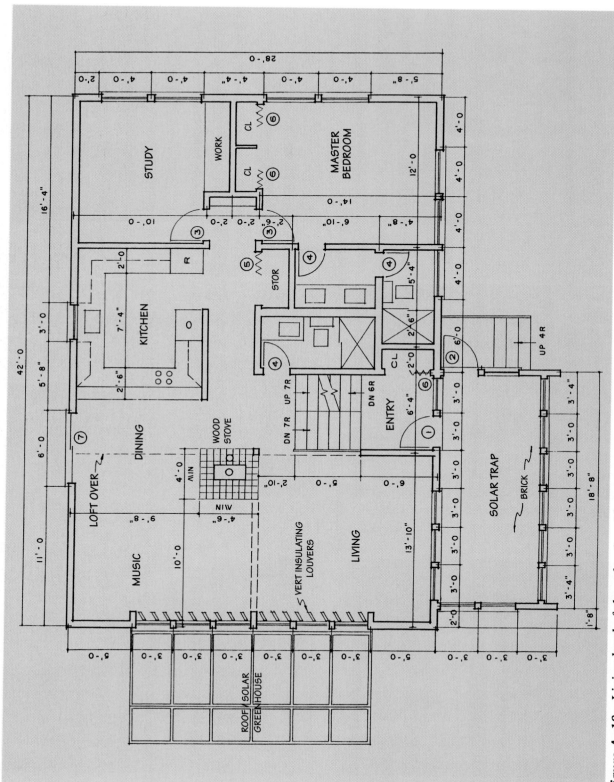

Figure A–18 Living level of the solar S residence.

Figure A–19 Sleeping loft level of the solar S residence.

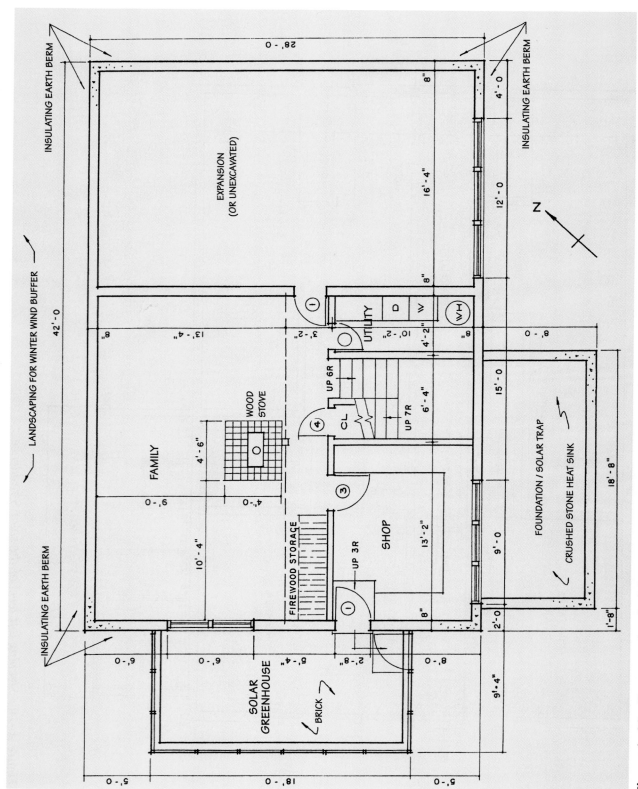

Figure A–20 Basement level of the solar S residence.

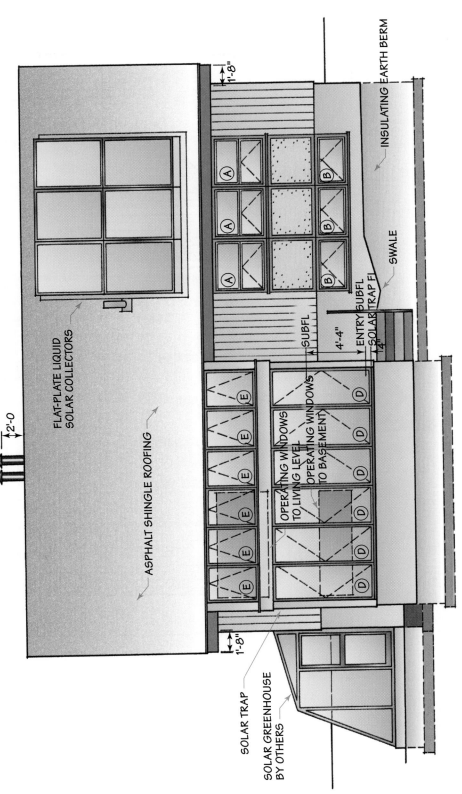

Figure A–21   Front (southeast) elevation of the solar S residence.

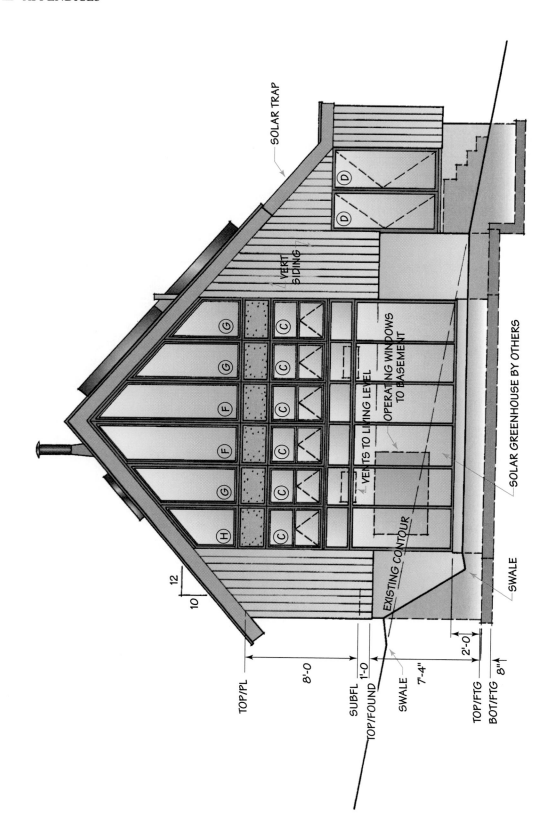

Figure A–22 Left (southwest) elevation of the solar S residence.

Appendix B
Windows

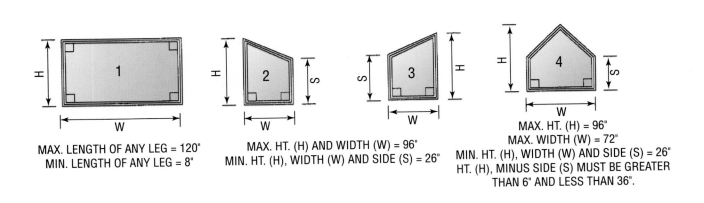

MAX. LENGTH OF ANY LEG = 120"
MIN. LENGTH OF ANY LEG = 8"

MAX. HT. (H) AND WIDTH (W) = 96"
MIN. HT. (H), WIDTH (W) AND SIDE (S) = 26"

MAX. HT. (H) = 96"
MAX. WIDTH (W) = 72"
MIN. HT. (H), WIDTH (W) AND SIDE (S) = 26"
HT. (H), MINUS SIDE (S) MUST BE GREATER
THAN 6" AND LESS THAN 36".

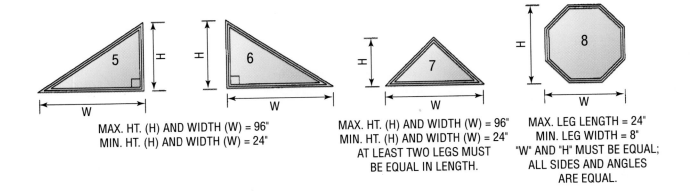

MAX. HT. (H) AND WIDTH (W) = 96"
MIN. HT. (H) AND WIDTH (W) = 24"

MAX. HT. (H) AND WIDTH (W) = 96"
MIN. HT. (H) AND WIDTH (W) = 24"
AT LEAST TWO LEGS MUST
BE EQUAL IN LENGTH.

MAX. LEG LENGTH = 24"
MIN. LEG WIDTH = 8"
"W" AND "H" MUST BE EQUAL;
ALL SIDES AND ANGLES
ARE EQUAL.

Figure B–1 *Pella® fixed frame clad windows—size parameters.*

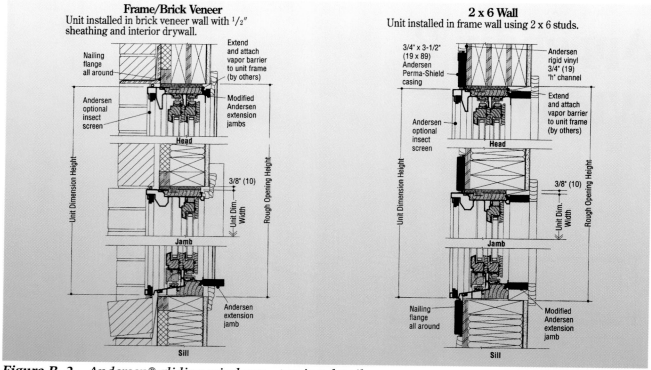

Figure B–2 *Andersen® gliding windows—tracing details.*

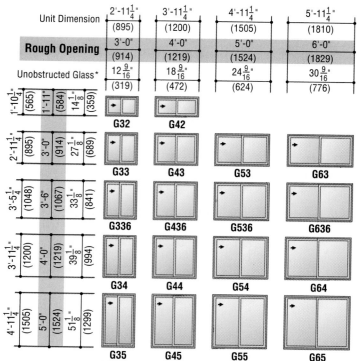

Figure B–3 *Andersen® gliding windows—table of sizes.*

* Unobstructed glass measurement is for single sash only.

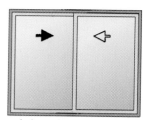

Active sash   Passive sash

Venting Configuration

NOTE: All venting indicated as viewed from the exterior.

Dimensions in parentheses are metric equivalents, shown in millimeters.

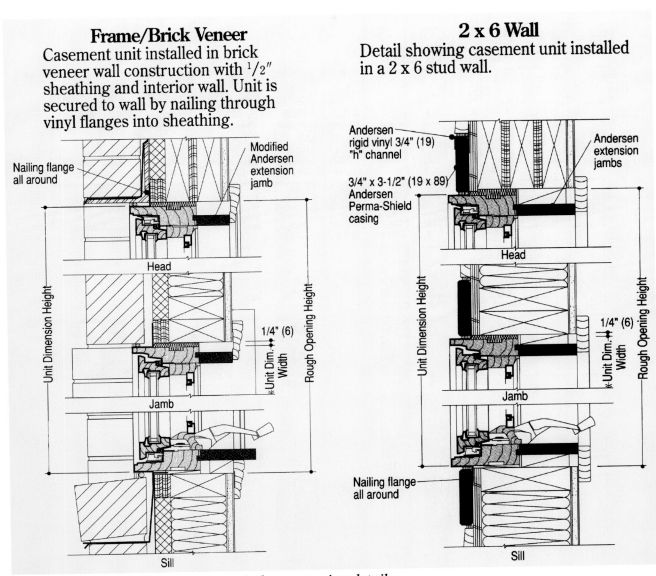

Frame/Brick Veneer
Casement unit installed in brick veneer wall construction with $1/2''$ sheathing and interior wall. Unit is secured to wall by nailing through vinyl flanges into sheathing.

Nailing flange all around

Modified Andersen extension jamb

Head

Jamb

1/4" (6)

Unit Dimension Height

Unit Dim. Width

Rough Opening Height

Sill

2 x 6 Wall
Detail showing casement unit installed in a 2 x 6 stud wall.

Andersen rigid vinyl 3/4" (19) "h" channel

3/4" x 3-1/2" (19 x 89) Andersen Perma-Shield casing

Andersen extension jambs

Head

Jamb

1/4" (6)

Nailing flange all around

Unit Dimension Height

Unit Dim. Width

Rough Opening Height

Sill

Figure B–4   *Andersen® casement windows—tracing details.*

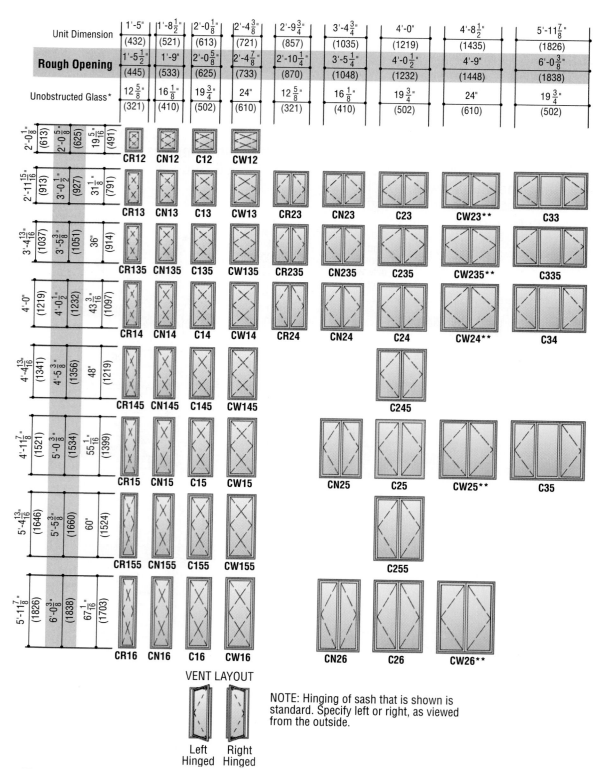

Figure B–5 *Andersen® casement windows—table of sizes.*

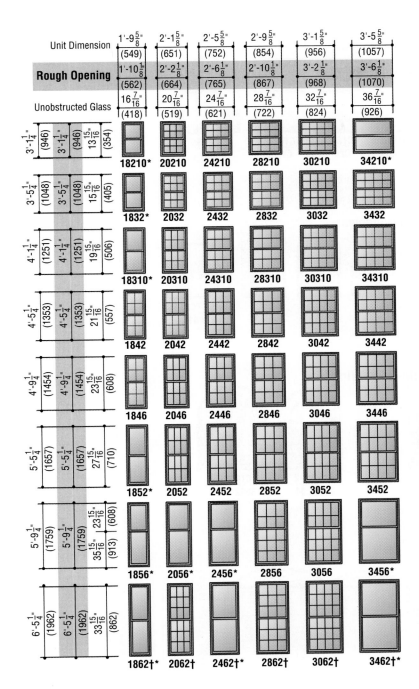

NOTE: Unobstructed glass measurement is for single sash only.
* Check with your local Andersen supplier for availability of these units.
† These units have restricted sash travel.

Figure B–6 *Andersen® Narroline double-hung windows— table of sizes.*

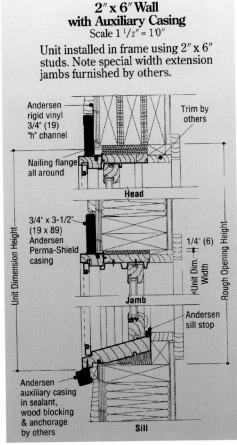

2″ x 6″ Wall with Auxiliary Casing
Scale 1 1/2″ = 1′0″

Unit installed in frame using 2″ x 6″ studs. Note special width extension jambs furnished by others.

Andersen rigid vinyl 3/4″ (19) "h" channel

Trim by others

Nailing flange all around

Head

3/4″ x 3-1/2″ (19 x 89) Andersen Perma-Shield casing

1/4″ (6)

Unit Dimension Height

Unit Dim. Width

Rough Opening Height

Jamb

Andersen sill stop

Andersen auxiliary casing in sealant, wood blocking & anchorage by others

Sill

Figure B–7 *Andersen® Narroline double-hung windows—tracing details.*

Circle Top Windows Combined with Casement Windows

CTQC1	CTC1	CTQCW1	CTCW1	CTQA3	CTC2	CTCW2	CTC3
C14	C14	CW14	CW14	CP303	C24	CW24	C34

Circle Top Windows Combined with Awning Windows

CTQC1	CTC1	CTQA3	CTC2	CTC2	CTC2	CTC2	CTC3
A21	A21	A31	AN41	A41	A42	AP421	A61

Circle Top Windows Combined with Double-Hung Windows

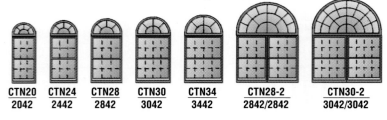

CTN20	CTN24	CTN28	CTN30	CTN34	CTN28-2	CTN30-2
2042	2442	2842	3042	3442	2842/2842	3042/3042

Typical Elliptical Window Combinations

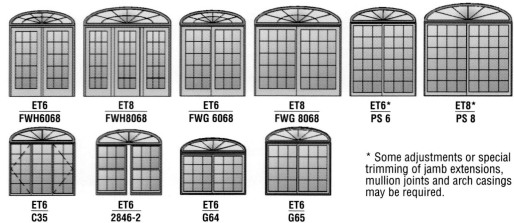

ET6	ET8	ET6	ET8	ET6*	ET8*
FWH6068	FWH8068	FWG 6068	FWG 8068	PS 6	PS 8

ET6	ET6	ET6	ET6
C35	2846-2	G64	G65

* Some adjustments or special trimming of jamb extensions, mullion joints and arch casings may be required.

Figure B–8 *Andersen® circle top and elliptical window combinations.*

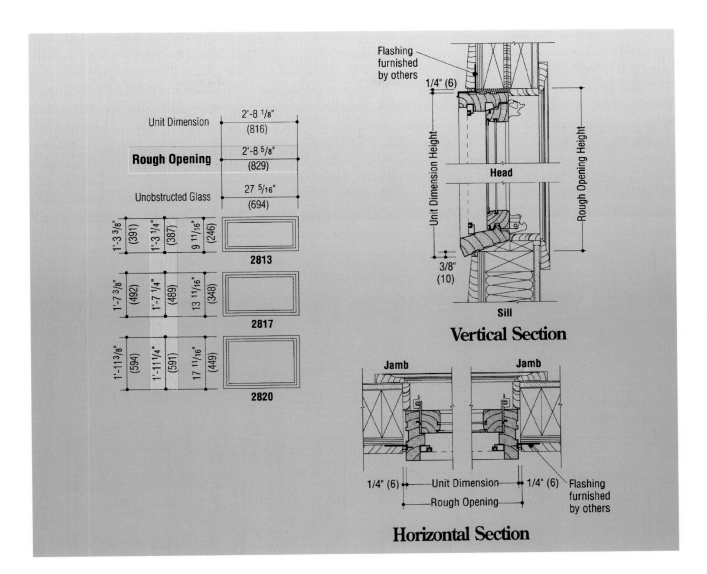

Unit Dimension
2'-8 1/8"
(816)

Rough Opening
2'-8 5/8"
(829)

Unobstructed Glass
27 5/16"
(694)

1'-3 3/8" (391)   1'-3 1/4" (387)   9 11/16" (246)
2813

1'-7 3/8" (492)   1'-7 1/4" (489)   13 11/16" (348)
2817

1'-11 3/8" (594)   1'-11 1/4" (591)   17 11/16" (449)
2820

Flashing furnished by others

1/4" (6)

Unit Dimension Height

Rough Opening Height

Head

3/8" (10)

Sill

Vertical Section

Jamb   **Jamb**

1/4" (6) — Unit Dimension — 1/4" (6)
Rough Opening

Flashing furnished by others

Horizontal Section

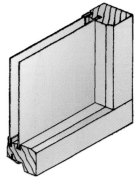

Corner section of sash with Removable Double Glazing panel applied.

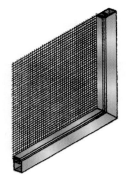

Corner section of aluminum framed screen.

Figure B–9   *Andersen® basement windows.*

Appendix C
Doors

TABLE C–1 *Units Made in Sizes as Shown:*

Thickness		Width		Height
$1\frac{3}{4}$"	×	2'-6"	×	6'-8"
$1\frac{3}{4}$"	×	2'-8"	×	6'-8"
$1\frac{3}{4}$"	×	3'-0	×	6'-8"
$1\frac{3}{4}$"	×	3'-0	×	7'-0

Some doors available in additional sizes.

M—Designates Pine Doors
F—Designates Fir Doors

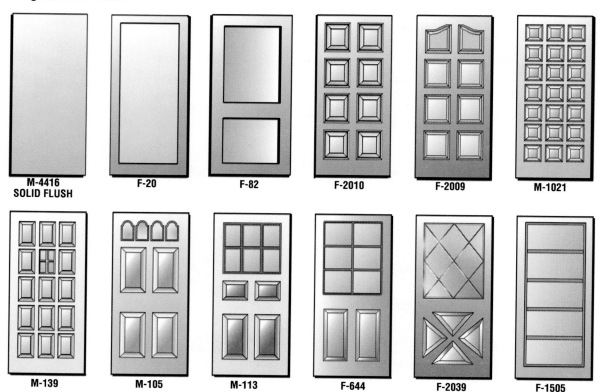

M-4416 SOLID FLUSH   F-20   F-82   F-2010   F-2009   M-1021

M-139   M-105   M-113   F-644   F-2039   F-1505

Figure C–1 Some Morgan exterior door styles.

TABLE C–2   *Units Made in 6'-6", 6'-8", and 8'-0 Heights and in Widths as Shown:*

2-door Units (M-2FD)		4-door Units (M-4FD)	
Width of Doors	**Jamb Opening Width**	**Width of Doors**	**Jamb Opening Width**
$11\frac{11}{16}"$	2'-0	$8\frac{11}{16}"$	3'-0
$1'-1\frac{11}{16}"$	2'-4"	$11\frac{11}{16}"$	4'-0
$1'-2\frac{11}{16}"$	2'-6"	$1'-2\frac{11}{16}"$	5'-0
$1'-3\frac{11}{16}"$	2'-8"	$1'-5\frac{11}{16}"$	6'-0
$1'-5\frac{11}{16}"$	3'-0		

M—Designates Pine Doors
F—Designates Fir Doors

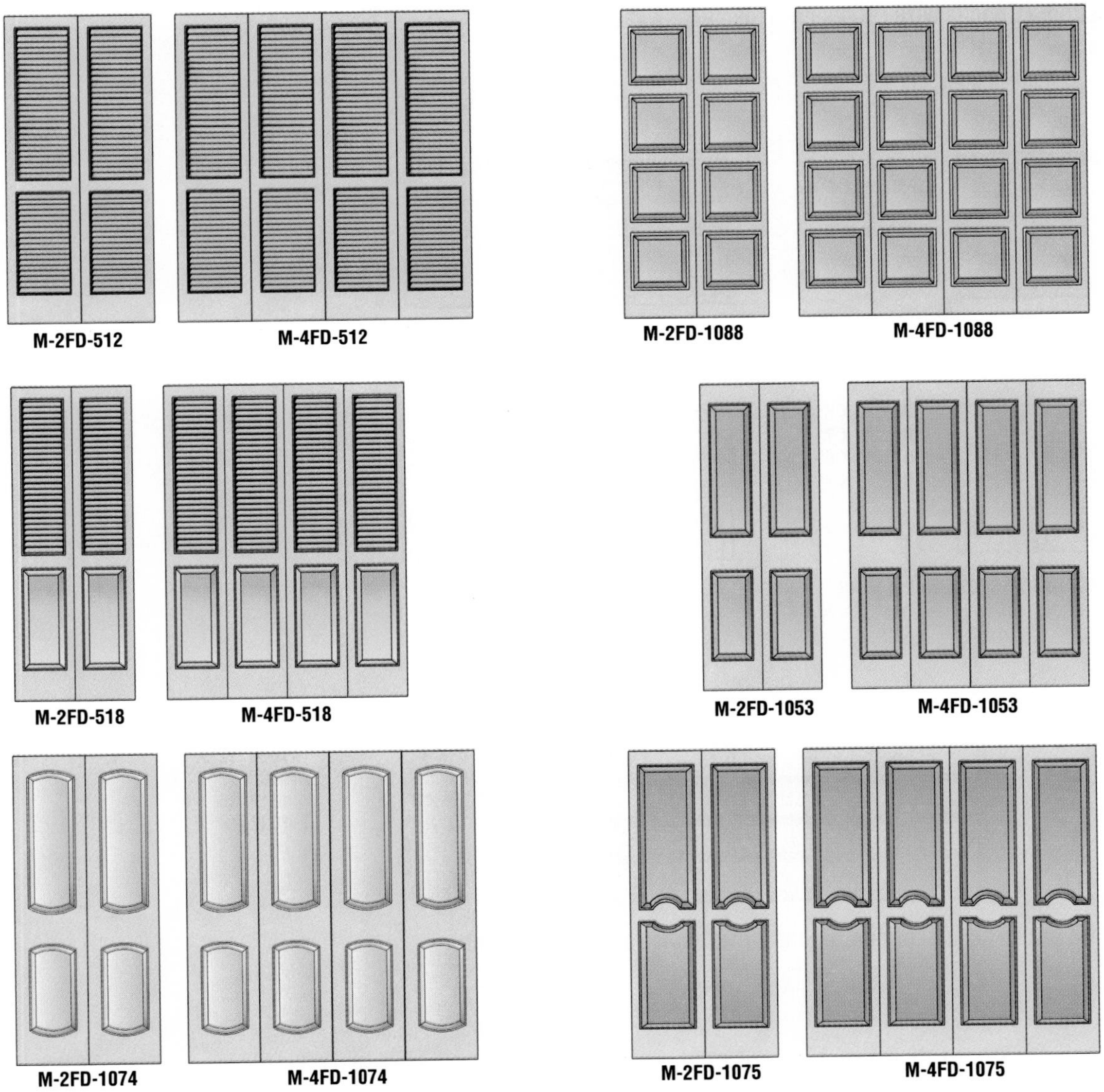

M-2FD-512   M-4FD-512   M-2FD-1088   M-4FD-1088

M-2FD-518   M-4FD-518   M-2FD-1053   M-4FD-1053

M-2FD-1074   M-4FD-1074   M-2FD-1075   M-4FD-1075

Figure C–2   Some Morgan folding interior door styles.

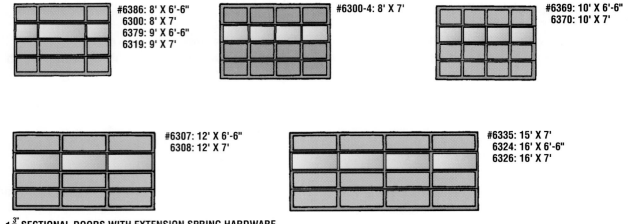

#6386: 8' X 6'-6"
6300: 8' X 7'
6379: 9' X 6'-6"
6319: 9' X 7'

#6300-4: 8' X 7'

#6369: 10' X 6'-6"
6370: 10' X 7'

#6307: 12' X 6'-6"
6308: 12' X 7'

#6335: 15' X 7'
6324: 16' X 6'-6"
6326: 16' X 7'

1 ⅜" SECTIONAL DOORS WITH EXTENSION SPRING HARDWARE

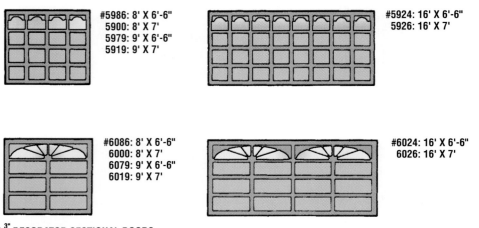

#5986: 8' X 6'-6"
5900: 8' X 7'
5979: 9' X 6'-6"
5919: 9' X 7'

#5924: 16' X 6'-6"
5926: 16' X 7'

#6086: 8' X 6'-6"
6000: 8' X 7'
6079: 9' X 6'-6"
6019: 9' X 7'

#6024: 16' X 6'-6"
6026: 16' X 7'

1 ⅜" DECORATOR SECTIONAL DOORS

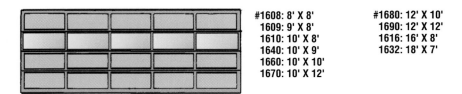

#1608: 8' X 8'
1609: 9' X 8'
1610: 10' X 8'
1640: 10' X 9'
1660: 10' X 10'
1670: 10' X 12'

#1680: 12' X 10'
1690: 12' X 12'
1616: 16' X 8'
1632: 18' X 7'

1 ⅜" LIGHT COMMERCIAL SECTIONAL DOORS WITH TORSION SPRING HARDWARE

Figure C–3   *Some Frantz sectional garage doors—styles and sizes.*

Appendix D
Architectural Spelling

These words are commonly used in architectural drafting and are often misspelled. Anyone seriously interested in drafting should know the proper spelling of the words of the trade.

Acoustical
Acre
Aisle (*isle* is an island)
Alcove
Aluminum
Appliance
Asphalt

Barbecue
Bathroom (one word)
Batten
Batter board (two words)
Bedroom (one word)
Bevel, beveled
Bracing
Brickwork
Bridging
Built-in
Built-up roof

Cabinet
Calk, calking (preferred over *caulk*)
Canopy, canopies
Cant strip
Cantilever
Carport (one word)
Casement
Center (*centre* is British spelling)
Centimeter (*centimetre* is British spelling)
Channel
Chromium

Cleanout door
Clerestory
Colonnade
Coping
Cornice
Corridor
Corrugated
Creosote
Cross section (noun)
Cross-section (adjective)
Cupola

Dampproof (one word)
Data processing (noun)
Data-processed (adjective)
Dining room (two words)
Dishwasher (one word)
Disposal unit
Double-hung
Dovetail
Downspout
Downstairs (one word)
Draft, drafting, drafter (*draught* is obsolete)
Dry wall (two words)

Eave (eve is evening)
Enclose, enclosure (*inclose* is used by land surveyors)

Fascia, facia
Fiberboard
Fiber glass (*Fiberglas* is trade name)
Fieldstone (one word)
Fire brick (two words)
Fireplace (one word)
Fireproof (one word)

Flagstone (one word)
Flue (*flu* is a disease)
Fluorescent
Formica
Freezer
Furring

Gable
Gage (preferred over *gauge*)
Galvanized
Game room (two words)
Gray (preferred over *grey*)
Grill (a grid for broiling)
Grille or grill (a grating for protection)
Gypsum

Handrail (one word)
Hangar (for airplanes)
Hanger (for hanging)
Horizontal

Jalousie (*jealousy* is resentment)

Kiln

Lanai (a Hawaiian porch)
Lath (*lathe* is a machine tool)
Lavatory (a sink; *laboratory* for experiments)
Level, leveled, leveling
Linoleum
Lintel
Living room (two words)
Loggia (a roofed, open porch)
Louver (*Louvre* is an art museum in Paris)

Mantel (*mantle* is a cloak)
Masonry
Meter (*metre* is British spelling)
Millimeter (*millimetre* is British spelling)
Miter, mitered
Molding or moulding
Mortar
Mortgage
Movable
Mullion
Muntin

Nosing

Ordinance (*ordnance* are artillery)
Oriel (*oriole* is a bird)

Paneled, paneling
Parallel
Perpendicular
Playroom (one word)
Projector

Rabbet (*rabbit* is an animal)
Receptacle
Remove, removable

Sheathing
Sheetrock (one word)
Siding
Solder (*soldier* is a member of the military)
Stile (of a door)
Story, storey
Style (of architecture)
Subfloor (one word)

Template
Terrazzo

Upstairs (one word)

Veneer
Vertical
Vinyl

Wainscot, wainscoting
Wallboard (one word)
Waterproof (one word)
Weatherproof (one word)
Weatherstripping
Weep hole (two words)
Woodwork (one word)
Wrought iron (means "worked iron")

Zinc

Appendix E
Abbreviations

Abbreviations must often be used by architectural drafters to fit notes into the available space. However, a list of the abbreviations used should be included on each set of drawings so that the meanings of the abbreviations are perfectly clear to all reading the drawings. The abbreviations shown in this section have been approved by architectural, computer, and engineering societies. They are based upon the following rules:

1. Capitals are used almost universally
2. Periods are used only when necessary to avoid a misunderstanding (like the use of *IN.* in place of *IN*)
3. Spaces between letters are used only when necessary to clarify the abbreviation (such as *CU FT* in place of *CUFT*)
4. The same abbreviation may be used for the singular and plural

Abbreviation	ABBREV
Acoustic	ACST
Acoustical plaster	ACST PLAS
Actual	ACT.
Addition	ADD.
Adhesive	ADH
Aggregate	AGGR
Air conditioning	AIR COND
Alternating current	AC
Aluminum	AL or ALUM
American Institute of Architects	AIA
American Institute of Steel Construction	AISC
American National Standards Institute	ANSI
American Society for Testing Materials	ASTM
American Society of Heating, Refrigerating, and Air Conditioning Engineers	ASHRAE
American Standards Association	ASA
American wire gauge	AWG
Amount	AMT
Ampere	A or AMP
Anchor bolt	AB
Angle	∟
Apartment	APT
Approved	APP
Approximate	APPROX
Architect, architectural	ARCH
Architectural terra-cotta	ATC
Area	A
Asphalt	ASPH
Assemble	ASSEM
Assembly	ASSY
Associate, association	ASSOC
At	@
Atmospheric pressure	ATM PRESS
Automatic	AUTO
Avenue	AVE
Average	AVG
Balcony	BALC
Barrel, barrels	BBL
Basement	BASMT
Bathroom	B
Beaded one side	B 1S
Beam	BM

Bedroom	BR	Channel	C or [(old designation)
Bench mark	BM	Check	CHK
Better	BTR	Cinder block	CIN BL
Between	BET.	Circle	CIR
Beveled	BEV	Circuit	CKT
Blocking	BLKG	Circuit breaker	CIR BKR
Blower	BLO	Class	CL
Board	BD	Clean-out	CO
Board feet	BD FT or FBM	Clear	CLR
Board measure	BM	Closet	C or CL or CLO
Book shelves	BK SH	Coefficient	COEF
Bottom	BOT	Cold water	CW
Boulevard	BLVD	Column	COL
Bracket	BRKT	Combination	COMB.
Brass	BR	Common	COM
British thermal unit	BTU	Compact disk	CD
Bronze	BRZ	Company	CO
Broom closet	BC	Computer-aided design or computer-aided drafting	CAD
Brown & Sharpe gauge	B&S GA		
Building	BLDG	Computer-aided design and drafting	CADD
Building Officials and Code Administrators	BOCA		
Built-in	BLT-IN	Computer-aided manufacturing	CAM
Bulletin board	BB	Concrete	CONC
Button	BUT.	Concrete block	CONC B
Buzzer	BUZ	Concrete floor	CONC FL
By	× (as 2' × 4')	Concrete masonry unit	CMU
		Construction	CONST
Cabinet	CAB.	Construction Specifications Institute	CSI
Calking	CLKG		
Candela	cd		
Candlepower	CP	Constructive solid geometry	CSG
Carpenter	CARP.		
Casing	CSG	Contractor	CONTR
Cast iron	CI	Copper	COP or CU
Catch basin	CB	Counter	CTR
Cathode-ray tube	CRT	Countersink	CSK
Ceiling	CLG	Courses	C
Cement	CEM	Cover	COV
Cement floor	CEM FL	Cross section	X-SECT
Central processing unit	CPU	Cubic	CU
Celsius	C	Cubic feet per minute	CFM
Center	CTR	Cubic foot, feet	CU FT
Center line	CL or ₵	Cubic inch, inches	CU IN
Center matched	CM	Cubic yard, yards	CU YD
Center to center	OC	Cylinder	CYL
Centimeter, centimeters	cm		
Ceramic	CER	Damper	DMPR
Cesspool	CP	Decibel	DB
Chamfer	CHAM		
Change	CHG		

Deep, depth	DP	Facsimile	FAX
Degree	° or DEG	Family room	FAM R
Detail	DET	Fahrenheit	F
Diagram	DIAG	Feet	' or FT
Diameter	φ or DIA	Feet board measure	FBM
Dimension	DIM	Feet per minute	FPM
Dining room	DR	Feet per second	FPS
Direct current	DC	Figure	FIG.
Dishwasher	DW	Finish	FIN.
Disk operating system	DOS	Finish all over	FAO
Distance	DIST	Finished floor	FIN FL
Ditto	" or DO	Fire brick	FBRK
Division	DIV	Fire extinguisher	F EXT
Door	DR	Fire hose	FH
Dozen	DOZ	Fireproof	FP
Double-hung	DH	Fireproof self-closing	FPSC
Dowel	DWL	Fitting	FTG
Down	DN	Fixture	FIX.
Downspout	DS	Flange	FLG
Drain	D or DR	Flashing	FL
Drawing	DWG	Floor	FL
Drawn	DR	Floor drain	FD
Dressed and matched	D&M	Flooring	FLG
Drinking fountain	DF	Fluorescent	FLUOR
Dryer	D	Foot	' or FT
Dry well	DW	Footing	FTG
Duplicate	DUP	Foundation	FDN
		Free-on-board	FOB
Each	EA	Front	FR
East	E	Fuel oil	FO
Edge grain	EG	Full size	FS
Elbow	ELL	Furnace	FURN
Electric	ELEC		
Electronic mail	E-MAIL	Gallon, gallons	GAL
Elevation	EL, ELEV	Galvanized	GALV
Elevator	ELEV	Galvanized iron	GI
Emergency	EMER	Gauge	GA
Enclosure	ENCL	Glass	GL
Engineer	ENGR	Glass block	GL BL
Entrance	ENT	Glue-laminated	GLUELAM
Equipment	EQUIP	Government	GOVT
Equivalent direct		Grade	GR
radiation	EDR	Graphical user interface	GUI
Estimate	EST	Grating	GRTG
Excavate	EXC	Gypsum	GYP
Extension	EXT		
Exterior	EXT	Hall	H
Extra heavy	XH or XHVY	Hardware	HDW
		Hardwood	HDWD
Fabricate	FAB	Head	HD
Face to face	F to F	Heater	HTR

Heating, ventilating, air conditioning	HVAC	Light	LT
Height	HT or HGT	Limestone	LS
Hexagonal	HEX	Linear feet	LIN FT
Hollow metal	HM	Linen closet	L CL
Horizon, horizontal	HOR or HORIZ	Lining	LNG
Horsepower	HP or HP	Linoleum	LINO
Hose bibb	HB	Liquid crystal display	LCD
Hot water	HW	Living room	LR
Hour	HR	Long	LG
House	HSE	Lumber	LBR
Hundred	C		
		M-shape steel beam	M or JR (old designation)
I beam	S or I (old designation)	Machine	MACH
Inch, inches	" or IN.	Manufacture, manufacturer	MFR
Information	INFO	Manufactured	MFD
Inside diameter	ID	Manufacturing	MFG
Insulation	INSUL	Mark	MK
Interior	INT	Masonry opening	MO
International Business Machines	IBM	Material	MATL
Iron pin	IP	Maximum	MAX
		Mechanical	MECH
Joint	JT	Medicine cabinet	MC
Junior beam	M or JR (old designation)	Medium	MED
		Metal	MET
		Meter, meters	m
Kalamein	KAL	MicroSoft's disk operating system	MS-DOS
Kelvin	K	Millimeter, millimeters	mm
Kilobyte	K	Minimum	MIN
Kilogram	kg	Miscellaneous	MISC
Kilowatt	kw	Model	MOD
Kitchen	K	Moderate-weather (a common brick grade)	MW
Kitchen cabinet	KC	Modulator/Demodulator	MODEM
Kitchen sink	KS	Molding	MLDG
		Mole	mol
Laboratory	LAB		
Ladder	LAD.	National	NATL
Landing	LDG	National Building Code	NBC
Latitude	LAT	National Electrical Code	NEC
Laundry	LAU	National Lumber Manufacturers Association	NLMA
Laundry chute	LC	No-weather (a common brick grade)	NW
Lavatory	LAV	Nominal	NOM
Leader	LDR	North	N
Leader drain	LD	Not applicable	NA
Left	L		
Left hand	LH		
Length	LGTH		
Level	LEV		
Library	LIB		

Not in contract	NIC	Quantity	QTY
Number	# or NO.	Quart, quarts	QT
Oak	O	Radiator	RAD
Octagon	OCT	Radiator enclosure	RAD ENCL
Office	OFF	Radius	R
On center	OC	Random-access memory	RAM
Opening	OPG	Random length and width	RL&W
Opposite	OPP	Range	R
Ornament	ORN	Read-only memory	ROM
Ounce, ounces	OZ	Receptacle	RECP
Outside diameter	OD	Rectangle	RECT
Overhead	OVHD	Redwood	RDWD
		Reference	REF
Page	P	Refrigerator	REF
Painted	PTD	Register	REG
Pair	PR	Reinforce, reinforcing	REINF
Panel	PNL	Reinforcing bar	REBAR
Paragraph	PAR	Required	REQD
Parallel	|| or PAR	Return	RET
Partition	PTN	Revision	REV
Passage	PASS.	Revolutions per minute	RPM
Pedestal	PED	Right	R
Penny (nail)	d	Right hand	RH
Per	/	Riser	R
Percent	%	Road	RD
Perforate	PERF	Roof	RF
Perpendicular	⊥ or PERP	Roof drain	RD
Personal digital assistant	PDA	Roofing	RFG
		Room	RM
Pi (ratio of circumference to diameter of circle)	π	Rough	RGH
		Rough opening	RO
Piece	PC	Round	φ or RD
Plaster	PLAS		
Plate	℔ or PL	S-shape steel beam	S or I (old designation)
Plate glass	PL GL	Schedule	SCH
Platform	PLAT	Screw	SCR
Plumbing	PLMB	Second, seconds	s
Plywood	PLYWD	Section	SECT
Point	PT	Self-closing	SC
Polish	POL	Service	SERV
Polyvinyl chloride	PVC	Severe-weather (a common brick grade)	SW
Position	POS	Sewer	SEW.
Pound, pounds	lb or #	Sheathing	SHTHG
Pounds per square inch	PSI	Sheet	SHT
Poured concrete	P/C	Shower	SH
Prefabricated	PREFAB	Siding	SDG
Property	PROP.	Sill cock	SC
Push button	PB	Sink	S or SK

Slop sink	SS	Tongue-and-groove	T&G
Socket	SOC	Tread	TR
Soil pipe	SP	Typical	TYP
South	S		
Southern Standard		Ultimate	ULT
Building Code	SBC	Underwriters Laboratories	UL
Specifications	SPEC	Unfinished	UNFIN
Square	□ or SQ	Uniform Building Code	UBC
Square foot, square feet	SQ FT	U.S.A. Standards Institute	USASI
Stairs	ST	U.S. standard gauge	USG
Standard	STD	User coordinate system	UCS
Standpipe	ST P		
Station	STA	Vanishing point	V
Station point	SP	Vent or ventilator	V
Steel	STL	Ventilate, ventilation	VENT.
Stirrup	STIR.	Vertical	VERT
Stock	STK	Vestibule	VEST.
Storage	STOR	Video gate array	VGA
Street	ST	Virtual reality	VR
Structural	STR	Volt, volts	V
Structural clay research	SCR	Volume	VOL
Substitute	SUB		
Supersede	SUPSD	W-shape steel beam	W, **WF** or WF (old designations)
Supplement	SUPP		
Supply	SUP	Wall cabinet	W CAB
Surface	SUR	Wall vent	WV
Surface 1 side	S1S	Water	W
Surface 2 sides	S2S	Water closet	WC
Surface 4 sides	S4S	Waterproof	**WP** or WP
Surface all sides	S4S	Watt, watts	W
Surface 1 edge	S1E	Weatherproof	**WP** or WP
Surface 2 edges	S2E	Weight	WT
Surface 1 side 1 edge	S1S1E	Weep hole	WH
Surfaced and matched	S&M	West	W
Suspended ceiling	SUSP CLG	White Pine	WP
Switch	S or SW	Wide flange	W, **WF** or WF (old designations)
Symbol	SYM		
System	SYS	Width	WTH
		Window	WDW
Tar and gravel	T&G	Wire glass	W GL
Technical	TECH	With	W/
Tee	T	Without	W/O
Telephone	TEL	Wood	WD
Television	TV	World coordinate system	WCS
Temperature	TEMP	World Wide Web	WWW
Terra-cotta	TC	Wrought iron	WI
The Radio Shack	TRS		
Thermostat	THERMO	Yard, yards	YD
Thick, thickness	T or THK	Year	YR
Thousand	M	Yellow Pine	YP
Thousand board feet	MBM		
Thread	THD	Zinc	Z or ZN

Appendix F

Glossary of Architectural and Computer Terms

Accordion door: A folding door containing many narrow leaves.

Active solar system: A solar heating system in which the heat flow is forced by some mechanical means.

Active window: The computer window currently in use.

Adjustable rate mortgage: A mortgage in which the interest rate depends upon some economic index and may change each year.

A-frame: Any frame in the shape of an inverted V.

Aggregate: Material such as broken stone, gravel, cinders or slag used as one of the constituents of concrete, the other constituents being sand, cement, and water.

Alcove: A recessed space connected with or at the side of a large room.

Americans with Disabilities Act (ADA): A civil rights statute requiring that all public and commercial buildings must be accessible to impaired persons.

Analog computer: Computer that processes and shows information in a smoothly varying form.

Anchor: A metal piece used to attach building members to masonry.

Anchor bolt: A threaded rod used to fasten the sill plate to the foundation.

Angle iron: A metal bar, L-shaped in section.

Apple Computer: A leading personal computer manufacturer best-known for the *Macintosh* line of computers.

Appliqué: See *pressure-sensitive transfer*.

Apron: The finish board immediately below a windowsill.

Arcade: A series of arches supported by a row of columns.

Arch: A curved structure that carries the weight over an opening.

Architect: A person who plans buildings and oversees their construction.

Architectural lettering: Commercial Gothic lettering altered to suit a drafter or office style.

Architectural styling: A building's exterior appearance as characteristic of a particular period.

Architectural terra-cotta: Terra-cotta building blocks having a ceramic finish.

Areawall: The wall of an areaway.

Areaway: A subsurface enclosure to admit light and air to a basement.

Artificial intelligence: Computer program that simulates human thinking.

Ash drop: A small door in the inner hearth for removing ashes from the fireplace.

Ash pit: The area under a fireplace's ash drop, used to collect ashes.

Ashlar masonry: Masonry composed of squared units laid with horizontal bed joints.

Asphalt: An insoluble material used in waterproofing.

AutoCAD: Popular computer-aided design software developed by AutoDesk, Inc.

AutoDesk, Inc.: A leading computer-aided design software publisher best-known for the AutoCAD program.

Awning window: A window whose frame is hinged at the top and swings outward.

Axonometric projection: An orthographic projection of an object tilted with respect to the picture plane.

Backfill: Earth replaced around a foundation.

Balcony: A platform projecting from the wall of a building, above the ground.

Balloon frame: A type of building frame in which the studs extend from sill to eaves without interruption.

Balusters: The small vertical members of a railing between the bottom and top rail.

Banister: A handrail.

Baseboard: The finishing board covering a wall where it meets the floor.

Basement: The lowest story of a building, partially or entirely below ground.

Batten: A strip of board for use in fastening other boards together.

Batter: Sloping a masonry or concrete wall.

Batter board: A horizontal board nailed to posts and used to lay out the excavation and foundation.

Bay: Any division or compartments of an arcade, roof, building, space between floor joists, or other area.

Beam: A horizontal structural member that carries a load.

Bearing partition: A partition supporting any vertical load other than its own weight.

Bearing plate: A support member used to distribute weight over a larger area.

Bench mark: A reference point used by surveyors to establish lines and grades.

Bent: A rigid, transverse framework.

Bevel weld: A butt weld with one mitered member.

Bibb: A threaded faucet.

Bid form: A sample bidding letter from the bidder to the prospective building owner, containing blank spaces to be completed by the bidder.

Bidet: A toilet-like bathroom fixture with spigots used for bathing lower extremities.

Bit: Binary digit (0 or 1).

Blocking: Small wood framing members.

Bluestone: A hard, blue sandstone.

Board foot: The amount of wood contained in a piece of rough, green lumber 1" thick × 12" wide × 1' long.

Board-and-batten: An exterior finish of vertical boards having joints covered by battens.

Bond: Mortar bond between mortar and masonry units; structural bond between wythes; pattern bond for decorative effect.

Bond beam: A reinforced concrete beam used to strengthen masonry walls.

Book match: A veneer pattern to alternate sheets turned over as are the leaves of a book.

Boot: Start a computer by loading the operating system.

Box beam: A hollow, built-up structural unit.

Brick bond: Adhesion of mortar (mortar bond), overlapping masonry units (structural bond), or decorative pattern of masonry units (pattern bond).

Brick veneer: A brick facing laid in front of frame construction.

Bridging: Cross-bracing between floor joists to add stiffness to the floors.

Brownstone: A brown sandstone.

Browser: Computer software that runs on the Internet, either text-only browser or graphical browser.

Btu: British thermal unit, a unit used to measure heat.

Bug: An error in a computer program.

Building board (also wallboard): Boards made from repulped paper, shredded wood, or similar material.

Building codes: Regulations that give minimum standards for building design, material, construction, and maintenance to protect the health, safety, and welfare of the public.

Building line: An imaginary line on a plot beyond which the building may not extend.

Building paper: Asphalt-saturated paper used between subflooring and finished flooring, and between roofers and roof covering.

Built-up beam: A beam constructed of smaller members fastened together with the grains parallel.

Built-up roof: A roofing composed of several layers of felt and asphalt, pitch, or coal tar.

Bundling: Computer software and hardware sold at one combined price.

Butt: See door butt.

Butt weld: A weld of members butting against each other.

Button: Small circle representing a choice to be made in a window on a computer screen.

Byte: Number of bits (usually eight or sixteen) that represent one character.

Cablesoft: Interactive television.

Canopy: A sheltering roof.

Cant strip: A form of triangular molding.

Cantilever: A beam or girder fixed at one extremity and free at the other. To cantilever is to employ the principle of the lever to carry a load.

Carbon steel: A basic structural steel containing carbon and manganese as main alloys.

Carpenter ants: Destructive ants that chew nests in moist wood.

Casement window: A window sash hinged at the side and swinging outward.

Caulking: A waterproof material used to seal cracks.

Carport: A garage not fully enclosed.

Casement: A window whose frame is hinged at the side.

Casing: The framing around a door or window.

Catch basin: An underground structure for surface drainage in which sediment may settle.

Catenary: The shape of a chain hanging freely between two supports.

Cavity wall: A masonry wall having an air space of about 2".

CD-ROM: A compact laser disk with read-only memory having more computer storage capability than a diskette. It reads only information provided by supplier of the disk. (compare with WORM).

Cement: A masonry material purchased in the form of a highly pulverized powder usually medium gray in color. The approximate proportions for portland cement are as follows:

Lime (CaO)	60%–67%
Silica (SiO_2)	20%–25%
Iron oxide and alumina	7%–12%

Centering: Form work for poured concrete floor and roof slabs; temporary form work for the support of masonry arches or lintels during construction.

Center to center: Measurement from the center of one member to the center of another (noted "oc").

Center-hall: A hallway near the middle of a floor plan.

Central air conditioning: A single-source air conditioning system, distributed through ducts.

Central processing unit (CPU): Computer hardware that processes and stores information: the computer's "brain."

Ceramic veneer: Architectural terra-cotta having large face dimensions and thin sections.

Change order: A document that changes the original construction document after the contract has been awarded.

Channel: A standard form of structural rolled steel, consisting of three sides at right angles in channel form.

Check: A lumber defect caused by radial separation during seasoning. Also see *door check.*

Chord: A principal member of a truss.

Circuit: The path for an electric current.

Circulation: The pattern of traffic between rooms and areas.

Clapboard: A narrow board, thicker at one edge, for weather-boarding frame buildings; siding.

Classic revival: Architectural style of the nineteenth century that imitated early Greek and Roman styles.

Clean-out: An accessible plumbing fitting to allow cleaning the house drain and sewer; a small door to the ash pit for removing ashes.

Clerestory: A window between roof planes.

Client: A person who employs an architect.

Closed plan: A floor plan of *rooms* rather than areas.

Closed riser: A stair having both treads and risers.

Cohesive construction: A structural system consisting of materials shaped while plastic and allowed to harden into cohesive members.

Cold-water supply: The water distribution system that supplies cold water from the water main to bathroom, kitchen, and laundry fixtures.

Collar beam: A horizontal member tying two opposite rafters together at more or less a center point on the rafters.

Colonial styling: Any architectural style developed by colonizers.

Column: A vertical supporting member.

Command: The operator's directions to a computer.

Commercial Gothic lettering: A type of Gothic lettering approved by ANSI for use on engineering and architectural drawings.

Compact disk: A large-capacity laser disk that stores information digitally. Most computer CDS are "read only" because the operator cannot store additional information on the disk.

Compaq Computer Corporation: A leading manufacturer of portable IBM PC-compatible computers.

Common brick: $3\frac{3}{4}" \times 2\frac{1}{2}" \times 8"$ brick used for general construction.

Composite wall: A masonry wall of at least two adjacent wythes of different materials.

CompuServe: A consumer information service accessible by computers with modems.

Concrete: A masonry mixture of portland cement, sand and aggregate, and water in proper proportions.

Condensation: Water formed by warm, moist air contacting a cold surface.

Conductor: A vertical drainpipe; material permitting passage of electric current.

Conduit: A pipe or trough that carries water, electrical wiring, cables, and so forth.

Configuration, rule of: The drafting principle that the shape of a plane remains about the same in all views, unless it appears as an edge.

Conifer: See *softwood*.

Construction loan: A loan obtained during the construction of a building.

Contemporary styling: Any building erected in a present-day style. Also, any *modern* style.

Contour line: An imaginary line representing a constant elevation on the earth's surface.

Contractor: A builder.

Control joint: An expansion joint in a masonry wall formed by raking mortar from a continuous vertical joint.

Convector: A heat transfer surface that uses convection currents to transmit heat.

Coping: A masonry cap on top of a wall to protect it from water penetration.

Corbel: A bracket formed in a wall by building out successive courses of masonry.

Corner bead: A metal molding, built into the plaster corners to prevent the accidental breakage of the plaster.

Cornice: That part of a roof which extends or projects beyond the wall; the architectural treatment thereof, as a *box cornice*.

Corrosion: Rusting of ferrous metals caused by moisture.

Counterflashing: A flashing used under the regular flashing.

Course: A horizontal row of bricks, tile, stone, building blocks, or similar material.

Court: An open space surrounded partly or entirely by a building.

Crash: Computer becoming inoperable due to hardware failure or software error.

Crawl space: The space between the floor joists and the surface below when there is no basement. This is used in making repairs on plumbing and other utilities.

Cricket: A roof device used at intersections to divert water.

Cupola: A small structure built on top of a roof.

Curb cut: A ramp between a walkway and a street.

Cursor: A movable symbol on a computer screen that indicates where data is being input.

Curtain wall: An exterior wall which provides no structural support.

Cut stone: Stone cut to given sizes or shapes.

Cutaway model: A model having surfaces removed in order to show interior features.

Cybernetics: Comparative study of human systems and computers.

Cyberspace: World-wide electronic mail; virtual reality.

Damper: A movable plate to regulate the draft in a chimney.

Dap: A circular groove used for split rings and shear plates.

Database: Information stored in a computer.

Dead-end room: A room not used as a throughway.

Deathwatch beetles: Destructive beetles that feed on moist wood.

Debug: Remove an error from a computer program.

Decay: Disintegration of wood through the action of fungi.

Decibel: A unit of measuring the relative loudness of sound.

Deciduous: See *hardwood*.

Dedicated: Computer hardware performing a single function (e.g., a dedicated word processor performs only word processing).

Default: The computer option offered if no other option is ordered.

Dehumidifier: A device to decrease relative humidity in a building.

Desktop publishing: Using a computer to typeset and print documents.

Digital computer: Computer that processes and shows information by numerical digits.

Digitizer: An electronically sensitized tablet that can sense and display the location of a puck.

Diskette: Flexible computer storage disk that may be inserted and removed from the CPU.

Disk operating system: An operating system for microcomputers.

Door buck: A door frame (usually metal).

Door butt: A hinge.

Door check: A device to slow a door when closing.

Doorstop: A device to prevent a door from hitting the wall when opening.

Dormer: A structure projecting from, or cut into, a sloping roof, usually to accommodate a window or windows.

Double-faced fireplace: A fireplace having two fireplace openings.

Double-hung window: A window having top and bottom sashes each capable of movement up and down in its own grooves.

Down payment: An initial amount given as partial payment at time of purchase.

Downspout: A vertical drainpipe for carrying rainwater from the gutters.

Drafting media: Tracing paper, vellum, and film used for drafting.

Drag: Move an image on a computer screen using a puck or mouse.

Drain: A pipe for carrying waste water.

Dressed size: See *finished size.*

Drip: A molding designed to prevent rainwater from running down the face of a wall or to protect the bottom of a door or window from leakage.

Drafter: A person who prepares plans using drafting instruments or computer equipment.

Dry rot: A dry, crumbly wood rot.

Dry wall: A wall finished with wallboard in place of plaster; stone wall built without mortar.

Dry well: A shallow well used for the disposal of rainwater.

Duct: A sheet-metal conductor for air distribution.

Eave: The lower portion of a roof which extends beyond the wall.

Eclecticism: A mixing of various historical styles in a single structure.

Edit: Examine and modify a computer file.

Efficiency, house: The percentage of a building's area actually occupied by rooms.

Efflorescence: An undesirable white crystallization that may form on masonry walls.

Elbow: An L-shaped pipe fitting.

Electric arc process: A welding process that uses an electric arc to fuse both members.

Electronic mail: Messages sent by computer rather than by mail.

Elevation: An orthographic projection of the vertical side of a building.

English styling: Houses fashioned after the type built in England before the eighteenth century.

Entourage: The surroundings (such as trees, human figures, and vehicles) included in an architectural rendering to increase realism.

Escalator: A moving stairway.

Excavation: A hole formed by removing earth.

Expansion joint: A separation in a masonry or concrete wall to permit wall expansion due to temperature and moisture changes.

Facade: The front or face of a building.

Face brick: A special brick used for facing a wall. Face bricks are more uniform in size than common bricks and are made in a variety of colors and textures.

Faced wall: See *composite wall.*

Facing: Any material, forming a part of a wall, used as a finished surface.

Fascia (also *facia, fascia board*): A flat banded projection on the face of the cornice; the flat vertical member of the cornice; the flat surface running above a shop window on which the name of the shop may be displayed.

Fenestration: The arrangement of windows in a wall.

Fiber glass: A material composed of thin glass threads used for insulation or with resin for a finished surface.

Fiberboard: Sheet material of refined wood fibers.

Fieldstone: Building stone found loose on the ground (field) regardless of its exact variety. Don't confuse with Flagstone.

Filigree: Fine, decorative openwork.

Fillet weld: A butt weld with the weld metal filling an inside corner.

Finish lumber: Dressed wood used for building trim.

Finished flooring: Tongue-and-grooved hardwood laid upon the subflooring.

Finished size: The *nominal size* is the size of rough lumber. After planing, the actual *finished size* is about $\frac{1}{2}$" smaller than nominal. The difference between nominal and finished size will vary depending upon the size of the lumber, ranging from $\frac{3}{4}$" for lumber over 6" to only $\frac{1}{4}$" for 1" lumber. The variations between nominal and finished size of American Standard Lumber are:

Nominal Size	Finished Size
1"	$\frac{3}{4}$"
2"	$1\frac{1}{2}$"
4"	$3\frac{1}{2}$"
6"	$5\frac{1}{2}$"
8"	$7\frac{1}{4}$"
10"	$9\frac{1}{4}$"
12"	$11\frac{1}{4}$"

Firebrick: A brick made of a refractory material (fire clay) that withstands great heat; used to line furnaces, fireplaces, and so on.

Fire cut: An angular cut at the end of a joist framing into a masonry wall.

Fireplace liner: A prefabricated metal fireplace form consisting of sides, back, damper, smoke shelf, smoke chamber, and air ducts.

Fireproofing: Any material protecting structural members to increase their fire resistance.

Fire rating: The ability of a building to withstand fire (from one to four hours).

Fire stopping: Obstructions across air passages in buildings to prevent the spread of hot gasses and flames; horizontal blocking between wall studs.

Fire wall: A wall extending from foundation through the roof to subdivide a building in order to restrict the spread of fire.

First-angle projection: A multiview projection in which the object is located in the first quadrant, resulting in its plan below its front elevation.

Fitting, plumbing: A pipe connector such as *coupling, elbow, tee,* or *valve.*

Fixed window: A window designed so that it cannot be opened.

Fixture: A piece of electric or plumbing equipment.

Flagstone: Flat stone used for floors, steps, and walks.

Flashing: The sheet-metal work used to prevent leakage over windows and doors, around chimneys, and at the intersections of different wall surfaces and roof planes.

Flat roof: A horizontal (water-film) roof, or a roof slightly sloped for drainage.

Floor plan: An orthographic projection of the floor of a building.

Floppy disk: Diskette.

Flue: A passage in the chimney to convey smoke to the outer air.

Flue lining: Terra-cotta pipe used for the inner lining of chimneys.

Folding door: A door of two leaves hinged together, one leaf being also hinged to the door jamb, the other leaf sliding in a track.

Footing: The bases upon which the foundation and posts rest.

Footprint: Amount of desktop area needed for computer hardware.

Formica: A plastic veneer trade name.

Fossil fuel: Combustible material such as coal, oil, and gas derived from the remains of former life.

Foundation: The supporting wall of a building below the first-floor level.

Framing: Lumber used for the structural framing of a building.

Frost line: The depth of frost penetration in soil.

Furring: Wood strips fastened to a wall or ceiling for the purpose of attaching wallboards or ceiling tile.

Gable: The triangular portion of an end wall formed by a sloping roof.

Galvanic corrosion: Corrosion when different metals come in contact and in the presence of an electrolyte.

Gambrel: A gable roof, each slope of which is broken into two planes.

General conditions: An important portion of a construction document containing additional contractual-legal requirements not covered by other contract forms.

Geodesic dome: A double-faced dome formed of members of nearly equal length.

Georgian styling: The architectural style in England during the reign of the Georges.

Gigabit: One billion bits.

Gigabyte: Approximately one billion bytes (precisely 2^{30}).

Girder: A large horizontal structural member, usually heavier than a beam, used to support the ends of joists and beams or to carry walls over openings.

Glazed brick: Brick finished with ceramic, clay-coated, or salt glaze.

Glulam lumber: Usual term for glue-laminated lumber.

Gothic architecture: The architectural style of cathedrals in Western Europe between 1150 and 1450.

Gothic lettering: A lettering style having all strokes of equal width and no serifs.

Grade or grade line: The level of the ground around a building.

Graduated payment mortgage: A mortgage having lower interest payments in early years but higher payments in later years.

Granite: A durable and hard igneous rock.

Graph paper: Paper produced with a network of lightly-printed lines.

Graphical user interface (GUI): A computer program using a mouse or puck to manipulate icons and windows.

Green efflorescence: An undesirable green stain that may form on masonry walls.

Ground cover: Usually roll roofing laid on the ground in crawl spaces to reduce moisture.

Ground source heat pump: A heat pump that transfers heat between the building and the ground (rather than air).

Grounds: Wood strips attached to the walls before plastering, serving as a plaster stop and nailing base for trim.

Grout: Mortar of pouring consistency.

Guardrail: A rail serving as a guard at the side of a balcony.

Gunite: Sprayed concrete using a dry mix and water.

Gutter: A trough or depression for carrying off water.

Gypsum board (also *plaster board*): Board made of plaster with a covering of paper.

Half-timbering: A frame construction where the spaces are filled in with masonry.

Handrail: A rail serving as a support at the side of a stair.

Hanger: An iron strap used to support a joist or beam.

Hardboard: Sheet material of compressed wood fibers.

Hard conversion: The changing of both the English dimensions and size of a product to a rational metric size.

Hard disk: Large-capacity computer storage disk built into the central processing unit.

Hardware: In computer usage, the computer's physical equipment.

Hardwood: Wood from trees having broad leaves in contrast to needles. The term does not necessarily refer to the hardness of the wood.

Header: A beam perpendicular to joists, into which they are framed; a masonry unit laid horizontally with the end exposed.

Headroom: The vertical clearance in a room or on a stairway.

Hearth: The masonry portion of a floor in front of a fireplace.

Heartwood: The dead, inner layer of a tree formed from former sapwood.

Heat pump: A heating and cooling system that uses a refrigerant to transfer heat between the building and the outside air or water.

Heat-treated steel: A high-strength steel that has been quenched and tempered.

High-strength bolt: A medium-carbon or heat-treated alloy steel bolt.

High-strength steel: A high-strength, low-alloy steel.

Hip roof: A roof with four sloping sides.

Hopper window: A window whose frame is hinged at the bottom and swings inward.

Hot-water heating: A heating system using hot water circulated through pipes to radiators or convectors.

Hot-water supply: The water distribution system that supplies hot water from the water heater to bathroom, kitchen, and laundry fixtures.

House drain: Horizontal sewer piping within a building that receives waste from the soil stacks.

House sewer: Horizontal sewer piping 5' outside the foundation wall to the public sewer.

Housewrap: Plastic sheet used between sheathing and finished exterior wall covering.

Humidifier: A device to increase relative humidity in a building.

I beam: A steel beam with an I-shaped cross section.

Ice dam: An undesirable build-up of ice on the edge of a roof causing water to leak into a building.

Icon: Computer terminology for "symbol."

Insulation: Material for obstructing the passage of sound, heat, or cold from one surface to another.

Interface: Communication between computer and operator (such as the screen images seen and used), between software programs, between hardware, or between software and hardware.

Interference-body bolt: A high-strength bolt with raised ribs on the shank.

International Business Machines (IBM): Largest manufacturer of mainframe computers, minicomputers, and microcomputers.

International style: A nontraditional building designed for function rather than decoration.

Isometric drawing: A drawing similar to an isometric projection, but having the principal edges drawn true length rather than foreshortened.

Isometric projection: An axonometric projection in which the object is tilted so that all three principal faces are equally inclined to the picture plane.

Jack rafter: A short rafter placed between a top plate and hip rafter or between a ridge and valley rafter.

Jacuzzi: Trademark for whirlpools manufactured by Jacuzzi, Inc.

Jalousie: A type of window consisting of a number of long, thin, hinged panels.

Jamb: The inside vertical face of a door or window frame.

J-groove weld: A butt weld with one gouged member.

Joist: A member directly supporting floor and ceiling loads and in turn supported by bearing walls, beams, or girders.

Joystick: A computer pointing device consisting of a handle that can be moved in different directions or rotations.

Kalamein door: A fireproofed door covered with metal.

Keyboard: A computer keyboard similar to a typewriter keyboard, but with additional keys such as function keys.

Keystone: The last wedge-shaped stone placed in the crown of an arch.

Kiln: A heating chamber for drying lumber (pronounced "kill").

Kilobyte: 1,024 bytes.

Kip: 1,000 pounds.

Knot: A lumber defect caused by an embedded limb.

Kraft paper: A strong, brown paper made from sulphate pulp.

Labor and material bond: A guarantee to the materials suppliers and subcontractors that their bills will be paid.

Laitance: An undesirable watery layer found in the upper surface of curing concrete.

Lally column: A steel column.

Lamella roof: A roof formed of short members assembled in diamond-shaped patterns.

Laminate: To bond together several layers of material.

Lanai: A roofed living area or passage with open sides.

Landing: A stair platform.

Laserjet: Brand of laser printer manufactured by Hewlett-Packard.

Lath (metal): Sheet-metal screening used as a base for plastering.

Lath (wood) (also *furring*): Thin wood used to level a surface in preparation for plastering or composition tiles.

Lattice: Openwork made by crossed or interlaced strips of material.

Lavatory: A washbasin or room equipped with a washbasin.

Ledger: A wood strip nailed to the lower side of a girder to provide a bearing surface for joists.

Lift-slab: A precast concrete construction method of casting all slabs on the ground and lifting them into final position.

Lightpen: A computer positioning device consisting of a light-sensitive detector painted directly on the screen.

Limestone: A sedimentary rock of calcium carbonate.

Lintel: The horizontal member supporting the wall over an opening.

Lobby: An entrance hall or reception room; vestibule.

Lookout: A short timber for supporting a projecting cornice.

Lot line: The limit of a lot.

Lotus Development Corporation: A leading computer software publisher.

Louver: A ventilating window covered by sloping slats to exclude rain.

Low-E glass: Low-emissivity glass manufactured with a microthin transparent coating that reflects heat.

Lumber: Wood that has been sawed, resawed, planed, crosscut, or matched:

Boards: Lumber less than 2" thick and more than 1" wide.

Dimension: Lumber from 2" to 5" thick and more than 2" wide.

Dressed size: See *Finished size*.

Finished size: The size of lumber after shrinking and planing; about $\frac{1}{2}$" less than the nominal or rough size.

Nominal size: The "name" size by which lumber is identified and sold.

Rough lumber: Lumber that has been sawed but not planed.

Structural lumber: Lumber over 2" thick and 4" wide, used for structural support.

Timber: Lumber over 5" in least dimension.

Yard lumber: Lumber of all sizes intended for general building purposes.

Macintosh: Personal computer created by Apple Computer.

Manhole: A sewer opening to allow access for a person.

Mansard: A hip roof, each slope of which is broken into two planes.

Mantel: The shelf over a fireplace.

Marble: A metamorphic rock used for building.

Masonite: A hardboard trade name.

Masonry: Material such as stone, brick, and block used by a mason.

Mastic: A waterproof material used to seal cracks.

Meeting rail: The horizontal rails of double-hung sash that fit together when the window is closed.

Megabyte: 1,024 kilobytes.

Member: A part of a building unit.

Membrane roofing: A thin layer of plastic, rubber, bitumen, or aluminum used as a roofing material for flat roofs.

Meter: SI unit of length equal to 39.37".

Microcomputer: Home computer with a single integrated circuit (microprocessor) used by one operator at a time.

Microsoft Corporation: The largest computer software publisher.

Millimeter: 1/1000 of a meter.

Millwork: Woodwork that has been finished (*milled*) in a milling plant.

Minicomputer: A computer larger than a microcomputer but smaller than a mainframe computer used by 10 to 100 operators at a time.

Miter: A beveled cut.

Modem: Equipment for transmitting computer data by telephone.

Modern styling: A contemporary, functional building designed without regard to historical styling.

Modular brick: 4" $\times$ $2\frac{2}{3}$" $\times$ 8" brick.

Module: A standardized unit of measurements.

Molding: Strips used for ornamentation.

Monitor: A cathode ray tube (computer screen).

Mortar: A mixture of cement, sand, and water used as a bonding agent by the mason.

Mortgage: The document used to hold a building as security for a loan.

Motif: The basic idea or theme of a design.

Mouse: A computer's pointing device that controls the cursor by rolling the device on a surface.

M shape: A lightweight structural steel I-beam.

Mullion: The large vertical or horizontal division of a window opening.

Multitasking: Operating several computer programs almost simultaneously.

Multiview projection: An orthographic projection of an object with its principal faces parallel to the picture plane.

Muntin: The small members that divide the glass in a window frame.

Netscape Communications Corporation: A leading computer software publisher.

Network: A set of connected computers.

Newel or newel posts: The post where the handrail of a stair starts or changes direction.

Niche: A small recess in a wall.

Nominal size: The "name" size of rough lumber. After planing, the actual *finished size* is smaller than nominal. See table at *finished size*.

Norman brick: 4" $\times$ $2\frac{2}{3}$" $\times$ 12" brick.

Nosing: The rounded edge of a stair tread.

Novell, Inc.: A leading computer software publisher.

Oblique projection: A parallel projection in which the projectors are oblique rather than perpendicular to the picture plane.

Old English lettering: An attractive, but seldom used, lettering style.

On center: Measurement from the center of one member to the center of another (noted "oc").

One-point perspective: A perspective projection in which the picture plane is parallel to a face of the object.

Open plan: A floor plan of *areas* rather than rooms.

Open-riser: A stair having treads without risers.

Open-web joist: A steel structural member fabricated with a top chord, bottom chord, and web members.

Optical scanner: See *scanner.*

Orientation: The compass location of the various rooms of a house to make best use of sun, topography, breezes, and views.

Outlet: An electric socket.

Overhang: The horizontal distance that a roof projects beyond a wall.

Owner-architect agreement: A document specifying the architect's services and the owner's obligations.

Palladian window: A window in the form of a central arch with a narrower rectangle on each side.

Palm top: Miniature handheld microcomputer.

Panel: A flat surface framed by thicker material.

Panelboard: The center for controlling electrical circuits.

Parallel projection: A projection having all projectors parallel to each other.

Parapet: The portion of a wall that extends above the roof.

Parging (also *pargeting*): Cement mortar applied to a masonry wall.

Parquetry: An inlaid floor in a geometrical pattern.

Particle board: Wood fiberboard.

Partition: An interior wall. (*Wall* refers to an exterior wall.)

Passive solar system: A solar heating system in which heat flows naturally rather than being forced by some mechanical means.

Patio: A courtyard, adjacent to a residence, and used as an outdoor living-dining area.

Penny: A term for the length of a nail, abbreviated "d." Originally, it meant the price per hundred nails (i.e., 8 penny = 8 cents per hundred nails).

Penthouse: A housing above the roof for elevator machinery.

Performance bond: A guarantee to the owner that the contractor will perform all conditions of the contract.

Perimeter heating: A heating system in which heat is supplied to the perimeter walls of a building.

Peripheral equipment: Accessory equipment to the CPU such as monitor, keyboard, digitizer, and plotter.

Personal digital assistant: A small hand-held computer.

Perspective projection: A projection having all projectors converging at a point (the observer's eye).

Pier: A rectangular masonry support either freestanding or built into a wall.

Pilaster: Specifically, an attached pier used to strengthen a wall.

Piracy: Illegal copying of computer software.

Pitch: A term applied to the amount of roof slope. It is found by dividing the height by the span. Also, a liquid material used in roofing; the center-to-center distance between bolts.

Pivoted window: A window that revolves on two pivots, one at the center of the top of the sash, and the other at the center of the bottom of the sash.

Pixel: Individual dot shown on a CRT, from "picture element."

Plank: Lumber 2" and over in thickness.

Plaster: A mortar of gypsum, sand, and water used to finish interior walls and ceilings.

Plate: A horizontal member in a wall framework on which rafters, joists, studs, and so forth rest, or to which they are secured, as in *sole plate, sill plate, top plate.*

Platform framing: A commonly used type of residential construction in which each story is built upon a platform.

Plotter: A device that draws on paper by moving technical pens as directed by a computer.

Plumb: Vertical.

Ply: The number of layers of roofing felt, plywood veneer, or other materials.

Plywood: Wood made up of three or more layers of veneer bonded with glue.

Poché: To darken in a wall section with freehand shading.

Pointing: Filling of joints in a masonry wall.

Points: The bank's service charge for a mortgage. Each point equals 1 percent of the mortgage.

Post-and-beam: A type of building frame in which cross-beams rest directly upon vertical posts.

Postmodern styling: An architectural style that applies the ornamentation of past classical designs to the clean lines of contemporary buildings.

Powderpost beetles: Destructive beetles that feed in dry wood.

Precasting: A casting in a mold that is not located at its final position in the structure.

Presentation model: A finished scale model showing *exactly* how a proposed building will appear and function.

Pressure-sensitive transfers: A thin, transparent plastic sheet with a printed pattern, and coated on one side with adhesive. Also called *appliqué.*

Prestressing: A method of compressing concrete members so that they will not deflect when in position.

Primary auxiliary view: An auxiliary view obtained by projection from a principal view.

Priming: The first coat of paint, mixed and applied so as to fill the pores of the surface preparatory to receiving the subsequent coats.

Primitives: Basic geometry shapes used to computer-produce solid-model perspectives.

Printer: A device that draws on paper by means of ink jets or laser beams as directed by a computer.

Projected window: A window whose frame swings open and slides at the same time.

Prompt: Computer request for a command.

Prototype: An example to be copied fully or partially.

Puck: A digitizer's palm-sized positioning device.

Pumpcrete: Pumped concrete.

Purlin: A horizontal roof framing member, laid perpendicular to main trusses and supporting the roof.

Radiant heating: Heating by radiating rays without air movement.

Radon: A harmful, radioactive gas that may leak into buildings.

Rafter: A member in a roof framework running from the eave to the ridge. There are hip rafters, jack rafters, and valley rafters.

Rail: A horizontal member of a door, window, or panel.

RAM (Random-access memory): Computer's main memory for storage and retrieval of data. Compare with ROM.

Ranch styling: A rambling, single-story contemporary residence with low-pitched roof.

Random match: A veneer pattern of sheets randomly placed.

Raster: Scan pattern of horizontal lines composed of pixels that fill the CRT screen.

Rebar: Reinforcing bar.

Reflective insulation: Sheet material with a surface of low heat emissivity used to reduce heat loss.

Regency styling: The architectural style in England during which the Prince of Wales was regent of England.

Reinforced concrete: Concrete containing more than 0.2 percent of reinforcing steel.

Relative humidity: Ratio of the amount of water vapor in air to the maximum possible amount at the same temperature.

Renaissance architecture: The architectural style between 1420 and 1700 characterized by the rediscovery and imitation of classical styles.

Rendering: A perspective that has been finished to show a building's completed appearance.

Renegotiated rate mortgage: A mortgage in which the interest rate may be adjusted every few years.

Resistance number: See *R value.*

Retaining wall: A wall designed to resist lateral pressure of earth.

Retemper: To replace water evaporated from wet mortar.

Return: A molding turned back to the wall on which it is located.

Reveal: The depth of masonry between its outer face and a window or door set in an opening.

Ribbon: A wood strip let into the studding to provide a bearing surface for joists.

Ridge: The top edge of the roof where two slopes meet.

Ridge cap: A wood or metal cap used over roofing at the ridge.

Ridgepole: The highest horizontal member in a roof. It supports the heads of the jack rafters.

Riprap: Stone placed on a slope to prevent erosion.

Riser: The vertical board of a step. It forms the front of the stair step.

Rocklath: A flat sheet of gypsum used as a plaster base.

Roll roofing: Roofing material of fiber and asphalt.

Rollerball: A computer pointing device that controls the cursor by rolling the device on a surface.

ROM (Read-Only Memory): Computer's memory containing permanent instructions. Computer can read data from ROM, but user cannot store data in ROM. Compare with *RAM*.

Roman brick: 4" × 2" × 12" brick.

Roman lettering: A lettering style having strokes of different widths and with serifs.

Roof boards (also *roofers*): The rough boarding over the roof framework on which is laid the roof covering.

Roof window: A glazed roof opening to admit light and air.

Rubble: Irregularly shaped building stone, partly trimmed.

R value: A classification number that indicates the effectiveness of insulation.

Saddle: A small, double-sloping roof to carry the water away from the back of chimneys. Sometimes called *cricket*.

Salvaged brick: Used brick.

Sandstone: A sedimentary rock of cemented quartz.

Sandwich wall: A wall of at least two adjacent and connected panels, usually reinforced concrete panels protecting an insulating panel.

Sapwood: The living layer of a tree surrounding the heartwood.

Sash: A framing for windowpanes. A sash window is generally understood to be a double-hung, vertically sliding window.

Scab: A small member used to join other members, fastened on the outside face.

Scanner: An optical device using a movable head that electronically reads and records drawings, printed lettering, handwriting, or bar codes.

Scarf joint: A joint made by tapering the ends of each piece.

Schedule: A list of parts (as a *window schedule*).

SCR brick: 6" × 2½" × 12" brick developed by Structural Steel Products Research for use in 6" solid, load-bearing walls.

Scratch board: A board with a white surface coated with black ink. When scratched with a stylus, the white lines are revealed.

Scratch coat: The first coat of plaster. It is scratched to provide a good bond for the next coat.

Seasoning: Removing moisture from green wood.

Secondary auxiliary view: An auxiliary view obtained by projection from a primary auxiliary view.

Section: An orthographic projection that has been cut apart to show interior features.

Semiconductor: Material such as silicon that is neither a good conductor nor insulator, used to manufacture inexpensive electronic devices.

Septic tank: A sewage-settling tank.

Serifs: Small lines at the ends of every stroke in Roman lettering.

Setback: The distance between a building and its front or side property lines.

Sewage disposal: The system of waste pipes that conduct the waste water to the public sewer or disposal field.

Shade: A surface turned away from light.

Shadow: A surface from which light is excluded by a shaded surface.

Shake: A hand-split shingle; a lumber defect caused by a natural separation of the annual rings.

Shear plate: A metal connector for timber-to-timber and timber-to-steel construction that distributes the load over a greater area.

Sheathing: The rough boarding on the outside of a wall or roof over which is laid the finished siding or the shingles.

Shed roof: A flat roof pitched in only one direction.

Shim: A piece of material used to true up or fill in the space between two surfaces.

Shingles: Roof covering made of wood cut to stock lengths and thicknesses and to random widths. Also fiber glass shingles, asphalt shingles, slate shingles, and tile shingles.

Shotcrete: Sprayed concrete using a wet mix.

Sidelight: A narrow window adjacent to a door.

Siding: The outside layer of boards on a frame wall.

Sill: The stone or wood member across the bottom of a door or window opening. Also the bottom member on which a building frame rests (sill plate).

Skylight: A glazed roof opening to admit light and air.

Slate: A metamorphic rock used for roofing and flagstone.

Sleeper: A wood member placed over a concrete slab to provide a nailing base for a wood floor.

Sliding door: A door that slides along a horizontal metal track screwed into the door frame head.

Sliding windows: Windows that run in pairs upon horizontal tracks.

Slip match: A veneer pattern of sheets joined side by side.

Slump block: A concrete block resembling stone.

Smoke chamber: The portion of a chimney flue located directly over the fireplace.

Snap header: A half-brick header.

Soffit: The undersurface of a cornice, molding, or beam.

Soft conversion: The changing of the English dimensions of a product to metric dimensions without changing the size of the product.

Software: Computer programs stored on disks.

Softwood: Wood from trees having needles rather than broad leaves. The term does not necessarily refer to the softness of the wood.

Soil stack: A vertical pipe in a plumbing system that carries the discharge from a toilet.

Solar heat: Any heating system that uses the sun's rays as the heating source.

Solar orientation: The orientation of a building to make best use of the sun.

Soldier: A masonry unit laid vertically with the narrow side exposed.

Sole: The horizontal framing member directly under the studs.

Sound absorption: Reduction of sound to prevent echoes or lower noise.

Sound amplification: Strengthening of sound as required in large rooms.

Sound masking: Overriding undesirable sound by introducing pleasant sound.

Sound reflection: Bouncing of sound from one or more surfaces.

Sound transmission: Passage of sound through walls, ceilings, and air.

Sovent: A single-stack, self-venting sewage disposal system.

Space frame: A three-dimensional truss system.

Spackle: To cover wallboard joints with plaster.

Span: The distance between structural supports (i.e., the length of a joist, rafter, or other member).

Spandrel: The area between the top of a window and sill of the above window.

Spandrel wall: An exterior wall that provides no structural support.

Specifications: The written description accompanying the working drawings.

Split block: A fractured solid concrete block laid with the split face exposed.

Split-level styling: A contemporary building having floor levels differing by half-stories.

Split lintels: Two lintels placed side by side in a wall.

Spreadsheet: A ledger of numbers arranged in rows and columns. A computerized spreadsheet can be used for calculations such as totals or differences between numbers.

Sprinkler system: A network of water pipes at ceiling level and provided with nozzles called "sprinklers."

Square: 100 sq. ft. of roofing.

S shape: A structural steel I beam.

Stack: A vertical pipe.

Standpipe: A vertical water pipe with fire-hose outlets at each floor.

Step: Verticle distance from one tread top to another.

Stile: A vertical member of a door, window, or panel.

Stirrup: A metal U-shaped strap used to support framing members.

Stool (also *water closet*): The wood shelf across the bottom and inside of a window.

Stop: See *doorstop*.

Story (also *storey*): The space between two floors, or between a floor and the ceiling above.

Stressed-skin panel: A hollow, built-up panel used for floors, roofs, and walls.

Stretcher: A masonry unit laid horizontally with the long face exposed.

Stringer: The sides of a flight of stairs; the supporting member cut to receive the treads and risers.

Stucco: A face plaster or cement applied to walls or partitions.

Stud: The vertical member that forms the framework of a partition or wall.

Stud welding: An electric arc welding process used to weld threaded studs to structural steel.

Study model: A model constructed by designers to help them investigate the function or appearance of a building.

Stylus: A digitizer's pen-like positioning device.

Subfloor: The rough flooring under the finish floor.

Sun Microsystems: A leading computer workstation manufacturer.

Superinsulation: Heavier-than-usual insulation.

Switch, electrical: Device used to control electric switches.

Tail beam: Framing members supported by headers or trimmers.

Tee: A structural steel member in a shape of a **T**.

Tempera: Opaque water-based paint.

Tempered hardboard: Water-resistant hardboard.

Template: A digitizer tablet menu card.

Templates, planning by: House planning by first using furniture templates to obtain room designs, and then using room templates to obtain a preliminary layout.

Tensile bond strength: Ability of mortar to adhere to masonry unit.

Termite shield: Sheet metal used to block the passage of termites.

Termites: Highly destructive insects that feed on wood.

Terra-cotta: Hard-baked clay and sand often used for chimney flues.

Terrace: A raised flat space.

Terrazzo: Floor covering of marble chips and cement ground to a smooth finish. Metal strips are used to separate different colors and create designs.

Thermal insulation: Any material that retards the passage of heat and cold.

Thermopane: Brand name for a hermetically-sealed double glazed window.

Thermostat: An instrument that automatically controls the heating plant.

Third-angle projection: A multiview projection in which the object is located in the third quadrant, resulting in its plan above its front elevation.

Three-point perspective: A perspective projection in which the picture plane is oblique to all of the object's principal edges.

Threshold: The stone, wood, or metal piece directly under a door.

Tie bar: A tie rod.

Tie beam: A framing member between rafters.

Tie rod: A steel rod used to keep a member from spreading.

Tilt-up construction: A method of precasting members horizontally on the site and lifting into their final vertical positions.

Toenail: To drive nails at an angle.

Tongue: A projection on the edge of a board that fits into a groove on an adjacent board.

Tool box: In *AutoCAD for Windows*, a display of icons.

Topographical orientation: The orientation of a building to take full advantage of the land contours.

Track: A horizontal member in metal framing, as in *top track*, *bottom track*.

Trackball: A computer pointing device consisting of a movable ball set in a fixed case.

Traditional styling: Any residential style that copies, with certain modifications, the kind of building built previously.

Translucent: Having the ability to transmit light without a clear image.

Transom: A small window over a door.

Transparent: Having the ability to transmit images clearly.

Trap: A device providing a liquid seal to prevent passage of air and odor.

Tread: The horizontal part of a step.

Treillage: An ornamental screen.

Trim: The finish frame around an opening.

Trimmer: A joist or rafter around an opening in a floor or roof.

Truss: A braced framework capable of spanning greater distances than the individual components.

Trussed rafter: A truss spaced close enough to adjacent trusses that purlins are unnecessary.

Two-point perspective: A perspective projection in which the picture plane is parallel to one set of the object's edges (usually vertical edges).

U-groove weld: A weld with one gouged member.

Umbra: The space from which light is excluded by a shaded surface.

Unfinished bolt: A low-carbon steel bolt.

Unit air conditioning: A multiple-source air conditioning system.

User-friendly: An easily understood and easily operated computer program.

Valley: The trough formed by the intersection of two roof slopes.

Valve: A device that regulates the flow in a pipe.

Vapor pressure: Air pressure caused by wet air attempting to mix with drier air.

Vapor retarder: A thin sheet used to retard the passage of water vapor.

Vault: A curved surface supporting a roof.

Vee weld: A butt weld with both members mitered.

Veneer: A facing material not load-bearing.

Vent pipe: A small ventilating pipe extending from each fixture of a plumbing system to the vent stack.

Vent stack: A vertical pipe in a plumbing system for ventilation and pressure relief.

Ventilation: Vents and fans in a building to reduce water vapor or introduce fresh air.

Veranda: A roofed, partially enclosed porch.

Vestibule: A small lobby or entrance room.

Victorian architecture: The style of architecture during Queen Victoria's reign (1837 to 1901), characterized by lavish ornamentation.

View orientation: The orientation of a building to obtain the best possible view of the surroundings.

Virtual reality: Realistic, three-dimensional, and interactive computer simulation of an environment.

Virus: A computer program that disrupts other programs.

Wainscot: An ornamental covering of walls often consisting of wood panels, usually running only part way up the wall.

Wall: An exterior wall. (*Partition* refers to an interior wall.)

Wallboard: A large, flat sheet of gypsum or wood pulp used for interior walls.

Wall tie: A metal piece connecting wythes of masonry to each other or to other materials.

Warm-air heating: A heating system using warm air circulated through supply ducts to room registers; cooler air is pulled through return ducts for reheating.

Warp: A lumber defect; twisting.

Waste stack: A vertical pipe in a plumbing system that carries the discharge from any fixture.

Watercolor: Transparent water-based paint.

Water distribution: The system of supply pipes that conducts water from the water main or other source to lavatories, bathrooms, showers, and toilets.

Waterproof: Material or construction that prevents the passage of water.

Water vapor: Water evaporated to a gas state.

Weathering steel: A high-strength steel that is protected from further corrosion by its own corrosion.

Weatherstrip: A strip of metal or fabric fastened along the edges of windows and doors to reduce drafts and heat loss.

Web: A set of related documents stored on computer systems.

Web site: A computer system that publishes Web documents.

Weep hole: An opening at the bottom of a wall to allow the drainage of moisture.

Well opening: A floor opening for a stairway.

Whirlpool: A jetted tub.

Wind bracing: Bracing designed to resist horizontal and inclined forces.

Wind orientation: The orientation of a building to take advantage of prevailing summer breezes and to block off winter winds.

Winder: A tapering step in a stairway.

Windows: A computer program providing icons to simplify commands. The icons are in areas called *windows* and are usually selected by moving a pointer with a mouse. Each window can be sized and moved about the screen. Also, multiple windows can appear on screen and be operated simultaneously. Windows programs were initially developed by Microsoft Corporation.

Wireframe: A computer-produced perspective showing all lines as visible.

Word processing: Using a computer to type and reproduce documents.

Working drawing: A drawing containing information for the workers.

WORM: Write once, read many laser disk that can read information supplied by a computer operator. Compare with *CD-ROM*.

W shape: A structural-steel, wide-flanged beam.

Wythe (also *withe*): A masonry partition, such as separating flues.

Zoning ordinances: Building regulations that regulate the size, location, and type of structure in each building zone.

Index